BRITISH AVIATION

The Adventuring Years
1920–1929

HM King George V, attended by Sqn Ldr Tom Harry England, inspects the new Gloster Gamecock fighter at the 1924 RAF Display. (*Courtesy Sir Roy Fedden*)

BRITISH AVIATION

The Adventuring Years
1920–1929

HARALD PENROSE

PUTNAM
LONDON

By the same author

BRITISH AVIATION: The Pioneer Years 1903–1914
BRITISH AVIATION: The Great War and Armistice
1915–1919

In tribute to
CHARLES GREY GREY
Engineer designer, Salesman, Journalist, Editor, Friend.
1875–1953

ISBN 0 370 10016 6
Printed and bound in Great Britain for
Putnam & Company Limited
9 Bow Street, London, WC2E 7AL
Filmset in Photon Times 11 pt. by
Richard Clay (The Chaucer Press), Ltd., Bungay, Suffolk,
and printed in Great Britain by
Fletcher & Son Ltd, Norwich

First Published 1973

CONTENTS

ACKNOWLEDGMENTS

The author wishes to express his gratitude to the following for permission to quote selections from their books and/or articles. C. H. Barnes, from his books *Bristol Aircraft since 1910* and *Shorts Aircraft since 1900*; Sir Arthur Bryant, from an article appearing in *The Illustrated London News*; Faber and Faber Ltd, *D.H. – An Outline of de Havilland History* by C. Martin Sharp; Wing Cdr Norman Macmillan, OBE, MC, AFC, AFRAeS, from his book *Wings of Fate* (G. Bell & Sons Ltd) and from articles appearing in *The Aeroplane* and *Air BP*; John Murray Ltd, *The Story of the British Light Aeroplane* by Terence Boughton; The Estate of Nevil Shute Norway, Wm Heinemann Ltd and Wm Morrow & Co Inc for *Slide Rule* by Nevil Shute.

PROLOGUE FROM THE WINGS

'My craft is that of trying to convey through the printed word a fleeting image of what some other period in history was,' wrote Sir Arthur Bryant in discussing the difficulty of that object. 'Unless one deals exclusively with some small and limited section of the past for which ample documentary material exists, the most one can hope to give is an imperfect and impressionistic picture which, however carefully one has studied one's sources, would be faulted in many particulars by anyone who had actually lived in the period and could recall from memory the age through which he had lived. I have been forcibly reminded of this by reading some of the recent reconstructions of events by young writers whose temper of mind is so different from that of England of half a century ago that, to anyone who remembers that time and the men of that time, the picture they have drawn from their reading is so utterly unreal as to amount to a fantastic travesty of the truth.'

I count myself fortunate to have experienced the events of fifty and more years ago, but that seems to make it even more difficult to compress within the limitations of a single volume the procession of aeronautical progress of just one of those decades. Commentaries, reports, descriptions, quotations, conversations, have had to be abbreviated and many matters eliminated. Nevertheless the story of the 1920s is here in broad essence – aeroplanes, events, policies, problems, achievements, disasters; and its people are those who step from the pages of the previous volumes of *British Aviation* and herald others who presently make their mark. Whereas *The Pioneer Years* and *The Great War and Armistice* were written from the viewpoint of a youthful onlooker of most of the events and aircraft described, *The Adventuring Years* sees the beginning of close personal participation in the industry – so the story is based on even closer recollection and experience, and confirmed, as previously, by much re-reading of every aeronautical journal of that decade, R & Ms and official reports, Hansard extracts, patent specifications and many private documents as well as discussions with pioneer participants. Though the book is part of a time-continuity, it can be read as a self-contained episode needing no reference to earlier history.

For those in particular who were pilots, the 1920s and the first few years of the 1930s were always remembered as the golden age of flying. Horizons beckoned with a glamour never found again. All the world was waiting to be explored by air. The spirit of pioneering was abundant – whether trail-blazing air routes that did not become established until years later, extending the boundaries of power and speed, conquering greater heights, or making aeroplanes and even airships, their engines and equipment, more efficient and reasonably reliable. With each new aircraft concept there was little realization that it was a further step towards the reality of war; if anything, we thought more in terms of purposeful development that would afford better and yet better communications between the peoples of the world, and thus aviation would lead to universal brotherhood. But the goal was never specific: it was a march forward to undiscovered ends, and despite tragedy and loss it was light-hearted and tremendously enthusiastic, for all were young – even the pioneer originators of the industry were scarcely middle-aged.

Because C. G. Grey was the most famous, most pertinacious aeronautical commentator of that time, I have quoted from him more extensively than from others in attempting to portray the atmosphere of those days. As that outstanding editor and aeronautical author, Maj Oliver Stewart has said: 'Charles Grey was a great man. He was witty, incisive, challenging, thought-provoking, altogether incorrigible, and even people who hated every opinion he expressed found themselves stimulated by his style and thought.' Because of these qualities most of us were more impressed by the sparkling editorials of *The Aeroplane* than the staid reflections and reports of *Flight* which made its mark with better technical articles. Much of C. G. G.'s perception of the shape of things to come, whether aircraft or international matters, has proved remarkably accurate; but it is the pungency of his comments on the contemporary scene which was, and still is, so enlivening. His sources of inside information were as endless as they were diverse, and used with subtle ingenuousness before any official release of what had been *Most Secret*.

Nevertheless, almost all old sources of information require reinterpretation in the light of modern experience. Even Martlesham reports cannot be accepted at their face value, for they were seeking explanation of what were then new phenomena. The further back in time, the more doubtful are the casual assumptions woven into aeronautical history. 'It is not imaginable to such as have not tried, what labour an historian (that would be exact) is condemned to,' warned that delightful virtuoso John Evelyn in the reign of Charles II. 'He must read all, good and bad, and remove a world of rubbish before he can lay the foundation.' That has been the aim, though I regard myself more as a witnessing chronicler of events than an uncompromising historian who holds the view that unless a matter was documented at its current time it cannot be a fact. Yet as John Smith, the earliest compiler of Gloucester history, wrote two hundred years before Evelyn: 'Much lye hid in the records of the Kingdom, into some of which places I have pryed and others saluted afar off, but I confess to have drawne the better halfe of these relations out of some of these honourable and spatious fields where still remain many fragant and fair flowers.'

CHAPTER I

LOW EBB AND DISASTER 1920

'I feel perfectly confident that this noble art of flying will soon be brought home to man's general convenience, and that we shall be able to transport ourselves and families, and their goods and chattels, more securely by air than by water, and with a velocity from 20 to 100 miles per hour.'

Sir George Cayley (1810)

1

Fourteen months had passed since the Great War ended, but from Switzerland to North Sea shores jagged lines of deserted and weed-grown trenches remained visible evidence of what had been the world's most titanic struggle. The first easy confidence of peacetime had vanished in the crowded competition of 'civvy' life. More than 700,000 men were out of work, and most women had been dismissed from factories. But in defeated Germany anxiety was worse; a great Army of Occupation was ensconced and poverty was widespread. Not only jobs were scarce but food was scanty, for the starvation blockade had not been lifted until summer of the previous year. Yet bad as these things were, the impact of revolution on our late ally Russia shocked the League of Nations Council at its inaugural meeting on 16 January when it was revealed that 30 millions were starving and help was imperative.

With the concept of a League of Nations against aggression safeguarding the future, many a Briton had expected a post-war Utopia. Instead, it was disconcerting to find that the pre-war £1 was worth only 9s 6d, that there were huge loans requiring interest and repayment, and that the nation owed the USA tremendous debts for goods of war – for taxes throughout hostilities had paid only for 30 per cent of the Government's expenditure. Despite the trade boom of 1919, dreams of easy peace-time prosperity were beginning to fade.

Flashing his verbal sword, the 45-year-old monocled editor of *The Aeroplane*, C. G. Grey, wrote: 'Last year was one of the extremes in the history of aeronautics. It saw the aircraft industry reduced from one of the greatest and most important to one of the smallest as regards output, and one of the least important as regards the existence of the Empire. Service flying ceased at home and lost interest for the public. Even the sport of aviation, though it flourished before the war, became a boring farce – yet the greatest sporting events in the history of aeronautics took place: namely, the Transatlantic flights by aeroplane and airship, and the flight to Australia.

'Firms which made hundreds of thousands of pounds building aeroplanes during the war shut down their aircraft departments and dismissed their skilled staffs, though others which had struggled gamely through bad times before the war were now spending their hard-earned and over-taxed profits in perfecting aircraft for peaceful purposes. Men who had come into the aircraft business during the war, and were given titles and honours for their services, turned their backs on aircraft in favour of their normal commercialism; yet the pioneers of aviation, though they received nothing better than a few OBEs, returned to their work of pioneering aeronautical developments in new directions. Financiers buttoned their pockets tightly against all appeals to support aeronautical undertakings; but a few firms and several individual pilots made highly profitable business in pleasure-flying on absurdly small expenditure, and although the Press and MPs had been wildly excited over the integrity of a single Air Force and the maintenance of a separate Air Ministry, they saw the Government reduce the Air Force to a smaller number of Squadrons than existed before the war.'

A White Paper entitled *Synopsis of Progress of Civil Aviation in Foreign Countries* presented early in 1920 explained:–

'Aircraft and agents have been despatched by British firms to various parts of the globe, but hitherto no official foreign missions have been despatched by the British Government owing to shortage of available Service personnel and a reluctance to confuse Civil and Service enterprise, and because of the traditional British policy that an industry should be left to develop, as far as possible, on its own merits.' This was also the traditional British way of avoiding official responsibility.

The report gave forthright warning that the French were making every endeavour to secure foreign markets. Complying with her usual policy of encouraging new industries, France was strongly subsidizing civil aviation with bonuses on distance flown, whether airline, charter, or private, adding a premium on tonnage carried and a special bonus up to 25 per cent per annum of the value of any machines which might be of military type and available in emergency for use by the State.

Unlike France and Britain, the USA had not embarked on foreign sales missions, and in a measure aviation seemed to be drifting into unimportance there. The great companies which had been so massively brought into aircraft production had reverted to other activities, leaving only a few pioneer firms; but as early as December 1918 a noteworthy start had been made with subsidized postal services for which $8,000,000 was currently

First British airliner in series production was the Vickers Vimy Commercial using Vimy bomber wing structure, rear fuselage and tail – later adopted by the RAF as the Vernon troop carrier. (*Courtesy Air Marshal Sir Ralph Sorley*)

demanded for extension, though civil aviation there was not yet controlled by a centralized body.

Of all possibilities it was China, vast of space and lacking communications, which seemed to offer greatest potential. It was reported that Vickers had secured orders for forty twin-engined Vimy Commercials, thirty training adaptations of the Gunbus (for which Disposal F.E.2ds were

The Vickers Vimy Commercial was probably the first transport aeroplane with built-in passenger steps. (*Flight*)

Prototype of the civil conversion O/7 of the war-time Handley Page O/400 bomber. Fuel tanks in the nacelles replaced those originally above the bomb bay.

substituted) and twenty-five reconditioned Avro 504Ks. Handley Page Ltd were also on firm ground with a signed contract for twelve O/400s adapted with ten-seat cabins, known as Type O/7, and the first had been tested at Peking on 6 December, 1919. Meanwhile Japan became the immediate target of the French, who aimed at supervising organization and training of the Japanese Flying Corps. Greece was similarly interested in building up its air arm, and already a British Mission was training the Greek Naval Air Service.

Ominously the White Paper stressed that Germany was a serious competitor: 'Partly to avoid the Peace Treaty terms, and partly in an early bid for foreign markets, large numbers of war-time aeroplanes and engines have been sold at low figures to Norway, Sweden, Holland, Denmark, and Switzerland, and Germany has secured a substantial footing. In her endeavour to develop civilian airships and aeroplanes, Germany believes that the leading commercial air power will have the strongest reserve Air Force in event of war; and Herr Euler, President of the Imperial Air Board, has stated that the Government intends boldly to foster civil aviation.' In fact the way was clear. The makers of the Treaty of Versailles, intent on formulating crushing military clauses, had failed to comprehend the potential of civil aviation, so that after a period of six months Germany was free to build small aircraft and in 1926 could build large ones: thus her aircraft industry had every hope of survival.

In Holland that *bête noir* of Allied airmen, 29-year-old Anthony Fokker, had established a factory to which he transported 200 of his German war-time aeroplanes, 400 engines, and much material from Germany under the very noses of the Allied War Commissioners, so was likely to become the focal point of the Netherlands' aircraft manufacture. As a large proportion of Dutch merchants was pro-German, it seemed unlikely that British products could find outlet there, and chances were further reduced because Koolhoven, who had achieved renown in Britain with his Armstrong Whitworth and B.A.T. aircraft, had also returned to Holland with intention of manufacturing there. His recent B.A.T. aeroplanes, pending decision on that firm's future, were housed at Hendon in charge of Chris Draper. 'This Company,' said a wit, 'strikes one as being rather like Hamlet without the Prince of Denmark.'

At the annual staff dinner of Handley Page Ltd on 24 January, the tall and heavily built, balding H.P., looking much older than his 35 years, ebulliently described his firm's activities in England as only a small part of what he hoped would be world-wide operations in the future, and Cricklewood would become a kind of Jerusalem from which would proceed the Gospel of aviation! He had cause for optimism: recently Handley Page Ltd had recapitalized greatly to its owner's benefit. The share prospectus indicated a war profit of £440,000, but new accountants whittled it to a mere £50,000 as no allowance had been made for taxes and depreciation. Nevertheless, in justification of his belief, H.P. had already sent demonstration O/400 re-builds to South Africa, the Argentine, China, Greece, India, Peru, Poland, Scandinavia, Spain, and the USA. Even an aerial mail contract between Brazil and the Argentine had been secured, and India and South Africa were similarly interested.

Except for Vickers, the sporadic efforts of other companies were unimpressive because concern with husbanding resources led to sales attempts savouring of tentative essays to see whether it was worth spending more. In general the answer was 'no', though Sopwith and Martinsyde had despatched demonstrators to several European countries; Fairey seaplanes were being demonstrated in Scandinavia; and the British Nieuport team was endeavouring to foster Indian military interest in the two-seat Nieuhawk.

In front of the O/7 folded wings are C. W. Meredith the Handley Page works manager and Capt G. T. R. Hill, MC, (later Professor), the H.P. test pilot and aerodynamicist.

Quiet Robert Blackburn, at 35, was hopeful, for his company had not unduly expanded, and recently had appointed as chief engineer Maj F. A. Bumpus, BSC, ARCS, a man of charm and intellectual attainment, a year younger than Blackburn, who had been the Air Board's representative at Leeds during construction of the G.P. Seaplane and Kangaroo. Because the Air Ministry had a requirement for a carrier-based torpedo-carrying replacement of the war-time Sopwith Cuckoo which Blackburns had built, the 1918 Short Shirl and Blackburn Blackburd having insufficient superiority, Blackburn saw this as a good chance of a commercial break into the otherwise black future, and determined to build a private venture machine with his war-time profits. Powered with a 450 hp Napier Lion personally loaned by Montague Napier, the new machine, T.1, later known as the Swift, was slightly larger than the Cuckoo, its estimated empty weight of 3,550 lb equalling the fully-laden Cuckoo. Unique in appearance because the forward decking sloped steeply to a slightly upturned nose, it had a specially strong tubular-steel framed central fuselage portion to withstand the heavy bumps of deck landing. With luck, Blackburn hoped to have it ready for the first post-war Olympia Aero Show envisaged for the summer.

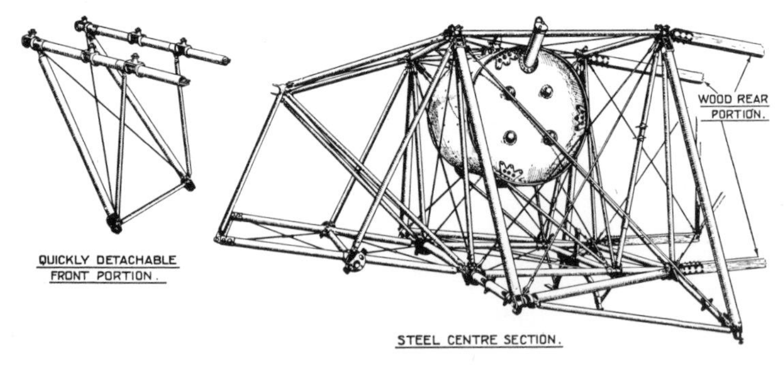

Slowly the change from wood to metal began with steel-tube structure such as the forward fuselage portion of the Blackburn Swift.

That was also where the austere Oswald Short, now 37, would reveal to the world his secret all-metal design, of which the great feature was a beautifully shaped, elliptical-section, stressed-skin monocoque fuselage of wrapped duralumin sheet butted and riveted on L-section transverse frames and longitudinal intercostal stiffeners. Nothing quite like this had ever been made, although metal-covered slab-sided fuselages and metal-skinned wings had been extensively developed in Germany by Dornier, Junkers, and Rohrbach. Unfortunately the Director of the Air Ministry's Technical Department could see no future for a material which he considered would be attacked by corrosion, nor did he think it had unique value as a new method of construction, because 'you have only replaced the plywood skin of conventional monocoque with a piece of metal' he told its inventor. Yet by the faith of a visionary engineer, here was the real beginning of the world's eventual system of aircraft construction, although

Dornier's 1918 Jagdflugzeug D1 biplane could be construed as a near contender.

Comparative tests by Short on corrosion of untreated duralumin and mild steel immersed in the Medway tides had demonstrated the superiority of light alloy; but at Boulton & Paul, research by tall J. D. North, not yet 27 but elderly in appearance and with seven years' executive design experience, resulted in such successful methods of protecting thin gauge fabricated steel-strip aircraft structures that in June he secured Patent 160,107 for methods of 'depositing a coating of zinc on steel, Sheradizing or Cosletizing it, or alloying with a high percentage of chromium, finally painting with organic substances such as esters.'

Not only had North's directors been so impressed by his talent for light steel work that the aircraft department had scrapped its woodwork plant and re-tooled with folding and drawing machinery for steel, but the Air Ministry was discussing with him a tentative design for a steel-structured, corrugated skinned, 'Postal' biplane, having twin engines within the fuselage. Though it soon became obvious that fabric must replace the weighty metal skin, here was North's great chance to prove his ability at pioneering metal aeroplanes.

2

Currently there was widespread preoccupation with the Government's great Civil Aircraft Competition advertised to begin on 1 March, and offering prizes totalling £64,000; but it had been strongly criticized by the Society of British Aircraft Constructors (SBAC) who emphasized that the rules failed to encourage design and production of sound commercial aircraft, particularly as size limitations were set too tightly.

Eventually the Air Ministry relaxed requirements of classification, disqualification, and method of marking points, and put back the date to 3 August. Instead of limiting the two groups to two-seaters and fifteen-seaters in the landplane section, the division was broadly extended to comprise a first group of any capacity between two and six passengers, and a second, whether twin or single-engined, carrying more than six. This opened the door to Vickers with their Vimy, and Sopwith with a cabin version of the Wallaby recently entered for the Australian flight; Alliance proposed a new version of the ill-fated *Seabird* originated for the 1919 Atlantic competition; Austins decided to build to minimum requirement a two-seat metal-framed biplane designed by Kenworthy; Tinson at Sages proposed a similar small machine; Tilghman Richards of Beardmore proceeded to design the six-seat single-engined W.B.X with duralumin structure and fabric covering, and also a twin-engined transport; Barnwell at Bristol made several tentative designs of a six-seater named Grampus, as well as modifications of the four-engined Pullman triplane to get requisite low-speed performance, but finally ran out of time and as a stopgap a three-bay derivative of the Bristol Fighter was entered. At Westland, Bruce made shrewd assessment that the best chance was a bigger machine than his three-passenger Limousine, and after investigating a twin-engined, seven-seat biplane, was designing a single-engined six-seat version with three-bay wings of 54 ft, utilizing a 450 hp Lion for maximum performance.

Significantly Martinsyde did not propose to enter; nor did Shorts, Parnall, Blackburn, Boulton & Paul, or Armstrong Siddeley, for they were preoccupied with targets aimed at Service contracts; and there was silence from Airco, though it was believed that their designer, Geoffrey de Havilland, was getting a D.H.18 ready. However Alliott Roe, now universally known as 'A.V.', decided to enter a triplane which had become his personal interest – for his young designer, Roy Chadwick, who recently had been taught to fly by their test pilot Capt H. A. Hamersley, had been gravely injured on 13 January when he crashed the Avro Baby demonstrator G-EACQ while approaching Hamble aerodrome low down in gusty weather, a vicious bump causing the machine to drop onto the garden of Roe's younger brother, the Rev Everard Verdon Roe. Though this accident delayed re-design of the Baby in two-seat form with fuselage lengthened by $2\frac{1}{2}$ ft, it gave A.V. personal opportunity to indulge in design again. With minimum cost in mind he attached triple 504K wings, using cut-down interplane struts, to a fuselage reminiscent of his pre-war Military Trials machine, standardizing 504K metal fittings wherever possible. Initially powered with a Disposal Board 160 hp Beardmore, the concept expressed the same frugality enforced upon Roe with his earliest machines when he used to search for dropped nuts and bolts and split-pins which had cost money he could ill afford. If given somewhat greater power the triplane seemed capable of meeting competition in the small class, so Roe entered a duplicate with 230 hp Siddeley Puma.

In the amphibian class, Vickers were building an improved version of the Viking flying-boat which John Alcock, victor of the Atlantic, had fatally crashed the previous December; Supermarine were re-designing one of their Channel flying-boats to take a folding undercarriage and greater power; Fairey was altering his ever adaptable twin-float IIID to the same end. But these were variants of existing machines, and only Sam Saunders, the oldest constructor of all, was building a novel contender, the Kittiwake, designed by Percy Beadle who had recently left the expiring Gosport Aviation Co, holders of Porte's flying-boat patents, and was now assisted by Bob Gravenell, the Saunders aircraft superintendent, and George Porter, a quiet little man long responsible for draughting their motor-boats, who had gained early aircraft experience in conceiving the hull of Sopwith's pre-war amphibious Bat-boat.

Though so radical a departure from conventional, the Kittiwake was designed in six weeks; but construction, true to the tradition of experimental aircraft, was behind schedule. A daring I-strutted structure was employed for the high aspect ratio 68-ft wings necessary to give climb with the high power loading of 20 lb/hp, and they featured a striking variable-camber device formed by knuckle-jointed full-span leading- and trailing-edge flaps which afforded a smooth unbroken wing-section when deflected – for which Patent 163,853 was granted to Beadle on 18 March. As eye-catching as the wings was the body assembly with hydrodynamic hull built as a separate unit bolted beneath a two-storey fuselage accommodating pilot and mechanic in the lower portion and five passengers on top. Long mid-gap balanced ailerons were hinged to the two outboard struts, and except for these and the tail surfaces the entire machine was surfaced with wire-sewn Consuta ply. Each inboard pair of struts carried an A.B.C. Wasp of 170 hp in a nacelle. Wiseacres shook their heads: with so many unconventional features there was bound to be trouble.

The Handley Page W.8, despite its apparent novelty as a new design, was regarded at the factory as 'a modified O/400' and built from sketch drawings.

Though the majority of entrants felt this Air Ministry competition had every possibility of being their last fling, Handley Page, with foreknowledge of the rules, felt confident that in his splendid W.8, revealed just before Christmas at the Paris Aero Salon, he had not only a potential winner but the precursor of an airliner fleet, though he did not disclose that it was cleverly constructed from O/400 bomber parts in the main. Since inauguration in September 1919 of his London—Paris service, using cabin adaptations of the O/400, his firm had carried 618 passengers and nearly 17,000 lb of freight, totting up 34,600 miles; and on his London—Brussels run had air-lifted 251 passengers and 26,000 lb of freight with a distance of 25,895 miles; additionally over 3,000 people were flown on joy-rides and charters.

His immediate British rival on the Paris run, Aircraft Transport & Travel Ltd, would not disclose the number carried in their D.H.16 four-seaters, but accomplished 200 out of 272 scheduled flights and covered 55,520 miles at an average speed of 102 mph. Its parent company, the huge Aircraft Manufacturing Co, known as Airco, operated by the ailing Holt Thomas, was tottering. Shareholders of the Birmingham Small Arms Co (B.S.A.) had just received a letter stating: 'Your directors beg to inform you that they have made arrangements for amalgamation with Airco by exchange of shares. That company has been one of the pioneers in aviation, and during the war, in addition to manufacturing large numbers of aeroplanes, designed a great proportion of the aeroplane, airship,

The D.H.16 was an expedient by Geoffrey de Havilland's right-hand man, Charles Walker, to accommodate four passengers by widening the fuselage of a D.H.9A and fitting a coupé lid. (*Flight*)

and accessories built by other factories in this country and the USA. It therefore holds and, if desirable, can retain a unique position in all matters connected with aviation.' Geoffrey de Havilland, now 38, who had his new eight-seat D.H.18 almost ready for flight, realized the significance in that phrase 'if desirable'. Equivocally the letter continued: 'Since the Armistice, and consequent cessation of Government orders, Airco has turned to manufacture of bodies for the motor trade, and is using its extensive premises at Hendon for this purpose. While your directors recognize the value of the commanding position of Airco in aviation matters, this has been largely ignored in deciding the present amalgamation. As they are satisfied that the transaction will provide our Company with great and important additional manufacturing facilities capable of producing sound and profitable business, your directors are glad to say that Mr Holt Thomas, the founder, main proprietor and chairman of Airco, will join the Board of B.S.A.' But what would the aeronautical future be even though Airco's profit for the past year, after allowing for taxation, including Excess Profits Duty, was £119,652 with a balance of £207,680 carried forward?

There were echoes of other changes. The splendid Ham Works of Sopwith Aviation, where thousands of Camels and Snipes had been constructed during the war, had just been taken over by Leyland Motors; a trickle of aviation companies were liquidating; others, such as Short Bros and Fairey, were registering debentures on their land and buildings in order to re-form. Even British & Colonial, after ten years of tremendous pioneering activity, was being wound-up to reappear as the Bristol

Aeroplane Co Ltd, paying creditors in full, and increasing capital from £1,000 nominal to £1,000,000, of which about half comprised assets of the original business. Meanwhile many foreign air forces were potentially interested in the war-time Bristol Fighter, and sales of its civil conversion, the Tourer, seemed hopeful.

Aviation questions were beginning to impact the judicial courts. Setting precedent was an echo of the Tarrant triplane crash of May 1919, with a claim before Mr Justice Roche made by Mr Thomas Dunn – as executor of Capt Dunn, one of the pilots killed – against the Aviation Insurance Association for a sum of £4,000. Associated as plaintiff was Mr W. G. Tarrant, constructor of the machine. It was a precedental case on the rights of insured persons in aeroplane accidents, the insurers contending the accident occurred before the machine left the ground, although the risk had been accepted by them 'from the time of the first flight'. The judge sided with common sense in deciding that the risk of flight begins on the ground – though this seemed to yield promise of later legal complexity. Hastily the Air Ministry absolved itself of all liability when aeroplanes were submitted for airworthiness tests under the Air Navigation Act, ruling that machines

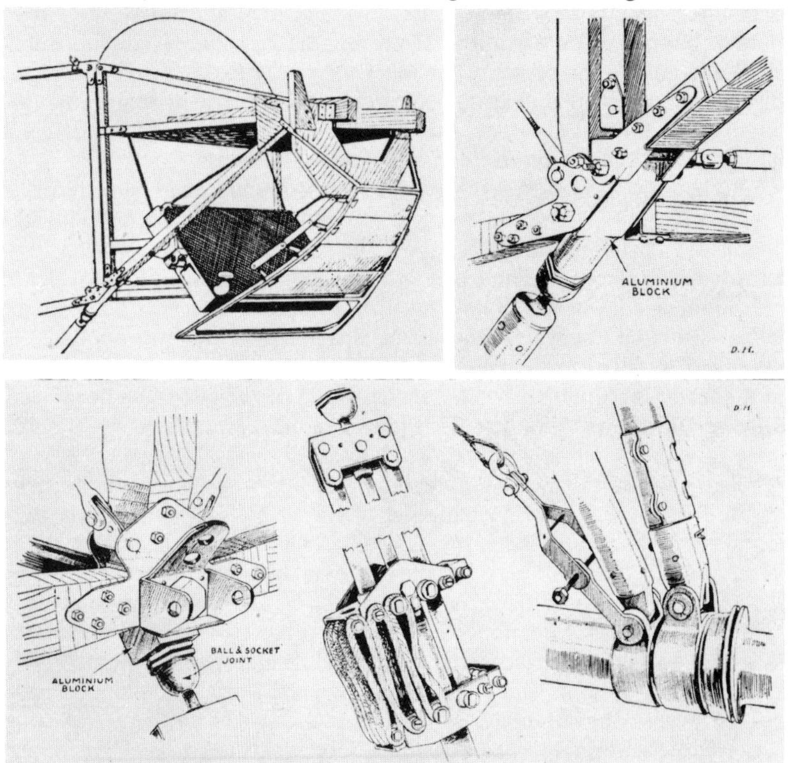

Details of the D.H.18 showing detachable engine mounting with integral radiator, fuselage joints at mounting and undercarriage strut, and undercarriage details. (*Flight*)

would be entirely at the applicant's risk not only during flight trials by RAF personnel but throughout the time they remained at a Royal Air Force station as well as during transport to and from it. In typical bureaucratic manner a clause was slipped in stating: 'The report of the trials is not to be published without consent of the Director of Research.'

3

There was strong hope that when British air routes were developed to the fullest degree the men and aircraft employed would form a hidden reserve of personnel and matériel for any future world war, though clearly airliners could not be converted to bombers overnight but should prove invaluable as strategic transports. Largely because of this significance a lecture to the Royal Geographical Society by Maj-Gen Sir Frederick Sykes, Controller-General of Civil Aviation (CGCA), on *Imperial Air Routes* attracted a big audience and was honoured by the presence of the Prince of Wales, Field Marshal Lord Haig, and Mr Churchill the Secretary of State for War. Sir Frederick had always proved strikingly adept at forecasting aviation development. His emphasis on our Protectorate Egypt as the great junction for aerial routes to India, Australia and South Africa therefore commanded attention. However, the Egyptian revolt in March 1919 and continuing pressure for independence, subject to free use of the Suez Canal, dictated the wisdom of an alternative base at Malta to avoid alighting on foreign territory, for even the route from England to Egypt had complications owing to international disagreements.

'We have charted the earth, we must chart the air,' said Sir Frederick trenchantly, 'but it must be recognized that the British aircraft industry cannot live unsupported. Direct assistance is a necessity. Subsidized competitors are in the field. The Imperial Dominion Governments must define and adopt a considered policy towards aviation.' The first route to be helped would be Egypt to India, for a Cairo to Cape route was not likely to pay commercially for the present, and because of its length the transatlantic flight to Newfoundland and Canada would be the last to be developed. However the Cape route of 6,223 miles had been ground surveyed and a chain of aerodromes established, but it was still, in effect, 'darkest Africa', although four aeroplanes were at this moment flying competitively from England to Cape Town in emulation of the great flight which had been completed to Australia the previous year. C. G. Grey put his finger on the spirit of it all: 'RAF officers and men in Asia and in Africa have been the real pioneers of our Imperial air routes,' he wrote. 'We know that Capt Ross Smith, under direction of General Borton, surveyed the route to Australia, and thereafter acquired much wealth and credit by competitively flying it with a civil Vimy. But who were the unnamed pioneers who trekked across the Arabian Desert in Ford cars? Who were they who sweltered down the Red Sea coastline, surveying the route to Aden, and who pushed into the African bush to build aerodromes along the upper reaches of the Nile and in the region of the Great Lakes? Surely they deserve mention in the official despatches of the peace-time RAF? Few have done more distinguished service for Imperial aviation than these anonymous pioneers of the air routes.'

Though the distance to Cape Town was so much shorter than the 11,500 miles to Australia it posed more problems, not the least of which were inadequate communications and aerodromes. The South African Government wanted the honour of its nationals making the first flight from England to the Union and for that purpose allocated a Vickers Vimy to be piloted by Lieut-Col Pierre Van Ryneveld, DSO, MC, and Maj C. J. Quintin Brand, DSO, MC, DFC, accompanied by two mechanics; some said General Smuts was personally responsible for this flag-waving attempt. At 7.30 on the drizzly morning of 4 February, the silver-painted Vimy, named *Silver Queen*, duly left Brooklands. It was expected to reach Cairo within a week, but not until 20 March was Cape Town achieved, and then only by Van Ryneveld and Brand completing the last stage from Bulawayo in a D.H.9, for their first Vimy had crashed on 11 February not far from Wadi Halfa some 530 miles onward from Cairo, and a replacement Vimy came to grief in attempting to take-off from Bulawayo.

The *Silver Queen* G-UABA, with uniformed Van Ryneveld and Quintin Brand awaiting the engine run at Brooklands before taking over.

On the same day as the *Silver Queen*'s departure, the adventurous Flt Lieut Sidney Cotton, inventor of the Sidcot flying suit, piloting the large and handsome Lion-powered D.H.14, with Lieut W. A. Townsend as second pilot, also left Hendon for the Cape, but badly damaged his machine on 26 February in a difficult forced landing in the dusk at Messina.

Concurrently *The Times* chartered a Vimy Commercial on what was regarded as a mystery flight, carrying the secretary of the Zoological Society, Dr P. Chalmers Mitchell, CBE, DSC, LL.D, FRS, and piloted by

In the last months of the war, construction of the big D.H.14 day bomber with Napier Lion was begun. It was completed by Airco in 1919, and subsequently loaned to Sidney Cotton for the attempted flight to the Cape. (*Hawker Siddeley*)

willowy Capt Stanley Cockerell, chief test pilot of Vickers, together with his tough assistant Capt F. C. Broome who had served under him in France with No. 151 Squadron. Though the object was an independent test of the practical utility of the Cape—Cairo route, light-hearted journalists alleged Mitchell was hunting for a Brontosaurus, taunting that he had cautiously taken train to Cairo, where he arrived at 4 pm the same day that the Vimy reached Heliopolis. Three days later, with all aboard, the Vimy set off for Khartoum, following the Nile, but on 8 February a forced descent due to a cylinder water-jacket leak was skilfully made in wild desert, and after repairs an engine failed on the 27th when taking off from Tabora in Tanganyika, causing a crash in which both undercarriages were forced through the lower wings; yet the undaunted crew managed to get the big machine mended and even succeeded in reaching the Cape only three days after Van Ryneveld and Brand.

Almost as mysterious was the secret departure of a Handley Page O/400 chartered by the *Daily Telegraph* for their aviation correspondent Charles Turner and piloted by Maj H. G. Brackley, DSO, DSC, with Capt Freddie Tymms as navigator. Earliest intimation that it was heading for Cairo was arrival at Brindisi on 6 February – but at Cairo there was a sad accident when a mechanic pulling away the chocks was fatally injured by one of the propellers; and beyond Wadi Halfa there was engine trouble, but though Brackley managed to glide to an emergency landing ground he landed with considerable drift and the machine collapsed.

Perhaps spurred by these events the Air Ministry decided the RAF might benefit from an African exercise, and on 25 February announced: 'A Vimy machine left Cairo this morning at 7.30 am en route for the Cape. The crew consists of Major Walsh and Capt Halley and two mechanics.

The machine is not flying in competition with the Civil aeroplanes, and its start has been purposely delayed to give others every opportunity to gain the honour of being the first to transverse Africa by air.' But it seemed that Africa was resisting with every means the inroads of civilization as typified by aircraft, for within two days of leaving Cairo the RAF survey Vimy had crashed at Wadi Halfa.

These crashes emphasized that liquid cooling systems, adequate for northern climes or flying at high altitudes in hot countries, were insufficient with heavy loads at low altitudes under the climatic conditions of the Cairo—Cape route, and this was exacerbated when machines had to take-off from an altitude of 4,000 feet as at Bulawayo where the temperature was 100 deg F. The uninformed but important 'man-in-the-street' could only assume these much publicized failures confirmed the uselessness of aeroplanes compared with ships and trains. They could not see these attempts as triumphs of endeavour; but the skies at last enabled great areas of the unknown to be explored from new perspectives where man could gain truer conception of his heritage the earth, and of the overwhelming vastness of the oceans. No longer need he imagine nations as colours on a map, but from the air could study the outline of their shores, the great rivers, mountains, deserts, the cities and the cultivated vales. Surely this knowledge, this new facility of journeying to another people and swiftly onward to yet another, must help in drawing mankind together in better understanding? Thus thought the dreamers – but to the men who flew, flight itself was the reward.

These pilots of the air had lived the brief years when aeroplanes became mature under the duress of war; they knew the romantic world of aerial solitude, of towering dazzling clouds, of sun-filled space, of remoteness from politics and struggle. In those high places there could be challenge, schooling of nerve and mind in face of peril – but it was matched by exuberance of freedom, joy of the whistling slipstream and buffeting of wind, the sonorous rhythm of the engine, and the buoyant lifting of wings slanted at the air. Yet the sum of their experience was negligible. Some had left the RAF with a bare 100 hours flying: to have 250 hours was average, and 400 hours counted high. It was unique that a pilot, for all the advantage of extensive test-flying at Martlesham, should attain the 695 hours of Capt Gerald Gathergood who had just given up piloting after 4½ years and 75 basic types in his log. Few pilots had flown more than half-a-dozen different types of aircraft.

Though many young men, lately released from the RAF, were determined to continue as civilian pilots, it was the RAF which had by far the biggest group who hoped to continue flying – yet that Service was so little evident in Britain that it seemed the aerial arm had ceased to exist. So when on 8 February, 1920, the *London Gazette* revealed that the RAF had won its first peace-time war with a mixed force of arms under Grp Capt Robert Gordon, DSO, a new vista of economic territorial control was opened. 'By the aid of an Air Force contingent using twelve D.H.9As, operations have

The D.H.9A had continuous RAF service in the Near and Middle East for a decade, loaded with extra radiator, extra fuel tank, thorn-proof tyres, spare wheel, water supply and even bedding. Handley Page slots were ultimately fitted to ameliorate the overladen condition. (*Ministry of Defence*)

been concluded in the space of three weeks against terrorists,' stated the Secretary of State for Air. 'In this time the Mullah in Somali has been reduced from a power in the land with many armed and aggressive followers, rich in stock, to a fugitive accompanied by a faithful few. His desultory war on the British Protectorate Government for 17 years, causing expenditure of millions of money, has been dealt with at minimum cost and practically no casualties. The general plan of the air campaign was prepared under the direction of Wg Cdr Frederick Bowhill, Chief of Air Staff.'

But if that was revolution in warfare, requiring only a few battalions in mopping-up operations instead of an army, Handley Page, lecturing at King's College, London, revealed that a mighty revolution in aircraft design seemed imminent, for he said: 'Our experiments have been directed to finding out whether it is possible to fly with smaller wings, and one of our discoveries indicates we shall be able to use planes of at least half the present area.' A little uneasily the management of other aircraft firms dismissed this as one of H.P.'s publicity-seeking exaggerations, for few had spotted Patent 157,567 issued to him on 24 October the previous year, describing transverse slots along the wing-span in which 'the opening on the under surface is in advance of that on the upper surface' forming a venturi 'to prevent "burbling" or eddying of the air above the leading edge of a plane set at a considerable angle of incidence. The auxiliary plane may be adapted to be moved into contact with the mainplane in order to close the slot, and for this purpose may be carried on a number of arms connected to the mainplane by vertical pivots which can be swung transversely in the manner of a parallel ruler.' His daring target was a system of adjustable slots in sequence across the entire wing chord, and it was on this that he prophesied half the present area for the same landing speed – though forbore to mention the enormously increased drag. Said one of his admirers: 'It seems more than possible that Mr Handley Page's technical and experimental department has done for aviation what Dunlop with his pneumatic tyre did for road vehicles. If so, we are on the verge of a new era in aviation, leading to a boom in aerial transport and aeroplane production alongside which the war boom in aviation will seem but a circumstance. One knows enough of the underlying principles and of Mr Page himself to say that future developments will be better worth watching than anything in aeronautics since the beginning of the war.'

Of greater immediate importance to the industry were the Air Force Estimates for 1920–21 introduced on 11 March by Maj H. C. Tryon, recently appointed Under-Secretary of State for Air. War-time expenditure of £1 million a day had dropped to £1 million a week in 1919, and in this second peace-time year would total less than £23 million including war debts of £8 million, necessitating cuts in RAF personnel from the 22,000 officers of 31 March, 1919, to 3,280 during 1920, with other ranks reduced from 160,000 to 25,000 and civilian staff from 14,000 to 6,000 including local nationals abroad; by the end of the year the WRAF would cease to exist, though a year ago it had 22,000 women. Encouragingly, Tryon stated that 'the work of rebuilding under that distinguished and able officer, Air Marshal Sir Hugh Trenchard, has been going on, and despite economies the RAF has engaged against the Bolsheviks in north and south Russia, the Afghans, the Pathans on the Indian frontier, and the Mullah; and in Mesopotamia the Civil Commissioner has referred appreciatively to co-operation of the RAF in maintaining order and communications, making maps, and even collecting revenue.'

Of paramount importance was the allocation of £1,389,950 for purchase of aeroplanes, seaplanes, engines and spares. It gave hope of constructional contracts for experimental aircraft, although the RAF was well equipped with an effective residue of war-time types, such as the Sopwith Snipe fighter, D.H.9A general purpose bomber, Vickers Vimy twin-engined bomber, and Avro 504K trainer, and for its marine services had Short and Fairey seaplanes and the Felixstowe F.5 flying-boat. Indicative of sales pressure by Vickers was an allocation of £984,000 for completion of airships and £37,000 for new purchases. Warned a critic: 'Airships are expensive luxuries when a nation is on the verge of bankruptcy.' Nevertheless a clique, who might financially or politically benefit, hoped the completed airships would be diverted to civil use on Empire routes.

For the first time the Estimates also revealed initial steps towards unification of all three Services in a single Defence Force as far as supplies were concerned, for the Navy was co-ordinating with the RAF in ordering rations, marine craft, torpedoes, machine-guns and ammunition, clothing, petrol, oil, maps, compasses; and the Army was helping with storage of guns, inspection of explosives, lands, barracks, equipment, and training personnel for medical, sanitary, and fire services. Maj Tryon presciently said: 'The Air Service goes forward united, preparing for peace or war. The officers and ranks constitute a Force that will maintain by its discipline, its training, and its skill, the pre-eminent position this country has attained. Looking back at the early years we see that the greatest triumph of all is the courage of those early pioneers, who, in the face of terrible losses, built the foundation of the RAF of to-day.'

The Estimates were not without critics, among whom Maj-Gen J. E. B. Seely, the previous Under-Secretary, in a general attack emphasized: 'The air industry in this country is dying; it is withering away, and it is most sad it should be so, and it is also very dangerous. Of the great firms which were producing aircraft and which had large design staffs – and it is on that our future in the air depends – nearly all have gone out of business. There remains one good big design staff, almost as big as before, one not greatly depleted, another reduced to only fourteen. There have never been more than a dozen first-class aeroplane designers in this country. Their design staffs consisted of ordinarily competent engineering draughtsmen, at £4 or £5 a week, a number of "stress merchants", some bright lads who design detail fittings able to take the stresses imposed on them, and a horde or tracers and blue-print makers. Altogether there were thousands.'

Winston Churchill, as Secretary of State for War and Air, wound up the Debate in a sweeping, convincing, and politely impolite reply to critics, stressing that decisions had to be wrested from other Departments of State, for which Cabinet sanction and Treasury sanction could only be obtained after prolonged negotiation. Thus construction and maintenance of airships had been taken over from the Admiralty; the Departments of Design, Production and Supply, including Research and the Experimental and Test stations, were acquired from the Ministry of Munitions; the Meteorological Office had been obtained; Civil Aviation had been rescued from the clutches of the Board of Trade and Ministry of Transport. In words long remembered: 'Civil Aviation must fly by itself; the Government cannot hold it in the air. The first thing the Government must do is get out of the way, and the next is to smooth the way. Any attempt to support artificially with floods of State money would not produce a really sound commercial aviation service which the public will use.' That certainly was not what the struggling embryonic air transport business hoped.

'During the year since we formed the new Air Council I have allowed no changes which could be avoided,' said Churchill. 'The same men are in the same seats exercising authority over the same sphere, and are doing so with increasing recognition that it is worthwhile to make plans and economies, not for the current year, but for next year and the year after; so they and their subordinates are increasingly feeling that they belong to a stable and solid institution where men are not jockeying each other for personal advantage or advancement, and where they may do their work in confidence day after day and month after month. I urge strongly on those who have the welfare of the Air Force at heart, that whatever Minister

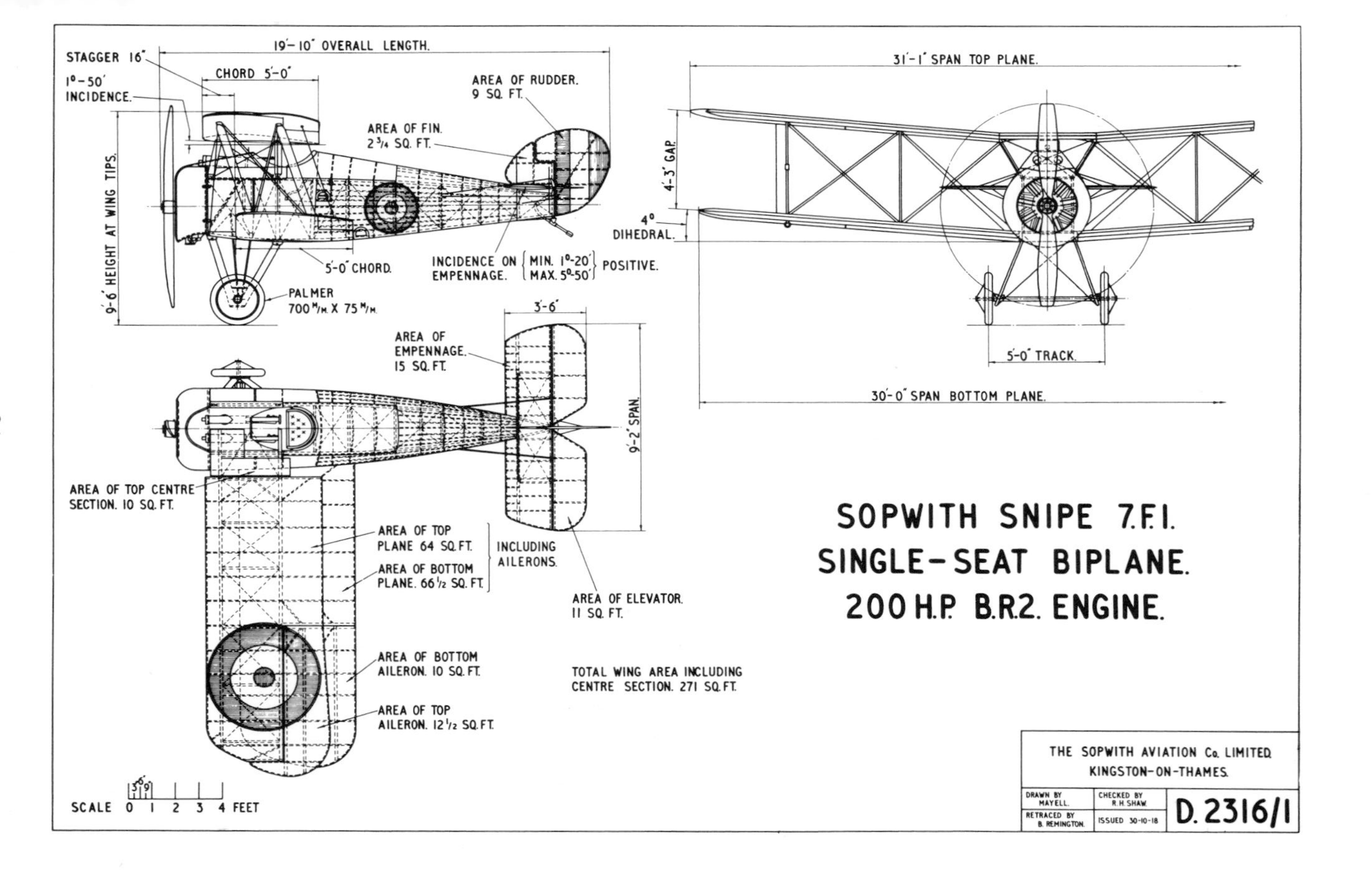
19'-10" OVERALL LENGTH.
STAGGER 16"
CHORD 5'-0"
1°-50' INCIDENCE.
AREA OF RUDDER. 9 SQ. FT.
AREA OF FIN. 2¾ SQ. FT.
9'-6" HEIGHT AT WING TIPS.
5'-0" CHORD.
INCIDENCE ON EMPENNAGE. MIN. 1°-20' MAX. 5°-50' POSITIVE.
PALMER 700 M/M. X 75 M/M.
31'-1" SPAN TOP PLANE.
4'-3" GAP.
4° DIHEDRAL.
5'-0" TRACK.
30'-0" SPAN BOTTOM PLANE.
3'-6"
AREA OF EMPENNAGE. 15 SQ. FT.
9'-2" SPAN.
AREA OF TOP CENTRE SECTION. 10 SQ. FT.
AREA OF TOP PLANE 64 SQ. FT.
AREA OF BOTTOM PLANE. 66½ SQ. FT.
INCLUDING AILERONS.
AREA OF ELEVATOR. 11 SQ. FT.
AREA OF BOTTOM AILERON. 10 SQ. FT.
TOTAL WING AREA INCLUDING CENTRE SECTION. 271 SQ. FT.
AREA OF TOP AILERON. 12½ SQ. FT.
SCALE 0 1 2 3 4 FEET
SOPWITH SNIPE 7.F.1.
SINGLE-SEAT BIPLANE.
200 H.P. B.R.2. ENGINE.
THE SOPWITH AVIATION Co. LIMITED
KINGSTON-ON-THAMES.
DRAWN BY MAYELL.
CHECKED BY R.H. SHAW.
RETRACED BY B. REMINGTON.
ISSUED 30-10-18
D. 2316/1

may, for the time being, be in charge, he must establish in this organization those conditions of stability and discipline without which no fine results, no real and certain success can possibly be achieved.'

Unfortunately there was little stability in world affairs. Ominously on 4 March a militarist revolution broke out in defeated Germany, and a few days later, the Germans penetrated the disputed Ruhr territory mandated to the French who immediately occupied Frankfurt, Darmstadt, and Hanau. Nor were tensions less on the eastern Mediterranean, for the Turks had massacred 7,000 Armenians in Cilicia (Syria), and Allied troops were rushed to Constantinople: war with Turkey seemed imminent. How effective would the League of Nations be? As a gesture to Empire solidarity, the dashing, immensely popular Prince of Wales concurrently departed on a voyage to Australia and New Zealand.

4

By far the biggest aviation deal of that time was announced on 15 March, whereby the entire stock of the Aircraft Disposals Board was purchased from the Ministry of Munitions by a syndicate registered as The Aircraft Disposal Co Ltd, with Handley Page Ltd as technical adviser and sole selling agent – though the latter arrangement was hardly surprising as H.P., his brother Theodore, and Lord Reading's brother Godfrey Isaac, of Marconi, were first directors. They stated: 'This new company will continue the active propaganda initiated by Handley Page Ltd, who have already sent missions to many countries overseas to demonstrate the merits of their own and other British machines. The development of this policy and the fact that all syndicate machines would be in perfect condition will go far towards ensuring a sound future for the British aircraft industry. The 10,000 aircraft taken over include the Vickers Vimy, D.H.10, D.H.9, D.H.9A, D.H.6, Sopwith Pup, Camel, Dolphin, and Snipe; the Avro 504K, Bristol Fighter, Martinsyde, and Government designed S.E.5, F.E.2b, B.E.2e. In addition there are large and small flying-boats of the F, H, and N.T. types, and a few Handley Page O/400 two-engined aeroplanes with Rolls-Royce or Liberty. The 35,000 engines include Rolls-Royce Eagles and Falcons, Napier Lions, Siddeley Pumas, Wolseley Vipers and Adders; French 200 and 300 hp Hispano-Suiza, Renault, R.A.F., Fiat, Anzani, A.B.C., Le Rhône, Clerget, B.R.2, Monosoupape, and others.'

There were also between 500 and 1,000 tons of ball-bearings, 350,000 sparking plugs, 100,000 magnetos, and vast stores of nuts and bolts. Value of the entire stock of aircraft, engines, equipment and miscellaneous items was £5,700,000, for which the syndicate paid £1,000,000 financed by the Imperial & Foreign Corporation, with whom Lord Balfour of Burleigh was closely associated. Half the profits on sales were to be repaid to the Government.

This was splendid for Handley Page, but what of other manufacturers? With tongue in cheek he said: 'A fixed rebate will be given to those aircraft firms who desire to buy back any machines of their own design. Owing to the existence of these stocks, the aircraft industry has to face a period when little manufacturing will be required, but modifying and renovating Government aircraft means much work for the industry. We hope that British firms will participate by contracting to the syndicate all this work and co-operate with us in establishing beyond question the supremacy of British aircraft in the world's markets.' The industry gritted its teeth and wondered how to make the best of things.

If H.P. was riding the crest of the wave, the *doyen* of all, 50-year-old George Holt Thomas of Airco, had come to the end of his ambitious undertakings. Hugh Burroughes, who left Airco at the end of 1919, told me years later: 'Holt Thomas had always a weakness in his throat and about the end of 1916 it began to get worse, seeming to impair his judgment. Immediately the war ended he wanted to go into all kinds of commercial work without adequate time to learn them, and it was over this that I had serious differences of opinion, with the result that I resigned – but he was very generous in paying me compensation. Pioneers vary as much as any other individuals – idealists, sheer adventurers, shrewd guessers: Holt Thomas came into the latter category, but it does not make him any less a pioneer.'

Recently a throat weakness had been confirmed as cancer: little could be done about it. Though Holt Thomas did not lack courage, there seemed little point in further struggling with the problems of aviation. Yet it took most people by surprise when *The Times* disclosed in March that: 'Mr Holt Thomas has resigned his chairmanship of the Aircraft Manufacturing Co Ltd, the future of which is now at stake. Mr Holt Thomas, one of the pioneers of British aviation, made a statement yesterday in which he clearly brought out the absolute necessity of State support. He explained his resignation as follows:–"By the amalgamation recently announced, The Birmingham Small Arms Co Ltd (B.S.A.) acquired control of my company. Their interest lies in our large factories, which are adaptable for productions such as motor bodies and engineering; naturally their first step is to foster these productions, and cut expense not likely to be remunerative in the near future. Could I honestly advise my co-directors, in view of the Government's apathetic attitude, to continue an expensive technical department devoted to aircraft design? Yet I could not, from the national aspect, regard the disintegration of a staff second to none in the world without considerable misgiving. I concluded that the only step was to leave the Board to decide matters without me, since in 18 months no official encouragement whatsoever has been given, nor is there prospect that Airco will have orders for experimental machines, say for £100,000, which would be sufficient to keep the staff together in the next 12 months. The utter failure of the authorities to view in proper proportion the importance of the Air Force, compared with Navy and Army, is the root of the matter. Minister after Minister has publicly stated that we must retain our lead. We are far from doing so. This is a new science, still in its infancy.

Ordering aeroplanes in quantity could not be expected, but £1 million a year spent on experimental machines distributed among design firms, would be a national insurance and maintain the technical staffs." '

Sombrely C. G. Grey said: 'It has been left to newcomers in the aircraft industry to shut down one of the oldest aircraft firms. Perhaps Mr Holt Thomas made an error in allying himself with the B.S.A.-Daimler group, but anybody might have made the same error. Happily there are other groups of directors who do not regard their duties as so strictly limited as those who now control Airco. The old original firms are strong in their belief and are maintaining aircraft departments even at the expense of more lucrative sides.'

The D.H.18 was de Havilland's next logical step in attempting speculative civil aircraft requirements, and was based on existing drawings of the proved Lion-powered D.H.14. (*Courtesy Air Marshal Sir Ralph Sorley*)

Geoffrey de Havilland, and his friend and colleague Charles Walker, who was five years older, had for several months been aware that Airco would probably cease operations. Encouraged by the successful first flight of the D.H.18 at this juncture in the hands of Frank Courtney, D.H. was quietly determined to carry on by starting a small new business. Holt Thomas was fully in sympathy. For the time being Airco remained an entity, and advertised de Havilland's machine as 'the Airco 18 Saloon Air Liner carrying pilot and eight passengers capable of a speed of 121 mph at 5,000 ft with full load, and a climb of 10,000 ft in $20\frac{1}{2}$ minutes.' The untrumpeted merit was that of design by a man who knew flying intimately, and therefore not only gave his creation good performance and robust structure but ensured safe and pleasant handling; nevertheless there was initial criticism that the pilot's location far aft gave appalling landing and taxi-ing view, for it was only possible to see by craning sideways because directly ahead the broad fuselage top and low centre-section made a bulky blind area.

As de Havilland was tied by transfer of his agreement from Airco to B.S.A., Holt Thomas advised him to approach the new directors, who significantly said they had realized he might wish to have his engagement cancelled, for if he remained with B.S.A. it must be on non-aircraft work. For a nominal sum they offered all design rights, jigs, and work in hand, including two D.H.14s partially constructed and repair work on D.H.9As. Now the problems began. It would take time to raise finances for a new business. As a start D.H. invested £3,000, and wrote to possible investors – just as in 1909 when, at 27, he left the Motor Omnibus Construction Co of Walthamstow and sought financial supporters so that he could construct his first aeroplane. Walker contributed £250; a motoring friend offered £1,000, but it was Holt Thomas who made things possible by promising £10,000 in instalments on condition that Airco's austere financial director, Arthur Turner – previously the master mind of contract finances at the Air Ministry – be appointed chairman of the new company.

5

The Tarrant crash had proved that civil action for liability was a contingency for which air transport operators must be prepared despite rigorous licensing control of flying personnel, ground crew, and aircraft. To clarify strength requirements, Report No. T 1425 of the Advisory Committee for Aeronautics issued a schedule of load factors mandatory from 15 March for future designs of civil aircraft, but not applicable to those already approved, though further revision from time to time 'will doubtless be necessary in order that it may remain in accordance with the demands arising from improvements in the constructional methods and design'. Stresswork methods were defined in the latest *Hand-book of Strength Calculations* (H.B. 806), based on the war-time booklet of Pippard and Pritchard, and must include a 'wire cut' case at not less than half the scheduled factor. Centres of pressure must be determined by wind-tunnel tests at 60 ft/sec on appropriate model wings of 6-in chord. Henceforth the calculators and aerodynamicists were very much in business, for aircraft design had at last become scientific by inescapable mandate. It was now in the interest of every manufacturer to establish his own wind-tunnel and testing laboratory whatever the endorsive help from the RAE. Only Handley Page and Airco had these during the war, but at Barnwell's insistence a wind-tunnel was built in June 1919 at Bristols, and Rex Pierson was pressing for one at Vickers, though to other manufacturers it still seemed doubtfully beneficial expenditure.

Urged by the Air Ministry, the new Bristol Aeroplane Co was considering an even bolder extension of activities – and here they were powerfully persuaded by the dynamic 35-year-old Roy Fedden, designer of the nine-cylinder radial Jupiter and three-cylinder Lucifer engines, whose employers, the Cosmos Engineering Co, had suffered serious setback unrelated to his aero department, and went into liquidation early in February. Only two engines had been delivered to the Air Ministry and paid for, but they were most promising, and at a weight of only 662 lb, or 1.4 lb per bhp, the Jupiter was the lightest aero-engine of its day. Brig-Gen J. G. Weir, earlier the Controller of the Air Ministry Technical Department, had written to Fedden: 'I wish to impress upon you that the changed conditions from war do not affect the importance of your experimental and development work, and it is most desirable you should continue your efforts. I need not enlarge upon the high hopes which I have regarding the Jupiter, beyond stating it is of very great importance to the nation that you should press on in all haste to perfect this engine which I feel sure has a considerable future for commercial aviation.'

Encouraged by such favourable support, Fedden obtained the Receiver's agreement to keep the splendidly equipped Cosmos Works at Fishponds and its design staff ticking while he attempted to find financial backing for a take-over. To that end he was vigorously attempting to interest not only Bristol but Vickers and Siddeley. Though Capt Peter Acland, the managing director of Vickers Aircraft Department, and General Festing his co-director were keen, their main Board, as the result of unfavourable pre-war and war-time engine experience, decided against. However, the fiery John Siddeley, intent on developing his rival Jaguar radial, was almost disposed to buy Cosmos to eliminate his competitor. At Bristols, essentially a family concern, Henry White-Smith, nephew of the firm's founder, greatly favoured acquiring Cosmos, but the managing director G. Stanley White, and Samuel White the current chairman, disagreed; convinced that the Jupiter was important, the Government began to put pressure on them to take over Cosmos.

Meanwhile the Bristol shops were almost empty of aeroplanes, for all contracts had been cancelled. Even the prospect of sales in the USA

Though not as big as the Airco empire, the Bristol establishment had increased tenfold and employed over 3,000 at the Armistice. Barnwell's original office was in the central row of villas still standing in the present factory.

vanished when that country imposed import duty on aircraft, but Herbert Thomas, youngest nephew of the founder, with outstanding effectiveness as works manager, retained his employees with carpentry of bus bodies for the Bristol Tramways Co and saloon bodies for Armstrong Siddeley cars. Barnwell's star assistant, W. T. Reid, who had designed the war-time metal-structured M.R.1, was given the side-line of producing a single-seat light car, and six prototypes were built, powered with a motor-cycle engine, but it soon transpired that miniatures had their problems. At Avros the ever inventive A. V. Roe was discovering similar snags with a monocar of his design, but though he personally tested one for 30,000 miles it was soon evident that the public was not interested. The car market in general was not encouraging, for the entire weekly production from all manufacturers was a mere thousand. There were only 150,000 cars on the road, and of these the Ford 'Tin Lizzie' swept the market for cheap vehicles with a price of £185 compared with £465 for a two-seat open Morris. Horses remained the mainstay of transport in countryside and town, numbering over a million, so that jingling harness was still the music of the roads and streets.

Every aircraft company was seizing any kind of work to keep going. Like Bristol, Shorts were building bus bodies, also specializing on marine craft from electrically-driven canoes to sea-going motor barges of 500 tons. Martinsydes had been building powerful sporting motor-cycles for nearly a year, but they were not proving profitable. Sopwith was making A.B.C. motor-cycles designed by the same Granville Bradshaw of the unfortunate war-time Dragonfly radial. Blackburn was another who turned to car and motor-coach bodies, but discovered a more lucrative if modest trade in nuts, bolts, turnbuckles, and aircraft general sundries (AGS parts) for the industry as a whole – but largely was sustained by hope that the private venture torpedo-carrier, the Swift, would prove a winner. Vickers, with less need to find ancillary manufacture, specialized on petrol cocks and pumps, but primarily were busy on their Vimy Commercial production for the Chinese Government, and an encouraging sign was the purchase of one by the shipping firm of S. Instone & Co following successful operations between their Cardiff, London, and Paris offices by Capt F. L. Barnard with a D.H.4. Vimys too had helped Westland consolidate with a contract for 25, but Bruce was an accomplished pianist and this led to the specialized task of constructing pianos for a German company. Fairey, like the others, had diversified and, while completing a contract for his splendid IIID seaplanes, was building motor bodies for Daimler through the aegis of a separate company known as Fairey & Charles Ltd. Even Handley Page, reassured by the splendid prospect of sales from the Aircraft Disposal Co, found his factory space excessive, so was cashing in by letting a large portion, of which a part was taken by an ex-barrowman called Smith who was packaging fried sliced potatoes under the title of 'Crisps'.

Unfortunately the B.A.T. Co received a bad blow on 23 March, when Chris Draper, their test pilot, put the demonstration Bantam into a spin

The civil registered Nieuport Nieuhawk was virtually a war-time Nighthawk with a much modified but still unreliable Dragonfly engine. (*Flight*)

which became too flat for recovery. Luckily he escaped with bad concussion and, though first reports were alarming, was soon out of danger, thanks to Graeme Anderson, the facial surgeon who had performed miracles for war-time pilots. But it finished the company. Lord Waring sold the remaining aircraft to Ogilvie & Partners, retaining British Nieuport as his sole aviation interest – and although 31-year-old Henry Folland, of S.E.5 fame, was its bright engineer, the future seemed doubtful for only a handful of Nighthawk and Nieuhawk derivatives had been sold and there was no interest in his civil designs.

6

There was much talk of the 'new poor', and that war profiteers had usurped the old aristocracy; certainly the younger generation had to keep more modest households than their parents in the days before the war. Speaking on the Air Estimates Churchill explained that the £15 million allocated to the RAF amounted to little more than £6 or £7 million pre-war. The Budget did nothing to encourage hope, for it embraced what seemed the vast sum of £1,340 million. Excess Profits Duty was increased yet again, and a Corporation Tax imposed on limited liability companies. Up went the tax on spirits and beer; postage was increased to 2d, telegrams to 1s; and to the discouragement of travel, the railways added still more to fares already 50 per cent greater than pre-war. Yet spending continued unchecked and, though some prices seemed high, others were bargains, for the sale of surplus stores had begun, handled by such people as Mr Mallaby Deeley, erstwhile MP, who bought for cheap re-sale hundreds of thousands of surplus suits of clothing originally intended for men on demobilization. Others got hold of great quantities of boots and masses of army shirts which were re-sold for a few shillings.

Compared with pre-war days the craze for entertainment enormously increased. There were dance clubs everywhere, with superabundant jazz. Golf and tennis clubs multiplied. Theatres were packed; *The Beggar's Opera* was a superlative hit. At cinemas, full houses watched the antics of Charlie Chaplin or swashbuckling films of Douglas Fairbanks and romances of Mary Pickford 'the world's sweetheart'. The young generation had felt little impact of the war in which they grew up, and even their elders half believed the pre-war world of their youth was recovering its former ease and tranquillity; but the unemployed were nearer despair.

International affairs remained the target of much ill-informed talk. The sea-moated British were strangely intolerant of the French on whose ravaged territory the war had been waged; nor for our American ally were there thanks. The £1 sterling had dropped below $4 on the exchange, and New York was scooping the world's gold: why could not those Yanks cancel the great war debt since they had done so little of the actual fighting? And Russia was dismissed as a bunch of useless Bolsheviks – yet there were less than 800,000 avowed Communists there.

It was Ireland where the deepest bitterness was springing. Ever since the 'coupon election' at the end of the war the Sinn Fein members, who had won every seat outside Ulster except four, refused to attend Westminster and met in their own self-constituted parliament in Dublin Mansion House intent on making Ireland a Republic. The American-born Eamon De Valera, who had recently escaped from Lincoln prison, was President of the *Dail*, but activities centred on a group who banded themselves into a Republican Army (IRA) under Michael Collins, armed with smuggled rifles and home-made grenades. At Easter, civil war broke out with IRA raids on barracks, post offices and tax offices all over Ireland. Hastily the Royal Irish Constabulary (RIC) was reinforced with English mercenaries who were paid 10s a day and wore khaki uniforms with RIC black belts and caps, thereafter known as the 'Black and Tans'. For a few months conditions were similar to the earlier struggle with the Irish hotheads; but presently a bitter spirit crept in, and on both sides there were ambushes, arson, hostages and murder.

But Easter in England passed quietly enough – too quietly for many in the aircraft industry, though it saw the opening venture of Cobham & Holmes Aviation Co with a flying fortnight of exhibitions and passenger-carrying at Warrington. So successful had been 26-year-old Mr Alan Cobham's winter flying with the former Berkshire Aviation Co that he proposed making a permanent career in aviation. Concurrently Avro began their joy-riding season at a number of centres, hopeful they could repeat their 1919 success when 30,000 passengers had been flown in four months; among their pilots were Capts E. D. C. Herne, F. G. M. Sparks, and H. A. Brown. Various other local joy-riding ventures were also operating, including one of Mr Handley Page's Avros, piloted by ex-Sgt Pilot Gordon P. Olley, which forced landed with engine trouble in Southwark Park, turning over in a shallow pond from which it was rescued overnight by the hard-driven H.P. team.

Joy-rides by many operators with Disposal Avro 504s had several years' success. Among them the scarlet G-EBIZ operated by Capt Percy Phillips became one of the best known and in ten years carried 91,000 passengers.

After Easter a major re-arrangement of the Air Ministry was made when the tall and statesmanlike Marquess of Londonderry, KG, was appointed Under-Secretary of State for Air in place of Maj Tryon who became Under-Secretary of State for Pensions. Other changes followed. Winston Churchill appointed that sporting pioneer Lieut-Col J. T. C. Moore-Brabazon, MP, as his Parliamentary Private Secretary for Air Ministry affairs, and Walter F. Nicholson, CB, formerly Assistant Secretary of the Admiralty, became Permanent Secretary of the Air Ministry in succession to Sir Arthur Robinson, KCB.

The Air Ministry now decreed that Government contracts for new aircraft and engines would be based on a general specification of requirements sent to major designing firms, entitling them to supply preliminary competitive designs within a time limit of ten to twelve weeks. These would be opened by the Contracts Department who would hand them to the Department of Research in such manner that the origin could not be identified by those who judged – though this overlooked the distinctive hand of every designer. After selection of the two most promising designs, Contracts would issue an order to the successful firms for prototypes at a sum intended to cover all expenses. Providing both proved successful, the better would be chosen as the 'Type design'. The Government would then purchase a set of master tracings for which a non-exclusive licence must be granted so that any selected firm could manufacture these machines, but as a sop it was stated that orders would be placed with the designing firm provided their production prices and deliveries were satisfactory. Manufacturers might growl, but from the Treasury standpoint competitive tendering at least should prevent price inflation likely with monopoly.

The Air Ministry was prepared to back steel-strip structures as exemplified by the Boulton & Paul Bugle of similar appearance to the aerobatic Bourges. (*Courtesy Air Marshal Sir Ralph Sorley*)

On this basis, Bristol and Boulton & Paul each submitted a design for an aeroplane described as 'Spares Carrier', featuring a multi-engine room with outboard transmission shafts to the wings. The Bristol Tramp was a triplane, based on the Braemar, powered by four central Siddeley Pumas, and the tender price was £23,000 of which £7,500 was for engines and transmission gears. Boulton & Paul entered a large metal-framed version of their proposed 'Postal' biplane having the general aspect of the Bourges,

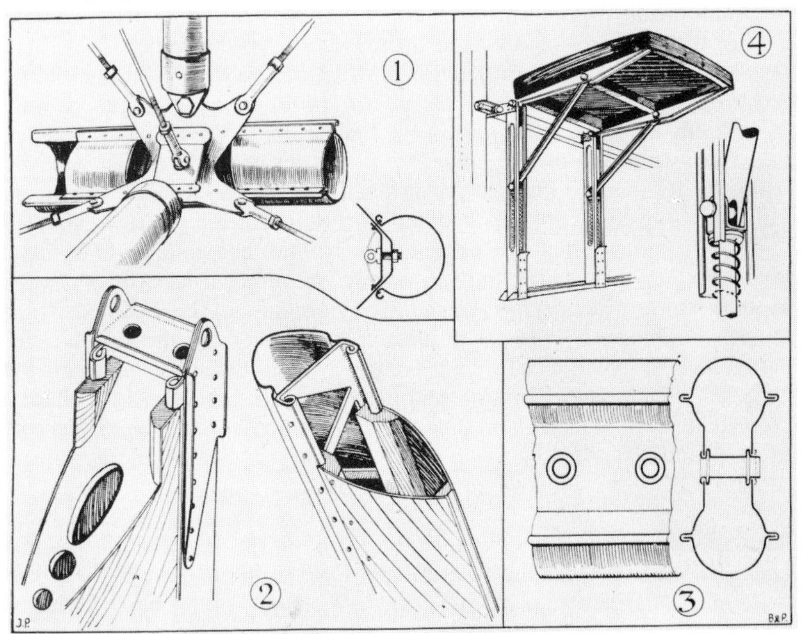

1, Boulton & Paul patented longeron and spacer joint. 2, metal fabrication of strut nose with plywood tail. 3, spar construction. 4, the gunner's folding seat was typical of care taken with all detail design. (*Flight*)

but with twin Napier Lions in the central engine room, and quoted some £20,000.

By contrast, at the Avro Hamble factory, convalescent Roy Chadwick was intent on a long-range single-engined heavy bomber, the Aldershot, to specification 2/20 utilizing the 650 hp Rolls-Royce Condor III. It was an impressive three-bay biplane of 68-ft span with all-up weight aimed at 10,000 lb. The fuselage of metal-tube girder construction was Avro's first attempt in this medium, but the contract was splendid acknowledgment of the regard with which Ministry officials accorded the young designer. His competitor was none less than de Havilland, who on low priority had been scheming the D.H.27 of a few feet less span than Chadwick's bomber.

Vickers were invited to tender for a long-distance twin-engined bomber (DOR Type 4), and 29-year-old Rex Pierson was draughting virtually an enlarged Vimy of 86-ft span, though it had some resemblance to Handley Page's outclassed V/1500 because the top wing was flat and only the lower had dihedral. Napier Lion engines were specified. The cost seemed remarkably low, for an initial order was placed for two airframes at £13,250 each. The second firm selected for this requirement was Armstrong Siddeley who offered a very similar biplane, the Sinaia, powered with twin 600 hp vee liquid-cooled Tigers.

Short Bros remained somewhat disinterested in such specifications, largely because Oswald was intent on the rapidly advancing construction of his all-metal *Swallow*, for he was building with obstinate determination against the strictures of Air Ministry metallurgists who were backing high tensile, thin-walled steel girder frame structures with conventional fabric covering for fuselage and flying surfaces. Like others, the Short shops were remarkably empty; the last nine of the Felixstowe F.5 flying-boats had just been completed for the Japanese Navy, but an order had been received from the Portuguese for a re-conditioned F.3, and a wealthy Australian, Lebbaeus Hordern, ordered another as a personal air yacht. Registered G-EAQT, it was tested at Rochester on 28 May and shipped to Sydney, but there languished, the hull ultimately becoming a fisherman's beach shelter. At low priority, was the unfinished structure of the Porte-type 113-ft span Cromarty flying-boat ordered to Admiralty specification N.3 towards the end of the war concurrently with a Vickers equivalent, the Valentia, which had a Consuta hull made by S. E. Saunders Ltd. Neither seemed needed, for RAF marine interest centred on a development of W.O. Manning's Phoenix flying-boat, the Kingston, which had a rounded section, mahogany-planked Linton Hope hull.

For those who pioneered the air before the war, a reminder of those days was the passing, on 17 April, of Alfred Kirby Huntington who had been Professor of Metallurgy at King's College but was an early ardent balloonist, later associating with Capt J. W. Dunne on problems of inherent stability, purchasing his original powered tailless biplane, and subsequently designing and building with him a curious tandem-like three-winger which he test flew at the age of 60, accepting like his younger contemporaries the risks attendant on personal experiment with untried aeroplanes.

On 29 April a further setback occurred to the development of super-large aircraft, for the Felixstowe Fury triplane flying-boat stalled and crashed when taking-off, and its pilot Sqn Ldr 'Rolly' Moon, an affluent officer of endearing character, was drowned. Designed by the late Col J. C. Porte, and his assistant Maj J. D. Rennie, the hull had been criticized by the maestro of NPL tank testing, G. S. Baker, when he received its model only after the machine was completed – for he found it liable to porpoise at 27 kt and it was too late for changes. Affairs at Felixstowe were currently

More than a decade passed before any other British flying-boat attained the weight-lifting capacity of the Felixstowe Fury at 33,000 lb all up, carrying 24 passengers, fuel for seven hours, and 5,000 lb of ballast.

chaotic, for Rennie had left to join Bristols, and the Fury was erroneously loaded, bringing the C.G. fully aft, where at the extreme weight used there was little available power for acceleration once clear of the water. Moon tried to hold the porpoising machine airborne when it bounced off, whereupon it dived into the sea and was wrecked. This, and the Tarrant crash of the previous year, put an end to several ambitious schemes for extra-large aircraft.

It was the sheer financial risk of building aeroplanes on speculation which was concerning the pioneers. From the Sopwith enterprise, Herbert Smith, their designer, said that in his opinion 'the day of the large aeroplane has not yet come, and much remains to be investigated and determined concerning multiple-engined machines. The Sopwith Co are at present pinning their faith to single-engined aeroplanes of moderately high speed. They are constantly receiving enquiries, but orders are not exactly as plentiful as they desire, and the Air Ministry does not provide the help and encouragement expected of them.' All they had in hand as the result of orders placed by the Larkin-Sopwith Co of Australia were a few Gnus and the two-seat Dove derived from the Pup. A more recent prototype was the 100 hp Anzani two-bay Grasshopper trainer, but with Disposal Avros available it seemed needless despite Hawker's valiant sales talk that it would inspire 'utmost confidence in the most nervous pupils'.

This metal Siskin II, powered with a Jaguar, was photographed at Martlesham Heath. Though generally similar to the war-time experimental Siddeley S.R.2, lineal descendant of the S.E.5, was very different in detail. (*Courtesy Air Marshal Sir Ralph Sorley*)

Yet there was still a place for new fighters if vigorously pushed. In the latter part of the war a young mathematician John Lloyd, now 32, who joined Siddeley-Deasy with his boss, Maj Green, on the virtual disbandment of Farnborough technicians, designed an attractive S.E.5-like fighter, the Siskin, with minimum help, using as drawing office a small back room above a laundry at Coventry. Armstrong Whitworth of Newcastle and Siddeley-Deasy had combined in 1919 to form the Armstrong Siddeley Development Co Ltd, and from it sprang the recently registered Sir W. G. Armstrong Whitworth Aircraft Ltd, though it had only a small corrugated iron shed on Coventry's London Road. The Siskin prototype, constructed of wood, had first flown with the impracticable 340 hp A.B.C. Dragonfly early in 1919, as evidenced by photographs in *Flight* and at the RAE, but now development trials recommenced at Radford aerodrome, Coventry. It seems probable that this machine was a subsequent prototype with Jaguar radial, and steel-tube structured girder fuselage with spacer struts secured by clips and lugs and braced with swaged rods as patented by Lloyd and Green (Siddeley-Deasy 139,034 & 151,085 of March 1919). In the hands of tall Frank Courtney of the wavy black hair, clipped moustache and *pince-nez* glasses with cord over his ear, the Siskin revealed handling characteristics which he considered superior to any fighter in contemporary service.

7

The future of the Advisory Committee for Aeronautics had been under surveillance by an investigating committee appointed when Lord Weir of Eastwood was Secretary of State for Air, and met under the chairmanship of the fluently discursive Sir Richard Glazebrook. Their recommendation was reconstitution in the form of an Aeronautical Research Committee (ARC) performing rather different functions, including executive duties; and that a Department of Aeronautics for advanced instruction should be established at the Imperial College of Science, London, directed by the Zaharoff Professor of Aviation, Leonard Bairstow. Glazebrook was re-elected chairman of the ARC, and there were to be two representatives from the Dept of Civil Aviation, two from the Dept of Supply and Research, two from Imperial College, two from the SBAC, with Alec Ogilvie representing the Royal Aeronautical Society (RAeS) and J. L. Nayler of the NPL as secretary. There would also be five scientific representatives: Horace Lamb, Professor of Mathematics at Manchester University; W. E. Dalby, Professor of Engineering at City and Guilds; B. Melville Jones, Professor of Aeronautical Engineering, Cambridge; G. I. Taylor, Lecturer in Mathematics, Cambridge; and Col H. T. Tizard, Lecturer in Natural Science, Oxford. Sub-committees would be appointed for accidents, air inventions, aerodynamics, engines, meteorology, and navigation, with others as required.

At this point Imperial College was battling for power to grant its own degrees – for students wishing to take a BSc had to sit for the examination of London University, and to an engineering training must add subjects alien to their special requirements. 'On the whole,' wrote C. G. Grey, 'one feels that those who aspire to become engineers will do well to avoid having more than is necessary to do with the imagination-cramping efforts of the professional scholiasts who control most of the established universities' – but these academic matters were scarcely of interest to an industry wondering how to survive. Consolidation was the target. Thus negotiations long underweigh between Crossley Motors and A. V. Roe & Co now resulted in amalgamation of the two companies, producing another sour comment from C. G. Grey: 'It is to be hoped that the Crossley directors esteem aviation more highly than do the directors of the Daimler and B.S.A. concerns, who have shown so little appreciation of the Aircraft Mfg Co's possibilities.' The new administration comprised Alliott V. Roe, OBE, and John Lord his dynamic little business assistant, together with Henry Fildes, MP, appointed by Crossleys as chairman, with W. M. Letts, OBE, as managing director, and H. E. Shuttleworth as third director giving voting majority. Sir Kenneth Crossley stated: 'Mainly with the object of manufacturing our own bodywork we have acquired a controlling interest in A. V. Roe & Co Ltd who have excellent factories in Manchester and at Hamble.

The former could hardly be better suited, but require considerable working capital to produce best results. Nevertheless we are anxious not to hamper development of aeroplane manufacture, for commercial aviation is certain to come sooner or later, and no one in Great Britain did more valuable work as an aeronautical pioneer than Mr A. V. Roe. His was the first machine to fly in England. But we cannot afford to run the aeroplane business as a philanthropic institution, so until the Government makes up its mind how to act, the provision of a large quantity of what we hope will be profitable bodywork for the Roe Co will enable it to retain the best of its labour and improve still further its staff and organization.' To consolidate their position Avro meanwhile obtained an interim injunction restraining Aircraft Disposal from selling any aeroplanes or goods as Avro products unless of Avro's manufacture because many other companies had produced the patented 504K components during the war, though there was no objection to these machines being sold 'as of Avro type'.

Avro Baby G-EACQ was bought by H. J. 'Bert' Hinkler in April, and after return by easy stages from Turin it attracted great attention at the Olympia Aero Show. (*Flight*)

Sir Roy Dobson told me in later years: 'Poor old Alliott! He seemed a bit in the dumps over this take-over and building bodies for Crossley cars – but he cheered up a bit when on 31 May little Bert Hinkler, an Australian who had done some flying for us after leaving the RNAS, took-off from Croydon in the Avro Baby, heading for Turin, which he reached in $9\frac{1}{2}$ hours, having covered 650 miles on 20 gallons of petrol. Those were the days!' Italy was well known to Bert, for in 1917 he had flown as Peter Legh's observer, before becoming a Camel pilot there, and the story was told that his skill as an engine tuner, coupled with his very light weight, enabled him to keep formation with pilots using 130 hp Clergets though his own had merely a 110 hp Le Rhône. His great ambition was to fly to Australia, for which Tom Sopwith, now 32, loaned a Sopwith Dove, but the start was so long delayed by lengthy arguments of the International Air Convention that by the time permissions were received the Dove was rescinded for an export order.

On the opposite side to the Sopwith sheds at Brooklands, the next Viking had been completed at the Vickers factory. *The Aeroplane* reported: 'Capts Cockerell and Broome, who carried out the flight tests, have expressed pleasure at the behaviour of the machine and its performance. The Viking is very light on controls, and easy to fly. It gets off after a very short run, and pulls up quickly on landing. The improved retractable undercarriage has stood up well, and tests of its retractability in the air indicate that the device works well.'

By way of exploring very different possibilities, the airship R.33, with Flt Lieut C. M. Thomas as skipper, and the US crew of the R.38 aboard, was flown with a suspended Sopwith Camel which had a rubber-covered safety petrol tank devised by Farnborough. With engine running, the pilotless Camel was released, and after an erratic flight crashed half a mile

Encouraged by the pilotless launch of a Sopwith Camel from R.33, a much later experiment was made with two piloted Gloster Grebes from the same airship. (*Flight*)

away, but no fire broke out when the tank burst on impact. Pundits shook their heads: such tanks were too heavy, for payload was more rewarding than safety.

At Farnborough a fascinating array of aircraft might be found. Said a visitor: 'One sees a B.E.2c with large backward stagger, a two-seat S.E.5 and a two-seat Snipe. There were three Bristol Fighters, one with a single wing-bay, another with two bays, and a third with three. One could find an S.E.5b, R.E.9, Sopwith Bulldog, Bristol Badger, D.H.9 with Napier Lion, and numerous Nieuport Nighthawks undergoing engine reliability tests. One gathered from the experimental pilots that the Nighthawk together with the B.A.T. Bantams are the most popular, and a forced landing with the Lion-engined D.H.9 is unknown.'

The chief test pilot, 26-year-old Sqn Ldr R. M. Hill, MC, AFC, had been making the first ever analytical investigation of aeroplane flying characteristics, for by anomaly these qualities became empirically developed before their nature could be realized and coherent idea formed of them. 'When they come to be analysed a multitude of obscurities is found,' wrote Hill in his report. 'The attempt to investigate, define and classify them is involving years of research. Early aeroplanes, such as the Wright biplane, must have taxed human skill to the uttermost in satisfying the elementary requirements of flight; in other words, while just rendering the maintenance of equilibrium possible, they were bad. The ideal aerial gun platform for a single-seat Scout should be susceptible to sensitive intentional movements, without appreciable time lag. If rigid, it is easy to keep the sight on a target, but not to bring it there. Instead it must respond evenly to the pilot's hand but not so suddenly as to disturb his aim. In short, the pilot wants to identify the nose of the aeroplane with one of his limbs, and not regard it as a separate agent, to be forced, coaxed, or juggled into the correct position. The highly unstable aeroplane makes a big call on the pilot's energy the whole time, runs away with him if he leaves the controls, and renders him liable to get into extremely awkward positions, which is very vulnerable. The highly stable aeroplane, on the other hand, while it will fly itself at most speeds with a trimming tail, is usually less manoeuvrable, stiff and unresponsive in a dive, and generally more unwieldy.

'Take the Sopwith Camel and the S.E.5, two conceptions diametrically opposed. The S.E.5a is stable with elevators free, the Camel unstable with them fixed. The Camel is more lightly loaded and has, with exception of the rudder, more powerful controls. In a dive the Camel is flicky, due to lighter loading and excessive longitudinal instability; the S.E.5a is very steady, but dull to small intentional movements. In a zoom the Camel improves greatly owing to its lighter loading and instability; the S.E.5a is inclined to become languid, and its stability near stalling draws down the nose so that a large backward stick movement has to be made. In a Camel the pilot has always to make small movements of controls to pick up steady speed which is difficult to maintain. Going into a dive, compared with the S.E.5a, the elevators of the Camel work the reverse way: the stick, though initially pushed forward, has to be pulled right in again, but the S.E.5a, if the stick is pushed forward, would drop its nose, creep up to trimming speed, and stay there. The S.E.5a is impossible to fly inverted; the Camel may remain so unintentionally. Nevertheless there is no doubt that the conception of the Camel as a fighting aeroplane made an irresistible appeal to a certain class of pilot, but it never could be comparable with the S.E.5a because, even assuming that it possessed certain qualities equal or even superior, the difference in performance – from 7 to 10 mph at 15,000 ft – and the view, gave the S.E.5a a superiority that only fighting pilots really understand.'

Though these were outmoded machines, replaced by the Sopwith Snipe, the stability and manoeuvrability factors were fundamental. Hill was no less illuminating on large types, describing the D.H.9A as excellent because of its general handiness and straightforward control; the Handley Page O/400 would only feel nose-heavy in a dive if badly out of trim, despite lacking an adjustable tailplane; a Vickers Vimy felt very nose-heavy as it always diverged from its unstable trimming speed and at 120 mph the force was so considerable that the pilot had difficulty in pulling up the nose; the D.H.10 although satisfactory in flight was tricky during take-off, for even if the engines were opened with deliberation the pilot had to pull the elevators up fully to counteract the marked nose-pitching defect due to the relatively high thrust axis.

Less serious was the combined lateral and directional stability problem. The S.E.5a was considered excellent, though with rudder held straight the machine described small lateral oscillations indefinitely; but the Rolls-powered Bristol Fighter was so poor that it was difficult to keep an accurate compass course when in cloud, and the Puma version was worse with the nose swinging emphatically from side to side if not checked by the rudder. 'Some aeroplanes are extremely sensitive to changes in position of the load, others can be loaded in the most extraordinary way and yet can be quite comfortable to fly. The Avro training biplane and the Handley Page O/400 are examples of the latter; the R.E.8 and Vickers Vimy of the former.'

Hill stressed the danger of multi-engined aircraft when an engine failed during take-off or initial climb. The out-of-balance swing caught out many a pilot, but: 'The Handley Page O/400 was flown for a considerable time on service without serious trouble because relative to the power the aeroplane was of large dimensions. If one engine lost revolutions the turning tendency was not violent, the whole aeroplane being comparatively sluggish due to its large moment of inertia; the fact that it had over-balanced rudders, though sometimes commented on, was not seriously complained of, though they caused the aeroplane to swing from side to side when taking off, and if left free in the air took charge and set up a violent swing; however, little force was required to prevent them doing this. It was with aeroplanes of smaller dimensions of the same power that the earliest troubles arose following failure of one engine. The Vimy and D.H.10 would both swing violently, the latter more quickly than the former owing to its small inertia. If the pilot had not time to gain sufficient speed to fly straight on one engine, the only thing was to throttle the other right down, landing wherever he was.'

He warned that more powerful aeroplanes of future design might well be fraught with difficulties making their characteristics unsuitable even in so simple a manoeuvre as landing. 'For the inexperienced pilot, a medium sized aeroplane of 400 to 500 sq ft area (such as the Bristol Fighter), which is reasonably handy without large control forces, presents the smallest difficulty in handling. The essence of a pilot's problem with the smaller Scouts is that his mental response is frequently not commensurate with their inherent quickness; with twin and multi engined aeroplanes it is his physical response to opportunities of control that is too feeble. Photographs of early aeroplanes getting off frequently show one wing dangerously dropped. The saving condition of these aeroplanes was their light loading. A medium sized, under-powered, lightly-loaded aeroplane is unpleasant to get off; a similar under-powered and heavily-loaded one is worse; even more so is an under-powered, heavily-loaded aeroplane of large size – which in commercial aeroplanes it is most essential to avoid. The attainment in a flat calm of a definite height at a definite distance from rest with a good margin of speed and consequently of control should be rigidly insisted on, and this height and distance should be such that the pilot is quickly freed from the embarrassment of the ground. Nothing must stand in the way of safety.' Clearly there were problems ahead for every designer additional to those of structural development and increasing use of metal. Not only must wind-tunnels be used to get the aerodynamics right, but it might be necessary to engage a permanent test pilot as a yardstick for the attainment of satisfactory handling qualities of each new design rather than use the temporary services of a 'free-lance'.

Crashes were still all too common. At the end of May, Air Commodore R. M. Groves, CB, DSO, AFC, who had taken temporary command from Sir Geoffrey Salmond in the Middle East, crashed in a Bristol Fighter at Almaza, Egypt, with fatal injuries. As Brigadier-General he had been Deputy Chief of Air Staff under Maj-Gen Sir Hugh Trenchard at the amalgamation of the RNAS and RFC, and re-assumed that post when Sir Hugh returned to the Air Ministry. Though a great administrator, those who knew his methods as a pilot had been astonished he survived so long.

8

Single-engined airliners in the guise of D.H.16s or 18s, and the French Breguet 14, together with an occasional twin, such as the blue and silver Vimy Commercial of S. Instone & Co, now used the new London Terminal aerodrome at Croydon instead of Hounslow as of old. Originally part of London air defences, it comprised two adjacent airfields of Wallington and Waddon used as one, with hangars on the first and flying from the latter, the two being separated by Plough Lane, across which the aeroplanes taxied while road traffic was held up by a man with a red flag. Maj S. T. L. Greer was aerodrome manager, and the ex-RAF living quarters were improvised as airport buildings, dignified with Customs facilities and even a lounge having a dining room and bar operated by Trust Houses Ltd. Said a reporter: 'Cutting the grass is progressing as fast as possible, but there is a large bump on the far side of the aerodrome which is fatal to tailskids. One hopes there will soon be marked improvement in the wireless at Le Bourget: machines usually arrive at Croydon long before the message that they have left Le Bourget, and those announcing safe arrival are delayed to a ridiculous extent. This is very annoying for all concerned.' Airline pilots chose their own flight course from start to finish, dependent on quick eyesight to avoid other aircraft, and with low cloud base often flew just above the trees of the undulating countryside between Croydon and the Channel. They were given somewhat uncertain weather forecasts by the Air Ministry before starting, and on returning to English shores circled Lympne, near Folkestone, where ground signs of canvas indicated visibility and wind direction across southern England. All too often it resulted in long detours dependent on the pilot's knowledge of hills and valleys. To fly as passenger was fraught with the unwitting peril of adventure.

July opened with the unpublicized arrival at Pulham of the largest of the Zeppelins, L.71, surrendered under the Peace Treaty. Piloted by Capt Heine with a crew of one officer and 21 men together with three British officers, it left Alhorn in Germany the previous evening despite a 20 mph head wind. The airship's arrival was entirely unexpected because the wireless was out of action, and the RAF personnel were still in pyjamas! After cruising over the Norwich district most of the morning it returned to Pulham at midday, landing without assistance in front of the shed built for

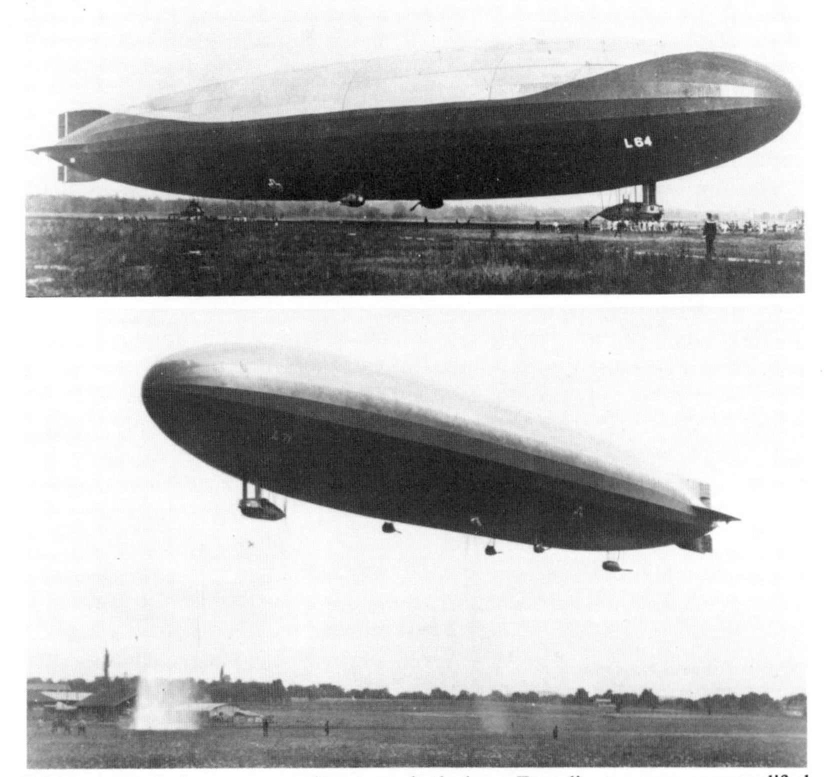

British airship designers were anxious to study the latest Zeppelin structures as exemplified by L.64 and L.71, but though percentage structure weight had been steadily reduced in the course of the war the Germans knew all-round improvement was possible.

R.34: no handling lines were dropped; when at ground level the landing party simply walked it into the shed. Admiration for the skill of the German crew was second only to the great question whether this last word in the art of airship construction could be continued as the beginning for Empire airship services – and to that end Vickers Ltd hoped to acquire it. A smaller sister, L.64, would be handed over at Pulham later in the month.

Already the Great War in which these ships participated had become history. On 9 June, King George V opened the Imperial War Museum in Lambeth. A large section was devoted to British aircraft, exemplified by the B.E.2c, R.E.8, Sopwith Camel, Salamander and Triplane, Nieuport Nighthawk, Parnall Panther, Bristol Fighter, the actual Westland-built Short seaplane used at Jutland, and a Norman Thompson flying-boat afloat in a pond – but there were also German aircraft such as the Friedrichshafen bomber, Roland, all-metal Junkers monoplane, Gotha bomber, reconnaissance L.V.G., and an 8-ft wheel from the POL Giant intended for a flight to New York to drop propaganda.

Adding to the summer's aeronautical interest was an announcement that a Royal Air Force Tournament would be held at Hendon on 3 July in aid of the RAF Benevolent Fund. The public imagined a gymnastic display, but it was soon made clear that this was to be a huge flying show to demonstrate the work and efficiency of the RAF. It was part of Trenchard's method of building up the *esprit de corps* of his Service, implementing the great tradition earned in aerial fighting.

Hendon was the subject of bitter contention between the Air Ministry and Claude Grahame-White, for the RAF proposed retaining the aerodrome because of its communication proximity to London; but Grahame-White wanted it back, hoping to resume his leading position in private civil aviation. His extensive factory no longer built aeroplanes, but was hopefully making furniture. True he had established the splendid premises of his pretentious London Flying Club, but members did little flying for the aerodrome could only be used by permission of the Air Ministry. A recent visitor reported: 'There stood a row of sheds bearing pathetically reminiscent names, such as The Aircraft Co Ltd, Deperdussin, Beatty School of Flying – but all were filled with thousands of overpowering leather hides in storage.' Meanwhile the Treasury claimed that Grahame-White owed some £120,000, and he counterclaimed £400,000 in respect of cancelled contracts and rental of the aerodrome. It was deadlock.

A crowd of 40,000 streamed to Hendon for the Tournament. From the same white-painted iron fences of the pre-war enclosures when week-end racing was all the vogue and the Aerial Derby of 1914 attracted an equally huge attendance, they saw what seemed an armada of aeroplanes, though in fact not more than 60 were present, grouped in flights of three and five, with a sprinkling of special types such as the S.E.5b. There were only 25 operational and 11 training squadrons in the entire Air Force at home and abroad, with five for naval use, compared with 500 squadrons in November 1918. Nevertheless the display at Hendon of khaki-painted

Aeroplanes of the first RAF Tournament – later known as the RAF Display. They comprised Avro 504Ks, Sopwith Snipes, Bristol Fighters, Handley Page V/1500s, and a few experimental aircraft. (*Flight*)

Avro 504Ks, Sopwith Snipes, Bristol Fighters, Vickers Vimys, and three great four-engined Handley Page V/1500s made an impressive sight. The 'trade' somewhat insignificantly backed them with an O/400 airliner, the repaired D.H.14 with which Cotton had tried to fly to the Cape, a smart pale blue Martinsyde two-seater, a grey Vickers Vimy Commercial, and a Sopwith Grasshopper.

There were aerial fights between a Fokker D.VII and a masterly Martinsyde F.4; a Bristol Fighter put up a splendid defence against two Snipes piloted by Flt Lieut A. Coningham, DSO, MC, DFC, and Flg Off G.

Visitors' cars were much more varied than the aeroplanes, and were numerically almost as limited, but the foot-slogging crowd packed the length of the aerodrome. (*Flight*)

E. Gibbs, MC. Even more breath-taking was an incredible display by Flt Lieut Jack 'Oojie' Noakes, AFC, MM, one of the ex-sergeant D.H.2 pilots of the McCudden era, with what seemed impossible skids and flat turns and imminent crashes in what he called 'crazy flying'. Said an onlooker: 'Mr Noakes is a very fine pilot indeed, and if he set out to be an exhibition pilot, which is quite different from a test pilot or a stunt pilot, he would find few to beat him.' Yet when Flt Lieut Walter Longton on a Camel followed with a supreme aerobatic display of leisurely loops and rolls it was clear that few could beat him either.

Following, came a display by the Handley Page V/1500s. 'The three vast beasts trundled across the rough grass aerodrome looking like performing elephants. Then they turned and swept towards the crowd as if to demolish the enclosure. It was a terrifying sensation as the twelve Rolls-Royce engines roared towards the railings. One confesses to a sigh of relief when Sqn Ldr Sholto Douglas's tailskid passed innocuously some 15 feet above one's head. The "super-Handley" is not exactly a machine for trick flying, but several times the three did turns with distinctly perceptible banks, and changed direction within the confines of the aerodrome.' From one of them Miss Sylvia Boyden descended with a red-white-and-blue *Guardian Angel* parachute, and a Nieuport Nighthawk and a Sopwith Snipe flew around and over the Handley Pages, like porpoises playing round whales. The crowd roared applause at the pluck of Miss Boyden – but to the discerning this drop by a woman revealed the glaring omission of parachute equipment for the RAF.

When five Bristol Fighters 'equipped with wireless telephones so that the pilots were manoeuvring to word of command like a platoon of infantry' demonstrated formation flying there was amazement at their precision. This 'wireless' was going to become important, it seemed. By contrast, a succeeding formation of Snipes lacking radio communication seemed less impressive, though their simultaneous looping and Immelmann turns were very pretty.

Powerful Martinsyde F.4 and lower powered Fokker D.VII represented the contrasting conventional wire-braced British design compared with cleaner German cantilever at the end of the war. (*Flight*)

Even for Bristol Fighters to take-off in formation seemed a daring exploit and certainly looked dangerous. (*Flight*)

'It's Hawker now,' and cheers rang for the hero of the Atlantic attempt. A masterly exhibition followed with his Sopwith Swallow monoplane – though those who knew felt he should not be flying because recently he had been undergoing treatment for a back injury from a crash during the war, and was far from well. As he came in to land, a Snipe dived to attack an ancient kite balloon. Flames leapt from the envelope, and a dummy observer fell through the bottom of the basket with his parachute jammed. Horror! It seemed a real man struggling to free himself, and when the burning remnants fell behind the brow of the aerodrome the crowd in the shilling north enclosure made an excited rush to view the disaster. After that, an attack by Bristol Fighters in line abreast on a trench which they blew up, and the dropping of white smoke bombs from an O/400, seemed anti-climax, though amusement revived when a zig-zag-strutted 'Heath-Robinson' caricatured Avro 504K took off with what seemed eight passengers sitting along the length of the fuselage, while clouds of smoke poured from a chimney behind the centre-section. It flew like a jumping rabbit and landed with a hop-hop-hop to the delight of the youngsters.

The splendid success of this first RAF Tournament not only resulted in £6,000 gate money for the RAF Benevolent Fund but was an inspiring introduction to the first post-war International Aero Exhibition which opened at Olympia, London, on Friday 12 July with an inauguration ceremony by the Marquess of Londonderry. Said a commentator: 'The hereditary legislator, born and trained to the work, has distinct advantage over mere elected amateurs. People are apt to forget that Lord Londonderry is an hereditary millionaire who has run his vast estates and mines for 15 or 20 years with great success and much to the benefit of thousands of Irish workpeople. The Air Department is fortunate indeed both in its Secretary of State and its Under Secretary.' Perhaps it was by intent his Lordship referred to the aircraft industry as 'undertakers', but he told the assembled members that he regretted both the present unavoidable stagnation and that the public had not adopted aviation with enthusiasm. Supporting this theme, Sir Hugh Trenchard said he did not want anyone to

View from the balcony of Olympia, with Central Centaur in foreground and Austin Whippet beyond. The Supermarine Sea King can be discerned between the wings of the Handley Page W.8 and the Vickers Vimy Commercial. (*Flight*)

think the Air Ministry did not recognize the usefulness of civil aviation – but it was necessary to keep the aircraft industry alive by sound means and not merely by subsidies: he hoped to see it equal the shipbuilding industry in time.

In the great glass-roofed hall of Olympia every major manufacturer had a stand, and machines ranged from the big Bristol Pullman triplane, Vickers Vimy, and Handley Page W.8 down to a resurgent B.A.T. display featuring the Limousine F.K.26 re-named Commercial Mk I, the F.K.23 Bantam, and the miniature Crow which amounted to a motor-cycle side-car slung under a pair of wings, with tail carried on a pair of box-girder beams. A surprise was the unheralded Fairey Pintail seaplane typical of its family yet with an upward-tilted parallel-sided rear fuselage reminiscent of the German Brandenburg, and rudder below the tailplane to give clear field of fire aft. Blackburn's new prototype Swift had been built in only five

At Olympia the 20-ft high Bristol Pullman fourteen-seat triplane seemed sensational to the public, but when at Martlesham for trials in August the Service pilots so strongly objected to enclosure in a cabin that the machine was scrapped.

Shown at Olympia with long fin, small rudder, and straight wings, the Blackburn T.1 Swift proved so tail heavy on its subsequent first flight that sweepback had to be built in and the tail re-set to greater incidence. (*Flight*)

The prototype Fairey Pintail I, powered with a 450 hp Lion I, was an amphibian with retractable wheels mounted inboard on each float, but this narrowed the track unduly.

months and, though not quite complete, showed remarkable advance on his previous machines, its importance emphasized by the torpedo resting on the ground between the undercarriage legs because its suspension crutch was on the Secret List banning display of specialized naval equipment. Despite the difficulty of stagger, the Swift's wings folded, though on a different principle from the 1917 staggered Short S.313, which was the first on which this problem had been tackled.

The single-seat Short Swallow mail carrier, renamed Silver Streak, with metal monocoque fuselage and metal-covered wings, was the glittering object of dissident contention at Olympia.

It was Short Bros with their all-metal Swallow, renamed Silver Streak, which marked by far the greatest advance. But *Flight*'s editor dismissed it with the comment: 'The fuselage was a metal shell similar in construction to the wooden monocoque type,' and in *The Aeroplane*, Capt W. H. Sayers seemed nearly as cautious, saying: 'One "Swallow" is but little towards the summer of metal construction. There are other examples showing that experimental work on metal aeroplanes has progressed greatly since 1914, and there is no insuperable difficulty to the manufacture of complete aircraft of either steel or light aluminium alloy. Metal has many advantages as a structural material as against timber. It does not swell and contract and change its shape with alterations in temperature and humidity as does wood. But it does rust or corrode unless very special precautions are taken against such happenings. And, of the two evils, probably many will prefer the warping of wood. With the existing demand there is no question that the metal aeroplane costs more than it is worth as compared with the existing type.' Another reporter wrote: 'Other aircraft constructors are building machines of steel. Yes, steel! Not mere light alloys. They are weaving delicate structures of steel like spiders' webs!'

Only the editor of *The Engineer* seemed fully to appreciate the significance of Oswald Short's great development: 'A broad critical survey of the exhibits leaves certain conflicting impressions as to progress and promise for the future. In many important respects aeroplane construction seems to have become stereotyped and shows little essential advance over the methods and means of 1914. The machines do not fail in every instance to impress a sense of progress, but we think this is caused chiefly by non-technical details of design, body work, decoration, upholstering and general arrangements for comfort of pilots and passengers. In mere size there has been a great step forward, as illustrated by the Handley Page and Bristol Pullman, each capable of carrying nearly a score. But mere size is not the criterion of technical development, although a great amount of progress is implied in that such large machines can be built to fly with as much trustworthiness as smaller ones. The real advance cannot be properly estimated at the exhibition at all, for it lies in features which can only be demonstrated in the air.

'Were we to judge solely by the evidence at Olympia there would be disappointment that only in *one* instance is there radical advance in constructional methods. We see general adherence of designers to wood, linen fabric, glue, stitching, and dope; the same apparently flimsy type of wing frame and fuselage construction; and receive a feeling of factors of safety cut down to perilously small values. Our dissatisfaction comes not so much from intuitive distrust of these features as from regret that none of the exhibitors – with one exception – show any sign of having seriously studied alternative methods.

'We have mentioned an exception, our reservation relates to the all-metal Silver Streak freight-carrying biplane shown by Short Bros. This machine is by far the most interesting technically. It is an engineering structure in the proper sense of the term, a description we hesitate to apply to any other aeroplane exhibited, much as we admired the ingenuity of design and perfection of craftsmanship which all embody. The Silver Streak contains not one particle of wood or linen fabric. Propeller, engine, fuselage and wings are entirely of steel, duralumin, or aluminium – but what is of even more significance, the designer has not been entirely content to replace wood and fabric by metal, he has developed radically different methods of construction; methods which are properly suited to metal. We would say that it is a machine of very great promise, and that as a sign of the advance being made in aeroplane construction it dominates our general impression of the show'.

Oswald Short was unbusinesslike in failing to secure a patent before exhibiting the machine, for this weakened his later position, but as he explained: 'My patent agent, Griffith Brewer, had asked me to allow him to make a search, and this took considerable time. What he found as the state of the art was eventually described in my first Patent 185,992, in Clauses 70 to 85, but none of those projects were ever carried into practice successfully, for the inventors failed to realize that such structures could not safely be made. In any case if I had waited for a Patent to be granted I should have had nothing to exhibit in that Aero Show.' On the opening day, two USA officers informed Oswald that if he would send detailed drawings to Washington they would guarantee that the US Government would purchase the machine; but he refused in hope that the Air Ministry would soon realize the merit of his construction.

Though somewhat shaken by the Silver Streak's advanced exploitation of metal, the greater part of the industry tended to excuse its own lack of initiative by saying the Short 'is before its time', preferring to continue with the well-tried system of wire-braced wooden frames and simple sheet metal fittings on the score that it was cheap and light and was the only method familiar to the traditional carpenters who were readily available. As a result there were no other completely metal-structured aeroplanes in the show, though Boulton & Paul might have exhibited the striking little two-seat Lucifer-powered biplane revealed at the Paris Aero Show had it not been damaged in a forced landing following engine failure. Another most interesting example would have been the Beardmore W.B.X with metal-fabricated girder longerons and structural methods based on their airship experience, but it was still incomplete although entered for the impending Air Ministry Competition. Except for the Austin Whippet's metal-framed fuselage and wing struts, and the tubular-structured fore part of the Blackburn Swift and Fairey Pintail, no other basic use of metal was displayed except fittings, small components, undercarriages, and engine mountings.

Built to advertise the tottering Martinsyde company, the little Semiquaver had minimum frontal area, and in side elevation had the typical pretty shape of George Handasyde's creations. (*Flight*)

Though using the prevailing materials of wood, glue, and fabric, Supermarine hulls were outstanding examples of the boat-builder's art. (*Flight*)

As a schoolboy visiting this exhibition with its profusion of glittering struts, bright wires, drum-taut fabric and acid-drop perfume of dope and pungent fresh varnish, my interest centred on the beautiful, scarlet-painted Martinsyde Semiquaver based on the F.4 with wings a mere 20-ft span, though powered with a 300 hp Hispano-Suiza. Recently at Martlesham Heath it had attained an outstanding 161.43 mph. Almost as compelling were the slightly larger 35 hp Avro Baby, the two-seat Sopwith Dove so reminiscent of the Pup, and the single-seat Supermarine Sea King with its glittering, polished, boat-built hull which made the bigger slab-sided Consuta-planked Vickers Viking appear clumsy.

Though less eye-catching than the aeroplanes, the really important *marque* was the display of aero-engines, revealing the great strides Britain had made in the five years since pre-war days. Of considerable interest was the Jupiter engine exhibit on the Bristol stand, finally confirming that they had taken over the defunct Cosmos Co, its patents and rights for the big air-cooled radials which Roy Fedden had been developing. Years later he told me: 'After our concern at Fishponds went into liquidation, the Bristol Aeroplane Co, a few miles away, acquired the designs and prototypes at a book value of £60,000 for £15,000. With thirty close associates I joined Bristol to found the aero-engine division as chief engineer. I had a wonderful opportunity to start off from the green fields with plenty of space to build modern test equipment. It was not long before I appreciated the enormous advantage of situation on an aerodrome giving opportunity to see how our engines behaved under flight conditions, and which equally enabled pilots to appreciate the care and patience put into them. Nevertheless setting up a new factory was a tricky undertaking, and with established companies already in the field, our rapid demise was prophesied.' It was a measure of the importance with which the Air Ministry viewed this take-over that a contract of £25,000 was placed for ten experimental engines. The Bristol directors accepted Fedden's estimate that within two years at a total expenditure including plant not exceeding £200,000, he would establish the design at 500 hp and weight of 650 lb. L. F. G. Butler remained chief designer, and the department of seven draughtsmen started work in September. 'Bunny' Butler was 'one of those few who could create a design and supervise its execution to the last detail.'

With the failure of the A.B.C. Dragonfly radial, Bristol's greatest rival was the Siddeley 300 hp fourteen-cylinder Jaguar, over which Maj F. M. Green, as chief engineer, had repudiated responsibility and was concentrating on airframe design because the fiery little John Siddeley refused to consult him on details of the engine. Italian-born S. M. Viale, pre-war designer of aero-engines, who during hostilities was in charge of Gwynne-built Clergets, took over, but was more interested in new design than development. The latter was largely handled under direct instruction from Siddeley by F. R. Smith, who was in effect chief draughtsman, and had 'considerable practical experience but was overly cautious and without training in scientific engineering and unable to exert great influence on Siddeley' according to Robert Schlaifer. The Jaguar had been redesigned by Viale to overcome its basic fault of cylinder heads loosening due to differential expansion, and Siddeley had eliminated the troublesome supercharger. Revolutions could now be increased beyond 1,100 and it at last gave an effective 300 hp.

More soundly established than any other engine company was Rolls-Royce, who displayed at Olympia the developed form of all their war-time engines, including the Condor of 600 hp for a dry weight of 1,390 lb and fuel consumption of 39 gal/hour. Yet for almost two years Claude Johnson, the managing director, in assessing priorities had refused to divert manpower from development on the car; so the R-R aviation department had merely existed on overhauls, leaving the field open to Napier whose broad-arrow Lion, designed by the elderly and retiring A. J. Rowledge, had come right to the fore with a reliable 450 hp for a weight of only 840 lb and consumption of 30 gal/hour. 'It is of the super-compressed type, and not intended to be flown at full throttle at any altitude less than 5,000 feet' stated their publicity leaflet; but contemporary 'high-compression' was only 5.55/1. Napier's gave no indication they were developing a secret 1,000 hp Cub sixteen-cylinder double vee in the form of an X which the Air Ministry had ordered in 1919 at £10,000 each.

Other manufacturers were baulking at the cost of engine development, though they displayed at Olympia; among them were Beardmore, Gwynne, and Sunbeam who had the largest engine on exhibition, the Sikh, for which 800/900 hp was hoped. Competing against every manufacturer, whether aeroplane or engine, was the Aircraft Disposal Co (A.D.C.), with prices knocking the bottom from any equivalent the British aircraft industry could produce, including sparking plugs, instruments, magnetos, and carburettors, all available by the hundred thousand. It was a most significant pointer to the need for caution.

The ever critical C. G. Grey reported: 'As a show intended to draw the great British public into interest in aviation, Olympia was a complete washout, but as a failure it was probably the greatest success of the year. Military, naval, and commercial attachés of all countries visited Olympia over and over again: many spent days in the AID and Technical Department sections studying our methods of inspection, and went away convinced that a machine granted an airworthiness certificate after such tests must be worth twice as much as any built in other countries. For that impression we must thank Air Commodore Brooke-Popham and Colonel Bagnall-Wild and their enthusiastic staffs.' It was only a fortnight before the show that the Air Ministry was asked to co-operate, and it responded nobly – Air Vice-Marshal Ellington's Department of Supply and Research surpassing itself. Even C. G. Grey admitted: 'In this section the Royal Aircraft Establishment, under Mr Sydney Smith, once so ill beloved as the Royal Aircraft Factory and now one of the greatest helps to the aircraft industry, contributes a number of most interesting examples of its work. Equally the once execrated Aeronautical Inspection Department, under the personal direction of Colonel Bagnall-Wild and his assistants Colonel Outram and Flt Lieut Mansell, have arranged the most convincing show of methods by which their officials test, check, measure, gauge, and otherwise ensure the reputation of British manufactured products all over the world.'

That there could be no clear run for British manufacturers was forewarned by a perpetual injunction granted on 10 July in the New York Law Courts by Judge Thomas Chatfield against sale or use in the USA of any foreign-built machines embodying features which infringed the Wright patents on lateral control by warping – astoundingly interpreted as any form of horizontal ailerons, although these had been patented in Britain in 1868. It was aimed by the Wright Aeronautical Corp of Paterson, New Jersey, against the Inter-Allied Corp of New York – an offshoot of Handley Page Ltd. Similar action was opened against the Aerial Transport Corp of Delaware which had intended to supply surplus British aircraft.

9

On the heels of Olympia followed the Aerial Derby of two circuits of a triangular course of 100 miles starting from Hendon on Saturday 24 July, in a cold and blustery wind. The 16 entrants included a landplane version of the Avro 539A Schneider with 240 hp Puma, and the Dragonfly-powered landplane Sopwith Schneider known as the Rainbow; making its first public appearance was the Bristol Bullet with the latest 450 hp Jupiter; Martinsyde had entered their speedy little Semiquaver; darkest horse was a fundamental re-design by Henry Folland of the Nieuport Nighthawk fitted with single-bay wings of 20-ft span and 3 ft 10-in chord, having a single I-strut each side attached to solid walnut compression ribs from spar to spar, and using a standard fuselage beautifully faired with formers and stringers to almost circular-section. At Martlesham on 17 June it had attained a British record of 166.5 mph, beating the Martinsyde. What with two Avro Babys, a Martinsyde F.4 and an F.6, three Snipes, two Nighthawk derivatives, and a D.H.14A entered by Sidney Cotton, it promised a fascinating afternoon's entertainment at Hendon where between 10,000 and 12,000 were present.

Starting them in handicap order was the RAeC pre-war timekeeper, little Mr A. G. Reynolds, his spectacles glinting beneath his trilby, full-skirted raincoat blowing in the wind, and watches and papers pinned to a large board suspended from his neck with string. There was early excitement. Cotton set off on a course more for Oxford than Brooklands, but seemed indifferent to a few miles lost. The Snipe piloted by Longton choked as he opened up, so that he lagged behind the others. Because of its small size the Schneider Avro seemed to shoot off at an amazing pace but was not seen again, Capt Westgarth-Heslam having landed at Abridge saturated with petrol from a leak. Jimmy James in the Nieuhawk tried to take-off with petrol turned off, and the two Martinsydes shot away while he was trying to re-start. Hawker and Uwins on the Rainbow and Bullet made professionally smooth take-offs, but Frank Courtney on the Semiquaver, deputizing for Raynham who had injured his shoulder when he piled up an F.4 at Brooklands, made a bouncy take-off through unfamiliarity with his little mount. Last away was the Nieuport Goshawk, gaudy with blue and yellow checks, but a trail of smoke indicated that the A.B.C. Dragonfly,

The Nieuport Goshawk was basically a Nighthawk fuselage faired to circular section and was reminiscent of the I-strutted S.E.4 which Folland designed before the war.

By lengthening the Avro Baby fuselage 2 ft 6 in, Chadwick was able to add a passenger in the extended cockpit. The first of these two-seaters, G-EAUM, had several owners until scrapped in 1935.

despite weeks of work on it, was still temperamental, and soon it was reported that at the Brooklands turn the engine was missing badly and Tait-Cox swept back to the aerodrome and landed.

While later starters were heading towards Brooklands, the earlier ones were appearing from the north prior to their second leg. In a buzz of chatter everyone awaited the winner. Ten minutes after the last of the first circuit, a machine was discerned low on the horizon. It was Hamersley on the first production Avro Baby, thus winning the Royal Aero Club Handicap. Six minutes later came Bert Hinkler with the Baby prototype. Hawker swept in with his racer, but as he had missed one of the earlier turning points he avoided the finishing line and landed by the Grahame-White sheds. A few minutes, and then with nose well down the scarlet Semiquaver raced across the finish as outright winner, banked steeply, throttled, side-slipped down, touched, made two enormous bounds, fell on her left wing-tip, cart-wheeled onto her back, and lay with wheels spinning plaintively. Stunned silence: incredible that Courtney, 'the man with the magic hands', could have crashed. Then cheer upon cheer as coils of his 6 ft 3 in dropped from the inverted machine. Second was Jimmy James with the Nieuhawk, and Bristol's test-pilot, Cyril Uwins, who had been nursing his Jupiter, came third. Even C. G. Grey was prepared to admit 'The almost brilliant success of this meeting should assure even greater success for the next.'

But the omens were less favourable nationally. Ever since the Armistice prices had been rising fast, and in midsummer the index was more than three times that of July 1914. Wages were rising despite efforts of employers to hold prices, yet the slight general increase in income was nevertheless nullified by lower purchasing power, and this was causing trade unions to abandon their earlier Socialist policy of long-term nationalization and shorter hours in favour of collective bargaining for higher wages. To that end the Trades Union Congress established a General

Frank Broome (*left*) and Stan Cockerell (chief) were the two light-hearted pilots of the Vickers team, flying the twin-engined Vimy Commercial and the amphibious Viking III in the Air Ministry competition.

Council to unify policy and action in every sphere of manual work. But though the Unions were now closely associated with the reorganized Labour Party, there was increasing subversive pressure to establish a Communist Party with similar aims to the revolutionary Russian Government – but Labour would have nothing of it. Though in June the Coal Industry Commission, half of whose members were Labour, issued their final report in such form that it enabled trouble among the miners to be avoided for a while, the threat of strikes was growing among railwaymen, and the dockers and transport workers were ready to back them. While industry in general was struggling to make the grade, few manual workers perceived that their jobs were precarious because of the deteriorating economic position and consequent dwindling sales. In particular the aircraft industry was staggering at minimum subsistance level. Orders were too scanty for any firm to keep going long, yet the workers who remained were a hard core of enthusiasts to whom aeroplane construction was still a romantic adventure rather than merely a wage-earning job.

By now the Air Ministry Competition was about to begin at Martlesham, the RAF testing aerodrome near Ipswich. Airco had scratched because of the company's liquidation, and neither the Beardmore nor Saunders entry were ready. The Alliance and Sage businesses had

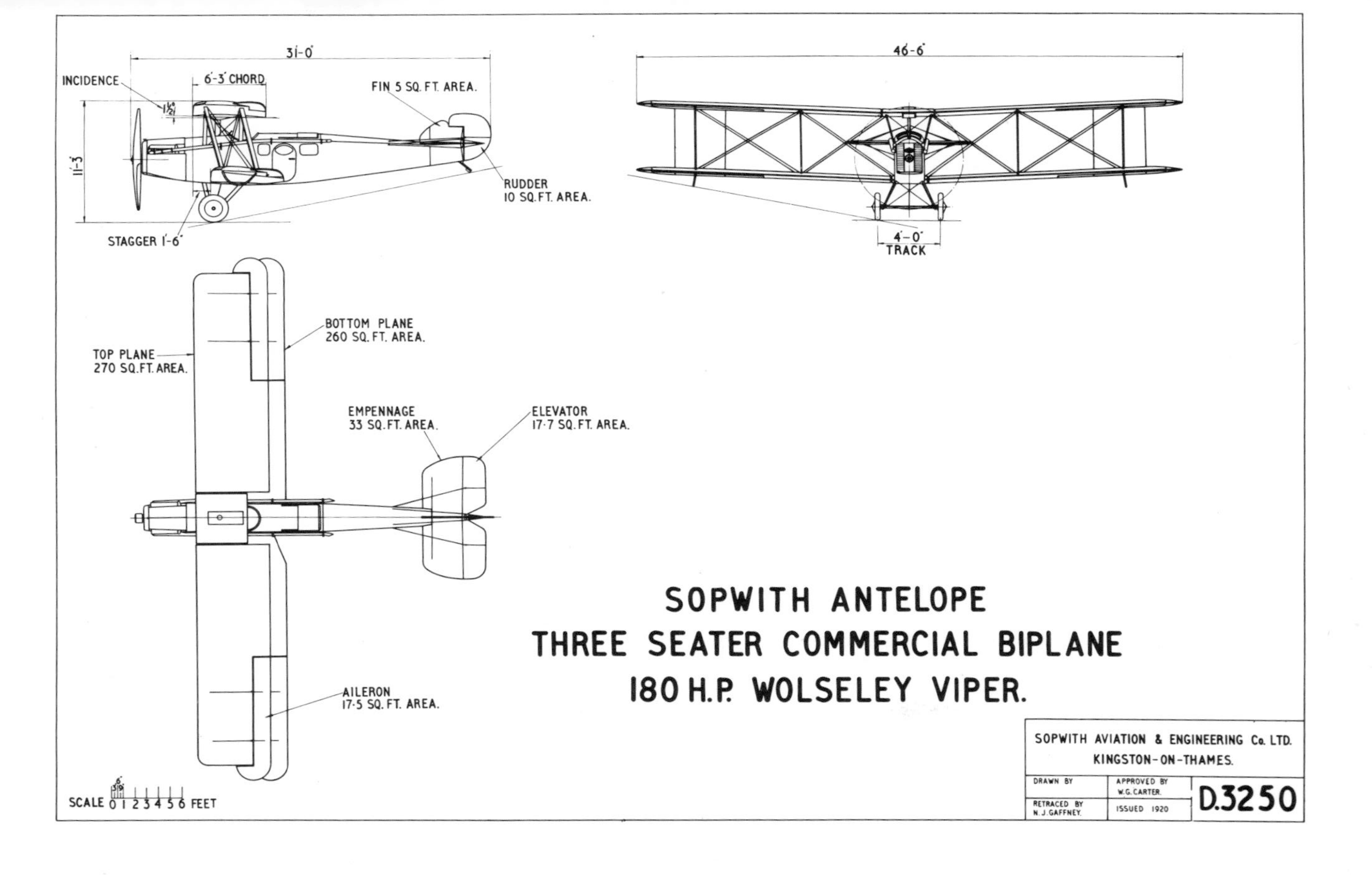
31'-0"
INCIDENCE
6'-3" CHORD
FIN 5 SQ. FT. AREA.
11'-3"
RUDDER
10 SQ. FT. AREA.
STAGGER 1'-6"
46'-6"
4'-0"
TRACK
BOTTOM PLANE
260 SQ. FT. AREA.
TOP PLANE
270 SQ. FT. AREA.
EMPENNAGE
33 SQ. FT. AREA.
ELEVATOR
17·7 SQ. FT. AREA.
AILERON
17·5 SQ. FT. AREA.
SCALE 0 1 2 3 4 5 6 FEET
SOPWITH ANTELOPE
THREE SEATER COMMERCIAL BIPLANE
180 H.P. WOLSELEY VIPER.
SOPWITH AVIATION & ENGINEERING Co. LTD.
KINGSTON-ON-THAMES.
DRAWN BY
APPROVED BY
W.G. CARTER.
RETRACED BY
N. J. GAFFNEY.
ISSUED 1920
D.3250

Based on the dimensionally similar Sopwith *Atlantic*, W. G. Carter devised the Antelope of half the power carrying three passengers in a small cabin. (*Courtesy Air Marshal Sir Ralph Sorley*)

closed down. Of accepted competitors, the Sopwith Antelope and twin-engined Central Centaur 2A failed to arrive on the opening Tuesday, but were there early next morning. Almost immediately the Centaur was in trouble, for though her airworthiness certificate had been granted on the estimated weight, it was found she was almost 1,000 lb heavier; so when loaded with 3½ hours' fuel there was no available payload for pilot or

Designed for a market that did not exist, the Central Aircraft Co's Beardmore-powered Centaur was low powered, old fashioned, and intended as a low priced eight-seat feeder airliner. (*Flight*)

passengers. As Harry Hawker smartly commented: 'She could only fly two of the specified tests – ability to stand unattended in wind and ability to fly uncontrolled!' However, it was decided to attempt the tests *hors de combat*.

Others soon had their difficulties. The Avro flew so wide of the *camera obscura* on slow speed tests that it could not be plotted. The Bristol Seely, flown by Uwins, insisted on a slow left-hand spiral in the uncontrolled test, and the Vickers Vimy Commercial, known as the 'Leaping Egg' due to her bouncing take-off, pushed into the 50-ft high balloon-suspended tape barrier instead of clearing it when demonstrating initial climb, and was quite unable to fly straight with starboard engine off, for sufficient rudder control could only be attained in a mild dive of 2,000 ft in 5 minutes – yet

The Puma-powered Bristol Seely was a hurried coupé lash-up of the F.2B Fighter, with deepened fuselage, and wings extended by an additional bay, resulting in doubtful stability. (*Courtesy Air Marshal Sir Ralph Sorley*)

with port engine off she would climb 500 ft. The Avro triplane was nearly as bad on take-off, attaining 1.18 ft at the barrier, and when it came to landing, ran well outside the 175-yard limit after a bumpy touch-down, but on a second attempt Hamersley landed even more heavily, and after the machine stopped it was found that the rear undercarriage struts were badly bent. As the Avro was in the way for Harry Hawker with the Sopwith, it was hauled to one side. Hawker glided in with brakes already on, landed, bumped, and there came three loud reports as his main tyres and one of the small forward ones burst – with the result that this was the only machine in the competition to stop within the stipulated distance, but unfortunately did not count as it had not landed intact! While onlookers crowded round to inspect the tyres 'there was a groan and a crash, and the Avro, which had been standing unattended, sat down flat on the belly of her fuselage. Everyone was laughing – and at that moment a RAF pilot on a single-seater landed nearby and completely wrote off his undercarriage, so there were three damaged machines sitting peacefully on the aerodrome in the morning sun,' wrote Capt Sayers, one-time designer of Admiralty aeroplanes at Grain but now technical editor of *The Aeroplane*.

The 160 hp Beardmore-powered Austin Kestrel designed by John Kenworthy had little chance as a profitable side-by-side two-seater but was conventionally pleasant to fly. (*Flight*)

Designed like the Austin Kestrel as a side-by-side two-seater, the wide-span Beardmore X was too handicapped by construction delay for it to be even tested before the Air Ministry competition.

Almost the same size as the Bristol Seely but a six-seater with twice the power, the Westland Limousine resulted from better appraisal of the rules by Capt Barnwell's friend Robert A. Bruce.

The Beardmore, with its interesting adaptation of airship metal structuring, was only finished in time to be sent from Glasgow without test, and when the engine was run the radiator built into the centre-section was found not to be functioning, and a replacement could not be obtained in time – though this would not have affected the competition as Capt Hamersley reported after an eventual flight test that 'somebody has mislaid the centre of gravity' and the machine was only safe to fly as a single-seater. This final stroke of fate decided Sir William Beardmore to close down the aeroplane department but continue with engines under 46-year-old Alan E. L. Chorlton, M Inst CE. A partly completed six-passenger amphibious biplane design, featuring four Adriatic engines in the hull driving two wing-mounted propellers through gears, shafts, and clutches, was scrapped. George Tilghman Richards resigned and set up as a consultant. But before the department finally closed, an astonishing design was patented in the names of Lord Invernairn (as Beardmore was soon to become) and A. Galbraith, revealing a large triplane reminiscent of the Tarrant Tabor, with tandem engines each side; and within the wide fuselage two further engines 'pivotally mounted on inclined air trunks which communicate with two air channels, the rear discharge from which may be controllable, or the discharge may be directed downwards through the trunks to assist take-off.' Despite ingenious control contrivances the multi-installation weight of this early STOL would have rendered it impracticable.

Trouble continued with others. Surprisingly the Austin Kestrel, regarded as pretty and very comfortable, had trouble with the Beardmore engine despite its great reputation for reliability; nevertheless after replacing a cylinder it managed with modest success. Even the well-tested Handley Page W.8 was immobilized the first ten days of the competition because of a damaged propeller, but when Maj Brackley flew her on the 14th he found that after 2½ hours the port upper trailing edge looked odd, and on landing discovered the top fabric stripping in the path of the port propeller slipstream. The H.P. crew re-covered the wing and had just finished doping it when one of the men fell from the high trestles bang through the lower wing and was so near being lynched that he was never seen again. But next day the machine was ready, and Brackley performed splendid take-off and landing tests, and with a level speed of 118 mph had the highest of the competition. Westland were flying equally well with their new Limousine. Particular comment was made of the determination of stocky Capt Stuart Keep, their test pilot, who after carrying out stability flights and landings, continued with two further flights of 3½ hours each. That was a rare enough feat for an onlooker to say: 'He did it without turning a hair, and was not even hard of hearing after landing. Such performance is not only high testimony to the pilot's skill and endurance, but to the design and construction of the Westland biplane and its Napier engine, for no pilot could have done as much on a machine which was heavy on her controls or tricky to handle.' A more caustic observer commented that: 'The really amazing thing was that the machine managed to lift Capt Keep's weight as well as the requisite load!' Since Keep was the only man who flew it, none knew that the ailerons were unsatisfactory.

It was soon clear that the Handley Page and Westland were the only possible winners, though some time must elapse before results were announced. One lesson at least was prominent, for as C. G. Grey put it: 'The competition was initiated with the idea of producing new types which would increase the safety, comfort, and economy of civil aviation. Actually it has collected at Martlesham a number of very nice aeroplanes which would have existed just the same had there not been a competition at all. So the Air Ministry might just as well have kept its £64,000 in its pocket but for the fact that "competition is good for trade", and presumably the winners in the different classes will profit thereby in one way or another, besides receiving the prize monies. Nevertheless to show any real advance over war-time practice we should have had new and original machines in the competition. Somebody ought to have had pluck to try a high lift wing for example: the only approach was in the Beardmore, but even that does not come anywhere near Fokker or Junkers wings for depth of canvas. To be of the highest interest we should have had the H.P. slotted wing at least in the form of an attachment to existing wings, if not in ultimate shape; and we should have had the Alula, or any other varieties which would increase efficiency and make for greater safety and economy.'

As displayed at Olympia the Handley Page W.8 had a lavish show finish, but it was pale pea-green by the time it reached Martlesham because H.P. had ordered all paint to be stripped to save weight. (*Flight*)

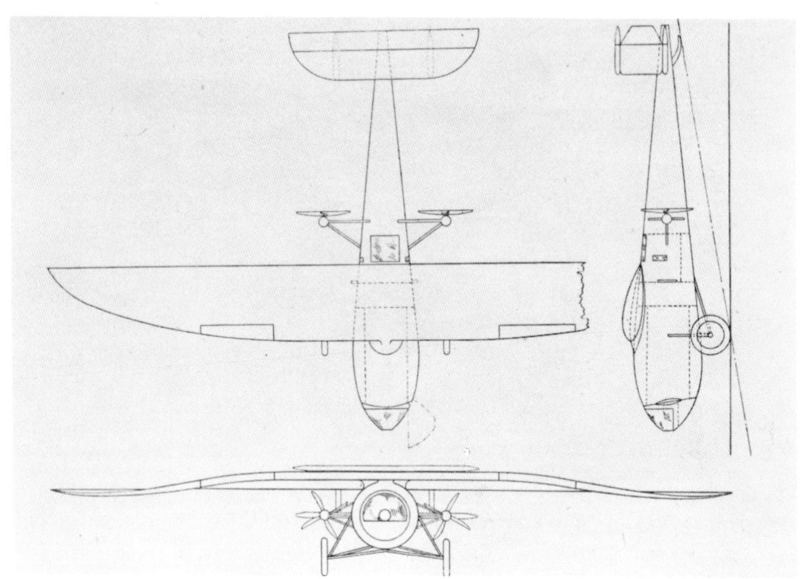

Though sponsored by Robert Blackburn, the Pelican was too advanced for its time, and some thought it a polite way of shedding the unruly Harris Booth from his staff.

The patented Alula wing (No. 167,086) was concurrently much publicized by Blackburn, and drawings were published of the 'Pelican' 4-ton aerial lorry version which Harris Booth and its inventor A. A. Holle had been designing in the form of a crescent-tapered high-wing monoplane of 146-ft span, its engine mounted within the beautifully streamlined wooden monocoque fuselage with shafts driving pusher propellers each side. Efficiency in terms of L/D, as measured in the small wind-tunnel at East London College, was 23. Designers and Farnborough scientists alike did not yet realize that there was the element of induced drag in a wing, and that with an aspect ratio of 15, whatever the plan form, there would be low resistance on climb, nor were the structural problems of making this long thin wing appreciated. Taking figures at their face value it had been happily assumed that the machine could be operated between London and Paris at a mere 1s 5d per ton mile. As a broad check on control and aerodynamics, a wing had been built and mounted on an obsolete D.H.6 fuselage. In July it was flown by Maj Veal and thereafter by Capt S. J. Clinch, but the machine refused to obey normal ailerons due to aeroelastic twisting, and was modified to use a hinged portion of the outboard leading edge instead.

Meanwhile trials of the three Competition amphibians began at Martlesham with take-off and landing tests. It was an hilarious affair. The Supermarine crew were true to marine tradition with heavy jerseys and grey trousers tucked in big sea boots, their pilot, Capt Hoare, applying an amphibious touch with a Norfolk jacket. His aircraft's registration G-EAVE was greeted with cheers for it echoed a well-advertised maker of uniforms who gave credit. Emulating their rivals, Capts Broome and

Though the Supermarine Commercial amphibian was placed second in the Air Ministry Competition it was preferred by RAF pilots, and in effect was the third metamorphosis of the war-time A.D. flying-boat, developing into the Seal of 1921 and, in production form, the Seagull.

Cockerell, the Vickers pilots, bought two little sailor hats and painted the name of their entrant Viking III in gold on the black hatbands. Because their machine often visibly rocked when climbing away, it was explained that the two pilots usually pummelled one another in a battle to be first pilot. However Lieut-Col Vincent Nicholl of Faireys, 'a man of charm who could look his tall chief in the eyes from the same height,' brought the real

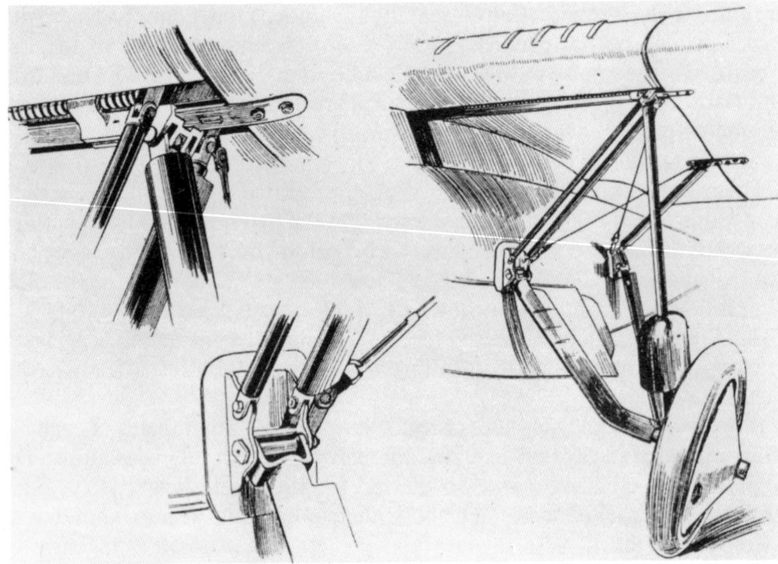
The hand-operated worm and nut, Supermarine retractable undercarriage for amphibians. (*Flight*)

marine atmosphere, for sea water was still dripping from the floats when he arrived after alighting 10 minutes earlier on the Orwell at Felixstowe, and his two passengers were too air sick on that gusty day to distinguish between the rough *terra firma* of Martlesham and riding the choppy sea.

'The fact that all competitors treat the affair as a very good joke does not in the least prevent them from competing very seriously, and working very hard to get the best out of their machines,' reported *The Aeroplane*. 'Up to the end of last week none had done any water tests as they wanted to complete land tests before chancing damage to their machines by awkwardly handled launches or a sudden squall.' However, the land trials successfully completed, the three arrived at Felixstowe, and while still splashing their way to success, the Air Ministry announced on 2 October: 'The Judges consider that the competitors for land aeroplanes show collectively less radical advance in design than anticipated, and though useful development and detail design have been produced, thus justifying the competitions, the award of the full prizes is not warranted.' In the large class first prize was therefore not awarded, but Handley Page secured the second of £8,000, while the third of £4,000 went to Vickers. In the small class, Westland won first prize of £7,500, Sopwith secured the second of £3,000, and the third of £1,500 was gained by Austin, whose relatively low-powered two-seater was the only one complying with the Air Ministry's earliest specification for this class. The announcement was followed by another on 11 October which stated: 'The Judges consider that the results achieved for amphibians show that considerable advance has been attained. The conditions fulfilled received little attention in the past and the competing firms deserve congratulation upon their enterprise. The Judges consider that the proportion of the monetary awards does not adequately represent the relative merits of the first two machines, and recommend an increase to the second prize.' Vickers gained first prize of £10,000, Supermarine had the second raised to £8,000, and Fairey as third achieved £2,000. Growled C. G. Grey: 'The results indicate no such disparity of performance as to justify a drop of £6,000 between second and third prizes.' But 33-year-old Dick Fairey made the best of it with a splendid series of advertisements impressing the tremendous reliability of his seaplane: 'She was delivered to the RNAS in April 1917; was still on service after the Armistice; bought back in May 1919 as a "Disposal" machine; used for experimental work during the summer of 1919; flown in the Schneider Trophy race in September 1919 and was only competitor to return intact under her own power; was used with experimental amphibian undercarriages during the early part of 1920; has survived nine different engines. Her performance equals that of the best modern flying-boat.' Meanwhile his new Pintail seaplane was receiving initial tests in the hands of Vincent Nicholl but proved tricky to handle when taking off.

Not until 11 September, far too late to compete, was it possible to launch the amphibian Kittiwake. Earlier, the King and Queen had been particularly interested, when they toured the extensive East Cowes marine

The Fairey entrant for the Air Ministry Competition had the classic III series structure adapted for a Lion engine and used the Pintail's modified amphibian undercarriage. (*Courtesy Air Marshal Sir Ralph Sorley*)

works of the dynamic and bearded Sam Saunders, to sit in the Kittiwake's cabin as their first experience of this kind. Several weeks after completing satisfactory static water tests, sea conditions were perfect for flight in the hands of Capt Norman Macmillan, but at 500 ft the cambering leading edge of the top wing sucked from its knuckle and blew off, and though he managed to put the almost uncontrollable Kittiwake safely on the sea a few hundred feet ahead, it hit one of the rocks at Egypt Point, damaging the hull. Thereafter tests were delayed by persistent trouble with the A.B.C. engines, one of which was hand-made and the other a production version. Subsequent flights by Warren Merriam, ex-RNVR and well known as a pre-war instructor, revealed the 'floating' ailerons ineffective at small angles, dangerously reversing their effect at big angles due to self-stalling, ultimately resulting in the machine being irreparably crashed by an Air Ministry pilot.

A Parnall amphibian, the Puffin, was also taking shape, built to the same specification XXII as the Fairey Pintail, but differed in having a single float. Key staff, discharged in March 1919 from the original Avery-owned Parnall & Sons, had returned to a new company wholly owned by George Parnall at the Coliseum Works, Bristol, retaining the mercurial Harold

Compared with steady evolution, it rarely paid to be unconventional, as typified by Beadle's design of the Saunders Kittiwake, nor were its chances helped by the unreliable Dragonfly engines.

Bolas as chief designer. He had been engaged since the previous November on this naval biplane, with Herbert E. Chaplin, MIMechE, as draughtsman designer, soon joined by R. J. Ashfield, the former Sopwith project designer, lately of Gosport Aviation where he had been draughting a two-seat sports flying-boat.

10

For a moment airships held the scene. At last the Vickers-built R.80 had been brought from her shed and given her first air trial on 19 July. The most streamlined British airship so far built, her length was 530 ft with diameter of 70 ft and capacity of 1,250,000 cu ft, affording a total lift of 38½ tons and disposable load of 17½ tons; maximum speed was 65 mph, with a cruising radius of 6,500 miles.

Designed by Barnes Wallis, the Vickers-built R.80 was a striking advance on its predecessors, with better streamlining greatly reducing drag, yet cruising speed was only 65 mph.

As early as March 1919, H. B. Pratt, the naval architect responsible for war-time airship construction at the Barrow shipbuilding base of Vickers Ltd, had written to his director Sir J. McKechnie urging: 'In considering the subject of international mail services the most experienced technical staff should at once be engaged for airships under a technical chief who has the confidence of the British Government.' He had in mind his 33-year-old chief designer Barnes N. Wallis, who now, a year later, was receiving a retaining fee of a mere £250 a year because the airship section had no work. Pratt proposed to Sir Trevor Dawson, the Vickers chairman, that: 'For the summer of 1920 I suggest we run sightseeing tours round the battlefields using R.80, which could carry 60 and R.33 which could carry 80, each making three trips a week,' and he gave operational details and costs. Since then he had investigated the possibility of using the big British rigids and recently-acquired Zeppelins on flights between Cairo and Karachi and eventually to Australia. Other proposals for small non-rigids to lift mining, survey, and exploration equipment likewise came to nothing. As Wallis was tapping his heels, it was with no surprise that he received from Commander Charles Craven, assistant to the chairman, cancellation of his retainer and termination of his services. In the next five months he took his BSc degree, and accepted a post as a mathematics master in Switzerland. Meanwhile Sir Trevor continued propaganda efforts by every means, including a lecture in September, *The Commercial Airship its Operation and Construction.*

Though the policy for airships was in doubt, interest in helicopters was more positive. The 67-year-old inventor, Louis Brennan, famed for the world's first guided weapon, the torpedo, and subsequently for gyroscopically-controlled monowheel trains and cars, was now engaged by the RAE to develop his secret 1916 Patent No. 281,735 for helicopter propelling, steering and balancing systems, which was extended in March 1919 by a further patent 'Relating to improvements in helicopters of the type described in Specification 281,735, wherein the car is hung by a universal joint from the gyroscopic centre of the lifting system and the direction of flight is controlled by imposing precessional forces on the lifting system.' Air Ministry decided to build a machine with a two-bladed rotor of 60-ft diameter having flying control flaps on the outer portion operated by pneumatic servos controlled by a cam linked to an inverted control column. Blade incidence could be varied collectively from full climb to auto-rotation in case of engine failure, and a swash plate enabled cyclic blade feathering during rotation by tilting its plane to ensure lateral stability.

Brennan, like Maxim, was an individualist who would go only his own way, and recognizing this, the Director of Scientific Research (DSR) gave him sole control, with exclusive drawing office and workshop, allocating the Balloon Shed for eventual erection and flight trials of the helicopter. There he was joined by R. G. 'Bob' Graham as assistant engineer at a salary of £350 per annum.

The beginning of Cierva's experiments in Spain with freely spinning rotors mounted on a pre-war Deperdussin fuselage.

Immersed in this fascinating and difficult project, Brennan may not have observed the issue in July of Patent No. 1657482 to J. de la Cierva y Codorniu, for 'An aircraft having the usual propelling means, and one or more horizontal airscrews supported from the fuselage and mounted to rotate freely in a plane which is slightly inclined upwards to the direction of motion.' In any case he would have ignored it. Was not his own rotor designed to rotate freely if the engine stopped? Nobody was aware that already this Spaniard had experimented with twin four-bladed, contra-rotating co-axial horizontal windmills on an old Deperdussin fuselage powered with a 60 hp Le Rhône driving a four-bladed tractor airscrew, and was sufficiently encouraged to design a further type of 'autogiro', as he called it, which would consist of a single three-bladed windmill with variable incidence blades to test a new idea for lateral control and stability.

The D.H.9 with full-span fixed slots had a taller undercarriage to give steeper ground incidence for landing, touched down to the detriment of the bottom of the rudder. (*Flight*)

Handley Page had provided a different and simpler solution to safe flight with his invention of the laterally venturi-slotted wing, though primarily he had been seeking greater lift. From wind-tunnel tests he proceeded to full-scale experiments using a Disposal D.H.9 with a continuous open slot full length along top and bottom leading edges. With true H.P. economy he had shortened the original nose ribs, and at intervals projected elongated solid ribs through the fabric cover to carry a tubular steel spar upon which was built the leading-edge aerofoil at a steep negative incidence.

Col Sholto Douglas, H.P.'s chief transport pilot, had left earlier in the year because Capt Geoffrey Hill, their test-pilot/aerodynamist, had been away ill and Douglas was asked to carry out urgent tests on the W.8 elevator and rudder, but he became nettled by Handley Page's terse refusal to pay extra for these duties which he was told were part of the game of making flying pay. At this he walked out, and rejoined the RAF with a permanent commission. Hill tested the new slotted wing and made comparative performance flights with a standard D.H.9 – even flying the

A benign H.P. and an interested D.H., in front of the slotted D.H.9, watch Geoffrey Hill climbing the standard D.H.9. (*Flight*)

slotted D.H.9 from the deck of HMS *Eagle* to demonstrate its much shorter run than conventional naval aircraft – but on the first public demonstration at Cricklewood on 21 October Maj Foot, another H.P. pilot, flew the slotted machine exploiting the take-off, slow flying, and slow landing, while Hill, undoubtedly with showmanship advice from Handley Page, flew the standard machine to prove it was visibly inferior, though in fact its drag was much less and therefore climb and speed were better. Of many journalists present one typically reported: 'It was obvious, particularly to those who knew the habits of the D.H.9, that the machine with modified wings got off at much lower speed with shorter run, and once off the ground climbed at an extraordinarily steep angle, though took as long in time to reach a given height. It could remain in the air with engine throttled at very low speeds and frightening angle of incidence. With the tail hanging down in a way which with a normal wing would certainly result in a stall, Foot turned and manoeuvred with certainty and precision, and in landing brought the machine close to the ground in what looked like an enormous "pancake", and then showed by dropping the tail appreciably and flattening out that it was still some way from stalling. A normal machine would have been entirely out of control. At present the wing has only been tested as a fixed affair, and the result is that top speed is reduced owing to the increased drag of the wings at fine angles, but there is no insuperable difficulty in retracting the nose piece to fit snugly against the wing, so that at high speeds the normal aerofoil can be used.'

Two methods of closing the slot had already been the subject of Patents 166,429 and 166,430 granted on 16 June. What Handley Page believed attainable was a multi-slotted wing of five or six narrow-chord aerofoils mounted in series, giving a tremendously high lift coefficient for an otherwise clean monoplane wing of very small area compared with biplanes of that time. For this he obtained Patent 172,109 on 31 August showing a mechanism closing all slots to give a wing with smooth top surface, though there was no hint as to how sufficient strength could be built into the slats to give a structurally integrated wing. Of the slot system in general, Dr Peter Thurston, one-time assistant of Maxim and later war-time co-ordinator of aircraft metal structural research, bitterly contended that his own 'rider' aerofoils, devised before the war, invalidated the Handley Page claim, though this seemed splitting hairs as they did not deliberately form a venturi but were analogous to aerofoils in tandem. This led him to interest Barnwell at Bristol in a patented method of lateral control using small rigid pinions at the wing-tips. Contention with H.P. was bound to follow.

11

Germany, though crippled by a punitive Peace Treaty and forbidden to import or manufacture military aircraft or operate a military air service, was taking full advantage of the Treaty makers' inability to foresee possibilities in civil aircraft. Professor Hugo Junkers was prompt to establish a factory at Dessau to build metal civil aircraft and during the summer demonstrated a commercial six-seat low-wing monoplane in the USA, hoping interest would warrant formation of a Junkers branch there. Claudius Dornier obtained a factory in Switzerland; and at the Zeppelin Works the giant Staaken four-engined, high-wing, 18-seat metal monoplane designed by Rohrbach had been completed. Within Germany the air

Symptomatic of Germany's developing techniques of all-metal aircraft was Rohrbach's E.4/20 built at Staaken. (*Courtesy Peter Brooks*)

transport company, Deutsche Luft-Reederei, was in full operation. Compared with the 'stick and string' conventional wooden construction of commercial passenger aircraft operated by Britain and France, the all-metal German equivalents were impressive indications of the future, though they seemed regarded by the British Air Ministry as fantastic experiments marking the end of Germany's war-time aspirations. Only at Shorts, with that beautiful monocoque metal fuselage, was there sign of potential superiority compared with the slab-sided German machines, though aerodynamically the latter were far and away ahead.

In Germany too there were strong signs of amateur flying enthusiasm indicated by a gliding competition in the Rhön district during July and August, organized by Oskar Ursinus, a war-time designer and now editor of *Flugsport*, who had found support with prize money from the German Air Ministry, directors of German aircraft concerns, and other aeronautical interests. 'It is interesting to note that poor defeated impoverished Germany,' said C. G. Grey, 'can thus maintain enthusiasm for the further conquest of the air, while we in this rich well-fed country cannot maintain a single aero club which does anything actively for the progress of aviation. The number of local clubs which entered this German competition is truly remarkable.' The principal winner was a clean low-wing monoplane with 'trousered' undercarriage like the 1911 *Monobloc* Antoinette, entered by the Aachen Aeronautical Society and designed by Dr Klemperer, which achieved 1,830 metres and duration of 2 minutes 20 seconds – but few remembered that Orville Wright in October 1911 kept aloft for 10 minutes with his biplane glider above the wind-driven sandhills of North Carolina.

In Britain things were going from bad to worse. At an auction of the Cambridge School of Flying an Avro 504K in perfect condition fetched only £50. Three D.H.6 trainers were sold for £2 5s, £3 10s and £6 10s. A 50 hp Gnome went for 35s and two propellers for 7s 6d each. Even a new acetylene welding plant with generator was knocked down for £7 10s. Manufacturers were faring as badly: neither they nor the RAeC were able to cope with the expense of sending seaplanes to the Jacques Schneider International Seaplane Race held at Venice on 19 September though both the Sopwith and Avro racers of 1919 were available.

While the commercial aircraft trials were underweigh at Martlesham the Nieuport & General Aircraft Co Ltd had announced they would have to close their works at Cricklewood and retire from making aircraft, at least for the present. Valiant efforts had been made to find overseas customers for Nighthawk fighters they had built in 1918 and subsequently bought in. Although the Receiver allowed the racing Goshawk to compete in the Gordon Bennett race held in France on 28 September, a series of delays due to erratic navigation by Jimmy James, who was piloting because Tait-Cox was ill, resulted in arrival at Étampes a day too late after numerous landings to enquire the way. However Martinsydes, of whom there were increasing rumours of financial difficulty, got the Semiquaver there in time, for it had been towed all the way by Fred Raynham and his mechanic, but to no avail. When he set out for his speed run the pretty little machine took less than half the take-off run of the others, giving every impression of being a winner. Minutes passed and no Raynham: 25 minutes had gone when at last he returned with engine misfiring. Oil was all over the machine – a fault common with Hispano-Suizas if the engine heated unduly, for the frothy sump oil escaped from the breathers. But Britain's chance had gone.

The 20-ft wings of the Martinsyde Semiquaver made pilots think it must be tricky, but in fact it handled well, though touch-down at 52 mph was regarded as too high for safety. (*Courtesy Air Marshal Sir Ralph Sorley*)

Winner was the famous French war-time pilot, thick-set, round-faced Sadi Lecointe, whose beautiful and slender little Nieuport biplane, powered with a 300 hp Hispano like the Martinsyde, attained 169 mph. Perhaps he was lucky, for his fellow pilot with identical machine landed with engine trouble after one lap, and the third French competitor was in similar difficulty with his Hispano-powered, stumpy, heavy looking but well streamlined Spad-Herbemont. Nor was the USA luckier with a powerful 600 hp Packard-engined Verville biplane and two advanced and rather terrifying 20-ft span high-wing Dayton-Wright monoplanes with 260 hp Hall-Scott engine, and undercarriage retracting into the fuselage by similar system to that patented by W. Brierley of the British Varioplane Co in December 1914, though the American's was the first seen in Europe. 'What complication and weight,' said British designers. 'Quite impracticable.'

Following this unsuccessful attempt of the Semiquaver, George Handasyde and Hamilton Fulton quietly resigned from Martinsyde Ltd, setting up a new business, Handasyde Aircraft Co Ltd – leaving their original partner H. P. Martin to continue with Martinsyde, and G. Tilghman Richards, until recently of Beardmore's, joined him as general manager. It was reported: 'There is no likelihood of the works closing down. Despite feeling the effects of the slump in the aircraft and motor industries, Martinsyde Ltd are in a good position financially, and when the industrial world is in more peaceable state they hope to have much profitable business.' Within days a Receiver was appointed by London County Westminster and Parr's Bank Ltd, confirmed by Court Order on 2 November. A financial commentator said: 'This is to be regretted, but appears the only way justice could be done to those who have conducted business with Martinsyde Ltd. Failing this, the Treasury would have seized the whole place to realize Excess Profit Duty, with the result that the bank and other creditors would have received nothing, for it would have been difficult to realize even the amount needed for EPD. This is typical of the way in which panic legislation, while on the face showing the country is able to pay its debts, does in fact strike at the root of the production of national wealth.'

Before this happened I had endeavoured to find how one could achieve a career in the aircraft industry. No schoolmaster could advise. Hopefully I wrote to several firms, but only Martinsyde and Central Aircraft were willing to consider an apprentice. Invited to an interview by Mr Martin at Weybridge, a great hall was revealed with a few motor-cycles on one side, many empty benches, and no aeroplanes; the works manager could promise only general engineering and no settled future. Before I could visit Mr R. Cattle of the Central Aircraft Co at Northolt, at the end of September disaster struck when their twin-engined Centaur 2A went into a left-hand spin after failure of its port engine at 500 feet, and in the crash their pilot, Lieut F. B. Goodwin-Castleman, and five passengers were killed. Cattle, grieved at the loss of life, explained that the crash had been financially disastrous as the machine was not insured, so recommended I continue at school another year, by which time it would be more evident whether the aircraft industry was worth entering. Both Bristol and Handley Page had even more clearly indicated there was no future.

Even more staggering than the demise of Nieuport and Martinsyde had been news on 11 September that the great Sopwith Aviation & Engineering Co Ltd had gone into voluntary liquidation. With closure first of Airco, and then Sopwith, it seemed a great blow had been dealt to the British aircraft industry. Nevertheless the decision was timely, for Tommy Sopwith knew his firm was still solvent and expected to pay creditors 20s in the pound. As a precaution he had already instructed agents to sell his Horsley Towers estate of 2,000 acres in Surrey, purchased from the Earl of Lovelace a short time previously. Although Sopwith Aviation was formed in 1913 with nominal capital of £26,000 almost wholly subscribed by Sopwith himself, the company had done so well that until a year ago it had reserves of £900,000, and even now the Receiver's statement showed assets to the value of £862,630 with liabilities of £583,510. 'It is sad that the name of Sopwith should disappear from places where they aviate or talk aviation,' mused C. G. Grey. 'Of course, if Parliament believed in a big Air Force and had ordered new machines to replace the hurriedly built products of war, there might have been enough to keep all *bona fide* aeroplane firms alive. Survival of the fittest is the law of nature, but there should be room for those who can last till next spring, when new RAF orders are expected in quite considerable quantities on a purely competitive basis. Perhaps then some far-seeing capitalist will reassemble the staff which made Sopwith aeroplanes so famous in the past, and we shall once more be able to include them among the world's best aircraft.'

That hope was shortly fulfilled. On 15 November there was registered the H. G. Hawker Engineering Co Ltd, a private company with capital of

£20,000 in £1 shares 'to acquire from F. I. Bennett all the patents, rights etc, relating to the manufacture of motor-bicycles, and to carry on the business of manufacturers of and dealers in cycles of all kinds, internal combustion and steam, motor-cars, aircraft etc. The first directors are F. I. Bennett, engineer; H. G. Hawker, aeroplane pilot; T. O. M. Sopwith, engineer; F. Sigrist, engineer; V. W. Eyre, engineer. Secretary: F. I. Bennett. Registered Office: Canbury Park Road, Kingston-on-Thames.' By this time the Sopwith chief designer, Herbert Smith, believing aviation at a standstill in England, had written to several Japanese companies, and the great shipbuilding yard of Mitsubishi seemed interested in discussing aircraft possibilities further, so he departed with equanimity from the scene of his momentous and vital war-time work.

Earlier, on 20 September, there had been an announcement of the formation of the De Havilland Aircraft Co Ltd, a private company with £50,000 capital in £1 shares, 'to carry on the business of manufacturers, inventors, designers, patentees, repairers of and dealers in aeroplanes, flying machines, airships, dirigible and other balloons, aeronautical apparatus etc. Subscribers are: A. E. Turner, director of companies; E. H. Neville, director of companies. First directors are: Geoffrey de Havilland, A. E. Turner, and C. C. Walker.' Holt Thomas's protégé, Turner, was duly appointed chairman. On 5 October, having paid Airco £834 for part-finished D.H.14s, 18s, and woodworking tools and benches, the new firm began to move into the two huts and two hangars of Stag Lane aerodrome acquired from Warren and Smiles, who had operated the war-time London and Provincial Flying School there but went out of business when their Fletcher-designed biplanes failed to secure a C of A for joy-riding in Ireland, to the disappointment of their youthful pilot, J. Wingfield Digby, whom they had taught after the war, and who now was at Cambridge studying aeronautics.

With de Havilland and Walker came D.H.'s early companion and brother-in-law F. T. Hearle as works manager, the Airco chief draughtsman Arthur E. Hagg, Wilfred Nixon as secretary, and Francis St Barbe who with Nixon had been Burroughes's assistant at Airco and now was made business manager of the new firm. 'As regards the company's policy, one does not expect to see it blosom forth into the organization of aircraft output on a quantity-production basis,' reported *The Aeroplane*. 'One is under the impression that the firm is to start operations with the care and maintenance of existing aeroplanes belonging to transport lines. If this is the intention, the logical result would be the production from time to time of new types of civil aerial transport machines based on soundness and cheapness of construction. This surmise is based on the fact that for some time Capt de Havilland and Mr Walker have been engaged on the design of a big passenger-carrying monoplane, with cantilever high-lift wings.' That seemed confirmed by the half-page distinctive but spartan advertisements which presently appeared declaring that as designers and manufacturers they were 'contractors to the Air Ministry'. It was more than worrying that by the date Hawker Engineering was registered, De Havilland Aircraft had an overdraft which could only be liquidated when Holt Thomas paid his £10,000.

12

Sponsored by the Air Ministry, an Air Conference held at the Guildhall in mid-October was full of good intent but largely preached to the converted. What was needed to impress business men was convincing evidence that it was by swift and safe communication that air transport could aid their success. Yet how remote was safety became evident in Sir Frederick Sykes's address when he stressed 'the great importance, in the interests of safety and regularity, of every commercial machine being equipped with wireless', and as an instance of its utility quoted the recent novelty of an aeroplane, flying with ten passengers, radioing for landing lights when forced to alight at St Inglevert after dark. As yet the vital questions of airway control, lighting, and radio had been largely side-stepped through lack of appreciation on the part of government officials of the difficulty and technique of aircraft navigation and the enormous increase in hazard when clouds were low, or there was mist, or it was dark.

Frank Searle, managing director of Aircraft Transport & Travel Ltd, and creator of the Daimler Car Hire Service, dryly commenting on Sykes's speech, said that civil aviation had been handled by aero-optimists; nobody could make a profit from two-seat or four-seat aeroplanes. Pilots could not fly by eye and without wireless; ground organization of communications needed establishing; it was impossible to fly by night, yet essential to use aeroplanes 10 hours a day to make them profitable; therefore the Air Ministry must light the air route adequately, turn-indicators should be compulsory, and height presented from ground level and not from sea level. He was surprised at the unpractical details of aeroplanes. Engines must do more than 3,000 miles without overhaul. Why use rubber connections in the petrol system? Why fix engine cowls with skewers instead of proper clips? Why use cables instead of rods? Why put magnetos where they were deluged in oil? Nevertheless he had a word of praise for Geoffrey de Havilland, pointing out that the D.H.18, which cost £5,000, could carry £160 of fares in the daily double journey to Paris and back: 'this showed a profit of £100 a day, so the machine could be written off in three months.'

Of all Conference speeches, that by Air Marshal Sir Hugh Trenchard on *Aspects of Service Aviation* seemed of most importance to the struggling upholders of the aircraft industry. It had all the stylistic signs of being written by that epicure of literature, Maurice Baring, his great war-time assistant. The Trenchard/Baring theme went to the crux of difficulties and considerations in establishing a third fighting arm. 'No sudden creation of aerial equipment to meet a national emergency is possible. It has been proved within the experience of every nation engaged in the last war that two years or more of high pressure effort is needed to achieve quantity production of aircraft, aircraft engines, and accessory equipment. Equally the training of personnel, including engineering, production, inspection, maintenance, and operating forces – covering some 50 trades and 75 industries – has itself proved a stupendous task when undertaken upon the basis of war emergency alone. Unfortunately no Air Force can provide a career within itself for all the young officers required, and some form of entry for comparatively short terms of service, to supplement older permanent officers, is a necessity.' As to the vital rank and file of tradesmen: 'We hope to enlist most of our requirements as boys for the trades requiring long apprenticeship; they will be given three years thorough training in theory and practice at Halton, and on completing this will do seven years colour service.' To many this seemed a pernicious system with risk of later unemployment for young men, but others saw it as a source of skilled labour for industry in general.

Although the Government had virtually decided a major war in Europe was impossible for ten years, Trenchard warned: 'Present needs are difficult to estimate in view of the universal unrest which causes the military storm centre to shift almost day by day. This inevitably makes for dispersion, which means inefficiency – for the RAF, being a highly technical service, is dependent on adequate workshop facilities, good provision of spares and technical supplies, and efficient supervision by higher ranks. We must, therefore, concentrate in as few centres as possible, with power to move quickly to any point required. This is most difficult to fulfil, for movement by air is complicated by the fact that to get anywhere from the UK we must pass over one or more European countries lacking essential ground organization, and all routes between Egypt and India cannot be used at present owing to Arab unrest.'

'Boom' Trenchard was an impressive man. The British aircraft industry was encouraged by his conviction of a purposeful and permanent Air Force. Not so the Navy. Admiral Sir E. A. M. Chatfield, Deputy Chief of Naval Staff, though protesting desire to see greater use of aircraft, clearly lacked faith in their reliability. 'His idea of a suitable naval aircraft appeared to be an apparatus possessing all the essential qualities of a battle cruiser and able to fly as well,' said one of those at the discussion.

Antagonistic though the Navy might be of RAF monopoly, Maj-Gen Sir Philip Chetwode, who had been Allenby's mainstay in the Palestine campaign, expressed in the clipped manner of a cavalryman his conviction that it was essential for the army to co-operate with the RAF. 'The last war was won by time, and it was the spade which gave us the necessary time, but the spade will never do it again.'

It seemed almost cynical commentary on the great Air Conference that in the House of Commons on 26 November, Mr. Churchill stated that as at 1 October the strength of the RAF was 2,802 officers and 23,862 men – figures which showed that the RAF had been reduced to less than one tenth its war-time strength – but at least it was greater than the US Air Service, where Congress passed a Bill approving 1,516 officers and 16,000 enlisted men as a 'separate and co-ordinate branch of the Army', though the US Navy proposed to retain its independent nucleus.

If enlistment was low, unemployment in Britain remained high with 691,000 out of work, and in November the Unemployment Insurance Act was extended to cover most employees. Though the prospect at home seemed depressing, there were hopeful signs that the Dominion and Commonwealth countries were growing in strength and independence. In February the Canadian Air Force had been initiated by Order in Council; in August the South African Air Force was formed; Australia was preparing to follow with an Air Force. Meanwhile on 16 November the Queensland & Northern Territory Aerial Services (Q.A.N.T.A.S.) was formed to operate air services, air-taxi work and joy-riding in Western Queensland. Sir Fergus McMaster was chairman, and a keen young wartime aviator, Hudson Fysh, the dynamo behind the venture, was appointed manager.

In England the London—Paris air services were not making the grade. Winter weather and cancelled flights were reducing income. Both AT & T and Handley Page Transport were sending canvassers to big London firms hoping to persuade them to send goods by air instead of rail and boat – but on 6 December a short article in the *Daily Mail* from its Paris Correspondent indicated that AT & T Ltd were bankrupt and had stopped operations; that Instone Air Line had laid up its machines for the winter; and Handley Page, instead of running a daily service, was only putting an occasional machine across the Channel if there happened to be enough freight or passengers to make it worth while. During the year H.P. had eight or nine O/400s in service, but they each averaged only half-an-hour's flying a day, and some 100 forced landings were made, due to bad weather or engine failure. Return fare to Paris was 18 guineas and single fare 10 guineas, with freight charges 2s per lb reduced to 1s 3d for large consignments.

Handley Page received a further knock when on 14 December Robert Bager, chief pilot of his airline, crashed into a tree after taking-off in the mist from Cricklewood with the now ancient O/400 G-EAMA, the machine bursting into flames on hitting the ground and killing him, his mechanic, and two of the six passengers. It was the first fatal accident of the company, for it had carried 4,000 passengers and many tons of freight in safety, covering a total mileage of 320,000 miles.

Only a week earlier Bager had written an amusing letter describing a typical cross-Channel run. 'We left Paris 12.35. Messrs Marconi had nothing of interest to tell until crossing the Channel when Lympne said that because of thick fog we must land at Cricklewood. Approaching Lympne we met the fog, and Cricklewood Marconi chipped in with "Fog clearing. Carry on." I grunted, and made for Cricklewood via North London. Passing Maidstone, Mr Marconi of Cricklewood said: "Thick fog at Cricklewood, proceed to Croydon." I didn't know exactly where Croydon was, but hoped to see the Crystal Palace – but the fog grew thicker and thicker. We went on, not recognizing a thing, until by my watch we should have been somewhere near, so I told the mate to ask Croydon to oblige with a rocket; but instead – what do you think? – they told us where we were! They said change 80 degrees for Croydon. I remembered how we used to be shot down in the big war when we got triangulated by listening posts, and thought this must be the same sort of thing, so altered course. Five minutes later we came plonk over Croydon. It couldn't have been done better if they had thrown us a line. We landed at 3.10 with five passengers and freight.'

At least Handley Page's background seemed secure, for he had managed to acquire the freehold of Cricklewood aerodrome from the Ecclesiastical Commissioners who had been claiming £43,575 against the Ministry of Munitions; but it was rumoured that H.P. had proposed an arbitrator – who settled for £31,320 to the company's benefit. He was equally well established at Croydon, where the aerodrome was owned by the Ministry of Munitions but held under licence by his Aircraft Disposal Co.

The end of the year closed with a hopeful note expressed by Maj W. E. de B. Whittaker, who wrote: 'Throughout 1920 the RAF has been at war in Russia, Turkey, Mesopotamia, Somaliland, Arabia, the Sudan, the Indian North-West Frontier, and all recorded so much gallantry and good work that it would be invidious to select any for quotation. It must suffice that in all war areas the RAF has maintained the highest tradition of the Army which gave it birth. At home the RAF has already lived down all reproaches made against it during the War and after the Armistice. The quality of its flying is as high as ever, and in all other respects has far surpassed its earlier levels. The smartness of the "other ranks", whether on pass or parade, their keenness on ceremonial, and the superlative excellence of their march discipline, is at once evidence of the high quality and enthusiasm of the airmen and of the good feeling they entertain towards their officers. The RAF is the last and least of the King's armed forces, but in no other respect does it now yield pride of place to the sister Services.'

CHAPTER II
GROPING THE WAY
1921

'In the lexicon of youth which Fate reserves for a bright manhood, there is no such word as Fail.'

Lord Lytton (1850)

1

In *The Times* controversy was mounting over naval shipbuilding policy, and prompted Rear Admiral Sir Reginald Hall to declare that the share aircraft could contribute in maritime battles of the future had been ignored, so intent were the deep sea protagonists on the respective merits of battle-ship and submarine. Though pinning faith on a trinity of 'thoroughly efficient air, submarine, and mining services', Hall was convinced 'air mastery alone can give the power of vigorous offensive'. Sir Sefton Brancker, adding his support amid the pleasures of temporary retirement, wrote: 'Some writers have stated the duty of the Fleet is to destroy enemy sea bases, coaling stations, fortifications and commercial harbours. Others contended that submarines have rendered such enterprise impossible. But they will *not* be impossible to aircraft. I am certain that, in the future, the Air Force must become *par excellence* the arm of offence against hostile commerce and territory. If this is admitted, then the proportion of money allotted to the Air Force compared to other Services must be very high. If this high allotment to the Air renders the allotment for the Navy in-sufficient for construction of capital ships then they *must* go. No nation can neglect power in the air to preserve a former defence so expensive in maintenance and so problematical in utility.' Brancker urged grouping the entire British defence organization under one head, involving creation of an Imperial General Staff which would deal with strategy without bias towards any particular arm of warfare. 'There is a growing desire in the Admiralty', he said, 'to wreck the Air Ministry and revert to the old non-progressive system of having separate naval and military air services. Such a retrograde step would spell ruin to the country and must be resisted at all costs.'

As though timed to emphasize the independence of the Royal Air Force, the King approved a design for its ensign comprising a flag of sky blue, with Union flag inset in the top quarter, and the quarter below and opposite bore the wartime-established roundels of royal blue, white, and central red.

Despite the cut-backs of the first year of peace the Air Ministry was fully alive to the importance of helping the ailing aircraft industry recover. Contracts had already been given to Avro for their Aldershot, Vickers and Armstrong Siddeley for twin-engined bombers, Bristol for the shaft-driven triplane with central engines, Boulton & Paul for the similarly centrally-powered biplane, and Fairey and Parnall for their two-seat seaplane amphibians, and Westland was completing a recent order for an extensive and ugly Air Ministry modification of the D.H.9A, converting it to a three-seater for naval use, powered with a Napier Lion. For Supermarine there was an order for a development, named the Seal, of their Air Ministry Competition single-engined commercial amphibian flying-boat; and from Blackburn the Air Ministry purchased the private venture Swift, which now bore RAF serial N139, instead of G-EAVN. On 23 December it had flown to Martlesham, but preliminary tests revealed the rudder too small, and Blackburns were making a larger one.

With a span of 50 ft 6 in and all-up weight over 7,000 lb, the Condor-powered, war-time-designed D.H.14 day bomber was a scaled-up version of the D.H.9A and handled equally well. (*Courtesy Air Marshal Sir Ralph Sorley*)

Engine builders were similarly encouraged with development contracts, notably for the Bristol Jupiter and Armstrong Siddeley radials, the huge 1,000 hp Napier Cub, and the 600 hp Rolls-Royce Condor of which the final form had at last been fitted to the D.H.14 day bomber as the first major job completed by de Havilland Aircraft. The aircraft industry was also receiving contracts for the repair of both aircraft and engines. Contract prices for the lot, including engines, totalled less than a quarter million – so it was not out of keeping that when the Air Ministry announced Cabinet approval to subsidize civil air transport it was limited to £60,000 for British companies operating approved aerial routes, paid on the basis of 25 per cent of the total gross revenue of each from 1 January onward. Settlement would be every three months provided the company concerned had carried out flights in both directions within the time limit of four hours from aerodrome to aerodrome (on the London—Paris route) on a minimum of 45 days in that period.

A potential contender for British passengers was the recently formed Koninklijke Luchtvaart Maatschappij voor Nederland en Kolonien (KLM), which would initiate services in April between Amsterdam and London, or Amsterdam to Paris at £10 8s using Anthony Fokker's revolutionary F.III high-wing monoplane which had a ply-covered cantilever wood mainplane and fabric-covered, welded-steel fuselage. Though cautiously regarded as unconventional, it was a pleasant aeroplane to fly. Capt Hinchliffe, distinctive with black eye patch, was appointed chief pilot as no Dutchman had sufficient experience, and Cyril Holmes (who later taught me to fly), Gordon P. Olley, and R. E. Duke became his captains. Behind the organization was the genius of Albert Plesman in the head office at The Hague. As KLM manager at Croydon he appointed the energetic and gay Spry Leverton who had been office assistant to Donald Greig, general manager of the demised Aircraft Transport & Travel Ltd, and in that capacity had encountered an irate Plesman in the latter part of 1920 on an occasion when he had booked by an AT & T aircraft only to find it unserviceable, so Spry had found him a seat with the Handley Page line – this had impressed Plesman so much that he gave him this job with KLM.

Designed to Fokker's instructions by Reinhold Platz, his chief engineer, the F.III proved amiable to fly, though pilot location alongside the engine caused initial surprise and criticism from airlines. (*Flight*)

For the moment no regular British air services were functioning, though the subsidized French were successfully operating Breguets, small Blériot-Spads, and twin-engined Farman Goliaths. This led to re-thinking by Sir Frederick Sykes as CGCA, for it was clear that the generous scale of France's subsidy would result in heavily undercutting current British fares. It was therefore decided to form a committee, including representatives of the aircraft industry and air transport firms, to devise an alternative method of applying the British subsidy.

Enthusiasm of airship devotees received a nasty setback on 27 January when the R.34, commanded by Capt Drew, on the first trip for nine months after overhaul and fitment of mooring mast nose-tackle, encountered intense fog at 800 ft and struck a hillside on the Yorkshire Wolds. With three of the four engines out of action, she managed to fly on at some 8 mph until dawn, when she was steered for Howden and skilfully piloted towards the ground where her handling ropes were taken by a landing party 300 to 400 strong, and she was moored ship fashion with three cables to triangulated ground anchorages. In an increasing wind she surged, slowly rose and fell, often hitting the ground, and presently bumped so violently that the front car was pushed into the envelope causing several structural rings to collapse and the nose crumpled and broke off. Playing it down, the official communiqué briefly reported: 'Owing to the damaged condition of the ship and gusty nature of the weather it was not possible to keep her under control during the night, and unfortunately the ship, to all intents and purposes, was wrecked. No casualties occurred.'

It was a bad omen for her slightly larger sister R.36, constructed by Beardmores and now, fitted with a passenger saloon, was almost ready for flight. To follow her from the National Airship Factory at Bedford was the R.37 built by Short Bros. What with these, the Vickers R.80, and the acquisitioned German L.64 and L.71, Britain's collection of expensive aerial monsters was slowly growing – but to what purpose would they be put?

There were questions too about the future of Winston Churchill whose current departure from his post as Secretary of State for the Departments of War and Air was regarded as something of a blow to British aviation. Sir L. Worthington-Evans replaced him at the War Office, and it became clear that the Government intended to establish the Air Ministry with a separate Secretary of State for Air, though mystery veiled his name. Most people assumed it would be Churchill's tall and distinguished Under-Secretary, Lord Londonderry.

'Recently Churchill has inherited a modest competence through the lamented death of Lord Herbert Vane-Tempest,' wrote the ever prescient editor of *The Aeroplane*. 'One hopes that as a result he will be able to devote his immense energy and brain power in future to pure and applied statesmanship. He has the personal charm and mental power of a leader of men. All that is needed is the will and time to use these qualifications. Given a party to back him in the House there is no reason why Mr Churchill, who is now only 46, should not be Premier during the Great Slav War in 15 or 20 years time, or at any rate War Minister over the sea, land, and air forces in a Labour government 10 years hence. Meantime aviation and all concerned therewith owe Mr Churchill a deep debt of gratitude for all that he had already done for aircraft.'

In his new office of Secretary of State for the Colonies, Churchill's initial interest was whether aircraft could be used for territorial control – a matter he originally proposed in March 1914 after massacre of the Somaliland Camel Corps by the Mad Mullah, and it was this that led to the successful expedition against that ruler in 1920. An unpretentious paragraph now recorded: 'The Rt Hon Winston S. Churchill, MP, Secretary of State for the Colonies, left London on Wednesday 2 March for Egypt. He was accompanied by Air Marshal Sir Hugh Trenchard. On arrival in Egypt they will discuss with representative military authorities, among other questions, the possibilities of the substitution of detachments of the RAF for military garrisons in the Middle and Far East.' It was to lead to complete revision of schemes for Imperial Defence.

But before he left, still as Secretary of State for Air, he introduced the Air Estimates. Excluding war liabilities of £1,471,000, the net total was £16,940,000 compared with the previous £14,998,230, so at least there was an additional £2,000,000 to purchase five further squadrons despite the generally increased cost of labour, material and transport, though consultation with the Admiralty led to decision suspending the RAF Airship Service 'as this would otherwise involve diminution of effort on services of which the fighting value has been more fully demonstrated'. Experimental, research, and civil aviation allocations remained the same, but for the first time there was provision for a 'Territorial Air Force' with the modest sum of £20,000. A critic noted with sour disquiet: 'Sir Hugh Trenchard now receives £3,375 in place of £2,725, the Director-General of Supply and Research is to have £2,702 instead of £2,302, and the Secretary of the Air Ministry rises from £2,000 to £3,000; Grp Capt Scott, OC of airships, is to have his salary increased by £550 to £1,500.'

On the important matter of training, Churchill said: 'At Halton we are going to teach 3,000 boys to be skilled mechanics, with eventual output of 1,000 a year. There and at Manston adult recruits are undergoing intensive technical education. At Cranwell we are training cadets to be officers, as well as a large number of boy-mechanics who will eventually be accommodated at Halton. At Upavon we are instructing in flying and practical engineering. At Netheravon and five other schools, one of which is in Egypt, we are training officers to fly. At Andover a school will be opened for air pilotage and night flying. Eastchurch will be a station where armament, aerial gunnery, and bombing are taught. At Gosport they will learn torpedo dropping from aircraft and observation for naval guns. At Flowerdown (near Winchester) there is an electrical and wireless school. At Farnborough there is photography; at Uxbridge an Air Force depôt. At Salisbury and Farnborough, artillery and infantry co-operation are taught. At Calshot, air navigation over the sea. At Felixstowe, and Leuchars in Scotland, naval co-operation is carried on. There are the great experimental stations of Martlesham, Grain, and Biggin Hill. Then there is the Royal Aircraft Establishment at Farnborough, an institution vital for general development of flying. There are also 28 fully formed Service squadrons, eight in India, one on the Rhine, one at Malta, three in Ireland, three working with the Navy, and one at home giving refresher courses to pilots.'

The scope of squadron dispersion reflected the strain Britain was undergoing at home and internationally. The Civil War in Ireland was continuing with bitter violence. At the end of 1920 The Government of Ireland Act set an invisible partition between the six loyal counties of Ulster and that of rebellious southern Ireland. North and South were granted separate parliaments under British administration, but the South was determined to maintain independence as a self-governing Irish Republic. Although Martial Law had been declared in eight of the Southern counties, the Black and Tans had managed in December to set the city of Cork alight. It seemed the forthcoming elections on 24 May would be another point of conflagration.

India too was agitating for self-government with campaigns of civil disobedience and boycott which had followed the Amritsar massacre of 1919. In Palestine there was intensifying Arab-Jewish conflict; and Egypt, victorious in revolt, was preventing settlement safeguarding Suez because independence seemed within grasp. As though incited by the vast killings of the Great War, murder, mass disobedience, sabotage, arson were everywhere growing commonplace.

Even the ebullient Lloyd George was finding it a problem to keep in step with France over reparations, which at the Paris Conference had been fixed at £11,300,000,000 payable over 42 years. On 28 January German representatives from the Supreme Council arrived in London for discussions, but their subsequent counter-proposals were rejected by Lloyd George on 1 March. There followed an Allied ultimatum stating that unless the Paris Agreement was signed by the following Monday measures would be taken to enforce it. Within days French and British troops occupied Düsseldorf, Duisburg and Ruhrort.

The position had not been made easier by the resignation of the urbane Arthur Balfour from the post to which Churchill had succeeded, and the Premier had also lost an important supporter in that melancholy Canadian-Scot, Bonar Law, who ascribed ill health to his retirement from politics. Chaos was mounting. Unemployment was increasing, particularly in shipbuilding, mining and heavy industries. For the first time, men and women were queueing for 'the dole' made available by the newly introduced Unemployment Insurance Act, but it brought in less than 30s for a small family. Whether in work or jobless, it was beyond comprehension of the masses that the trade recession was largely due to the upward spiral of product costs which had begun with pressure for more and more pay. One effect was the dismal failure of the promised housing programme, for many bricklayers and carpenters were daily adding to the list of unemployed.

Yet there remained a superficial air of prosperity. Many of the army of female factory workers dismissed at the end of the war returned to neatly uniformed domestic service. Milk and bread were delivered as in pre-war years by horse-drawn floats and vans with iron-shod wheels, and the postman made three calls a day, dressed like a Victorian militiaman with blue serge uniform and double-peaked hard hat. Tramways remained the principal mode of town travel, with the rarer motor 'bus and thousands of bicycles as alternative, though for special occasions there were high-roofed motor-taxis and slow old horse-drawn 'growlers', but the hansom cab had gone. The few private cars were regarded with speculative interest. For long-distance travel and 'commuting' the train was supreme and British railways were judged the best in the world as well as profitable. When it came to going abroad then the Channel steam ferry was the mode, and you were regarded as rather odd if you went by air.

2

The Report of the Advisory Committe for Aeronautics, now reconstituted as the Aeronautical Research Committee, issued in February, was the last under the old name, and revealed a number of aspects indicating the current direction of British aeronautical technical development. Primarily it aimed at establishing the parameters of successful if stereotyped design. Nevertheless the field covered was certainly wide, for seventy R & M technical reports were printed, though others had been omitted as too revealing. A profusion of research was being conducted on propellers, instruments, and engines. There had been experiments with a Wasp-powered Avro on the worrying question of fire-proof installations, but unfortunately, in typical fashion, the Wasp engine failed, due to a broken crankshaft, before completion of tests, though the proposed mandatory types of asbestos bulkhead, metallic petrol pipe joints, and so on, were at least tried. Nor were Admiralty needs overlooked, for the Fleet Air Arm required another aircraft-carrier to accommodate the new spotters and torpedo machines, exemplified by the Parnall Panther, Westland Walrus, and Blackburn Swift, so the NPL on behalf of the Director of Naval Construction had conducted tests with model ships to investigate the airflow over the deck. During the summer of 1920, full-scale comparative

In addition to Panther prototypes constructed by Parnall & Sons, the British and Colonial company became responsible for 200 production Panther two-seaters for the Fleet Air Arm.

The Westland Walrus was a literal interpretation of naval requirements which specified the D.H.9A converted to a three-seat spotter with prone defensive position, a paravane undercarriage and air-bags, and a Napier Lion engine to drag it along. Thirty-six were built with bigger rudders than the prototype.

trials had also been made on HMS *Eagle* which showed that 'the difference as regards the loss of direction of wind on model and actual ship are within experimental error, so that model work promises well as an aid to future design.'

Of immediate use to aeroplane designers was Hermann Glauert's R & M 710 summarizing knowledge to date on stability and control, the analysis of which had stemmed from Professor G. H. Bryan's work initiated at Bangor University as early as 1903, before the Wrights had flown a powered machine. Glauert's concern was to discriminate and list appropriate R & M references under five main headings of General Problem of Stability; Longitudinal Stability; Lateral Stability; Static Control; Dynamic Control. To these he added appendices dealing with stability derivatives, control surfaces, downwash, and slipstream. A note indicating further work required to correlate theory and practice stressed that: 'The most urgent problem of control is improvement of conditions at low speed, as it is of great importance in reducing landing speed. This is mainly a question of improved elevator and aileron control, the critical aspect being a reduction in the tendency of the aeroplane to spin when stalled.' Nobody questioned whether Farnborough was over-emphasizing this common cause of crashes, for contemporary aeroplanes had such light loading that they were subject to sudden change of incidence due to bumps which could be catastrophic when approaching to land at too slow gliding speed. At this point not even Handley Page appreciated that the solution was in his own hands, for he was absorbed with the idea that the important feature of slots was to produce high lift.

Of special interest to theoreticians was a report on the researches of Professor L. Prandtl at Göttingen. Abandoning classical theory which assumed that streamlines separated at the nose of an aerofoil and reunited at the trailing edge, he proved that relative 'circulation' around a wing must exist, though not in the sense of tracking a particle of air along its path until it returns to its starting point, nor could the particles be given a spin. Instead it was measured by the velocity of a fluid along the boundary multiplied by the length of the boundary, and became plus or minus depending whether the boundary be 'on the right or left of a swimmer moving with the current'. Taking the transverse section of an aerofoil as the inner boundary of the associated streamlines in flight, those above and below being opposite in sense, then 'because the upper one has higher velocity than the lower, an integration round the entire periphery shows a circulation in a clockwise sense, the outer boundary in effect being indefinitely remote. This is not to be confused with vortices formed as the result of a body moving through a fluid. In the vicinity of the aerofoil the main cyclic system is carried along with the aerofoil as the system whose energy is conserved, but the impulse distributed over the area of air swept by the aerofoil results in a vortex pair in the form of a wake rotating from the wing-tips.' Prandtl's was the mathematical development of the empirical theory originated by the British engineer F. W. Lanchester before the war. At Farnborough, the German's work became Glauert's speciality, and as a result he produced R & M 723 – an authoritative booklet of 22 pages explaining the vortex theory, and showed that the resistance of a wing comprised not only drag due to the shape of its profile but an additional proportion depending on aspect ratio.

Glauert's brother Otto, the mathematician at Boulton & Paul, immediately appreciated the value of this theory and converted John North to its acceptance, for flight tests had long proved the value of high aspect ratio wings empirically adopted for the twin-engined Bourges. As North said: 'In so far as the primary aerodynamics of wings are concerned, the use of the quantitative vortex theory has thrown into true perspective the whole mass of partially and imperfectly correlated data on wing sections, plan forms and multi-plane combinations.' It seemed typical that Lanchester's pioneer inductive reasoning should find acceptance in Britain only after intervention of a long war and through Continental sources; but much investigation must ensue before suitable wing shapes for particular purposes could be mathematically evolved.

'Other work is in progress on aerodynamic properties of thick aerofoils suitable for internal bracing,' stated the supplement to the Advisory Committee Report, and significantly continued: 'The German Junkers and latest Fokker employ this type of wing; so also does a new design under construction at the request of the Air Ministry. Before there is much flying in this country on such designs, it is desirable that model experiments should be made, and these are now in hand; in particular, it has been found that these monoplanes are very liable to spin, and it is hoped that model experiments will show that this tendency can be counteracted.' The monoplane being built was a Woyavodsky low-winger in which the aerofoil deepened in chord towards the root and blended into the fuselage sides. The elderly Robert Bruce of Westland had submitted its design as a single Napier-powered 'Postal' eight-seater with fabric-covered wing having multi-spars based on the Junkers which he and Barnwell had inspected the previous year. Of the twin-engined Woyavodsky proposal for which Harold Bolas had received an earlier design contract, Herbert Chaplin said: 'I took a large wind-tunnel model of it to the office of the International Steel-Wing Aircraft Syndicate in Kingsway, London, who patented the construction (No. 193,980), and that was all we ever heard of it again.' Bolas's ingenuity and brilliance were switched by the Air Ministry to a small triplane with the same theme of two shaft-driven propellers from a central engine installation as the prototypes known as the Tramp and the Bodmin which Bristol and B & P were building.

The late Col Porte's assistant, Maj W. F. Vernon, had for some months been on the staff of Bristol to develop a huge flying-boat version of their Braemar triplane unconventionally powered with steam turbines, but the cost of developing such a powerplant and its condensers proved impossibly high. Farnborough had been considering the parallel possibility of gas turbines, and investigated various existing industrial models, such as the Holzwarth, Karavodine, and Lemale, but concluded these were not yet a proposition for aircraft because of low thermal efficiency. A possible solution was thought to be a constant-pressure turbine using weak mixtures and large compressor with a Curtiss two-velocity-stage rotor of 3-ft diameter. 'The circumference ratio of the nozzles is rather high, but could only be reduced by increasing the rotor size,' stated the RAE report. 'In building up the rotor disc, Laval practice is followed with the disc equally loaded radially and tangentially. Taking the temperature inside the combustion chamber as 1,500°C, the speed within should be below 10 ft/sec. A cast-iron cylindrical chamber 3 ft diameter and 3 ft long would answer the purpose, placed axially in line with the turbine and carrying at one extremity four discharging nozzles. At the opposite end would be the air and fuel entry pipes. The fuel pump would have to overcome the pressure in the chamber of about 150 lb/sq in and handle 1,000 lb/hr of oil. The gasified fuel enters the end of the combustion chamber through a series of jets and is mixed thoroughly with incoming air. The compressor will be coupled direct to the turbine and run at 4,000 rpm.'

Unfortunately the final weight estimate revealed that this gas turbine would be about double that of a piston engine, with fuel consumption also double, so there seemed little point in developing the scheme – but it was an interesting example of this country's scientific pioneering doomed to be dropped.

Meanwhile the Bristol Tramp remained far from completion because of continuous trouble in testing the transmission system of the four 230 hp Siddeley Pumas interconnected through a system of clutches to shaft drive a propeller either side. Until there was semblance of a satisfactory solution for the Bristol and the somewhat similar transmission for the B & P Bodmin biplane, the Parnall proposal, using a single 450 hp Napier, was given low priority, and Bolas turned to several widely differing designs in the intervals of modifying his Puffin amphibian which, like the rival Fairey

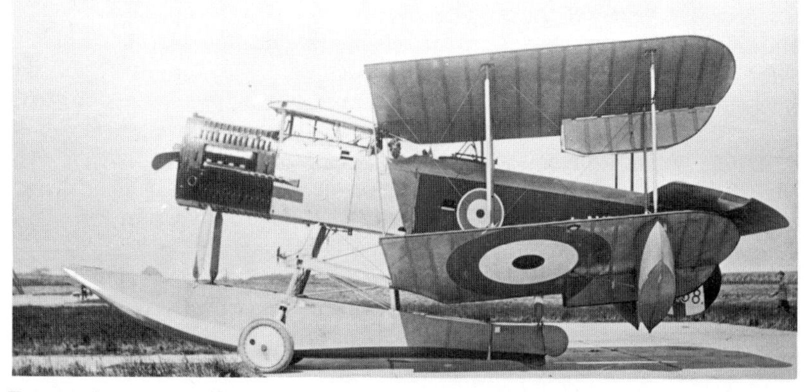

To meet the same specification as the Fairey Pintail, Bolas similarly placed fin and rudder below the fuselage for his Parnall Puffin, but used a single main float, and set the fuselage at mid gap of the staggered wings to ensure folded clearance.

Pintail, was having difficulties. 'We were in trouble over that central float,' Chaplin told me years later. 'When launched for its sea trials in the hands of Norman Macmillan the big punt-like float threw up so much water that it broke the propeller. I had always felt the forebody was too short, and made a model with much longer nose, upturned in front, and tested it in the swimming bath of the local school by towing with a drum cranked by my upturned bicycle, and found it gave a very much cleaner run.' The float was therefore modified to that shape, bracing the longer overhang with two forward struts; but when retested in a slight chop there was still tremendous spray. As there is nothing like having the designer aboard to witness defects rather than read pilot's reports, Bolas invariably flew with Macmillan on all these tests; as a result a false chine was fitted, increasing the beam and breaking the spray. With this the Puffin took off cleanly and quickly, though tended to get airborne too soon, so great was the hydrodynamic lift; but like the Pintail, there were also stability and control problems. The Puffin proved very tail-heavy before adjustment, and the

Unfortunately the Puffin's float did not run cleanly, and even after adding chines it was necessary to lengthen and uptilt the bows with detriment to air resistance.

rudder, although increased in size on the second prototype, was ineffective. Small rudders were a characteristic Bolas fault, inherited from the doctrine of his friend Harris Booth who considered they kept pilots out of trouble, though in fact this could make recovery from a spin impossible.

The Fairey Pintail, despite greater relative fin area than the Puffin and much bigger rudder, proved directionally unstable. Tests temporarily were interrupted by illness of Col Nicholl, the Fairey test pilot. Presently Capt Norman Macmillan took over, gaining unique opportunity to compare the rival amphibians. He found that even with additional small fins fitted to the Pintail outboard below the tailplane, directional stability remained unsatisfactory. Many years later he recorded: 'Both aircraft had a two-position adjustable seat for the pilot. Puffin view was so good that only the lower position was needed, and the adjustment could have been dispensed with. Pintail view was obstructed by centre-section and engine

The Pintail floats also had to be lengthened, and for the third aircraft the wheels were mounted centrally within the floats.

Foundation of Fairey fortunes was the Eagle-powered IIIC developed towards the end of the Great War.

whatever the seat position, and when down was definitely bad for formating or making direct approach to a given object, such as a Carrier's deck. The seat had to be raised to sight the gun, and I reported that the gap between fuselage and upper wing should be increased and the gun lowered to improve its accessibility.' This was done, but Service trials at Grain could find no special reason for ordering either of these unconventional machines, though it led to understanding that if an aircraft was designed for dual purpose as seaplane and landplane, only a weighty compromise

The Fairey offices and works at Hayes still used the small field (at right) for occasional flying in early post-war years, and still had room to expand.

could result; so reliance reverted to the easy-to-fly, long-tailed Fairey IIIC as the standard seaplane.

In view of the big production at Hayes an announcement that Fairey Aviation Co Ltd had gone into voluntary liquidation came as a great surprise. It was explained that registration of a new company with the same name was purely a reorganization 'designed to put upon a permanent basis a business which was commenced during the war, under somewhat uncertain conditions.' The last two words were perhaps a euphemism, for in the earliest years there had been trouble over the Clayton Road factory ownership which Fairey had forcefully settled, and in 1917 he had transferred to a new factory built in North Hyde Road, on the opposite side of the Great

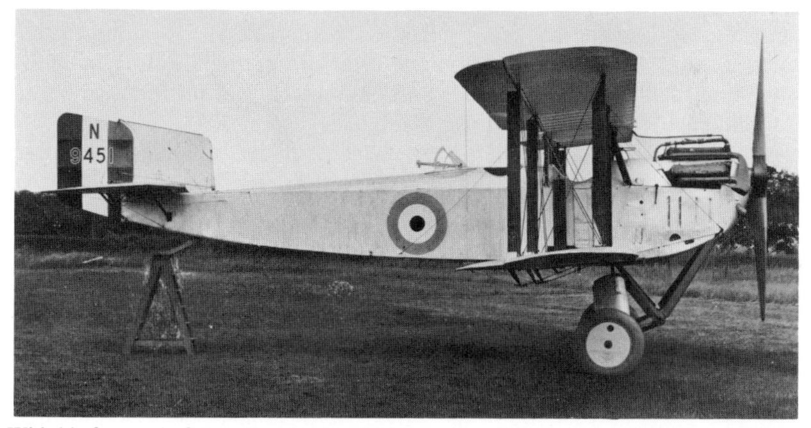

With his factory in full production for IIID landplanes (*illustrated*) and seaplanes, Fairey was in more fortunate position than most manufacturers.

Western Railway mainline, bounding what was always referred to as 'the aerodrome', though it was only a rough piece of ground of doubtful ownership. In the post-war slump he had been glad to allocate half this factory, under the name of Fairey & Charles Ltd, for a contract with the Daimler Co to build motor-car bodies. This proved so successful that his specialist partner in this work, a Frenchman, in conjunction with Daimler, tried to take over the whole factory. A bitter tussle followed, for Dick Fairey was devoted to his vocation of building aeroplanes, and cars had only been an expedient; but he had overwhelming personality and grim determination, and those pale blue eyes could be enormously intimidating. Precisely what happened has not been revealed, but he was soon in possession of the entire factory and car production was tapered off. As a consequence the new company, Fairey Aviation Ltd, was registered on 9 March as a private business with capital of only £100 in £1 shares. The subscribers were C. R. Fairey and his friend F. G. T. Dawson, both of whom were to be permanent directors together with the secretary C. O. Crisp, with remuneration of £300 free of income tax for Fairey as chairman, and similarly £250 per annum for his co-directors.

A contemporary of that time said: 'C.R.F. retained for himself the title of chief designer, in addition to chairman and managing director. Thus the most that anyone under him could achieve was the title "head of design".' The aloofly withdrawn F. Duncanson held this anomalous post, but the man who seemed to carry heaviest responsibility was the 50-year-old chief technician, P. A. Ralli – a genial Greek with a remarkable combination of pure mathematics and engineering science, who had been trained at the Sorbonne, and as an evacué joined Fairey during the war. He is remembered as 'completely devoid of self interest – just lived for the job and delighted in finding elegant mathematical solutions to avoid long tedious calculations'. Prominent among the draughtsmen was Marcel Lobelle, the wounded Belgian soldier who had joined Martinsyde under similar circumstance, but subsequently transferred to Tarrant, detail designing the ill-fated Tabor. After that huge triplane crashed he joined Fairey in 1920, attending night classes at one of the London technical institutes to polish up mathematics and his picturesque English.

On the constructional side, Fairey had an able lieutenant in Wilfred Broadbent who had been works manager since 1917, when he was 26. Building the factory in North Hyde Road had been one of his tasks, and he had supervised every prototype since the beginning of Fairey design, whether landplane, seaplane, or flying-boat. As assistants he had A. C. Barlow and W. Walmsley in charge of about 200 men at Hayes and Hamble.

3

Handley Page, in a lecture to the Royal Aeronautical Society on his new 'break-through' to greater wing lift, revealed some of the development steps between his initial idea of five chord-wise slots and the ultimate span-wise single and multiple slots, and predicted: 'It would appear from our recent experiments that a total lift/drag ratio on a complete aeroplane can be obtained of not less than 1 in 15 at top speed, and with this value and a propeller efficiency of 70 per cent a speed of 120 mph can be attained with 33 lb/hp,' but he did not reveal that this applied to a secret ten-seat high-wing multi-slotted monoplane which George Volkert, assisted by Harold Boultbee, was hopefully designing. Given with H.P.'s typical urbane salesmanship, and undoubtedly with an eye to securing some of the modest allocation for 'research' in the Air Estimates, the lecture was received with stoically British scepticism coupled with an uneasy feeling that H.P. 'might have something'. Professor Leonard Bairstow, the supreme pundit of classical aerofoil theory, voiced the prevailing doubt with the comment: 'It is very obvious that between the original inception and final results Mr Page has passed through a number of unsuccessful experiments' – and Maj F. M. Green, now chief engineer of the newly registered Sir W. G. Armstrong Whitworth Ltd, criticized H.P.'s contention that there were no particular difficulties in constructing a slotted wing and said they would be considerable; with which Dr A. J. Sutton Pippard, the structural specialist,

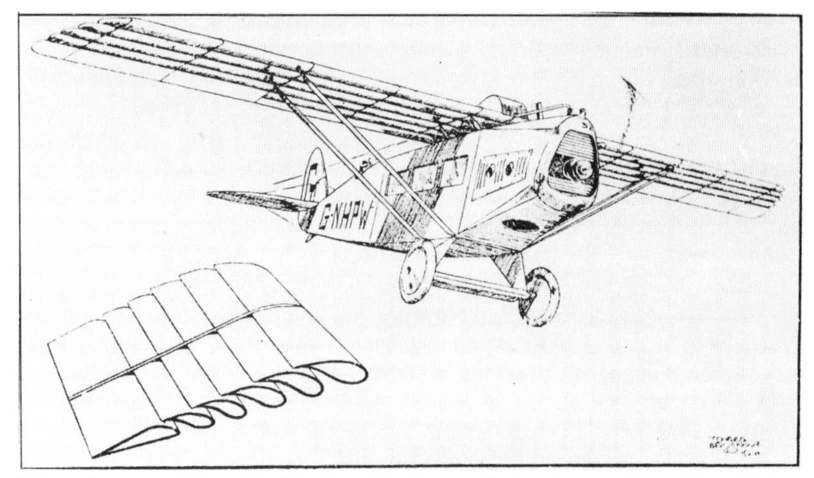

Schematic proposal for a Handley Page five-slot small-winged monoplane with equivalent lifting capacity to the big-span Fokker.

agreed. More contentiously Dr A. P. Thurston challenged: 'Although Mr Page's discoveries seem to indicate something in the nature of a revolution in methods of aerial support I think I have discovered something even better' – but did not explain that his patented wing-tip pinions had just been applied to a monoplane version of the Bristol Babe, which the Press reported 'was fitted with a curious form of wing'.

With a mischievous chuckle, Handley Page pointed out that the Director of Research was present, and hoped that he would profit by the instruction offered by the lecture. As to Dr Thurston's remarks on a new method of support, he thought such discovery most opportune in these times of unemployment! As this subject of anticipation of patent had been raised, he would show a slide claiming Continental priority of a wing devised by Herr Lachmann which had resemblance to his own, though the shapes of the sectional aerofoils were wrong because they gave too restricted a venturi: to prove that, he had tested a model and it gave only $\frac{5}{7}$ the lift that might be expected. It soon transpired that the German Patent Office would not accept Lachmann's application because he had made no wind-tunnel verification; but unknown to H.P., this refusal stirred Lachmann to join the German aerodynamic research centre at Göttingen where he was busily working on his slotted wing.

Nobody was aware that Handley Page's company was virtually bankrupt. All his airline O/400s were grounded at Cricklewood, and there was no hope of being able to replace them with the intended W.8s because he had no money with which to finance production, nor had he accepted the Air Ministry's offer to purchase the prototype G-EAPJ which had won the civil aircraft competition, for it was too expensive to build another *de novo* as the prototype had been largely devised 'on the job' from O/400 components, so complete re-drawing was required.

The next *Handley Page Bulletin* stressed that 'British civil aviation will cease to exist unless the British Government decides upon a radical modification of the proposed subsidy of £60,000, the terms of which are utterly hopeless in comparison with the French subsidy.' However, following consultation between Air Ministry and air transport firms, a temporary resumption of service was proposed for Handley Page and Instone Air Line. H.P. had only five pilots on the pay roll, with R. H. McIntosh succeeding Brackley as chief pilot, and the former publicity manager, Alex Cogni, a wealthy Pole, became general manager in place of Maj G. Woods Humphery. At Cricklewood only a handful of men and apprentices, among whom was G. C. D. Russell, were working under MacRostie as foreman, overhauling a few A.D.C. machines, including a Bristol Fighter for the King of the Belgians. It seems Handley Page Ltd had an overdraft of £400,000, and even H.P.'s associated Aircraft Disposal Co was affected, for he owed it £40,000 from sales, nor had he paid for his own shares of £336,000. There was a tremendous row, but the A.D.C. directors refrained from liquidating Handley Page Ltd in return for the right to nominate management of the main and transport companies, and H.P. agreed to relinquish £176,000 alleged to be due to him. His brother Theodore resigned, and the Bank of Scotland as main creditors appointed four new directors, with Lieut-Col J. Barrett-Lennard as general manager.

Realizing that his multi-slotted, passenger monoplane would probably not be built, and dismayed by the blank prospect, George Volkert, who had been with the company for nine years, decided to leave. Apparently Maj Brackley had breezed into his office to say good-bye because he was leaving England with the pilots and engineers of a mission to Japan organized by Colonel the Master of Sempill at the behest of Rear-Admiral Kobayashi, the Japanese Naval Attaché in London. Brackley, who was disappointed at the failure of British air transport to make the grade, expected employment in Japan to last several years, and told Volkert that the team was short of technicians and therefore advised him to contact Sempill. The H.P. designer knew the effervescently enthusiastic Sempill as the pilot who flew the first American-built O/400, and with that common ground he was promptly accepted as chief of design and inspection. Handley Page, genial as ever, gave his man as a parting gift all the bound volumes of the Advisory Committee Reports, but avoided a cash token of his regard, presumably believing that gratitude was due on Volkert's part since it was H.P. who had picked him among pre-war Northampton Institute students and had trained him.

Japan wished to model its fighting Services on the British. 'In 1920 official application was made for an aviation mission composed of personnel from the Royal Air Force,' recorded Sempill. 'Several Departments of State had a voice, and decision was long delayed. Eventually a reply was sent regretting that the British Government was not in position to render assistance. The Japanese were seriously considering applying elsewhere, for a French aviation mission had already been sent to reorganize the

The Gloucestershire Sparrowhawk was a simple conversion of the Nighthawk airframe to take a B.R.2 rotary engine. (*Courtesy Air Marshal Sir Ralph Sorley*)

military Air Service. However, so impressed was Japan with the pre-eminence of Britain in naval aviation that they decided to request the name of a suitable officer to command a privately organized mission, and I consider myself singularly fortunate in being given the first offer. In January 1921 I came to an agreement and commenced organization the following month. The Japanese placed no major restriction whatever in respect of equipment and training for their Imperial Naval Air Service.'

Sempill found that Avro 504Ks, 504L seaplanes, and F.5 flying-boats from Shorts had been ordered, and to these he added 50 rotary-engined Gloucestershire Sparrowhawks, redesigned from Nighthawks by Henry Folland who was still chief designer to Sir Samuel Waring's Nieuport & General, and also a batch of Supermarine Mk II Puma-powered Channel flying-boats, of which Japanese officials had been given a convincing

To the right are Japanese-built Sopwith Cuckoo torpedo-droppers. On the left is Herbert Smith's ultimate 300 hp version with wide-track undercarriage and different fin.

demonstration by Henri Biard at Southampton in a gale of wind with heavy seas. An order for twelve Parnall Panthers followed, subcontracted to Bristols. A few war-time machines, such as the S.E.5a, D.H.9, Martinsyde F.4, were included to broaden the scope, and a little later two Blackburn Swifts were purchased for experimental use, and Fairey succeeded in selling a redesign of the Pintail with long fuselage. For torpedo dropping, the Sopwith Cuckoo was built under licence by the great Japanese shipbuilders Mitsubishi with Herbert Smith as chief engineer – so his was a familiar task.

Glad though the aircraft industry was to receive such orders, there were dissident voices that it was wrong to supply armaments to a potential enemy. *The Aeroplane* warned that eventually this could lead to a Japanese–American war, though 'at present the combined flying Services of the US Army and Navy are strong enough to bomb Japanese arsenals to destruction if transported to Korea by the US Fleet; but in two or three years' time the position will be reversed if nothing happens to hinder Japan's preparations for war. A naval attack on the American Pacific coast might draw the entire USA Pacific Fleet to that area, and force a concentration of US aircraft for defence of San Francisco, Seattle and other cities, leaving the Mexican coast open for a Japanese landing assuming their Fleet could defeat or hold the US Fleet.' In due course came a reprimand from the Society of British Aircraft Constructors (SBAC), with implication that British companies might cease to advertise in *The Aeroplane* unless it supported British overseas sales whatever the country or the ethics.

Whatever the political fears, the boat was pushed out for the Japanese. When the Crown Prince of Japan and his entourage visited the RAF Station at Kenley on 16 May, he was received by the Duke of York, who was attended by Air Vice-Marshal Sir John Salmond, and the Secretary of State for Air presided at luncheon, after which an impressive flying display was given, which included aerobatics by Flt Lieut F. L. Luxmore, DFC. A month earlier the first of the British Naval Mission had arrived in Yokohama, having travelled by the N.Y.K. steamer *Suwa Maru*, and the second party of instructors and engineers followed within a week. Sixteen experts were also sent from Short Bros to spend two years in Japan, chiefly to instruct in construction and maintenance of F-Type flying-boats.

4

At the beginning of April came the surprise appointment of 45-year-old ex-Life Guardsman, Capt the Hon F. E. 'Freddie' Guest, DSO, CBE, to the post of Secretary of State for Air. As Liberal MP for East Dorset since 1911 he had been made Patronage Secretary to the Treasury, and was much in the entourage of Prime Minister Lloyd George, for he and his brother Lord Wimborne were relations of Churchill. At least the anomaly of a dual-headed War and Air Minister was eliminated, but many were disappointed that the Marquess of Londonderry had not been chosen; instead he accepted office in the Government of Ireland.

The advent of the new regime led *The Times* on 5 April to fire an opening shot in further contention between Admiralty and War Office: 'Military ideas have been supreme at the Air Ministry under Mr Churchill; every favour has been shown to the military side of aviation – barracks, elaborate aerodromes, high sounding titles. Meanwhile actual flying has languished. Little progress has been made in design of engines. Civil aviation has been systematically cold shouldered. Subordination of Air Ministry to War Office is to blame.'

Three weeks later Capt Guest made his debût in the Air Estimates Debate. A commentator reported: 'One thing was abundantly clear, namely that he knows very little about aviation, but it is to his credit that he does not pretend to know. His first startling statement was that during the war the science of aviation advanced a hundred years! We could demonstrate that it has been retarded at least five years.' Guest ascribed savings made the previous year on civil aviation to 'want of interest taken by the general public in civil aviation.' But already the future held a more cheerful note, for arrangements had been consolidated for daily air services between London and Paris by Instone Air Line and Handley Page Transport.

When Instone inaugurated its subsidized service with the Vimy Commercial *City of London* on 21 March, the occasion was adorned by the presence of appropriate dignitaries, including Lord Londonderry – but wily Mr Handley Page had forestalled his rivals two days earlier by cutting all ceremony and sending an O/400 piloted by W. L. 'Wally' Hope from Cricklewood to Paris. Nevertheless Instone plans were more ambitious and, with Capt F. L. Barnard as chief pilot, proposed additionally using three D.H.18s of which two were hired from the Air Ministry, the big Westland Limousine, and their D.H.4A. Less successful was their idea of equipping a commissionaire with a magnificent uniform adorned with silver braid and the company's initials. His duty was to shepherd passengers to their seats, but after a brief morning of such arduous work he was discovered at 2 pm in a state of utter inebriation, and instantly dismissed, followed by a hunt for garments with which to send the fallen hero home.

Explaining to the Commons how the subsidy would be provisionally allocated, Capt Guest said: 'The Air Council guarantee a clear profit of 10 per cent to each firm on gross receipts excluding subsidies, any excess to be returned to the Air Ministry. The subsidy will be £75 for each single schedule flight, provided the maximum payable does not exceed £25,000 to each firm for the trial period of seven months.' With glee, C. G. Grey commented: 'If a company cannot run a machine like the D.H.18 or the Handley Page at £75 per trip plus 10 per cent of its gross receipts then it is time the firm went out of business. Considering it is possible for a hard working joy-ride company to support six people year in year out from pleasure flights at a guinea a head with two Avro biplanes, as the Berkshire Aviation Co have done, it looks as if a properly organized London—Paris air service should be able to put the whole of that £75 into its pocket with the help of a little skilful camouflage.'

The recently produced D.H.18 Limousine, with 450 hp Napier Lion, was regarded as a big step towards profitable operation. On the basis of six flying 1,000 hours per annum, de Havillands began to advertise a gross cost of 4d per passenger mile 'by which is meant fuel, pilot, insurance, maintenance and depreciation of machine and engine, all overhead charges, and every expense in running an airline'. Pilots found the D.H.18 pleasant to fly, like its smaller progenitor the D.H.4, and despite earlier criticism of their being placed so far behind the wings, approach and landing view had not proved too difficult, and certainly there was a happy feeling that it was the safest place in a crash. Passengers at least had burstable fabric covers over escape hatches in the roof, but although O/400s had similar panels by the exit, instructions were on such little labels that they were not evident. The Westland Limousine had nothing except the pilot's cockpit for emergency exit, and the Vimy, Goliath, Spad, and Fokker were worse unless a window could be broken. No aircraft carried buoyancy apparatus, lifebelts, or rafts, though it was observed that some pilots discreetly carried life-jackets for personal use if forced to ditch – and one KLM passenger thoughtfully equipped himself with an inflatable belt worn over a long and large leather overcoat, but by no means could he have escaped through window or hatch. Cyril Holmes was the first to demonstrate the value of escape hatches, for the engine of his D.H.18 failed after take-off, and he made a crash-landing in a suburban garden near Croydon. The machine fell sideways and the door jammed, but the passengers escaped through the roof without too great a panic.

Good though the D.H.18 was, Geoffrey de Havilland's creative imagination had led him through a sequence of schematic designs for sporting and transport aircraft, including a Puma-powered cantilever monoplane, the D.H.26, somewhat like the Fokker, and thence to the D.H.29 – a bigger and more powerful version which interested the Air Ministry's Research Department sufficiently for a prototype contract to be awarded. In April a visitor to Stag Lane reported: 'The monoplane is rapidly taking shape and has its Lion engine in position and the undercarriage fitted. Pilot and navigator sit right forward immediately behind the engine, and then comes a saloon for 10 passengers. Behind it is a gunner's position, as the machine unfortunately is to be a Service machine and not for airlines as had been hoped.' All the same, it was clearly intended as a successor to the D.H.18, for this was the Air Ministry's method of encouraging and subsidizing new civil designs.

At Stag Lane Capt H. 'Jerry' Shaw was testing the 525 hp Rolls-Royce Condor powered D.H.14 which had been ordered in 1918 as replacement for the D.H.9A, but with cessation of hostilities all urgency had gone except for the sole Napier-powered machine which Sidney Cotton used for his attempted England—Australia flight in 1919. Shaw reported that the new D.H.14 was very like the D.H.18, and spoke highly of its performance. Numbered J1939, it was immediately delivered to Martlesham. The makers of the engine had just made a move of unimaginable importance by inducing the quiet, self-effacing chief designer of Napiers, A. J. Rowledge, to join them. His Lion was now the premier engine, and the clean design of its triple-banked cylinders was a big advance compared with the Mercedes-inspired ruggedness of the rival Eagle and Falcon. Too long had Rolls-Royce been content that their engines were the first to fly the Atlantic, to Australia, and to South Africa, though rightly they insisted that the future of flying depended on the reliability which was their hallmark.

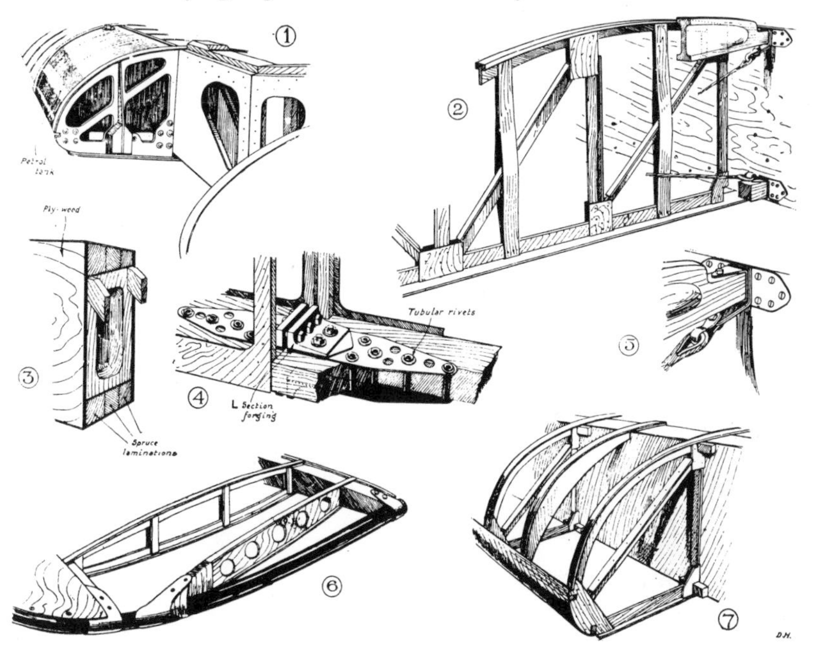

D.H.29 construction: 1, Centre-section and wing tank. 2, Rib and compression struts. 3, Box spar. 4, L-section forgings joined the spars centrally. 5, Top compression strut. 6, Wingtip. 7, Nose ribs. (*Flight*)

While Stag Lane always was a place animated by close-knit team-work and enthusiasm, Croydon now regained its old atmosphere of *camaraderie.* 'English once more can be heard in the Trust House, and Emile Bouderie and his Goliath aircraft are again getting their legs and undercarriages pulled. To make it still more like the days of last summer, some of the old pilots are returning in varying capacities,' wrote one of its devotees. Nevertheless, one of the operators, Handley Page Transport, were far from the end of their troubles. The old O/400s had great difficulty in taking-off from Croydon. One just cleared the Custom-house, but was unable to climb over the top of a hill towards Purley, and dropped onto a field whilst still flying furiously. Another, taking off downhill, had to land immediately after crossing the new Brighton road. Then came the knock: the C of A was reduced to authorize only five passengers. Their machines were hastily sent to Martlesham to be weighed, pending possibility of raising the number of passengers, but it was found they differed from one another by as much as 500 lb. Hence that sinking feeling, as one wit put it!

H.P. Transport Ltd came back to Cricklewood as starting place for their outward service, for though smaller it was flat. Meanwhile the Vimy Commercial with the same power and engines could take-off easily from Croydon with ten or a dozen passengers. Where were the W.8s, long promised as re-equipment? The prototype was in fact still the only example, and 'incessantly expected' at Croydon, but was quietly accumulating dust at Cricklewood, for its Lion engines had been returned to Napiers who had only loaned them. That something was up was first realized by the habitués of Croydon when Lieut-Col W. A. Bristow, of Ogilvie & Partners, appeared in the Handley Page office – for almost the first act of Colonel Barrett-Lennard, as manager of H.P. affairs, was to give him responsibility for the technical running of Handley Page Transport: few could be better, for Bristow had been in charge of all RNAS matériel for bombing operations in Flanders.

As a quick, cheap expedient to test the slotted monoplane formula, Handley Page mounted an experimental wing on one of his Disposal D.H.9A fuselages.

And what of those slotted wings? It was reported that Frank Courtney had been flying a 'parasol' creation at Cricklewood consisting of a standard Liberty-powered D.H.9A fuselage beneath a cantilever wing with full-span slotted leading edge – but it seemed a lash-up, adapted so that the lift struts could pick up the original lower strong points on the fuselage intended for the standard biplane wings. Some said it was the wing intended for the ten-seater which H.P. had abandoned. Within a few days it was crashed in too costly a manner for experiments to continue.

5

Industry was much disturbed by labour unrest. The end of March had seen a coal strike with grave consequences as it was our major export. Closure of manufacturing plants was threatened through lack of gas and electricity generated by steam power. A State of Emergency was proclaimed – for a conference between Government and miners ended in deadlock on 7 April. Next day the Air Ministry issued a Notice asking former pilots of the RAF to volunteer for strike duty. Thousands of ex-officers responded, and tens of thousands of the so-called middle classes volunteered for every form of duty. There was little sympathy for the euphemistically termed 'worker'. Transport crews and railwaymen answered with strike action supporting the miners, but though this workmen's triple alliance was withdrawn on the 15th, following conciliatory offers of more pay and better conditions, its actions had added to the increasing unemployment in mining, shipbuilding, and heavy industries. An optimist wrote: 'The arrival of an emergency like that which has just passed shows the nation is sound at the core, and that all the revolutionary talk and stirring up of class war which is the self-imposed task of a few Bolshevist agitators are only the vapourings of a noisy minority.' But the militant discontent of the miners continued. The Government offer was rejected, and the conference between owners and workers came to nothing. Not until 1 July did the coal strike end.

Ireland was causing even greater concern. Elections on 24 May for the Northern and Southern Parliaments led to Sinn Fein again winning all but four of the 128 seats in the South, and in the North the Unionists held 40 seats, with Sinn Fein and Nationalists twelve. Next day there was a great battle in Dublin, because Trinity College represented the four non-Sinn Fein seats of the South, and the Custom House was burned down. On 6 June the first sitting of the Ulster Parliament in Belfast was marked in London by Sinn Fein raids in the suburbs where telegraph and signal wires were cut. An appeal for a Truce was made by King George V when he opened the Ulster Parliament on 22 June. Two days later Lloyd George invited the rebel De Valera and the Northern Ireland Prime Minister Sir James Craig to a conference in London on 11 July. There, Dominion status was proposed for Ireland, with autonomy of taxation and finance, and power to maintain a military home defence force, though the North would have freedom to determine its own relationship. It was no good. Steadily Irish affairs deteriorated.

Come what may, that was no reason for England to stop her summer cricket, nor the delights of Ascot or Henley Regatta, nor such spectacular flying events as the RAF Pageant, as it was now termed. Though again no British seaplane challenged for the Schneider Trophy Contest, at least there was the Aerial Derby to look forward to, and it seemed there would be a British entry for France's Coupe Deutsch.

The versatile Prince of Wales, back from his Empire tour, spoke prophetic words at a dinner for the Dominion Premiers. 'There is no doubt,' he said, 'that the future of rapid Imperial inter-communication lies in the air, and I trust the day is not far distant when civil aviation will have built a great air organization on the same lines as our mercantile marine, and that delegates at the next Imperial Conference will travel by the Imperial air routes now being worked out. Present ship and rail communications are insufficient for a Commonwealth of Nations which extends over all parts of the globe. The British Empire has more to gain from efficient air communication than any other State in the world, and I feel sure no time will be lost in solving the problems connected therewith.'

Churchill, on his return from the Middle East, harangued Parliament to similar effect on 14 June, and after explaining the vital conclusions he had reached with Trenchard on the rôle of British aircraft in policing the Near East, added: 'Arrangements are being made for aeroplanes to fly regularly to and fro across the desert between Baghdad and Cairo. Once the route has been marked out the whole Air Force in Mesopotamia could be speedily transported to Palestine or Egypt, or *vice versa.* Arrangements could also be made to fly a certain number of commercial aeroplanes, which would carry mails, and possibly passengers, and could afford a most valuable link in the chain of Imperial communications, which might ultimately result in very great advantage in shortening communication with India and with Australia and New Zealand.'

Discovery of oil was the factor making it essential to stabilize the Near East area so that fuel could be exploited for the benefit of Europe and Britain in particular. Between Transjordan and the great rivers of Iraq spread 500 miles of desert. Those who had flown the desert observed how emphatic were the wheel marks of lorries used near camps; as a result the Churchill/Trenchard conference decided that the best guide for aircraft would be a double furrow ploughed the whole length of the route, with fuel reserves stored in five-gallon drums at suitable points. Led by Sq Ldr W. L. Welsh, work was immediately started by parties moving centrally from either end, but in all that barren extent only two sparse streams gleaned water from the adjacent rugged hills, and it was therefore essential to control them, particularly that key watering place for nomadic tribes, Rutbah Wells.

Already the Arab rebellion had cost the British some 2,000 casualties. Four RAF Squadrons were now transferred to Iraq, under the monosyllabic Grp Capt Hugh Dowding, in order to spot refractory tribes, drop warnings to lay down arms, and if they continued belligerent, to blockade their areas using pairs of aircraft in sorties day and night, ringing their tents and flocks with small bombs but avoiding injury. A similar air umbrella was extended to Transjordan and Palestine, mandated territory under League auspices, but the latter mandate, to the concern of many and relief of others, included an undertaking to establish 'a national home for the Jewish people.'

Meanwhile in England the Air Ministry issued a *communiqué* setting out yet newer terms of State assistance for commercial flying. Operating firms were to be 'approved', but this posed no immediate difficulty for there was no intention of discontinuing the services operated by Handley Page Ltd and S. Instone & Co Ltd under the temporary scheme. 'Orders will be placed by the Air Ministry for aeroplanes of modern commercial type to be hired out to approved firms who will be required to pay a monthly rental equivalent to $2\frac{1}{2}$ per cent of the cost of the machine. After 30 such payments the machine will become the property of the firms.' The entire personnel of each airline, including directors and shareholders, had to be British, and aircraft and engines of British design and manufacture; but though this exclusiveness should prove a boost to manufacturers, the operators were more intent on the fact that: 'With the approval of the Lords Commissioners of HM Treasury, a sum of approximately £200,000 per annum will be set aside from civil aviation votes for three years.' This sum had to cover cost of some £50,000 to date of the temporary scheme, as well as purchase of 'hire' aircraft, but the crux was that 'The Air Ministry will grant subsidies during the same period of three years on the basis of 25 per cent on an approved firm's gross earnings operating on any of the following routes: London—Paris, London—Brussels, London—Amsterdam. Further routes may be approved at a later date.'

To remind the country that airships might prove a better proposition, particularly for long distances, MPs and Press representatives were taken to Pulham for a flight with Maj Scott in the R.36. 'The mooring mast is only in the experimental stage, and possesses no means of ascent other than ladders – which require a certain amount of nerve to tackle,' recorded one of the passengers. 'However, by 7 a.m. I had safely negotiated the 120 feet climb, crossed the platform over what seemed nothingness, down the

The Beardmore-built R.36, with civil registration G-FAAF, moored on the mast devised by Major Scott. (*Flight*)

Major G. H. Scott (*right*), most experienced of all British airship commanders and a key figure in airship development.

gangway, through the opening in the airship's nose, descended the catwalk beneath the gas-bags, and finally was comfortably installed in the passengers' cabin – arm-chairs, tables and bunks for everyone, in addition to plenty of room to stroll. Never have I experienced such travel comfort as in this airship – no dust, no smoke, no sway, no draught, very little noise, and practically no vibration.' At 50 mph they cruised the countryside, passed over London, and circled Croydon aerodrome 'where a parachute was dropped with the newspaper correspondents' reports'.

A few days later one of the dangers of airship operation attained equal publicity, for on coming up to the mooring mast a gust of wind carried the weightless R.36 over the top, and the sudden arresting pull of the mooring cable strained the forward bays of the airship so badly that she had to be manoeuvred into the shed, only to encounter further disaster, for her bows and side hit the walls and were badly damaged. This led to startling revelation that: 'The Zeppelin L.64 has had to be wrecked to make room in the Pulham shed for R.36 as there was nowhere else to put either ship.' It seemed that airships might be an expensive business.

Nevertheless, on the night of 23 June, the R.38, the world's largest airship, made her initial flight at Cardington, where she had been built by the Royal Airship Works, previously Short Bros' factory. Commanded by Flt Lieut A. H. Wann, RAF, and carrying 48 crew and passengers, including Air Commodore Brooke-Popham, Air Commodore Maitland, Mr C. I. R. Campbell who was the Naval Constructor Superintendent of Airships, and the two American officers, Commander L. H. Maxfield and Lieut-Cdr Beig who were to deliver her across the Atlantic to her US Navy purchaser, she cruised six and a half hours over the home counties before turning back and mooring in the pale dawn.

Aeroplanes now took the stage with airships. It was to Hendon that all roads seemed to lead on 2 July when the great RAF Air Pageant was presented for the second year. Excitement had been stirred with promise of 'aerial fighting of the most realistic character in which machines will be "shot down" in flames. Two great Handley Page bombers will be attacked by Scout machines and "destroyed" as realistically as in war. A kite balloon will be attacked by an aeroplane, and onlookers will be thrilled by the sight of the observer leaving his burning balloon and descending by parachute. The *pièce de résistance* will be the complete destruction of a model village by machines belonging to No. 24 Squadron. This village is one of the most realistic affairs imaginable, built from old aeroplane "scrap" and complete even to a church with an important looking tower.' And so it happened, and much more. King George, Queen Mary, Queen Alexandra, and the Duke of York were present, attended by Air Vice-

The Royal enclosure at the 1921 RAF Pageant. (*Flight*)

Marshal Sir John Salmond who had organized the Display, assisted by Air Vice-Marshal A. V. Vyvyan, Air Commodore H. R. M. Brooke-Popham, and Grp Capts H. C. T. Dowding, T. C. R. Higgins, and A. M. Longmore together with six other officers, and Commander Harold Perrin representing the Royal Aero Club.

To emphasize Britain's prestige in airships, R.33 flew overhead, sometimes like burnished silver, at others dark and sinister like a shark, and then ghost-like and semi-transparent, fading into fleecy clouds. As a laughter-maker with serious purpose there was the old prototype B.E.2c, crazily painted, its wings bereft of stagger and fitted with apparently rickety struts and an additional undercarriage on the centre-section, though its normal rôle at Farnborough was to measure flight loads in the fuselage at the behest of Professor 'Bones' Melville Jones who had flown her after sawing through the longerons and re-connecting the top pair with hinges and the lower with springs.

This faked-up B.E.2c relic of the Great War flew to amuse youngsters many of whom probably became pilots in the next World War. (*Flight*)

Of those who performed most splendid aerobatics at Hendon, Flt Lieut W. H. Longton on a Sopwith Camel for the second year running was clearly the supreme artist at smoothness of every manoeuvre, though it was Flt Lieut 'Oojie' Noakes who caught the public fancy with his startling crazy flying in a red-painted Avro and afterwards was presented to the King. Another pretty exhibition was by Flg Off P. W. S. Bulman, MC, AFC, on Farnborough's B.A.T. Bantam which had been used for research on inverted spinning. Said C. G. Grey: 'Mr Bulman did everything that can be done so far as actual manoeuvring is concerned, but he has not quite acquired the art of showing off a machine to the best advantage. Those of us who saw the late Peter Legh and Major Chris Draper on the same machine in 1918 perceive the difference. Nevertheless Mr Bulman's performance gave a very fine impression of the aeroplane and of his ability as a pilot.'

6

Tragedy again: the British aircraft industry, as well as several private flyers and two French entrants, were busy transforming their machines into 'racers' for the Aerial Derby to be flown on 16 July over the old pre-

At the wheel of his racing car, the thirty-year-old Australian, Harry Hawker, who had joined T. O. M. Sopwith in 1912 and became Britain's pre-eminent test pilot. He was the first to exceed 100 mph in a 1,500 cc (A.C.) car.

war course of Hendon to Brooklands and Epsom, thence to the eastern turning point of West Thurrock on Thames side, and so northward to Epping and Hertford and back to Hendon on a 100-mile circuit flown twice. Twenty had entered – among them Harry Hawker flying the Nieuport Goshawk, powered with the fire-prone Dragonfly, and holder of the British speed record of 167 mph, though pundits considered that Henry Folland's latest design, the Gloucestershire Mars I known as the 'Bamel', flown by Jimmy James, was the likely winner.

Four days before the race Hawker was testing the Goshawk at Hendon. A couple of miles away at Stag Lane, Geoffrey de Havilland was conferring on the aerodrome with Charles Walker over difficulties in flying the new D.H.29 monoplane which D.H. had initially tried on the 5th. 'Looking up,' as C. Martin Sharp writes in *D.H.*, 'they watched an aircraft climbing at an unusually steep angle out of Hendon. It went to a good height and then seemed to get out of control and came down and they saw a cloud of smoke rise. They went over in cars, but could do nothing. Returning, de Havilland said, "Well that's that. Now we'd better get on with this one", and tests continued in no cheerful mood,' for it was Harry Hawker who had been killed. Regarded as the greatest of all British pilots, Sopwith's pre-war protégé had thrilled the world in 1919 as the first man to attempt flying the Atlantic non-stop, only to be forced down but spectacularly rescued through the long odds of a ship passing by. 'If ever there was a trier, Hawker was one,' wrote C. G. Grey. 'Once he made up his mind to do a thing, he would try, and try, and try again until he succeeded. Failure spurred him to new efforts. Within a small exterior he had a great and restless spirit, a driving force which made it imperative for him to be up and doing. He loved things that were worth while, and did them for the sake of doing, not for gain. When ill health should have kept him on the ground he continued to fly, not because he understood his illness, but because his restless nature would not let him take things easily. Had he been the wealthiest man in the world, Hawker would still have done the things he considered worth doing. If these carried with them an element of danger, he did not shirk them on that account; in his mind they were mere incidentals in the game he loved, the game which now claimed him.'

At the inquest Dr Gardiner of Weybridge made it clear that Hawker would have lived only a few weeks in any case. Eighteen months earlier he had seen a specialist about a pain in his back, which had been injured some years ago, and X-rays revealed a tubercular spine. Since then the injured vertebrae had been reduced to a shell, resulting in haemorrhage while testing the Goshawk, and paralysing his legs. An eye-witness saw a burst of smokeless flame under the machine and thought the pilot turned the aeroplane on its back as though to drive the fire from the cockpit by falling upside-down, then righted, made a rapid spiral and straightened out as if for a fast landing, but went straight into the ground.

Flight, with the mantle of prophecy, wrote: 'The name of Hawker deserves to live in the history of aviation, not for the performances popularly associated with his name but for his contributions to the increase of aeronautical knowledge. His recent venture into the motor trade was only a means to an end, for he desired to build better aeroplanes than anyone has built and hoped to make from trade the money with which to build them.' For the daily Press, Hawker's accident was the finest opportunity of the year for booming an aviation fatality.

Inevitably his passing detracted from the light-heartedness of the Aerial Derby, and despite splendid weather the attendance was poor. Even before the race started accidents reduced the field to only twelve. The great French pilot Sadi Lecointe nosed-over his Nieuport in landing and broke his arm; his fellow countryman Count Bernard de Romanet broke a wing-tip and tailskid on his special Lumière-de Monge racer and it was with-

The 25 ft 6 in span 230 hp Avro 539A was Roy Chadwick's idea of an inexpensive racer designed on an overlay of the Baby 534 drawings, but it proved 20 mph slower than the Semiquaver.

The pretty little Martinsyde Semiquaver became birdlike with its Alula wing, but though eventually flown it was too flexurally dangerous for serious investigation. (*Flight*)

drawn; Westgarth-Heslam, test pilot for Avro, overshot with the Lion-powered Avro racer derived from last year's Schneider, and disastrously ran into the railway cutting, again breaking his recently broken leg; Maj 'Tubby' Long broke his Martinsyde; the Wasp engine in Chris Draper's Bantam refused to function, likewise the venerable R.A.F. engine of Mr Curtiss's B.E.2e; and lastly the unconventional shoulder-wing monoplane entered by Mr Holle, comprising a small cantilever 'Alula' wing of his design mounted on the fuselage of last year's Martinsyde Semiquaver racer, became a non-starter. Frank Courtney, the winner when it was a biplane, took one look at the sparless, flexible, boat-built wing and decided it would be sheer foolishness to race this untried contraption, or even fly it at all. Holle tried to get Draper or de Romanet to take Courtney's place, but the former made himself scarce, and though de Romanet went to Northolt to look at the machine he too shrugged his shoulders and departed.

Said one of the aviation zealots: 'When I went to the sheds before mid-day on Saturday it looked as if there would be several other non-starters. The 'Bamel' and Nieuhawk appeared in process of building *in situ*, and the Bristol Bullet was without spinner. Sundry mechanics, whose enthusiasm was apparently being evaporated by the temperature, were screwing up odds and ends here and there until just before starting time. Finally the machines were ready with seconds to spare.'

The race was preceded by an event all declared a great success – an air race between Oxford and Cambridge, where many ex-Service pilots were studying for their degrees. The Royal Aero Club put up the money to hire six S.E.5as from Aircraft Disposal. Oxford was represented by A. R. Boeree, the ex-Martlesham pilot soon to become a schoolmaster, N. Pring, and A. V. Hurley; Cambridge had H. A. Francis, W. S. Philcox, and R. K. Muir. Oxford encountered great difficulty in finding a team because of their university's prohibition against flying and activities had to be *sub*

The University aeroplanes were Disposal S.E.5as made available 'by courtesy of Frederick Handley Page', and were flown by students who were war-time pilots.

rosa, but the more enlightened Cambridge authorities regarded the race with favour and allowed notices to be displayed asking for entrants, with the result that Cambridge Aeronautical Society was inundated with applications from S.E.5 pilots. Unfortunately Boeree, the inspirer of the race, relied on his stop-watch and moved over the starting line before the flag dropped, so had to turn and recross while the others took off. Cambridge climbed high, and Oxford flew low down, but on the second circuit Pring landed at Enfield with engine trouble. After 20 minutes the rest were seen returning in a bunch to Hendon, finishing within 30 seconds of each other, and Cambridge in the first three places.

The Oxford and Cambridge Air Race teams with Capt. A. R. Boeree to right of top most figure, Sir Sefton Brancker in front of him, Harold Perrin of the Royal Aero Club between Brancker and the pilot in a white sweater, Jack Savage of A.D.C. centre of front row, and Capt H. A. Francis, leader of the Cambridge team, third from right in back row.

While they were still flying, the Aerial Derby machines had been lined up, and Capt Tully on the two-seat Avro Baby was despatched on his handicap time, closely followed by a brand new single-seat Baby piloted by Bert Hinkler recently back from demonstrating one in Australia. Off went Forester-Walker's Sopwith Pup; then a wealthy young private owner, Alan S. Butler, piloting his Bristol Tourer; there was Tait-Cox with an Avro Viper, Hubert Broad with his Sopwith Camel; Longton and Ortweiler on S.E.5as, Noakes on the Nieuhawk, Foot on the Martinsyde F.4, Cyril Uwins on the clipped-wing Bristol Bullet, and finally James with the 'Bamel'. Hinkler and Tully forced-landed in turn at Brooklands. The Pup gave up, and tipped on its nose when landing; Ortweiler cut a turning point; and Foot came back with engine trouble. As predicted it was the 'Bamel' which won, gaining the Derby Trophy and £400, with an additional £200 as winner also of the Handicap.

Barnwell drastically revised the Bristol Bullet, reducing the wings from 195 sq ft to 180 sq ft and swelling the fuselage to the external diameter of the Jupiter engine. With these and lesser refinements speed increased from 155 mph to 170 mph at the expense of high landing speed. (*Flight*)

Although the 'Bamel' was designed for the Gloucestershire Aircraft Co by Henry Folland as a spare-time job, he was still employed three days a week by Lord Waring's company, Nieuport & General Aircraft, which was on the point of collapsing through lack of orders. Folland's old Homburg hat, glasses, winged-collar, spats, and somewhat untidily fitting clothes, gave the impression more of a retired city man than popular conception of a designer, but could not mask his self-assured expression, for he was certainly pleased and even a little conceited over the success of his new machine.

The 'Bamel' had been put together, rather than designed, in the mere space of three weeks by adapting one of the Gloucestershire-built Nighthawk fuselages and tail. A multiple bulkhead replaced the steel

Like the Nieuport Goshawk, the Lion-powered 'Bamel' was devised inexpensively from a war-stock Nighthawk fuselage and components using Folland's favoured I-strut interplane bracing. (*Flight*)

mounting plate of the standard radial, and from it sprang the supporting structure for the special Lion engine. Under it was a finned oil tank and a projecting rectangular honeycomb radiator. A completely faired cabane came next, with sides tangential to the rounded fuselage, containing gravity and water header tanks, but though its rear edge was immediately in front of the pilot the obstruction to view was remarkably small, for in a conventional tractor the pilot looks diagonally and not directly ahead, particularly in landing where the nose obliterates the vista. The slightly staggered biplane wings were similar to the Goshawk, with equal span and chord, though the top one had rounded tips and the lower was squared. Only a single Rafwire each side braced the front top spar in flight, though the rear spar had duplicated wires as the C.P. would be aft at speed. Single landing wires ran to the lower front spars from the top of the tank cabane, and to take tension at the cabane, cables ran within the fairing from front and rear

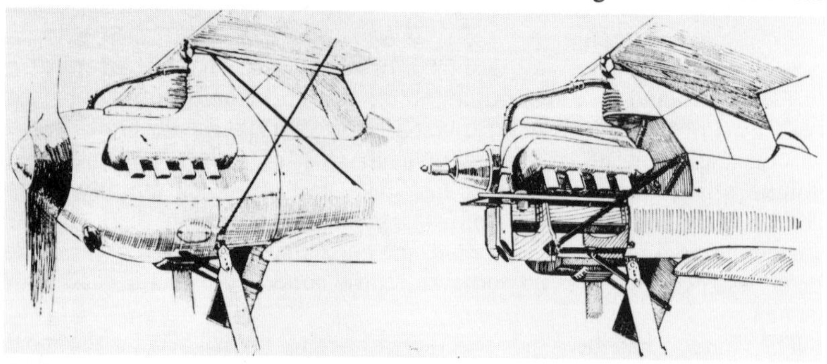

Metal spinners were apt to disintegrate but Folland's simple and novel solution was to enlarge and fair the propeller boss. Fuel was carried in the streamlined cabane. (*Flight*)

top root fittings to the rear leg attachment of the undercarriage. The wooden propeller, so often the subject of doubt of its efficiency, had been designed by Ogilvie & Partners' H. C. Watts, the foremost British propeller expert. It was expected that the 'Bamel' would soon be sent to Martlesham for an attempt at the speed record, and subsequently would represent Britain in the Coupe Deutsch to be held in France at the beginning of October.

There followed the long-expected announcement that Henry Folland had resigned from the Nieuport & General Aircraft Co to join the Gloucestershire team as chief designer and engineer. He had ideas for a commercial aeroplane powered by a Napier Cub. 'The Gloucestershire Aircraft Co Ltd may for the moment not be so well known,' reported *Flight*. 'This is a shortcoming which under Mr Folland's technical leadership will no doubt soon be remedied. The firm commenced business in 1915, building aeroplanes to other people's designs, and now that original designs are to be built one may expect to hear good accounts from Cheltenham. Probably the first new design will be for the Air Ministry, and of this nothing may therefore be said. In the meantime it is gratifying to learn that Mr Folland's services are not to be lost to the industry.' That the company was determined to become well known was immediately indicated by full page advertisements in *Flight* and *The Aeroplane*.

Hugh Burroughes, after parting from Holt Thomas, had returned to the Martyn and Gloucestershire Boards, representing them in London, and paying weekly visits to Cheltenham where David Longden, a most able man, was administrator. A year earlier Longden had shown considerable courage in buying all the Nighthawks they had made, storing them in the Winter Gardens Pavillion, and found immediate reward with a contract

The Nieuport Nighthawk might have rivalled the Martinsyde F.4 as a fighter had the Dragonfly engine proved reliable. Instead, orders were cancelled, though re-instituted for a few with Jupiter or Jaguar engines. (*Courtesy Air Marshal Sir Ralph Sorley*)

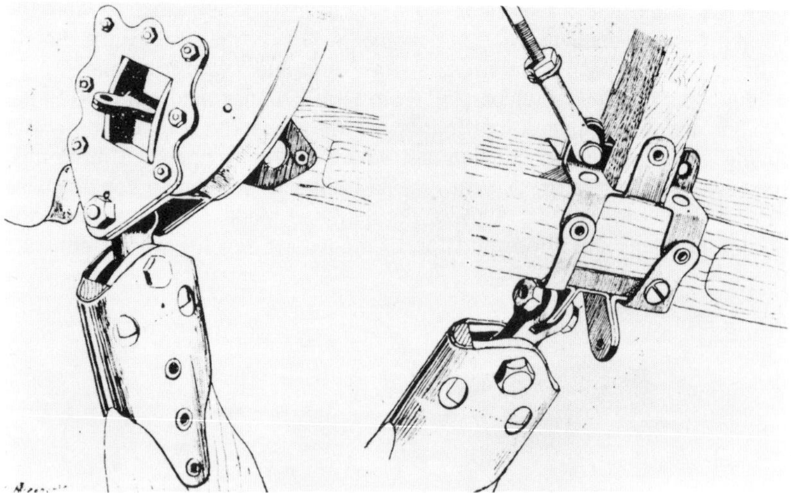

Typical fittings and structure of Folland's design which persisted in character until metal structures supplanted wood. (*Flight*)

from the Japanese for fifty Bentley-powered Sparrowhawk variants and forty in components as reserves. It was this, even more than the success of the 'Bamel' which led to enlistment of Folland and his able lieutenant, little H. E. Preston. The stamp of approval was wrung from the Air Board, and G.A.C. was in business for the first time as a fully-fledged design contractor – a great act of faith when other aircraft companies were only surviving with the aid of a prototype contract for each and a small amount of repair work. As Burroughes recorded: 'The G.A.C. carried on and paid its way with contracts for reconditioning Parnall Panthers, a small contract for

Nighthawks for the Chanak crisis, and manufacturing components of that invaluable war survival the D.H.9A – a stand-by for which several firms were duly grateful, or should have been. The 'Bamel' was promoted to establish G.A.C. with a reputation for high-speed aircraft, and to stake a claim with the Air Ministry in the single-seat fighter field. In the meantime the company pressed on by designing as a private venture another variant of the Nighthawk, first with a B.R.2, later with the Armstrong Siddeley Lynx and finally with the Armstrong Siddeley Jaguar.'

7

August seemed likely to match the torrid heat of the drought in the early part of July, but interest in flying during the holiday season seemed phlegmatic. However, Berkshire Aviation with their Avro in the hands of the then smoothly shaven O. P. Jones, handsome idol of the flappers, and smiling A. F. Muir with his Avro at Croydon had their fill of joy-riders, and the de Havilland Co's D.H.9s were continuously busy with photography and taxi flights. Private flying was negligible, though the Royal Aero Club was hopefully offering one B.E.2e and three Avro 504s for hire to its members at £3 an hour inclusive of petrol and oil, basing them at Croydon in charge of Muir. Occasionally there were moments of excitement such as when Alan Butler, full of amateur enthusiasm at gaining 3rd place in the Aerial Derby handicap with his Bristol Tourer, competed with

Bristol built some two dozen Puma-powered two- and three-seat Tourers, but it took two years to sell them. Alan Butler purchased G-EAWB and toured southern Europe with it between April and June 1921. (*Flight*)

Cyril Uwins to see whom could land it in the shortest distance. Butler's effort was described as sensational! Uwins then took him as passenger, and at 500 feet they could be heard in powerful discussion. In no wind, Uwins pulled off a pretty landing, stopping in some 70 yards, so Butler decided to put the machine away for the night, 'before we wreck it,' he explained. He taxied to the level-crossing on Plough Lane, turned far too quickly, and hit a ferro-concrete post with his wing, so that the machine swung violently round with the engine hanging over the fence. 'On getting out of the machine, Mr Butler proceeded to turn a cheerful somersault on the grass on the strength of it,' reported Geoffrey Dorman in *The Aeroplane*, and then cautiously added: 'Mr Butler seems to have the makings of a star pilot when he has had a bit more experience if his luck is kind. One hopes he will not take risks as civil aviation cannot afford to lose enthusiastic amateurs of his calibre.' But regardless, Mr Butler was now sold on aviation and wished to identify himself more closely with it. As an opening venture he collaborated with Sidney Cotton in capitalizing a seal-spotting company in Newfoundland, equipping it with the big Westland Limousine G-EARV, and a Martinsyde.

Sidney Cotton opened seal and fish spotting operations in Newfoundland with the big Westland Limousine III (*left*) and smaller Martinsyde F.6 on skis, presently adding three Limousine IIs.

While the world at large might be indifferent to flying, people avidly read the daily Press for the epitomized march of events: Ireland offered Dominion status with power to maintain a Home Defence Force; death of Signor Caruso; famine in Russia, thousands starving; railways decontrolled; death of King Peter of Serbia. On 17 August, devoid of valedictory headlines, a small paragraph announced: 'It is with deep regret that we record the death at Geneva, of Lieut-Gen Sir David Henderson, KCB, KCVO, DSO. His health had been failing for a long time, and his death had been expected for some months past. He was 59 years old.'

Though Henderson's span of activity with military aviation was only seven years, it was epoch-making pioneering. Under the *alias* of 'Henry Davidson' he had learned to fly in 1911 with Howard Pixton as instructor at the Bristol School at Brooklands when he was a Brêvet Lieut-Colonel. Appointed Director of Military Training at the War Office in 1912, he laid the very foundation of the Royal Flying Corps, with its naval and military Wings, and in September 1918 became the first Director-General of Military Aeronautics with rank of Brigadier-General. At the outbreak of the Great War he took the RFC to France with its historic three and a half squadrons, totalling 42 unarmed pushers and tractors, which helped to cover the great British retreat from Mons. From this nucleus grew and expanded the great fighting Service which became the RAF with 22,647 aircraft, 27,373 officers, and 263,410 other ranks, and nearly 700 aerodromes.

In October 1914 he had returned to soldiering, for he considered it his primary duty – but a month later he was recalled with rank of Major-General and appointed GOC of the RFC with Headquarters at St Omer, the atmosphere of which was brilliantly portrayed in Maurice Baring's classic *R.F.C., H.Q.* In 1915 it was decided he must operate permanently from the War Office as Director-General, and Trenchard took over his Command in France. Henderson's had been no easy course, and in many ways he was too gentlemanly to fight back with the acrimony of his opponents when he became victim of bitter attacks on the management of the RFC and his choice of aircraft, resulting in a Judicial Enquiry in the summer of 1916 which would have broken a weaker and less able man. Acceptance or rejection of the resulting committee recommendations took time, so affairs grew worse before improving with the appointment of William Weir as Controller, early in 1917, to speed production; but creation of the Air Ministry and appointment of Lord Rothermere as Air Minister led to further trouble. An Air Council was created, of which Sir David Henderson, knighted in the New Year's honours, became Vice-President. Maj-Gen Sir Hugh Trenchard was made Chief of Air Staff, and Maj-Gen John Salmond succeeded him in the Field. Within a short time Rothermere was in conflict with Trenchard who promptly resigned, and inevitably his friend Henderson backed him with his own resignation, particularly as Maj-Gen Sykes, whom he distrusted, was given Trenchard's position. Once again Henderson reverted to the army, but within months his son Ian lost his life in an aeroplane accident at Turnberry in Scotland, with deep effect on his father. Without disrupting the organization of Command, there was no adequate post of responsibility which Henderson could be given other than Area Commandant in Paris; but when the Peace Conference opened he was appointed Military Counsellor to the Supreme Council, remaining there until June 1919 when he became Director-General of the League of Red Cross Societies at Geneva – an administrative post for which he was eminently fitted. 'David Henderson was a great soldier,' wrote C. G. Grey. 'Above all things, he was a very great gentleman.'

A week later catastrophe overtook R.38. 'At intervals in the course of every human effort towards progress, moral or physical or mechanical, there befalls a disaster which arrests the attention of the world and causes the faint-hearted to regard it as the end of all things,' wrote a commentator. 'In reality these disasters cause a temporary check, but ultimately do good in that they act as a spur to further effort because the human man will not acknowledge defeat, and they provide valuable lessons in what to avoid in future.'

Under American ownership as ZR-2, it had been intended to fly this beautiful airship to the USA at the end of August; to that end her US naval crew had been some months in England studying operational techniques, and latterly had been conducting flights under instruction by the RAF. On

23 August they had been cruising all day between Howden and Pulham, but instead of landing that evening it was decided to continue overnight, and shortly after 5.30 on the 24th were nearing Hull when suddenly onlookers were horrified to see the great airship beginning to buckle in the centre, and then amid flames and smoke she broke in two, the stern portion dropping more slowly than the nose, each falling in the river. Two parachutes came floating down, one dropping into the flaming wreckage. The mass of smoke from burning petrol surrounding the wreck made rescue work so difficult that only five survivors were picked up, among whom was the injured captain, Flt Lieut A. H. Wann, RAF. Among those lost were Commander L. A. H. Maxfield, USN, his two Lieut-Commanders, three Lieutenants, and nine crew. Among the fatalities were the great airship protagonist, Air Commodore E. M. Maitland, who was AOC Howden; C. I. R. Campbell, the Superintendent Naval Constructor of Airships; and J. R. Pannell of the NPL, the skilled scientist specializing on airship wind-tunnel research who was aboard to check full-size results against his figures for the model.

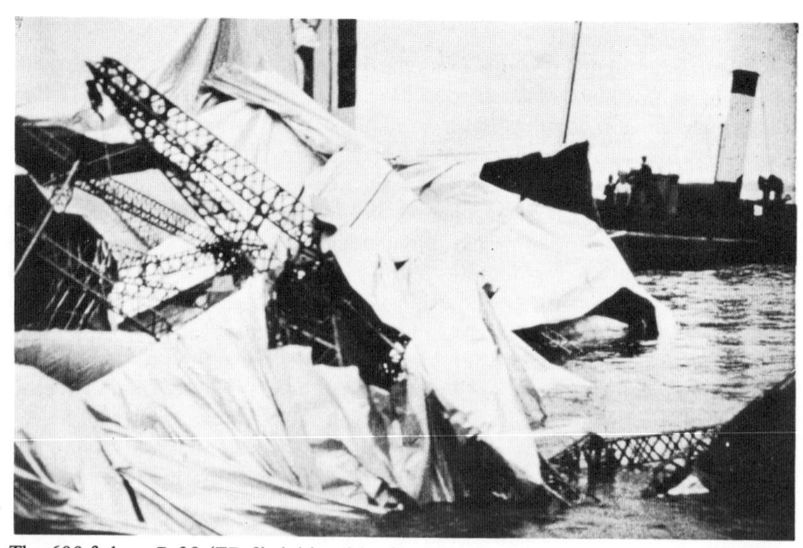

The 699-ft long R.38 (ZR-2), initiated by Short Brothers in June 1918, revealed control trouble on its first flight, instability and structural failure on the third, and catastrophe (*seen here*) on the fourth flight. (*Courtesy Philip Jarrett*)

The inevitable investigation by a Court of Enquiry followed, presided over by Sir John Salmond. Their report indicated grave doubt of the validity of airship design procedure: 'The system by which both construction of a ship and inspection of the work are centred in one head, as it was at the Royal Airship Works, Cardington, is unsound. Having regard to the great differences in requirements between HM Airship R.38 and previous British airships, the design should have been examined and discussed by an

The doomed crew of the airship R.38. (*Courtesy Philip Jarrett*)

official and competent committee before construction was commenced. There is no evidence that this was done, although opportunity arose after the Armistice when information on details of ships built in Germany became available.' The condition under which R.38 failed was certainly a limiting case where a minimal safety factor was permitted, for the rudder had been deliberately applied at full lock, flying at some 45 to 50 kt, and immediately reversed. Failure followed between frames 9 and 10, the forward section catching fire at the fracture due to a spark from electric wires. Although not expressed in the report, this was to lead to more intensive development of diesels to replace airship petrol engines.

The accident could hardly have come at more inopportune moment for airship gamblers. Government policy had been in the melting pot throughout the year, with strong bias towards scrapping the entire effort. Meanwhile Mr A. H. Ashbolt, Agent-General for Tasmania, continued to push the case for an Imperial Air Company operating six rigids between Great Britain and Australia, and not only had the ear of Prime Minister W. M. Hughes in the Australian House of Representatives but was strongly backed by Vickers Ltd.

The problem of airships, and particularly the cost of erecting mooring masts, providing bases, fuel supplies, upkeep and commissioning of the existing fleet, had been discussed at the recent Conference of Prime Ministers of the British Dominions as the result of a report favouring cessation of airship work which had been presented by the Committee of Air Communications instituted by Churchill before he changed office, and chaired by Capt Guest.

Interviewed by *The Times*, Mr Hughes, when asked if he was satisfied with the recommendations, replied: 'So far as the advice tendered has relation to the technical side, it is not for me to criticize it. But if you ask whether I am satisfied with the proposal to cast aside this instrument of Empire communication – which I may remind you cost some £40 million to bring to its present state, and which offers the most hopeful means of bringing this world-wide Empire of ours within narrower compass – then emphatically I am not satisfied. I hope nothing will be done by way of sale of airships, or spare parts, or machinery and plant so as to make any attempt at a successful experiment by Australia impossible.'

Concurrently the Australian Government was cautiously going ahead with internal airlines, and called for operational tenders covering routes from Sydney to Brisbane and Sydney to Adelaide. There were many signs that Australians realized to the full the vast possibilities of aircraft in opening up communications with their vast territory, and were equally determined to have aerial defence of their country, to which end an Air Force had been established on 31 March by proclamation pending passage of the Air Defence Act, and on 13 August the King approved the title of Royal Australian Air Force. With her great seaboard and small population scattered throughout an enormous area, Australia decided that the initial unit would be six Squadrons of twelve seaplanes which were ordered from Fairey Aviation, and the first machine was ceremoniously handed over on 12 August at a flying demonstration at Hamble attended by many distinguished guests warmed by Dick Fairey's hospitality to counteract the pouring rain.

8

Late in August, in reply to a letter from me seeking apprenticeship, de Havillands offered an interview. From the new tube extension to Edgware it was a brisk walk to the narrow, rough-surfaced Stag Lane. At its rural termination came green fields and a simple wooden bungalow office, and beyond were three canvas hangars and the roof of a very small factory. The sun shone brightly, and on the rough and sloping little aerodrome stood that thrilling D.H.29 monoplane of which I had read in *Flight* and *The Aeroplane*. I was astonished at the smallness of the set-up compared with the splendid Martinsyde factory I had visited, or the still bigger war-time factories, such as Sopwith and Bristol, which I had studied in photographs – but I was desperately anxious to be taken on as this seemed my only chance of entering the aircraft industry; and above all, Geoffrey de Havilland was my schoolboy ideal of a pilot-designer.

That bungalow had a full width room each end, one of which was shared by de Havilland and Walker, the other used by Hearle as administrative centre. Each side of the passage was a smaller room, the right shared between Nixon and St Barbe, and the left for the typist-telephonist and as waiting room. From there I was presently summoned to Mr Hearle, and found a tall quiet man who quickly calmed my anxieties and led me to talk of aeroplanes – yet took the gently pessimistic view that these were anxious times, and though they had just accepted another youngster as apprentice (whom I later discovered was R. E. Bishop, destined as chief designer in the unimaginable next World War) they might consider me if certain unexplained difficulties were overcome. He said he would have liked Mr Walker to have seen me, but unfortunately he was on holiday; I felt almost relieved there was no mention of interview by the fabulous de Havilland himself.

Perhaps amused by my shy enthusiasm, Hearle showed me round the shed-like, corrugated-iron factory with its shrieking saw cutting planks, and the planer showering shavings. There was the clean fresh smell of wood, the pear-drop perfume of fabric dope, the tang of new-brushed varnish. At one end a few men were filing fittings and others operating limited metal-working machinery driven by what I recognized as a Beardmore by its six copper-cased cylinders. 'We only use three of them' said my mentor, 'the others are blanked off.'

The D.H.29 with Scarff gun ring as accepted for the RAF at Martlesham, but though an ASI on the fin may have given minimum position error at cruise there must have been a big airflow change at throttled slow speed. (*Courtesy Air Marshal Sir Ralph Sorley*)

Behind the sliding doors that opened on the airfield were several D.H.9s and 9As being rebuilt – but what held my eyes above all was yet another imposing, ultra-modern looking, D.H. 29. In addition to the novelty of its deep and tapered, fabric-covered 54-ft wing, it had, for the first time in D.H. practice, a 'tumble home' fuselage with wider bottom to spread the undercarriage, and was ply covered not merely along the cabin portion, as with the well-proved D.H.18, but from end to end. 'The forthcoming trials of the D.H.29 at Martlesham possess a very large measure of interest because the machine marks a considerable advance, in theory at least, over anything which has gone before,' ran *Flight*'s editorial. 'It is a reversion to the monoplane, a type much to the fore previous to the war, but which fell into disuse owing to limitations of constructional knowledge of those early days. Since then the science of design and construction has progressed enormously and is represented in the D.H.29 by the adoption of a cantilever principle of construction in the wing design.' But de Havilland and Walker had become seriously concerned with the machine's behaviour – so much so that they were hastily designing a biplane replacement, the D.H.32, powered with the Rolls-Royce Eagle of which the price had recently been reduced to £1,000.

Originally the prototype D.H.29, J6849, looked very different from the development I saw, for it had a bulky nose radiator and a high top cowling ending in a massive 'conservatory' windscreen with sides impinging on the leading edge. The engine was a foot lower, and drove a four-bladed propeller instead of the eventual two-blader. The wing, despite its unique appearance for that time, was of conventional wooden practice except for the deep box spars with drag struts fitted independently to top and bottom flanges, cross-braced with cable transversely and horizontally to form a rigid structure similar to contemporary biplanes, thus avoiding torsional weakness.

On the first flight D.H. had a startling time. He seemed to be in the full blast of the slipstream which swirled viciously around the windscreen buffeting his head, and stove in the fabric-covered emergency hatches and panels as well as shaking the tail. Landings were erratic, particularly if the throttle was opened to reduce the rate of sink, for the change of trim became so great that the control column had to be pushed full forward, and when taxi-ing the rudder was completely ineffective. Yet a model had been tested carefully by Walker in the Airco wind-tunnel, to which they still had access, and it showed none of this. The whole future of the company seemed to the worried principals to devolve on the success of this their first machine designed and built at Stag Lane – particularly as it was built to order of the Air Ministry on whom they depended for approval if the company was to compete in the military field against the long established businesses of other manufacturers.

De Havilland thought part of the trouble might be because the maximum intensity of slipstream was in line with the wing leading edge, so he decided to re-design the entire nose, re-installing the engine with higher thrust line and using an upward sweeping under-cowling with adjustable slat louvres to control air flow to a radiator under the engine. The top fairing of the pilot's cockpit was remodelled and smoothed, and fitted with a small off-set windscreen lacking side panels. Anxiously Walker watched D.H. take-off for the next test and presently return. He could judge from his chief's expression that there was little improvement. The horn balance on the elevators was next removed, but the tail still shook at the stall, and it was now found that at the high incidences used on the climb there were signs of fuel starvation because the gravity head had become insufficient for the raised engine. A low-pressure windmill-operated header tank was therefore installed behind the pilot's cockpit, projecting three feet above it, and this cured the difficulty – but they were getting nowhere with the other problems.

The new Daimler Airway, which Colonel Searle was creating from the ashes of AT & T Ltd, was interested in purchasing eight of these impressive looking monoplanes with which to initiate their cross-Channel services in the spring of 1922. Delivery time was paramount. Baffled by problems of interference and buffeting beyond his experience, de Havilland decided to abandon the design as a commercial venture after handing the two prototypes to the Air Ministry, and proposed offering the D.H.32 to Searle. 'Typical of de Havilland's sanity in critical moments,' said Walker years later. Yet had they persevered and fitted a complete front cockpit enclosure undoubtedly the trouble would have been cured, judging by personal experience with similar cockpit position a decade later; but even in my time there was prejudice to be overcome, for RAF pilots were convinced they must have their heads in the open, and were even more adamant in 1921.

Despite longitudinal and directional aerodynamic problems, the D.H.29 revealed a notable advance in lateral control. A craftily drawn patent, No. 184,317 of 13 June, 1921, in Hagg's name, described a simple acute-angle bell-crank arrangement 'so that for a given movement of the pilot's control lever the upwardly moving aileron is displaced through a greater angle than the downwardly moving aileron.' It was outstandingly successful, at last solving a problem which had concerned every designer and pilot since the Wrights, for now the drag of the upturned aileron outweighed the other, and assisted instead of counteracting the turn.

For the D.H.32, the Air Ministry agreed a new specification, No.18/21, based on the valiant D.H.18, but with all eight seats facing forward instead of two with backs to the engine. The crew of two were located similarly to the D.H.29, but flow disturbance would be less because of the thin wing, and in any case this was where airline pilots preferred to sit in order to have better view when approaching to land. Because of low price the Rolls-Royce Eagle was initially selected, but as existing fleets had Lions it seemed inexpedient to stock spares for a different engine, so after inter-consultation between de Havilland, the Air Ministry, Daimler and Instone, a slightly larger, Lion-powered version, the D.H.34, was agreed, with all-up weight increased from 5,740 lb to 7,200 lb, though this gained only an increase of 100 lb on the original useful load of 1,540 lb because of the greater fuel requirement, and the landing speed was a frightening 70 mph compared with the already greatly criticized 61 mph of the D.H.18. Either way it meant disaster if force-landed in the small fields of England, though the lightly laden Handley Page O/400s and the Vimy Commercial could get away with it as they touched down at only 48 mph. To reduce fire hazard in a pile-up the D.H.34 was designed with fuel tanks fitted inboard each side under the top wings instead of in the fuselage.

In case time was too short for spring delivery of D.H.34s, the Air Ministry ordered two further D.H.18s as replacements for time-expired G-EARO and G-EAUF lost in a crash, brought up to date as the D.H.18B by covering the fuselage entirely with ply, like the D.H.29, and using a new system of engine starting by clutching in a hand-wound inertia fly-wheel.

During this time de Havillands, who were paying a rent of £1,500/annum for Stag Lane aerodrome, endeavoured to purchase the freehold, but the owners, Warren & Smiles, held out for £20,000, which far exceeded available capital. The first trading year, ending 30 September, showed £32,782 for contracts invoiced to date, with a profit of £2,387, but expectation ran high, for the Daimler contract increased the order book to some £90,000. Hard-headed Nixon was confident things were shaping well, and at the next Board it was agreed to pay £2,000 deposit to Warren & Smiles. It put de Havillands deep in the red. Now they must work with rigorous economy on an overdraft, yet somehow raise the balance for the aerodrome and extend the factory.

Under these circumstances Hearle wrote to me saying the way was not yet clear to take on another apprentice – and concurrently a letter came from Handley Page in which he too said that a career in aeronautics seemed a doubtful prospect, but if I still wished to pursue it then I would be wise to stay another year at school and then take an aeronautical course at the Northampton Institute of London University where he used to lecture. The die was cast. I stayed.

In later days Nevil Shute Norway, the author, told me he was at de Havilland's as an unpaid vacation student from Oxford that summer, working in a wooden drawing office hut with half a dozen draughtsmen. In *Slide Rule* he said: 'In so small an organization, which at the same time covered practically every branch of aviation, I had a magnificent chance to get a knowledge of all sides of the business, and I think I took advantage of it. Mr King and I worked side by side, the senior performance calculator and the junior assistant, and as the new projects came to life upon our graphs and columns of figures, de Havilland and Walker still spent long periods cogitating upon our drawing boards, to my immense benefit.'

But was there a future? That same September Capt Frank Barnwell resigned from the Bristol Aeroplane Co and went to Australia having accepted a commission in the RAAF, and was posted to the Aircraft Experimental Department at Randwick. That the designer of the famous Bristol Fighter should go seemed more than ominous. His final effort at

Basically the Bristol Bullfinch was a strut-braced parasol monoplane single-seater, but a 3-ft section for a gunner could be interposed behind the pilot and a small cantilever wing added below it for lift and trim. (*Courtesy Air Marshal Sir Ralph Sorley*)

Carrying three more passengers than the big Westland Limousine, similarly powered with a 450 hp Lion, the Bristol Ten-seater originally had a four-wheel undercarriage when tested at Martlesham. (*Courtesy Air Marshal Sir Ralph Sorley*)

Filton had been the Bullfinch, a single-seat, strut-braced, parasol monoplane which could be converted to a two-seat reconnaissance biplane by adding a smaller lower wing and an additional mid-bay in the fuselage. Evolved with considerable research in the new Bristol wind-tunnel, many variations of wing and fuselage shape had been tested, but in final form Barnwell had changed the circular-section fuselage based on the racing Bullet to slab sides and horizontal stern like the Bristol Fighter, and with small twin tail fins beneath and an astonishingly small elliptical rudder above, similar to the pre-war Bristol monoplanes designed by Coanda. To Barnwell it was the last straw that only three machines were ordered and his other military projects had been turned down.

Of the designs on which he had been working with Wilfred T. Reid, his dark and tall, elderly assistant, even the bulky new Ten-seater, of similar arrangement to the D.H.34, had been a private venture by Bristols. First flown on 21 June, it was now at Martlesham Heath; but with the de Havilland competitor ordered 'off the drawing board' Barnwell saw it had no future, particularly as his directors decided it must be produced only with the Jupiter as standard powerplant – and that engine was still a long way from production, although currently it achieved the distinction of first of any make to pass the rigorous mandatory 100-hour type test newly instituted by Colonel L. F. R. Fell, the Assistant Director Research and Development (Engines). This success was only just in time to prevent the engine department closing down, for exactly a year had gone since Roy Fedden joined the company, and he had already spent £197,000 with no return.

Reid now became chief designer. He was an ex-Farnborough man, and earlier had been trained as marine engineer at the Fairfield Shipbuilding Co. As a first co-operative step with the 'press-on' Fedden, he designed a cone swivelling mounting for the Ten-seater's Jupiter to reduce fire risk and enable the engine to be changed in the field in two hours – a matter of great interest both for civil and military work.

Fedden was not only a clever and dynamic engineer but an outstanding salesman, for it was clear that his main hope of survival must be to interest other manufacturers in his engine and not depend solely on Bristols. Tom Sopwith already had seen the merit of the Jupiter in his 1919 Schneider seaplane, and now proposed to use it for another prestige racer on which his new chief designer Capt Bertram 'Tommy' Thompson, BSC, ACGI, was engaged – though in later years Sir Thomas Sopwith had no recollection of how Thompson came to be chief designer, but remembered 'he seemed an amiable fellow'. Trained as an automobile engineer Thompson had served in the Admiralty during the war, and probably was introduced to Hawker Engineering by F. I. Bennett, their director-secretary who also was a motor engineer and currently chief enginner, with Fred Sigrist as managing director.

Thompson's earliest task was to complete the 'Humpback' fleet spotter design left unfinished by Herbert Smith and his chief draughtsman, W. G. Carter, when they left the Sopwith Aviation Co. A patent issued to Hawker Engineering and Thompson in November described a method of closing both the top and bottom apertures of a Handley Page leading-edge slot to reduce resistance at speed, and was probably intended for the fleet spotter. Of the Hawker Racer, on which he was engaged, no details seem to have survived, but there are several later patents of Thompson's dealing with car gear systems, including an early syncromesh.

9

Not until 3 June were pilots of the RAE allowed to fly the Short Silver Streak, though it had been delivered on 1 February. Now numbered J6854, it had initially been civil registered as G-EARQ, but a C of A had been refused on the paltry excuse that long-term behaviour of duralumin as a primary structure was an unknown – despite the fact that continuing use of duralumin was proposed if bigger and better airships were eventually built.

True, officials had cause for caution, because in August the previous year, when the glittering Silver Streak made its first flight in the hands of John Lankester Parker, the paper-thin aluminium wing-skin rippled and buckled, so it was replaced with duralumin. This took a long five months, as it meant completely rebuilding the wings, but on 27 January Parker flew it again, when, in common with the outlook on testing of that day, it was deemed that 15 minutes airborne was sufficient to prove the aircraft acceptable. Immediate delivery to Farnborough was attempted with that great character Oscar Gnosspelius as passenger, but fog necessitated landing at Croydon, and the flight was completed on 1 February. Thereafter the Silver Streak was long the subject of scrutiny by RAE 'specialists' who apparently expected it to shed the rivets or collapse from corrosion. Only after considerable pressure and a 'round-robin' from the pilots of the RAE, headed by Sqn Ldr Roderic Hill, was permission given for cautious non-aerobatic flight tests. On the first of these, on 3 June, Flt Lieut Noakes recorded climbing to 10,000 ft in 11 minutes and a top speed of 125 mph. In the next two days it was flown by Sqn Ldr Hill and Flt Lieut E. R. C.

The Short Silver Streak J6854 was delivered to Farnborough as a two-seater, and the thin aluminium wing skin had been replaced by duralumin.

Scholefield who agreed the machine was astonishingly good, easy to control and had excellent manoeuvrability.

No further flights were permitted, and Short Bros were not even allowed to show the machine to potential overseas purchasers. 'It was ignored by the Technical Department when it had flown except for making it ready for destruction tests,' Oswald Short told me. 'These were deliberately delayed so that a steel girder fuselage, not a streamline monocoque like mine, could be tested first. I was told it had priority; however, I got in touch with Air Commodore Brooke-Popham, Director of Research, and he sent a telegram ordering my machine to be tested forthwith. When the RAE got down to it, the tests showed the fuselage to have amazing strength, but when I sought a copy of the report from the Technical Department, I was told I could not have one. It was only in a roundabout way that I was able ultimately to obtain it. Thus was the dice loaded against me while the year passed.' As C. H. Barnes records in *Shorts Aircraft since 1900*: 'In the wing loading test, failure occurred at just above the calculated ultimate stress by buckling of the front spar outboard of the outer pair of struts, though the residual strength of the buckled spar was still adequate for normal flying loads. The cantilever tailplane withstood an overall loading of 57 lb/sq ft and the fin and rudder a lateral loading of 63 lb/sq ft, both well in excess of requirements for conventional wooden airframes; in a torsion test on the fuselage shell a moment of 2,000 lb-ft produced no visible distortion or permanent deflection. On completion of the static tests in September, the airframe was subjected to 100 hours vibration testing, at the end of which in November no sign of cracking or loose rivets could be detected.'

In his efforts to secure attention and a development order for his epoch-making system of construction, the iron-willed, outspoken Oswald Short undoubtedly angered crucial members of the Technical Department, who in turn obstinately closed their doors to him. Undeterred, Oswald began to enlist attention of politicians who might influence aircraft policy, though he had badly handicapped his case through that earlier failure to take a patent before revealing his machine. Not until 10 November, 1921, did he secure Patent 185,992 for 'Construction of fuselages and like structures for aeroplane flying machines in which it is desired to provide all-metal construction, and moreover a construction in which a light and strong metallic alloy, such as Duralumin, can be advantageously and safely employed to form the main part or body of the structure.

'According to this invention the shell or body of the fuselage or like structure, which may be of circular, oval, or elliptical form in cross section and is usually tapered towards the rear end, is composed of a plurality of sleeves of sheet metal, the sleeves being of such relative diameter that the end of one will enter and fit within the adjacent end of the next, and so on, so that the overlying edges of the sleeves can be riveted or otherwise fixed together, and thereby a fuselage can be formed composed of a plurality of sleeves.'

Though Britain thus secured an undoubted lead with the form of construction which eventually was to be primarily exploited by US technicians, it seemed at that time of less concern than the failure of the Royal Aero Club to find a British entrant for the Schneider Trophy held at Venice on 7 August. The only competitor against the Italians was Sadi Lecointe with his Nieuport-Delage seaplane, but it failed to pass the navigation test and buckled its chassis on attempting to take-off. Of the three Italians, Zannetti's M.19 had an engine seizure and caught fire, Corgnolino ran out of fuel, and de Briganti on the slowest machine, a Macchi M.7, won at 120.3 mph. As a race it seemed very trivial, but it established the Italians in a powerful position should this race attain prominence later. Had the 'Bamel' been put on floats it could well have won. Yet it failed to win the classic International speed trials of the year, the Coupe Deutsch de la Meurthe presented by the great patron of French aviation of that distinguished name to replace the Gordon Bennett Cup won outright for France the previous year by Sadi Lecointe. For the new contest, flown on 1 October, there were five French machines, one Italian, and the 'Bamel' for Britain. Certainly the Gloucestershire Co had cause for hope, for their hybrid yet attractive little racer had new mainplanes of 20 sq ft less area, using a high lift top wing section and high speed lower. She seemed so fast that Sadi Lecointe, the redoubtable, heavily built Nieuport pilot, 'thought bad of this apparatus'! On arriving in Paris, Folland hopefully obtained a pair of ovoid Lamblin low-drag radiators with which the 'Bamel's crew at Étampes replaced the English honeycomb high-drag radiators.

Setbacks had already befallen the French. Sadly enough, the charming de Romanet fatally crashed a week before the race when the fabric ripped from a wing of his aggressive looking Lumière-de Monge. The advanced-looking Hanriot, with aluminium skin covering its triangular-section steel-tube fuselage, was so late that the first test flight would have had to be the eliminating trial, so the builder wisely withdrew the machine. For the competition Sadi's 300 hp Nieuport-Delage 'Sesquiplan' took off at 10 a.m., attaining the first turning point at 207 mph, only to crash near Toury, rumoured due to burst fabric like de Romanet's. Next was Brackpapa on the big 700 hp Fiat, but after a fast first lap he also forced-landed at Ruau due to a leaking fuel system. In mid afternoon little Jimmy James started with the 'Bamel', but shortly after beginning the second lap was seen returning, landed up the hill and taxied to his shed. To the disappointed Gloucestershire team he explained that after the first lap, on opening the Napier wide, he saw the fabric ballooning on the upper wing from four ribs on the left and one on the right, and with de Romanet and Lecointe in mind, concluded it was on the point of bursting and came in. With Sadi Lecointe, Brackpapa and James out of the trial, proceedings lost interest. Lasne with a Nieuport biplane circuited in just under 70 minutes, followed by Kirsch on another 'Sesquiplan' who took five minutes less, and won at 278½ kph (173 mph), but landed down hill at 90 mph, ballooned, opened up and choked the engine, bounced furiously half a dozen times, stood on one wing, but the wheel that side hit a bump before the tip struck, and the machine bounced level, spun round and stopped. To the British it seemed to vindicate their passion for light wing loading and low landing speed, despite the move in the opposite direction by de Havilland.

Though this international participation proved no advertisement for British aviation, a great air-taxi tour of 5,000 miles by Alan Cobham, around the principal capital cities of Western Europe, obtained widespread publicity aided and abetted by his American passenger, Lucien Sharpe, who created mystery by initially withholding his name. Never before had so extensive a tour been made, for it lasted three weeks and seventeen cities were visited. 'In those early days Cobham was not rated as the best pilot that the company employed,' recalled Norway, 'but he had a fantastic capacity for hard work and organization; could work 18 hours a day, month after month, and was soon to prove it by a series of pioneering flights about the world that brought him a great reputation.' It was on one of his journeys to Spain that he encountered the ribald and golden-haired Charles Barnard who had been delivering D.H.9s from Croydon for A.D.C., and as the D.H. air-taxi business was swiftly expanding, thanks to the publicity efforts of Bob Loader of Lep Transport's Piccadilly office, Barnard was taken on as additional pilot.

Croydon, as Customs airport, was departure point for the scarlet-painted D.H.9 taxis. Only recently had a battery of petrol pumps been installed in the Customs enclosure by the rival companies Anglo-American and Shell, the latter in the charge of 'Jerry' Shaw, ex-AT & T; previously refuelling had been by hand-operated bowser and petrol cans. Furthermore, Croydon and Lympne now each had an aerial lighthouse visible for some 30 miles to help pilots through the dusk, though there was no official night flying yet.

Dignity was soon to be enhanced, for Instone Air Line ordered uniforms of naval cut for pilots and general staff who hitherto had worn a motley of civilian suits. But Croydon remained a gay band of airline and free-lance pilots, of whom most were still only in their mid-twenties. By any standard pay was modest, even to eking a precarious livelihood by flying new machines at a ridiculously low figure while waiting for a job in the aircraft industry to turn up which would offer a more definite prospect. For production testing a Disposal machine the fee was £1. To deliver it to Brussels gained £8, but the pilot had to pay his expenses, netting a mere £3 for a two hour flight and overnight sea-going return.

As spare-time enthusiasm, Croydon pilots arranged a week-end race meeting using joy-riding Avros and borrowed S.E.5s. Several sporting Service pilots joined in, and despite blustery easterly wind and drizzle, many notabilities, including debonair Sir Sefton Brancker, jovial Frederick Handley Page and the equally tall though much younger looking Richard Fairey, and even Air Vice-Marshal Sir Edward Ellington, came as a kindly gesture to civil aviation.

At Croydon it was also noted that: 'Mr A. S. Butler, who garages his private Bristol Tourer at the aerodrome, reappeared on Sunday afternoon. It's so long since he visited that it was thought he had forgotten he owned an aeroplane.' In fact he had been ill. His aeroplane had proved a splendid facility for travel abroad without the inconvenience of ship and rail. Now he would like something faster, more up to date. The editor of *The Aeroplane* recommended him to contact the vigorous new firm of de Havilland. 'A special design will cost a fortune, probably £3,000,' said D.H. when told of the enquiry; but when St Barbe called on Butler he accepted the price without question and, though still convalescent, went to Stag Lane to discuss requirements more fully, specifying a three-seater, with large luggage locker, and easily accessible Rolls-Royce under a car-like bonnet.

This darkly handsome young man, who looked much older than his 23 years, seemed rather interested in the expanding possibilities of de Havillands, judging it to be a very honest firm, yet it was with some embarrassment that he said on leaving: 'I have been considering investing in an aircraft company. Would you be interested in, say, £50,000?' Somewhat shaken, D.H. admitted they were needing further capital, but lest he appeared too anxious hastily added: 'But not as much as that.'

To bring in a single large investor would almost certainly require his appointment to the Board, and de Havilland was not anxious to extend the close-knit circle of directors, for they regarded this young ex-Guards officer with caution, though it was established he was a man of considerable substance, already a director of several companies in addition to the Newfoundland venture with Cotton, and a member of several leading London clubs. Yet the need for money was pressing; presently they agreed to invite him to join should he raise the matter again.

A fortnight later he visited with his uncle. 'Conversation turned to the matter of his aeroplane, details of estimated weights and performance, and it looked as if another pleasant meeting would end as casually as the first,' recorded Martin Sharp. 'However, when the time came to go, Butler said,

"Well, what about that investment? How much were you thinking of at this stage?" ' Nixon told him that they needed £7,500 in connection with their proposed expansion, diffidently adding: "The first £5,000 would be very useful in a matter of days if that is not inconvenient.' There and then he was handed a cheque for £5,000, and three days later Butler was appointed a director, with an allotment of 7,500 shares, for which he paid the balance of £2,500 just before Christmas. Nixon immediately contacted building contractors. *Flight* was quick off the mark, and reported: 'In these days of comparative quietness in the aviation world it is a matter of considerable satisfaction to record that at least one firm has found it necessary to expand their works to cope with the demand for machines. We are referring to the de Havilland Aircraft Co Ltd of Stag Lane, Edgware, who are busily engaged on levelling the ground and generally making preparations for laying the foundations for extension of the old works. As the work of constructing the steel buildings is in the able hands of Messrs A. Dawnay & Son, of Battersea, it may be expected that progress will be rapid, and the new works at Stag Lane should soon begin to take shape.'

10

To remind the Nation for all time of the million of the British Empire killed in the Great War, the King, on the anniversary of Armistice Day, 11 November, unveiled a dignified and simple Cenotaph designed by Sir Edwin Lutyens, which replaced a temporary memorial that was erected in Whitehall for Peace Day in July of the previous year. For all, it was deeply moving, and extended beyond the massed crowds lining Whitehall in stricken silence to every city, town and village. An endless procession slowly passed the Cenotaph in silent homage, and in the Abbey an unknown soldier from the battlefield was entombed at a Service of great solemnity and beauty, typical of the sadness of English pageantry. For many a day after, men passing the Cenotaph lifted their hats, and women bowed. They knew an unrepayable debt was owed to those many who never would return. Almost all believed that the Great War had been the War to end all wars, though a few politicians and military commanders saw clearly that the frailty of human relationships could presently lead to another and yet greater war against which the only safeguard was to establish a powerful and publicized deterrent Force which could not only defend but strike a devastating blow from the air. From every aspect the recent war shouted the paramount necessity of having sky supremacy – yet idealistic believers in the League of Nations imagined future quarrels could be amicably settled by discussion.

Meanwhile precautionary steps continued. Thus at Grain the Marine Experimental Establishment had throughout the summer been testing the relative merits of the Vickers-Saunders Valentia twin-Condor powered flying-boat with Consuta hull designed by George Porter – that modest man, unknown to fame, who had designed some of the fastest motor-boats in the world; the Short Cromarty with similar engines and improved Porte

The prototype Vickers Valentia had a biplane tail and fully cowled engines, but Saunders also built this variant with hollow bottomed hull and monoplane tail.

type hull; the Phoenix-erected four-engined N.4 Titania with flexible hull of the type devised by the late Maj Linton Hope, and built by Fife on the Clyde; and the Dick Kerr N.4 Atalanta with hull designed by yacht designer Charles E. Nicholson of Camper & Nicholson Ltd at Gosport, and built by May, Harden & May at Southampton using widely spaced inflexible web frames and light longitudinals compared with heavy longitudinals of F Type boats. The wing structure of both N.4s was designed by Fairey Aviation. Lecturing before the Royal Aeronautical Society on *Flying-boat Construction*, Capt David Nicolson, an early assistant of Colonel Porte at Felixstowe, who became the Air Ministry's chief production officer on flying-boats, compared relative merits of the diverse systems of construction in these boats, all of which were ordered in 1918 but only recently completed. His object was to secure 'class' rules for scantling sizes and constructional methods standardized in similar manner to those for

Completed three years late owing to contractual indecision, the Short Cromarty was an enlarged version of the Porte/Nicolson-designed 12,200 lb Felixstowe F.5, and with twice the power had a loaded weight of 20,000 lb.

Of the four-engined N.4s, the Fairey Titania with the later Condor III carried 1,100 lb greater load than Atalanta and had an all-up weight of 31,600 lb. Until the late 1940s the hull of N129 lay on the mud at Bawdsey Ferry, Felixstowe.

A conventional F-Type hull, built on wooden longerons, formers and spacers, required many trades working together as it was akin to aeroplane construction.

racing yachts. As there was no comparative trial data available, no clear-cut conclusion could be reached, though there was much to be said for the Atalanta method because of simplicity and strength, and it could lead to further evolution. In fact 'Messrs Beardmore have recently advanced to the next step by building a composite boat. This is a stage shipbuilders arrived at many years ago. The composite boat known as W.B.IX is on the longitudinal system, the main portion of the hull being of approximately circular form, with 17 continuous longitudinals, and the channels forming them are of .04-in duralumin 6-in apart, braced Zeppelin fashion with fluted duralumin crosses. Keel and chines are of rock elm, and the hull is skinned with two thicknesses of mahogany planking $\frac{9}{32}$-in at the bottom and $\frac{5}{32}$-in above.'

In forecasting that the Linton Hope hull, exemplified by Supermarine and Phoenix, was likely to be standardized Nicolson said: 'Construction is such that the structure is capable of resilient distortion, so that when alighting it can spring, reducing the shock. The hull cross-section is egg shaped, very light, possesses great strength, and is built of longitudinal stringers with bent hoop timbers inside and light frames outside the stringers, skinned with double planking, through-fastened together. No web frames or cross-bracing are required, and the hull is a continuous structure with steps externally added.

'With a hull of conventional F.5 type, such as the Cromarty, you start by criticizing it as a commercial proposition, for you run into such items as turnbuckles, bolts and nuts, wires, cables, sheet steel, steel tubing – in fact a hundred and one parts as though building a large aeroplane fuselage. You must employ not one trade but a number, such as boatbuilders, carpenters, sheet metal workers, fitters, machine hands, riggers – and are immediately in the midst of demarcation troubles in arranging the working squads. With a Linton Hope hull you need only one class of labour – boatbuilders; a small number of men and boys can be placed on the job, and if pieceworked under supervision the chances of holdups are small: there is no complication, and they carry straight through and finish their job. A standard Supermarine four-seat hull, 31-ft long, takes three men and two boys on an average $5\frac{1}{2}$ weeks to build, working a 47 hour week.'

Matters having great bearing on flying-boats and naval aircraft were discussed at the Washington Conference, opened by President Harding on 12 November, at which Mr Hughes, United States Secretary of State, proposed that the USA, Britain, and Japan should scrap most of their respective navies and cease to build further warships for ten years. In the case of aeroplane carriers the USA and Britain should be allowed a maximum tonnage of 80,000, and Japan 48,000 – though with the escape clause that carriers twenty years old could be replaced by all three countries with new construction. Luckily, because 'naval aircraft may readily be adapted from special types of commercial aircraft it is not considered practicable to prescribe limits for naval aircraft.' A. J. Balfour for Britain whole-heartedly accepted these restrictions and added proposals forbidding construction of submarines. Commented C. G. Grey: 'If the Navy is stopped building capital ships there will be all the more money to spend on aircraft – naval and land going. Thus the Navy's misfortune may well be the RAF's opportunity, for it is now more obvious than ever that the RAF must become our first line of defence. But it is by no means plain flying to reach that destiny. There are deep and influential and energetic intrigues being carried on by the senior Services to split the RAF and abolish the Air Ministry and get the remnants of the Air Force back into the Navy and Army.'

M Briand's contribution to the Conference made it clear that France remained implacably anti-German, for the German Army though so recently beaten had not been broken and the day of retribution was much in mind. Accordingly Balfour promised British support against any German aggression. Further, on 13 December the USA, Britain, France and Japan signed an 'Agreement for Preservation of Peace in the Pacific' – and to ensure fullest co-operation with the USA, the Anglo-Japanese Treaty would be revoked. Yet what could be done when a committee chaired by Sir Eric Geddes was officially investigating the possibility of slashing the cost of running the Air Ministry and RAF? Even the Chief of Imperial General Staff, Sir Henry Wilson, had openly said: 'It is for those who govern the world to ponder whether, if they want to limit the horrors of war, it would not be better to limit aeroplanes rather than submarines.'

The Paris Aero Show, opening on Saturday 14 November, might be a diverting respite for visiting technicians, but such was the state of the British aircraft industry and its lack of confidence in international sales that the only two British aeroplanes displayed were a torpedo biplane in skeleton, built by Levasseur to drawings of the Blackburn Swift but powered with a 370 hp Lorraine, and the other a Vickers Vimy Commercial, with power stepped up by installing two Napier Lions, displayed on the stand of Les Ateliers des Mureaux who had obtained a constructional licence for this type, and sold this Brooklands-constructed forerunner to the Grands Express Aériens for their London—Paris route. French manufacturers excitedly disapproved of such intrusion when they were so emphatically demonstrating their own interest in airlines with six new multi-engined prototypes, of which the four-engined Blériot-Spad, with single I-outboard strut each side, seemed most advanced and daring. Of other exhibits, a number exemplified metal construction – the most remarkable being the huge Latécoère fuselage, derived from the 'geodetic' construction of the war-time German Schütte-Lanz airship, with channel-section duralumin diagonals, running spirally to form a streamlined frame riveted at every intersection, and intended to be covered with thin-gauge aluminium plating.

Said one of the visitors: 'Considered purely as a show for the entertainment of a non-technical public, the Paris Aero Salon of 1921 is the most beautiful thing that has ever happened. The general scheme of uniform decoration is more beautiful than ever, the lighting more brilliant than ever, and the paint and polish on the aeroplanes more lavish than ever. Compared with the higgledy-piggledy arrangement, the go-as-you-please standard designs and the lugubrious atmosphere of Olympia, the Grand Palais is as the Russian ballet to an ordinary English music hall show. But considered as an exhibition of aerodynamic achievement and aeronautical engineering production it is merely lamentable.' That was harsh criticism, for though in Britain there was no production order for all-metal aeroplanes, in France Louis Breguet was exhibiting his entirely metal-structured Type 19A biplane which was to become the standard two-seat fighter for the French Air Force. This 'sesquiplan grande reconnaissance' with its big round fuselage, large span top wing and small lower having raking I-struts each side with lift wires taken from beneath them to the undercarriage, appeared far more advanced aerodynamically, and even constructionally, than any on British drawing boards except the Westland Woyavodski; making even the Short Silver Streak, with its many struts, appear dated.

It was the Bristol Jupiter which saved the day for Britain at Paris, and Paris saved the Jupiter, for Fedden later confessed: 'We were right at the limit of the development expenditure I had estimated to the Bristol directors, but in September the Jupiter had been the first to pass the 100-hour test and attained 450 hp. Then in November when exhibited at the Paris Salon it attracted such enormous attention that we gained immediate success, for the Société des Moteurs Gnome et Rhône purchased a constructional licence for France. It was an enormous boost to morale, and our position was consolidated in December when the British Government gave their first production order for 42 Jupiters together with a 100 per cent holding of spares.'

Undoubtedly the goal had been achieved only through the immense drive of Roy Fedden, though in turn he attributed it to the remarkable spirit of enthusiasm and loyal support from his little team of chief designer and seven draughtsmen. Behind Fedden was the influence of a good home, for he attributed much of his outlook to his father – 'a remarkable man, who practically gave up business to found the National Nautical School for Boys at Portishead.' The merits of leadership and discipline Fedden had learned from his housemaster at Clifton, where he played a prominent part in games and became a prefect. At Bristol Merchant Venturers Technical College it was Professor Morgan of the Automobile Engineering Department who taught him the value of theoretical approach, though perhaps Fedden learned the best axiom of all when apprenticed to the Bristol Motor Co where the master mechanic under whom he worked told him he 'must never dash off the moment the whistle blows'.

One more success came to Great Britain before the close of the year. The Gloucestershire 'Bamel' was fitted with wings reduced by a further 20 sq ft since the Deutsch de la Meurthe Cup, and with only 165 sq ft now had what was regarded as the stupendous wing loading of 15 lb/sq ft. The cowling had been reshaped to smoother contour; the interplane struts were

Painted pale blue with white wings and tail, the modified Gloster 'Bamel' looked elegant and fast, and was better still when one of the 'lobster pot' radiators was removed. (*Flight*)

faired into the wing with aluminium cuffs; one of the two radiators was removed; and to prevent the wing fabric lifting in the slipstream it was first sewn to wrappings round the ribs and then standard stitched through tapes and covered with the usual doped-on strip. Taken to Martlesham because of the wide area for landing, rough though it was, the 'Bamel' on 10 December, piloted by Jimmy James, covered a straight kilometre at 212 mph, thus apparently beating the world's record of 204 mph – but homologation by the Fédération Aéronautique Internationale (FAI) required the mean speed of four runs into and down wind, and as this averaged 196.4 mph it could only be regarded as a British record.

Although most of the sixteen airframe manufacturers could count themselves fortunate in having contracts to design and build one or more experimental aeroplanes at Government expense, the disastrous trend of general manufacturing, particularly heavy industry, was towards an increasing recession. Like a tidal wave, unemployment had swept to the enormous figure of 2,170,000 before it began to recede. Ships were laid up in every estuary to save dues. Steel works and coal mines stood still. Men stood apathetically around. Wretchedness was alleviated only miserably by the recent extension of the new Unemployment Insurance Act to provide that uncovenanted benefit dubbed 'the dole'.

Throughout the year there had been Government moves towards rigorous economy starting with a Cabinet decision 'That means be adopted to organize all scientific work which is of common interest to the Fighting Services of the Crown to ensure utmost economy of expense and personnel and due co-ordination of the various technical Naval, Military, and Aeronautical Establishments to avoid overlapping either with each other or with research organizations of the civil departments of State.' A Co-ordinating Board was instituted, and Henry Tizard, now 35 and Professor of Thermodynamics at Oxford, was appointed to the part-time post of assistant secretary at a salary of £1,360. But that was only one aspect. A Committee on National Expenditure had been established with Sir Eric Geddes as chairman 'To make recommendations to the Chancellor of the Exchequer for effecting forthwith all possible reductions in the national expenditure on supply services, having regard especially to the present and prospective position of the revenue.'

In May the Treasury had circularized all Government departments emphasizing the vital need for heavy slashes in the Estimates, which must be reduced from the current year's £603,000,000 to £490,000,000 for 1922/1923; but now the Geddes Committee was instructed to aim at economies saving a further £62,000,000, by cutting not only the allocations for the Fighting Services but for Education, Health, Labour, Unemployment, Old Age Pensions, and War Pensions. Yet the Cabinet did nothing to encourage and expand exports other than pass the defeatist Safeguarding of Industries Act imposing import tariffs as high as 33½ per cent to stem the flow of competitive goods mainly from bankrupt Germany. Faced with enormous reparations, pressed unendurably for payments by France, and somewhat shamefacedly by Britain, more and more paper Marks were being printed to maintain internal economy while gold gained by foreign exchange was despatched to her victors. The slippery path had started.

The whole world seemed involved in problems. Ireland fighting her way to an entire Republic; Islam and the European Powers in conflict; China in confusion, still making reparations from the Boxer War; India drowning in its tide of Nationalism; Britain struggling with her uncertain hold on Egypt, battling in the League of Nations, and adding administration of the former German African Colonies to her problems of the Dominions and Commonwealth. Famine through disastrous drought, followed by monetary collapse, was hitting Russia; only through an American Commission, headed by the famous Norwegian explorer Dr Nansen, was the population saved with wheat and supplies from the USA. Though Britain and Italy ratified a Trade Agreement with Russia, they did nothing to send relief to the starving. Italy in any case was beset with internal unrest – for its slide towards Communism, led by the Government under Signor Giolitti, was encountering violent opposition in the form of a huge black-shirt militia known as Fascisti, led by a heavily-jowled newsman called Mussolini whom the King invited to form a government.

Amid these vast global problems those of the aeronautical industry must be counted insignificant. Survival was the problem. Cautiously, economically, paying meagre salaries to their technicians and draughtsmen, the British aircraft manufacturers intended to keep going, counting on the enthusiasm and sense of vocation of their handful of men.

CHAPTER III

THE PALE DAWN
1922

'It must not remain our desire only to acquire the art of the bird, nay it is our duty not to rest until we have attained to a perfect scientific conception of the problem of flight.'

Otto Lilienthal (1891)

1

It was the auxiliary-sail steamship, *Quest*, voyaging southward through the Antarctic Atlantic, rather than devastating world affairs, that became the source of interest for the British public, for Sir Ernest Shackleton, the great British explorer who had succeeded Capt Robert Falcon Scott in popular esteem, died aboard on 5 January. The expedition had left Tower Bridge on the Thames the previous September, after loading a special two-seat 80 hp Le Rhône Avro Baby seaplane for polar reconnaissance under

Biggest and most powerful of the Avro Baby series was the rotary-engined 554 seaplane which ended its days as a seal spotter in Newfoundland. (*A. V. Roe and Co*)

the care of Major C. R. Carr, lately test pilot at Martlesham and famous earlier as Grahame-White's chief pilot. It was a pretty little biplane with only superficial resemblance to the original two-seat 534, and was referenced Type 554. The top wing had greater span than the lower, and for easy maintenance they were braced with tubular-steel struts instead of wires; but despite all forethought the machine was destined to return unused the following September, the voyage of the *Quest* abandoned.

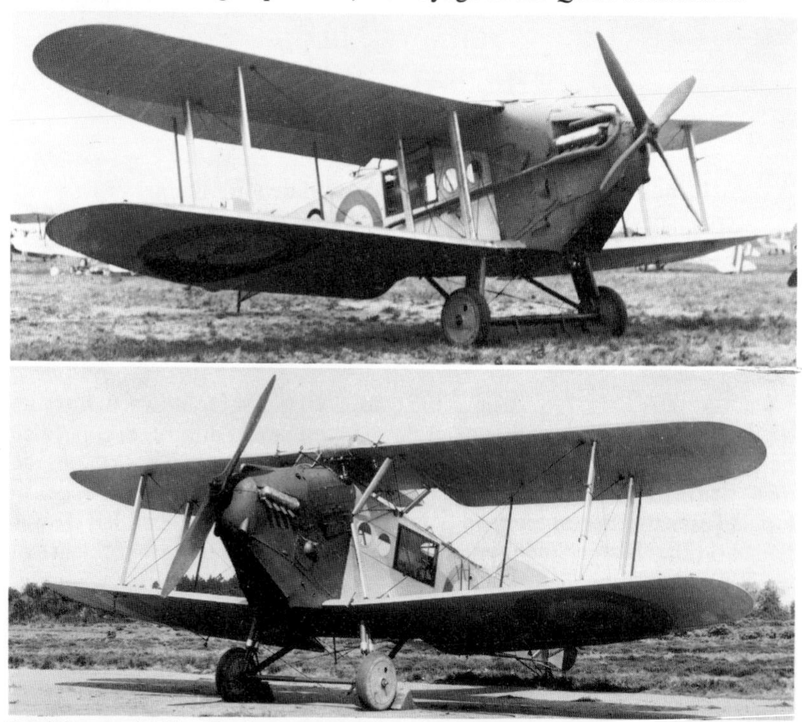

Reminiscent of Alliott Roe's pre-war Military Trials biplane was the Avro Bison I deck-landing Fleet Spotter of which two prototypes and twelve production machines were ordered. Mk II had the upper centre-section raised.

Bert Hinkler, already famous for flights in Australia with the prototype Baby, was now test pilot to A. V. Roe Ltd, and had recently flown their first post-war Service prototype, the somewhat ugly Fleet Spotter Type 555, Bison. The influence of Roe himself could be discerned, for there was strong family likeness to the fully-enclosed biplane he made for the pre-war Military Trials, though the spidery elegance of his famous 504K was entirely missing, the only identifiable similarity being the practically finless balanced rudder. The 555 was competitor to the Blackburn R.1, built to retrospective specification 3/21 (D of R Type 7A). For maximum deck landing view the Bison's pilot was similarly in front of the top centre-section, and the cowling sloped steeply to a low thrust-line Lion, the rectangular-section fuselage filling the entire wing-gap and forming a capacious cabin for gunner and navigator, who had a raised platform for access to the Scarff gun-ring on the top longerons. Like the Aldershot heavy bomber, still some way from completion at Hamble, the fuselage was constructed with steel tubes and the central portion boxed with plywood. Relatively thick, allegedly 'high-lift' wings had conventional front and rear spars, with struts and swaged Rafwire bracing. To meet deck-landing requirements soft shock absorbing was attempted using an oleo dashpot with compression rubber springing, and three claws were fitted to the axle to engage deck arrestor wires.

Unfortunately the first flight revealed similar troubles to those de Havilland had with his monoplane, due to interference between fuselage and wing exacerbated by the open cockpit at the leading edge: there was directional instability, and elevators and full-span ailerons required too much strength to operate easily. Martlesham confirmed the company's worst suspicions that the Bison in its prototype form would not be bought for production. There was small consolation that the bull-nosed Blackburn was even uglier and suffered similar interference, and, though its controls were more satisfactory, the elevator became ineffective at big incidences owing to blanketing. Both Avro and Blackburn were intended to replace the stopgap Westland Walrus biplanes which had suffered many crashes,

N150 was the Blackburn R.1 Blackburn prototype to the same requirement as the Bison, but basically was a Dart redesign with full depth monocoque mid-fuslage and pilot moved forward.

and even when taking off on delivery flights from Yeovil one had engine failure and stalled and two others could not overcome the down-draught on the low surrounding hills and were smashed beyond repair.

Blackburns had found encouragement in export orders from USA, Japan, Spain, and Brazil for the Swift Mk I. Trials with the private venture prototype at Martlesham early in 1921, and subsequent deck landings on the *Argus*, had led to an Air Ministry order for three, re-named the Dart, having slightly reduced span and powered with the latest Lion, the first of which, N140, went to the *Argus*, and on 12 January N141 was flown to Martlesham for official trials with more than six months' advantage over its secret competitor the Handley Page H.P.19 Hanley, the design of which

'Star' Richards had completed on promotion to chief designer following the resignation of Volkert. Ginger-haired Richards had the more stolid Paul Newell as his chief assistant. The outspoken C. W. Meredith was works superintendent, and in charge of the experimental department was Frank Meadows, rough and with a wealth of bad language, known as 'the man who was never wrong' – but the overwhelming personality was still H.P. in his office by the entrance.

Already there were good reports of the Dart, for its light loading of 9 lb/sq ft resulted in a gentle stall at 38 kt, and it was very easy to land, earning the *soubriquet* 'gentlemanly', and that was usually a guide to production orders. However, the Hanley, more conventional though it was at first sight, had the interesting feature of full-span leading-edge slots to give low landing speed because of the high-lift coefficient. The possibility that it might be safer by preventing stalling in turbulent conditions was not yet apparent.

For export this Blackburn torpedoplane retained the name Swift, but those ordered for the Fleet Air Arm were named Dart. It was regarded as particularly easy and safe to fly. (*Courtesy Air Marshal Sir Ralph Sorley*)

The dynamics of winds were still a qualifying factor in aircraft operation. Most aeroplanes were doing well to climb at 500 ft/min with an engine efficiently developing maximum power, but even mild deterioration could reduce climb by 20 to 30 per cent, and a down-current might be disastrous. Even the customary light wing loadings caused ground handling problems, and many a machine blew onto a wing-tip and then stood on its nose or turned over. Not even airliners were immune. *Flight* reported: 'On Friday last, with a terrific gale raging, Mr. R. H. McIntosh made another of his wonderful bad-weather flights. Two Handley Page O/400s left Paris together and with the wind behind made good time as far as the French coast. Here they were caught by the full force of the Channel gale. The one flown by Mr Wilcockson managed to land at Berck aerodrome, but after his passengers had alighted, and before anything could be done to protect the machine, it was caught broadside and lifted onto one wing and then the other, being badly wrecked, slightly injuring the pilot. Mr

McIntosh, who was about to land, saw the accident, and decided to carry on to Croydon where there would be proper staff to handle his machine. Although quite low, and within a few hundred feet of the houses in Berck, he put his engines full on and the machine rose vertically for about 3,000 feet. He had a tremendous fight to get across the Channel, and at times he and his mechanic had to hang on to the controls. When he finally landed at Croydon, only managing to escape a crash by fine piloting, he had been 4 hours 35 minutes on the journey from Paris.'

On another windy day an O/400 piloted by Wilcockson took well over six hours from Paris to Lympne where it was necessary to refuel, but at least the pilot was able to break the tedium of the flight by 'wirelessing' instructions so that the fuel was ready waiting. But worse could befall, for *Flight* mentioned that in yet another incident McIntosh 'was injured in the foot when his O/400 crashed the other day in fog near Le Bourget. He manages to hobble about fairly well on crutches. The muscles and tendons of his leg were badly wrenched, and it will be some time before he is able to fly again.'

All airlines were experiencing similar troubles. A special Fokker F.II piloted by Mr Duke for KLM took four hours from Amsterdam to Lympne, where he had to stay owing to darkness, arriving at Croydon next morning after a further 50 minutes, but while taxi-ing the wind got under a wing and tipped the machine on its nose with considerable damage. A few weeks later one of the Grands Express Goliaths was written off in a forced landing at Farnborough because the pilot was unable to see the ground owing to mist, and knocked off the undercarriage though 'he and three passengers escaped injury, and the latter were in fact rather intrigued by their experience'.

Even the indomitable Alan Cobham, after flying nearly 7,000 miles with his D.H.9, on one of his taxi-flights was caught out by fog when approaching the coastal aerodrome at Venice, and in attempting to avoid trees, jetties, ships and buoys, decided to ditch rather than crash into the obstacles, and was found with his passenger sitting on the top wing waiting for a rescue boat. For pilots, such accidents were all part of the risk of their vocation – but it was not surprising that only two or three passengers might be the complement of a twelve-seater airliner.

At least it was H.P.'s hope to scrap those converted war-time O/400s. His contract for the prototype Hanley showed that the Air Ministry believed in him and restored a measure of confidence to his creditors. Money at last became available to lay down two more W.8s, now referred to as the W.8b, but in consequence of an unsuccessful battle with Napiers for free loan of Lion engines, he turned to Rolls-Royce who proved more amenable, and 360 hp Eagles were therefore installed. Tactfully H.P. explained: 'In the case of the W.8, practical experience showed that not only is the excess of power provided by the Lions too great, but the speed of the machine so fitted is greater than necessary for work on the London—Paris line.' In fact top speed with the Eagles was 103 mph and the machine cruised at a very modest 90 mph. He continued: 'While a good margin of power is the best possible manner of securing regularity and safety, a point is ultimately reached where economical reasons dictate a stop – therefore the Eagles.' But did he mean the economics of flight or his own purse in purchasing the engines? As a contribution towards load-carrying capacity the engines were stripped of cowling, because 'by the time the weight of cowling has been added, and considering the extra complication and difficulty of getting at the engines, the gain would have been so small that it has been considered better to leave the engines bare'. Following the lead of the D.H.34, fuel tanks were placed on the top wing, for although officially a safety precaution it resulted in a simplified petrol system. For a stopgap, lest there be further delay in re-equipping with W.8bs, he managed to secure from Bristols the loan of two 400 hp Jupiters which he installed in one of the O/400s giving splendid take-off and climb, though it seemed doubtful that he would buy such costly engines.

In hope of stimulating interest in civil aviation, the Government held another Air Conference in February at the Guildhall, London. It was preceded by a visit to Croydon on a perfect day of blue sky and bright sunshine, with every type of airliner on display. Arriving by special train from Victoria, the visitors passed through the Aircraft Disposal Co building, with its stacks of aeroplanes tipped on their noses to save space, and were met by Sir Frederick Sykes, who explained that this area would eventually house airliners, and the temporary buildings by Plough Lane would be replaced by permanent offices and works. Flights and demonstrations followed, but in fact only served to accentuate that no real progress had been made with civil aircraft since the previous year's Conference. However, a pleasant party was had by all, and in that sense of *camaraderie*

Among many others displayed for the Air Conference visitors were the prototype Bristol Ten-seater with simplified undercarriage, and a civil registered, scarlet D.H.29. (*Flight*)

the session opened next day at the Guildhall under the chairmanship of Capt F. E. Guest. Despite the interest of lectures given by many specialist exponents, it was preaching to the converted and could do nothing to bring civil aviation closer to the masses needed to make air transport a practicable proposition. The populace not unrealistically clung to the idea that aviation was a very dangerous business. As *Flight* recorded: 'The Conference was an event which should have had considerable effect in letting the man in the street, the business man, the financier – in short, all who are not intimately connected with aviation – know the present position of aviation, what it has to offer, and at what price. It must be admitted that this object was not attained this year. The papers read were too long and too technical. Next year let us stick to matters of policy, matters of national and international interest, but for Heaven's sake let us present a more united front to the outside world than we did in the Guildhall. The impression left on the non-technical section of the audience must have been that the aviation world does not know what it can do, nor how much it will cost to do it.'

C. G. Grey said of that time: 'Nobody knew anything about civil aviation so nobody knew the direction in which it ought to be controlled or directed or fostered.' Fumbling for solution, the Cabinet appointed a Civil Aviation Advisory Board with Lord Gorell, Under-Secretary of State for Air, as Chairman, replacing the Advisory Committee on Civil Aviation which had been presided over by Lord Weir. In addition to members of the Society of British Aircraft Constructors (SBAC) there were the Controller-General of Civil Aviation (CGCA), the Director-General of Supply and Research (DGSR), representatives of the GPO, Associated Chambers of Commerce, Lloyds, the Royal Aeronautical Society (RAeS), the Air League of the British Empire, and the Royal Aero Club (RAeC). Said Lord Gorell: 'I am authorized to say that the Secretary of State has decided to refer to this Advisory Board, as the first subject upon which he desires its recommendations, the question of the cost and practicability of an Imperial air mail service.'

Committees are the great British solution to everything. Under direction of the Air Ministry they were already numerous:– The Aerodrome Committee, the Aeronautical Research Committee (ARC), Awards to Inventors Committee, Committee on the Future of Experimental Establishments, Contracts Co-ordinating Committee, Committee on Cross-Channel Services (Subsidies), Royal Air Force Fire Committee, Meteorological Committee, Permanent Building Committee. There was the Whitley Council intended to bring about better relationship between masters and men as applied to staff, and the industrial version dealing with manual labourers. There was a Medical Advisory Board to co-ordinate the work of the RAF Medical Service with other Services and with civil practice. There were even more sub-committees, and inevitably the ARC appointed sub-committees of eminent technicians, professors, researchers, and industrial representatives compartmented for aerodynamics, engines, materials and chemistry, meteorology and navigation, accidents' investigation, fire prevention, load factors, and air inventions. Undoubtedly committee work was comprehensive, and undeniably in aeronautical research Great Britain led the world – even though the aircraft of most manufacturers appeared stereotyped and conservative compared with war-time development in Germany, or the recent trend of dashing experimentation in France and the more specialized improvements in American high-speed aeroplanes and engines.

2

With publication of the Report of the Geddes Committee the 'axe' at last had fallen, but with sense of relief that proposed cuts were no greater than people feared. For the Navy it amounted to 26 per cent, the Army fractionally more; but the £5½ million reduction from £15½ million for the RAF amounted to 35 per cent, and of that £1 million was against technical equipment – and this sounded ominous. Yet throughout the clauses dealing with reductions in the Air Force Estimates there was clear realization of the value of the RAF. 'We have come to the conclusion that the costs of the defence of the Empire, so far as it falls upon the British taxpayer, must be considered as a whole,' ran the opening remarks of the Report. 'The necessity for this is much more apparent now than before the war, more especially because of the advent of the Air Arm, which has come so much to the front, either as an addition to the older fighting Services, or in substitution for them.'

In analysing the requirements of the Royal Navy, Army, and RAF, the Report touched on the never ending struggle for power between them: 'The Navy and the Army both urge that the most effective and economical use cannot be made of the Air Arm so long as the personnel is controlled by another Service, and they suggest in general terms that it might be possible to effect economies by utilizing Air Forces in place of, say, light cruisers in the Navy or cavalry in the Army, but no concrete proposals on these lines have been furnished to us. On the other hand, by dividing the Air Force between the two Senior Services duplication in experiment, design and supply would be inevitable, and duplicate flying schools would result. To avoid duplication the two Senior Services admit the need for a certain amount of joint organization of subordinate character, but have given no clear appreciation of what this joint organization should be.

'All the arguments of an economic character which have been urged for the absorption of the Air Force into the two older Services apply, in our opinion, also the fusion of all three Services under one Minister. Complete co-ordination in supply, transport, education, medical and other services would then be possible. There is, however, the further argument that without a separate existence there is a grave danger that the Air Service would be unable to work out developments which might in the next decade or so entirely revolutionize methods of attack and defence and so render

Maj-Gen Rt Hon Sir Frederick Hugh Sykes, PC, GCIE, GBE, KCB, CMG (*left*) – an outstandingly brilliant but uneasy man, originally commissioned in the Hussars, and as Bt Major became the first commander of the Royal Flying Corps. (*Flight*), and (*right*) Alfred Hubert Roy Fedden, an innate engineer of outstanding ability and tremendous personality.

possible very large economies in the cost of the Fighting Services as a whole, by substituting Air for Land or Sea Forces.

'We cannot say we have been convinced that the Air Force is less economically administered than the other Fighting Services, but we are impressed by the fact that public funds are admittedly being spent in both the older Services because of a certain overlapping with the Air. The question of Aerial versus Naval or Military command in operations in the future would doubtless cause difficulties; but economies to an increasing extent ought to result in the older Arms from the advent of the Air Force. We are particularly impressed with the very large savings which we are told can be realized in the Middle East as soon as the transfer of responsibility from Army to Air Force can be effected. By the use of aircraft in this region it has been found possible to reduce the estimates in respect of the Middle East from £27 million in 1921/22 to £13 million in 1922/23. It can no longer be denied that by the intelligent application of air power it is possible to utilize machinery in substitution for and not as a mere addition to man-power.'

The copy of the Report to which I had access was annotated by Bristol's Sir Henry White-Smith, who noting that the Committee proposed reducing the 32½ squadron strength of the RAF by 8½ squadrons, scribbled in untidy hand: 'And they suggest this when they know that France this year will have about 200 squadrons, and Germany can start in May building their so called *civil* machines! Surely the more we weaken our forces in

Egypt and Palestine the longer the trouble will last and will be far more expensive in the long run in money and men?'

The number of aircraft possessed or on order by the RAF was reported as follows:

AEROPLANES	Active Service	Training	Stored	Total
Snipe	96	39	397	532
Avro 504K	14	395	514	923
Bristol Fighter	360	227	503	1,090
D.H.9A	271	124	268	663
D.H.10	60	7	12	79
Vickers Vimy	—	14	71	85
Vickers Ambulance	3	—	—	3
Vickers Vernon	36	—	—	36
	840	806	1,765	3,411
MARINE AIRCRAFT				
F.2a & F.5	45	24	40	109
Fairey IIIC	12	—	10	22
Fairey IIID	42	12	—	54
Westland Walrus	24	24	—	48
Sopwith Cuckoo	24	6	15	45
Parnall Panther	36	3	51	90
Sopwith 2F.1 Camel	36	—	43	79
	219	69	159	447
Grand Total	1,059	875	1,924	3,858

Of which 36 Vernons, six IIIDs, and 22 Walrus not yet received.

In the current Estimates it was proposed to order 92 new aircraft; 65 to be converted; 361 to be reconditioned by contractors; 184 to be reconditioned in Repair Depots.

Because the number of new machines was far too scanty to maintain sixteen airframe constructors and four engine manufacturers, particularly as it included experimental aircraft taking two years to design and build, the future must largely depend on profits from reconditioning. Armstrong had a few D.H.9As, delivered two or three at a time, and a D.H.9A had been converted to the same specification as the Walrus and was known as the Tadpole; Avro had a number of 504Ks, a few of which were converted to Type L seaplanes; Blackburn had two or three Cuckoos; Bristol some Fighters; Boulton & Paul and Handley Page a few D.H.9As; Hawker a batch of Snipes; Vickers a couple of Vimys; Short and Supermarine an occasional flying-boat; and Westland, as parent firm for the D.H.9A, was responsible for all its many modification drawings and had sequences of two or three for reconditioning. With the unofficial connivance of AID and Air Ministry officials, the practice of every firm with these contracts was similar to the admirable arrangement which Samuel Pepys instituted for Charles II's Royal Navy, whereby obsolete ships were sent to the docks and so considerably rebuilt that they often emerged as quite different, brand new ships. Although aircraft contractors did not go quite so far, every item was removed to the last bolt and piece of wood, and replaced by new if inspection gave the slightest excuse, resulting in a gleaming re-assembly indistinguishable from new. No wonder the standard of work-manship of British aircraft was considered the best in the world!

Frank Cowlin, that early Sopwith stressman, recollected: 'The post-war slump broke the project on which I had been working with Blackburns for civil air routes between Brough and Scandinavia, and I found myself back in the Air Ministry in R.D.3 section of DTD dealing with stressing and performance of Service and civil prototypes, and airworthiness certifica-tion. R. J. Goodman Crouch was head, and I took over performance. H. B. Howard had the stressing section, and McFarlane dealt with airworthiness and licences. Howard's right-hand man was A. E. Hayward. The total staff was small and barely able to cope with the work, which was beginning to expand as civil aviation belatedly seemed to get into stride. Then the "Geddes Axe" fell. Bagnall-Wild, who was DTD, was instructed to reduce his H.Q. staff to one third, and adroitly solved this by sacking the least useful third and dividing the residue into two – half of which was to stay in London, and the other half to join the RAE. We, in R.D.3, formed part of this Farnborough contingent, and acquired the new name of "Airworthi-ness Department RAE". Initially we numbered about 10, but Crouch rapidly discovered that the Geddes Axe took no cognizance of RAE weekly paid staff, and as civil work was still increasing, began to recruit more men, profiting from the generally poor employment situation in industry. By this time I was Crouch's deputy as he spent considerable time on general aviation policy in London with the AID of which Outram was director, and with the Department of Civil Aviation. We were enormously busy. By the end of the year our numbers were nearly double, but still we could not cope. So Hayward and I visited three or four selected firms to see if they would operate a "delegated responsibility" stressing scheme – that is, accept responsibility for numerical accuracy on the basis that we gave approval to the method of stressing and the basic assumptions. In the following year this idea was adopted as a pilot scheme, and soon expanded to cover all established firms. Hayward sent visiting representatives to supervise; and as work increased, some of them became permanently resident, essentially as stressing and technical officers (RTOs). A feature at this time was the sudden enthusiasm for light aircraft, and for Aero Club race meetings Crouch and Rowarth became closely involved as handicap-pers, assisted by Dalton.'

In contradiction to the Geddes recommendation that the RAF strength of five squadrons in Egypt and Palestine be reduced to save £2,500,000 per annum, Winston Churchill in a statement to Parliament on 9 March stressed the great success and enormous economy of using the RAF in place of the Regular Army in the Middle East. Instead of cuts he had raised the strength to eight squadrons concentrated at Baghdad in a loop of the river. 'They constitute the principal agency,' he said, 'by which the local levies all over the country are supported, by which the authority of the Arab Government is rendered effective, and by which the frontiers are to a large extent defended. It is a powerful concentration. There is nothing like it elsewhere in the British Empire. The eight squadrons represent one third of the whole strength of the RAF. We propose during the coming year to place the military control and security of Iraq under the officer commanding the RAF. By the time the Air Force takes over control there will only be four battalions of troops left, and they will be there for the purpose of keeping order in Baghdad and securing the aerodromes from marauders, and not for the purpose of wandering about the country on expeditions.'

Commented Maj-Gen Seely, one-time Air Minister: 'The Secretary of State for the Colonies has announced a very remarkable policy, and one of entirely novel character. It opens vistas of economy compared with which anything suggested by the Geddes Committee is as nothing. If this experiment succeeds, you will be able to do the most astonishing things all over the world in the way of reducing the number of garrisons – but do not let any one think that the Secretary of State's statement will go unchallenged. I know it will be bitterly opposed by old fashioned people.' And he proceeded to emphasize the tremendously effective manner in which aircraft had been used to quell four little rebellions in Iraq where 'if it had not been for aeroplanes, we would have sent an expedition which would have taken several weeks to get there, and there would have been considerable casualties. But on each occasion aeroplanes flew over and dropped a warning. After appropriate interval they went back and dropped a more efficient warning in the shape of a bomb or two, and the very next day the insurgents came in and apologized and surrendered. I congratulate the Secretary of State on his imagination in utilizing the Air Force as policemen. This is a red-letter day in the military history of the world, because for the first time we see this new young Air Force taking charge of a big country with the older Service ancillary to it. I assure him that there is a great body of opinion in this House which will back him with all its power in adopting and developing this efficient and merciful means of maintaining our Empire.'

But when Mr Amery, Parliamentary Secretary to the Admiralty, introduced the Navy Estimates, Rear-Admiral Sir Reginald Hall as spokesman for the Navy made renewed claim for separate Air Services. Despite the old threadbare arguments, he was seconded by Viscount Curzon who had been chairman of the Air Board in 1916 at the time of the attacks on General Henderson and Colonel O'Gorman. Neville Chamberlain, a Birmingham business man of great integrity who was the son of the famous

Joseph and half brother of Austen, made detailed reply to the Navy's case, concluding: 'To sum up, the Government believes that to abolish the Air Ministry, to reabsorb the Air Service into the Army and Navy, would be a fatally retrograde step. Even if it removed a little friction and improved and facilitated co-operation between the Air Services and purely naval and military operations, it would unquestionably retard development of the Air Services in their own element, in which it may be that the future of national defence lies. The decision of the Government to establish a separate Air Ministry was based on war experience. What is now required to ensure the success of the present scheme is close and intimate co-operation, and that the three Services should regard themselves as the common servants of the nation in endeavouring to attain a single object. This cannot be achieved so long as the existence of the Air Ministry and the Air Force remains in doubt, and the Government think it right and fair to that Service and to the distinguished officers who are at its head, and no less fair to the other two great Services, that they should thus define their attitude so that all may know what is expected of them and what system they have to follow.' The die was cast. The RAF had won through, at least for a long period. Admiral Hall bowed to the storm and asked leave to withdraw his Amendment to the Navy Estimates.

3

The Vickers Vernons mentioned in the Geddes Report were a military ambulance version of the Vimy Commercial used by Instone Air Line, but the nose could be opened to load four stretcher cases; alternatively the cabin could take eight sitting cases, and in addition to pilot and navigator there was room for two medical staff. Climb in the burning heat of the desert route was weak, and take-off could be long and difficult, particularly as Amman was 2,600 ft above sea level, giving an air density equivalent of 4,000 to 5,000 ft. With the surrounding hills in close proximity, the margin of safety was slender. Several accidents occurred through failure to get airborne or sinking in down-currents stronger than the normal small rate of climb. Sq Ldr Roderic Hill, former OC of Farnborough's flying, had been appointed to command the squadrons of the Cairo—Baghdad air mail run, and he urgently recommended that the Vernons be fitted with Lion engines to give better performance. Following Rowledge's change to Rolls-Royce, Capt G. S. Wilkinson had been re-engaged by Napiers to re-develop the Lion. Pre-war he had been chief draughtsman, and in 1915 held similar position in the engine D.O. at Farnborough, working with Major Green and S. D. Heron on design of air-cooled engines, but in 1917 joined Airco to develop an air-cooled engine for de Havilland.

With no loss of time the Air Ministry instructed Napiers to supply Vickers with two Lions for Vernon J6884. Rex Pierson, grimly massive and now looking more than ever like his bald clergyman father, gave priority to installation drawings prepared by his chief draughtsman Paul Wyand, and the shops worked to such effect than in January the machine

Vickers Vernon Mk. Is with 375 hp Rolls-Royce Eagles were the transport mainstay of the Middle East when the RAF took over military control of Iraq.

was flown to Martlesham for evaluation, followed the next month by a similar Vernon fitted with 120-gallon overload outboard tanks, and an oleo-pneumatic undercarriage of improved shock-absorbing capacity replacing the bouncy 'bungee' springing used for every squadron machine in the RAF, whether Vickers or otherwise. So successful were the trials that all the Eagle engines of No. 70 Squadron were replaced, and as Hill recorded: 'The Vernons became animated with the fury of Lions. It happened almost inevitably that this Squadron with its higher performance aircraft took over the ex-Iraq mail entirely.' At that time D.H.9As of No. 47 Squadron at Helwan in Egypt and No. 30 Squadron at Baghdad had carried the mail, together with the D.H.10s and later Vimys of No. 216 Squadron at Heliopolis. Hill said: 'In those days there were no desert refuelling points, so the D.H.9As had to strap on as many as twelve 4-gallon petrol tins underneath their wings on the bomb racks, and when the petrol in their tanks was exhausted, they landed and filled up from the tins. They looked like veritable Christmas trees, and the external addition of tins adversely affected their performance.'

The Vickers shops at Weybridge now became very busy, for not only was there a production line of Vernon Mk IIs and several production Viking Mk IV amphibians with longer wings of high-lift T.64 section developed by the RAE, but there was also the prototype 88-ft span Virginia long-distance bomber assembled in skeleton form, and alongside was its troop-carrying version, the Victoria, with similar elliptical-section cabin to the Vernon. There seemed little hurry for these two big machines, for they had been only slowly taking shape for the past eighteen months. Instead, the works were full-out to complete six single-engined airliners, four of which had been ordered by Instone. Known as the Vulcan, this design had single-bay wings, a biplane tail, and an enormously deep elliptical-section fuselage with pilot's cockpit in front of the top wing. The Eagle-powered prototype of this rival to the D.H.34 had seats for eight passengers, though the D.H. and Bristol equivalents carried two more.

The Vernon Mk. II with Napier Lion II engines was a more effective military machine, but although originally cowled, the RAF found it more practical to have the engines bare. Note the Shilovsky-Cook turn-indicator pitots at wing-tips. (*Courtesy Air Marshal Sir Ralph Sorley*)

Compared with the Vickers Vulcan, the Lion-powered D.H.34 seemed more rational, and had adequate head room despite a smaller fuselage and the necessity of stooping to enter the low doorway (*lower illustration*) although it was wide enough to admit a spare engine. (*Flight*)

Design had started in February 1921, when the D.H.18 was its only functioning competitor, though Bristol had already started work on their prototype Ten-seater, and made such rapid progress that it was ready to fly five months later. Although the Vulcan prototype had been registered in February as G-EBBL, it was late on contract delivery, and the first flight was scheduled for April.

Despite the efforts of Percy Maxwell Muller, the efficient Vickers works manager, and Archie Knight his burly assistant, the Vulcan was beaten to the post by the smaller and somewhat anxious organization of de Havilland – for the scarlet-painted D.H.34 made its first flight at Stag Lane on 26 March piloted by Alan Cobham. It was his first experience of testing a new type, but he appeared delighted with the handling, having found no difficulty in adapting to a cockpit location very different from the machines he usually flew. Geoffrey de Havilland followed almost immediately with a confirmatory flight, and found control and behaviour conventionally normal; but so, in fact, was the machine, for it was a lineal descendant of the first B.E. he had designed a decade earlier. Yet, in this year of 1922, the D.H.34 seemed very advanced to her pilots, and the mechanics found it a big step forward to have an undistortable ply-covered fuselage and an engine quickly removable as a unit with internal radiator, and there were foldable built-in platforms for servicing or cranking the inertia starter.

Of the merits of wooden construction D.H. was convinced. 'I cannot quite agree that at present the all-metal machine is practicable,' he said. 'The life of a wood machine is longer than generally believed, and wood has the advantage that for experimental work it affords cheapness of construction. The life of a wooden machine could be tested very cheaply, and it might be useful to carry out such tests. It is sometimes claimed that fabric-covered wings get soggy, and that in consequence a machine loses its performance, but I do not believe this deterioration is as serious as generally thought. One difficulty which firms encounter is that as soon as a machine is finished it is taken away from the constructor, and thus he gets no opportunity of testing it thoroughly. I would plead for support from the Government to enable firms to do more full-scale flying tests.' Literally, there was no development flying unless there were very obvious faults, and not even a dive was attempted. Questions of controllability, manoeuvrability, stability, and suitability were left entirely to the discretion of RAF pilots. Thus the D.H.34 went through its brief airworthiness tests in a single day at Martlesham, and was delivered to Daimler Airway at Croydon on 31 March. Two days later, after brief handling flights by their pilots, it was flown on its inaugural service to Paris by Capt Hinchliffe, late of KLM, and carried a consignment of newspapers. *Flight* reported: 'The load was, apparently, too much for the machine; after an extraordinarily long run it only just managed to clear the hedge, and was actually so low that it disappeared from sight in the valley south of the aerodrome. When attempting to leave Le Bourget for the return journey, again with a heavy load, the machine stuck in the mud and was unable to get away. This sort of thing rather amuses the old hands, as they have been all through it years ago; but it would be better if new companies could benefit by some of their experience instead of beginning all over again and making the same mistakes. It will be remembered that this happened with the D.H.18s; for a long time their full load was at the most seven passengers, but when pilots and mechanics gained experience, the full load of eight passengers and quantities of baggage could be carried with ease.' Meanwhile the second machine, with silver-doped wings and blue fuselage, was accepted by Instone Air Line and flown to Paris by Capt F. L. Barnard on 2 April.

Though de Havilland and Walker were wedded to wooden construction, Major Green of Armstrong Whitworth was entrenched in the opposite camp, and insisted: 'So long as wood is used, the making of good aeroplanes will be a matter of art rather than science. The substitution of metal for wood may enable us to effect improvement in weight. Using steel of 45 tons/sq in tensile strength, it is possible to make aeroplane structures of at least no greater weight than wood. The difficulty is that in making structures suitable for aeroplanes the thickness of the material has to be small, and we have yet to find the best methods of making and using thin sheet materials. Many fundamentals have been established, and development of this sort of work calls for research of rather a practical nature.' With that, John North of Boulton & Paul certainly agreed, but stated that all-metal machines already could be built 10 per cent lighter than wood structures. Both these designers had devoted extensive research to thin steel fabrication, and North in particular had advanced considerably in the course of his Bodmin design, assisted by Harold J. Pollard who had previously been at Vickers after training at University College, Nottingham.

Frank Courtney taxies out the Boulton & Paul Bodmin at Mousehold, Norwich. Behind him the engineer looks from his side hatch in the stuffy engine-room.

In that design, every part made of wood in a conventionally trussed-girder constructed aeroplane – such as spars, ribs, struts, longerons – was built from high-tensile sheet steel, rolled, drawn, or stamped to the required section using the extensive machinery with which the factory was now equipped, and everything was held together with many small solid rivets. Even where standard tubes might have been used for longerons, North found it better to fabricate from rolled sheet in the form of a three-quarter folded circle in section, flanged at its edges, and riveted to a similarly flanged quarter segment, enabling right-angled fittings to be interposed to take attachments for the tubular-steel vertical and horizontal spacers and their swaged rod bracings. The main spars, whose variable linear loading was the chief problem, were of hour-glass section with flanged semi-circular top and bottom riveted to similarly flanged bell-topped webs, the sides being steadied laterally with a sequence of axial tubular spacers through which tubular drag struts projected with an overhang on which the built-up steel interplane struts were secured. Ribs were curved steel channel-section strips top and bottom and lattice braced as being the lightest construction. The enormous detail work of this delicate and costly fabrication was regarded as worthwhile, for North found that the Bodmin structural weight compared with wood was lighter not by 10 but almost 20 per cent – though this appears to have discounted the structure carrying the complex of shafts, gears, and tandem propellers which was independent of the wing to avoid problems of rigging, and formed a triangulated unit each side of the tandem Lion engines which were centrally mounted and supported by the top longerons.

At Armstrong Whitworth, Major Green and John Lloyd, as his chief designer, were somewhat less ambitious. The long delayed, wood-structured twin-engined Sinaia was at Martlesham. Recent patents indicated that a tubular-steel-structured girder fuselage was favoured both for the latest S.II version of the Siskin, and for a new big twin-engined troop carrier of 105-ft span, named the Awana, which was well advanced in design and construction. However the Awana had conventional wood-built wings, but elegant metal wings had been built for the Siskin as far back as 1920, employing rolled and corrugated built-up thin gauge steel spars of the type developed and patented in November 1920 by Major Hamilton N. Wylie, whom John Siddeley appointed consultant. Although the Siskin S.II has been reported as having wooden wings and steel-tube fuselage, it is also recorded as 'all metal', so one of the sample machines may have used wings of the type built eighteen months earlier, possibly with ribs changed from narrow duralumin tube to U-section thin steel.

Bruce at Westland was also involved in metal construction, but restricted the application to wing spars and cabin of the Woyavodsky Dread-

The cantilever Woyavodsky wing of 18-ft chord had a scaled-up, modified T.64 section, and was attached to fuselage bulkheads at four stations.

nought. Considerable stressing problems ensued due to departure from the classic two-spar system and a special mathematician, J. D. Williams, was engaged to cope with the six-spar arrangement. Influenced by the Junkers multi-spar wing which Bruce and Barnwell had inspected in 1919, the Westland design was the first British construction of this type. Tubular-steel booms top and bottom were separated by vertical tube struts and cross braced with swaged rods vertically in the plane of the spar and transversely and diagonally to the next spar. The central fuselage frames were rectangles of steel tubes, with deep triangulated structures each side taking the wing attachments and forming a central vestibule braced transversely and longitudinally with open portal frames. It was intended to have five seats in each wing root, and the sloping sides of the cabin in effect constituted a fillet running the whole length of the fuselage, though it was empirically shaped in the absence of any known mathematical approach.

Much over-all aerodynamic model testing had been undertaken by the RAE, using a wing-section very similar to T.64 blown up in depth to 6½ ft at the fuselage and 3½ ft at the root juncture, tapering normally to the tips. The leading edge was very heavily radiused on the top as far as the first spar, and had a sharp juncture to the bottom surface – for nobody was aware of the danger of early flow separation. Control hinge moments had been carefully investigated, and both rudder and elevators had inset balances of advanced design, and the ailerons were of new form devised by Leslie Frise of Bristol, whose Patent 194,753 stated that upward movement of the aileron 'causes its leading edge to project beyond the contour of the thick section of the wing, but downward movement is possible without producing any such projection or gap'. When the leading edge projected below the wing it formed an air passage helping to make the aileron more effective at the stall, but primarily added useful drag to the downward going wing. In combination with Hagg's differential it could have afforded the perfect aileron. Bruce was the first outside constructor, other than Bristol, to employ this form of aileron, largely because in pre-war days he had worked with Barnwell. In fact Handley Page had already patented a similar form of aileron with the advantage that a slot was formed when the aileron was raised, thus giving greater effectiveness, while the opposite deflected aileron closed the gap and so produced less drag. But constructors seemed a little chary of Master Handley Page.

For the rest of the structure, wood was mainly used except for engine mounting and undercarriage. The crew of two were located in an open cockpit ahead of the wings, immediately behind the 450 hp Lion. Design by now was sufficiently advanced for many parts to have been finished, and some of the structure was being assembled in the big main erecting building, known from its war-time origin as the 'Vimy shop' which still had only a gravel floor, such was the pressure of post-war economy.

As there were gaps of weeks in the fluctuating demands for D.H.9A repairs, Stuart Keep, the Westland test pilot, was taking his mandatory annual training period as a Royal Air Force Reservist. Having a cross-country to perform, he decided to look up his Yeovil friends. The aerodrome was relatively small, some 650 yards by 400, and somehow he managed to undershoot and hit the westward hedge, so that his Bristol Fighter turned over, and he catapulted from the cockpit. There was no ambulance in those days, or many a year after, and the only fire-fighting apparatus was a small hand-trolley, fitted with discarded aeroplane wheels, carrying four useless little extinguishers. By the time the alarm was raised, Keep had got up and was staggering around, then collapsed from concussion. The rest of his flying leave was spent in hospital.

4

On 1 April the Air Ministry announced that Maj-Gen Sir Frederick Sykes, GBE, KCB, CMG, had resigned as Controller-General of Civil Aviation on expiration of the term for which he was originally appointed. Sykes had written to the Secretary of State for Air explaining that: 'I have come definitely to the conclusion that, owing to the very small scale to which you have found it necessary to reduce the Department of Civil Aviation, there is no scope for such a position as mine, and I would not be justified in continuing even for a year to receive a salary for work which circumstances make it impossible for me to perform.' Capt Guest had already explained in the Debate on the Air Estimates that retraction of civil aviation was in no way the fault of the department over which Sir Frederick presided. 'It is the difficulties of the times in which we live that are responsible. It might be that in time to come this side of our work will expand,' he said. An immediate result was to abolish the office and title, but it was rumoured that in the reorganized Air Ministry there would be a new department known as the Directorate of Civil Aviation. Who would fulfil this post was any man's guess, but it seemed that Colonel Brabazon might be a good choice, or even Brancker. What to offer to that great public servant, Sykes, posed a more difficult question, for despite differences of opinion between him and Henderson, his services had been great and he had been the first CO of the infant RFC in 1912.

Though so much was going slowly in civil aviation, the upholders of airship potential, and Vickers in particular led by Lieut-Cdr C. D. Burney, were pushing new schemes for Empire air services using the redundant British rigids – but they chose an unfortunate moment, for the Italian semi-rigid *Roma*, owned by the US Army, concurrently was wrecked, only eleven of the forty-five on board surviving. *The Times* reported: 'The airship became unmanageable when at about 1,500 ft. Observers saw a huge kite-like structure under the tail cone swing loose at an angle of 45 degrees. The *Roma* took a nose dive and fell rapidly to the ground. It missed the chimney stacks of buildings at the army base, and fell upon high tension wires. A terrific explosion followed and the mammoth gas-bag turned over on its cabins, whose occupants were precipitated into the roaring flames.' Although the *Roma* was a semi-rigid airship, the fact that it crashed from structural failure added to the emphasis that insufficient was known about the aerodynamic forces on airships and the method of calculating stresses.

Nevertheless, at a meeting at Caxton Hall, London, on 30 March, Burney assisted by the enthusiastic A. H. Ashbolt, Agent General for Tasmania, with Major G. H. Scott advising on technical aspects, explained a new scheme backed by Vickers and Shell Petroleum for an operational company of £4 million capital – subject to the Government transferring free of cost all airships and material, the airship bases of Pulham, Cardington, and Howden, and supplying wireless telegraphy and meteorological services. Of the capital, Vickers and Shell would each subscribe 100,000 shares at par, and the British Government was asked to provide £91,000 per annum subsidy and guarantee the proposed 6 per cent dividend for ten years; Australia and India would contribute £40,000 each.

The Air Ministry duly replied: 'In the opinion of the Air Council, the scheme constitutes a notable advance upon any other utilization of airships in Imperial communications. They consider that, with certain additions, it offers a reasonable prospect of being able to operate ships satisfactorily between India and this country.' It was stipulated that the number of airships should be increased from five to six, and an airship base, complete with shed, must be erected in India, but added the ominous note: 'The Air Council are at this stage unable to make any statement regarding the financial aspect of the scheme.' Undoubtedly the Treasury would apply a brake since the national finances were precarious.

Assuming an average of 55 kt including halts *en route*, it was hoped to fly from London to Bombay in 5½ days instead of 17 by surface transport; Rangoon 7½ days instead of 22; Hong Kong 8½ days instead of 4 to 5 weeks; Australia 11½ days instead of 4 to 5 weeks. Certainly the saving of three weeks to Perth in Australia would be thought invaluable by business men and diplomats had not recent accidents reduced the public's confidence in airships to zero. Nor did the first mid-air collision in the history of air transport help, for a Farman Goliath belonging to Grands Express, flying at a few hundred feet in misty visibility flew into a Daimler Airway D.H.18, piloted by the popular R. E. Duke, near Grandvilliers in France with fatal results, though luckily both aircraft were almost empty. With no effective wireless systems integrated with ground control it was likely to happen again.

An immediate result was a meeting in the Handley Page office at Croydon of air transport representatives and pilots 'to plan routes and rules which would make the chances of air collision practically nil', *Flight* reported. 'Several resolutions were agreed, the principal being that pilots should hold meetings among themselves to plot definite routes, all marked by well known and easily distinguished landmarks, and that once these routes are fixed all aircraft must keep to the right of them. It was further resolved that all machines should carry wireless, and it is noted in this connection that the French representatives, whose machines were not fitted with wireless, were as emphatic on this as the British. Resolutions dealing with improvement of wireless communication and weather reports were forwarded to the Air Ministry for their consideration and one hopes necessary action.' But the problem was not so easily solved, for each pilot had his own favourite landmarks on certain parts of the routes, and none could agree which was the best for every kind of weather.

Nevertheless the first night flight over the British portion of the air routes was safely carried out on 7 April when 'an aeroplane which carried eight people, including a navigator, wireless officer, and Air Ministry officials responsible for lighting and wireless arrangement of the route, left Biggin Hill about 8.30 pm, flew to Croydon, and landed there. The pilot, who had great experience, reported that the aerodrome flood lighting of dispersed searchlight beams and an illuminated landing L were the best he had ever seen, and made landing as easy by night as by day'. Thereafter the machine flew to Lympne, guided by temporary aerial lighthouses at Catsfield and Cranbrook, then over the Channel towards St Inglevert aided by the marine lighthouse at Cap Gris Nez which was visible from above Biggin Hill, and soon the French aerial lighthouse at St Inglevert came in sight. The pilot next retraced his course, landed at Lympne, left again for Croydon, re-circled its mass of lights, and headed back to Biggin Hill, his landing lit only by wing-tip and ground flares. 'The general impression of those aboard,' said one of the passengers 'was that it is easier to find a course by night than day, and that provided the Continental ground organization is as good as ours there should be no difficulty in commercial night flying over the London—Paris route.'

Although not mentioned in the official report, the machine was a Handley Page O/400 used as a flying laboratory by the Instrument Design Establishment (IDE) at the South Camp of Biggin Hill, where the RAF Station had just been closed as a Fighter base on disbandment of No. 39 Squadron and its Sopwith Camels. Only five pilots were attached to IDE for routine testing of new devices such as D/F receivers and transmitters experimentally installed in an Avro 504K used for investigation of ground control approach using a narrow cone of sound from a klaxon at the focal point of a 20-ft mirror.

The scope of hopeful aerial endeavour now extended to two proposals to fly round the world – the first by the England—Australia flight winners, Sir Ross Smith and his brother Sir Keith, flying a Viking IV, with J. M. Bennett, who had been promoted to lieutenant, as engineer; the other by Major W. Blake and Capt Norman Macmillan as pilots, with Lieut-Col L. E. Broome as navigator, who hoped to fly a three-seat Disposal D.H.9 to Calcutta, a Fairey III seaplane to Alaska, a Canadian D.H.9 to New York and a Canadian F.3 flying-boat for the last stage across the Atlantic and home. Ross Smith intended keeping to the land masses nearest the equator, though, after Tokyo, his course would be unexplored to Petropavlovsk on the top of Siberia, then across the Bering Strait to Kodiak Island off the Alaska Peninsula, across Canada to Newfoundland and thence direct to

A Vickers Viking IV (*upper illustration*), but with open cockpit replacing the enclosure, was built for the ill-fated attempt of the Australia race winner, Sir Ross Smith, to fly round the world. Cockpit layout was unconfusingly simple (*lower*). (*Flight*)

England. His Viking IV had 4 ft more span than the Mk III amphibians, and used T.64 section for greater lift, but it was not a very good machine: there was considerable change in trim between engine on and off due to the high thrust-line, necessitating a hard push on the stick when suddenly throttled, and as the wing-section had a sharp stall it could be dangerous. For the round-the-world flight, Rex Pierson had eliminated the cabin enclosure to save weight, and fitted large tanks behind the second cockpit to extend the range from 420 miles to 925.

On the morning of Thursday 13 April, Ross Smith and Bennett arrived at Brooklands for a familiarization flight with Stan Cockerell, the Vickers test pilot, who O.K'd him and got out. Ross Smith took over, and the machine left the ground normally, but after climbing to 1,500 ft a wing suddenly dropped as though beginning a sharp turn but developed into a spin. A witness told reporters: 'After coming down to about 700 or 800 feet it looked as if Sir Ross was regaining control, but immediately the machine commenced spinning again until quite close to the ground, but it was then too late to flatten out, and the machine crashed into tall fir trees at the back of the Byfleet banking of the motor track, both occupants being killed instantly.' Thus, by awful coincidence, first the 1919 conqueror of the Atlantic and now the man who first blazed the trail to Australia were killed in the same type of Vickers machine.

Norman Macmillan and his crew, with their far inferior aeroplanes and ancient flying-boat, began their round-the-world attempt from Croydon on 24 May, but ended near Chittagong, then in India, on 23 August when water in the petrol caused engine failure and necessitated alighting on the rough sea. After taxi-ing some time, and even attempting to take-off again, the starboard float leaked so badly that the machine began to sink and during the night a heavy sea turned it turtle, leaving the crew sitting on the bottom of a float until late next day when they were rescued by the river launch *Dorothea* and taken to hospital.

Meanwhile Fairey Aviation had built three special IIID seaplanes for the Portuguese Government, whom, it was rumoured, intended to let two of their officers use one for a round-the-world flight. These machines were distinguishable from the standard Eagle-powered version by an additional bay in the wings giving a span of 60 ft to enable reasonable take-off with over-load long-range fuel tanks slung under the bottom wing at the inboard struts. Piloted by Capt Sacadura Cabral, with Capt Gago Coutinho navigating, a first attempt was made in April to cross the Central Atlantic from

Pilot and navigator of the Portuguese long-range Fairey IIID seaplanes had slightly staggered side-by-side cockpits abaft the fuel tank filling the forward fuselage bays.

The Netherlands were early purchasers of British aircraft. Here a Fairey IIID takes-off from Southampton Water on its delivery flight to Holland for service in the Dutch East Indies.

Lisbon, with stops at the Canaries and Cape Verde Islands – but it ended after eighteen days when the float undercarriage collapsed on alighting in too rough a sea alongside a Portuguese warship stationed at the microscopic St Paul's Rock, 500 miles short of the Brazilian coast, but luckily they were rescued through the forethought of their Navy. The next flight also was disaster. Not until their third attempt, using the last machine, did the gallant pair attain Pernambuco on 5 June.

5

England at least could give a sigh of relief at the end of April, when De Valera issued instructions to suspend all offensive operations. Though the IRA continued with sporadic outbursts, the final move towards establishment of the Irish Free State had been made – foreshadowed by an agreement Britain had signed on 28 February recognizing Egyptian independence. China also had changed status as the result of bitter civil war, and on 11 May Yuan-Hung assumed presidency of the Chinese Republic. Nevertheless, for the average Briton, these international affairs paled in contrast compared with the great assault being made on unconquered Mount Everest.

The end of April saw initial tests both of the W.8b G-EBBG, and the Vickers Vulcan G-EBBL. Almost immediately they flew to Martlesham for Type tests. Service and airline pilots alike were particularly pleased with the W.8b's low landing speed of some 42 mph, and from 4 May it was used in familiarization flights between London and Paris. When the Instone pilots flew the Vulcan at Croydon on 8 May they were equally happy with its short take-off and initial climb, though the Handley Page beat it into regular service by a fortnight, following official public appearance on 16 May when Sir Sefton Brancker, who had just accepted the post of Director of Civil Aviation (DCA), named with time-honoured champagne the first

Forward and rearward view from the Vulcan was considered ideal, once pilots got used to their perch, and the somewhat uncertain directional stability was improved with a small central fin above the tailplane. (*Flight*)

two *Princess Mary* and *Prince George*. Though his skill and judgment as a pilot was negligible, Sir Sefton's commonsense and enthusiasm for aviation was enormous, and despite his outspokenness, or even because of it, he was regarded with affection by all. 'In the past I have criticized the Air Ministry's aviation policy,' he said, 'and now with my appointment as DCA, it becomes my turn to be criticized – and I hope I shall be, for I always welcome criticism.' Removing and re-securing his glittering monocle, the dapper 'Branks' with richly crisp intonation added a quick face-saver: 'Unfortunately, as far as the present financial year is concerned, the policy for civil aviation is already settled, but next year I hope to frame the best possible.' Frederick Handley Page, towering benignly beside him, was seen to smile with gratification – or was it because he had stolen another march over the Instone Air Line?

Whatever the success of their civil aircraft and big bombers, the Vickers Board was determined to make greater impact in aviation commensurate with its armament and shipbuilding activities, and therefore proposed

As a dignified but cheap colour scheme for his new fleet of W.8bs, Handley Page stuck to standard silver dope, particularly as it was lighter and easier to repair, but edged it prominently with blue. (*Flight*)

tackling the market for smaller two-seat military aircraft which in a year or two should be needed worldwide to replace those built with earlier techniques. The requisite constructional trend seemed to be a fuselage framed with steel-tube and faired to attractive shape with stringers, and the simplicity of single-bay wings, even though 40 ft in span, would give a convincing modern appearance compared with the thinner multi-bay wings of most British Service aircraft, and for maximum performance and neat installation the well-tried 450 hp Lion was advisable. From these considerations, Pierson, with Camden Pratt as section leader, developed appropriate dispositions of surface and weight to provide an impressive looking biplane, the Vixen, aiming at a gross weight of 4,500 lb and empty weight of 3,000 lb, and potential speed of over 135 mph with climb to 10,000 ft in 10 minutes, and ceiling at 20,000 ft.

In Parliament Mr L'Estrange Malone, an ex-Colonel fighter pilot of great renown, asked with obvious inside knowledge whether the Air Ministry had been following helicopter experiments in various countries; what conclusions had been reached as to their practicability for war or civil use; and whether it was intended to make any practical experiments? Capt Guest replied that the authorities considered helicopters sufficiently important to warrant investigation and experiments were in progress.

'Is it intended to hold a State competition for helicopters?' pressed the MP. 'What prizes are offered; what are the conditions; and when will the competition take place?'

He was told: 'The Air Ministry have decided to offer a prize of £5,000. The conditions will be announced as soon as possible.'

A few privileged people had already seen the 'secret' helicopter which that elderly genius Louis Brennan had been building in the airship shed at Farnborough, where it was barricaded from prying eyes by canvas curtains made from an old Bessonneaux hangar cover strong enough to cushion the machine if it drifted during tethered trials. Its 60-ft diameter single rotor-wing seemed enormous, poised above two superimposed pyramidal central structures with a 230 hp B.R.2 in the rotating upper, and seat and controls in the lower. Blade incidence was varied collectively, and a swash-plate governed feathering by tilting the plane of rotation, but there was no tail to give directional equipoise. Power from the horizontal engine was transmitted by internal shafts to a large four-bladed wooden propeller at each rotor tip. The lower pyramid formed the base on which the machine stood and was suspended from the rotating structure by a universal joint on the C.G. immediately below the engine. Through this joint went the flying controls, pneumatics, and fuel pipes from a tank below the pilot's seat.

R. G. 'Bob' Graham, who eventually became Director of Aircraft Mechanical Engineering Equipment, told me: 'Toward the close of 1921, the aircraft was ready for engine runs and engagement of the rotating wing. A rig was provided in which steel cables were run up the centre of the shed along the roof, down the walls, and back to the centre, the helicopter forming a link in this system. It was prevented from drifting laterally but

Louis Brennan standing by his helicopter in the original Airship Shed at Farnborough, with Bob Graham aboard. The B.R.2 rotary was mounted horizontally above the rotor, and starting it proved a problem.

was free to ascend 20 ft, and shock absorbers were provided at top and bottom. Some difficulty was experienced in running the rotary engine on its side, and lubrication, induction, and cooling required modification; but when these were rectified and the engine running satisfactorily, trouble followed with engine starting. It was not possible to swing the tip propellers by hand and a Hucks starter had to be used, but was unsatisfactory because the thrust tended to override the chocks used to prevent rotation of the main rotor-wing and eventually a propeller was broken. This led Brennan to say we would have no more of the Hucks; we would have a mechanical starter, although no such thing existed for rotary engines, so he designed a light-weight mechanism on the lines of those now used for motor lawn mowers, which was loaded by hand, and when released by the pilot gave a pull of about 1,500 lb on cable round the propeller boss. This worked perfectly.

'About this time the old man said to me: "You have done the aerodynamics and stressing of this machine and know all about it. Don't you think it right that you should fly her?' From then on, I was at the controls for 98 per cent of the trials, although the Air Ministry appointed two RAE test pilots, Flt Lieut P. W. S. Bulman and Flg Off C. A. Bouchier as

BRENNAN'S COMPLETE SPECIFICATION No. 12327 DATED 8TH MAY 1923.

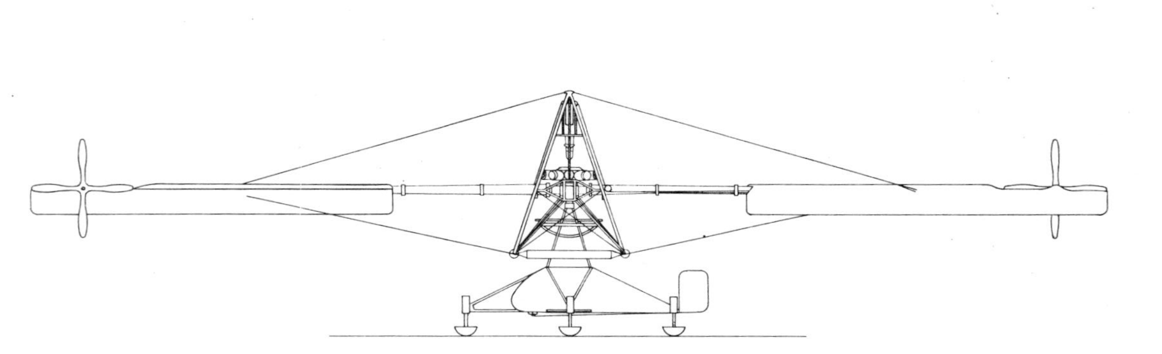

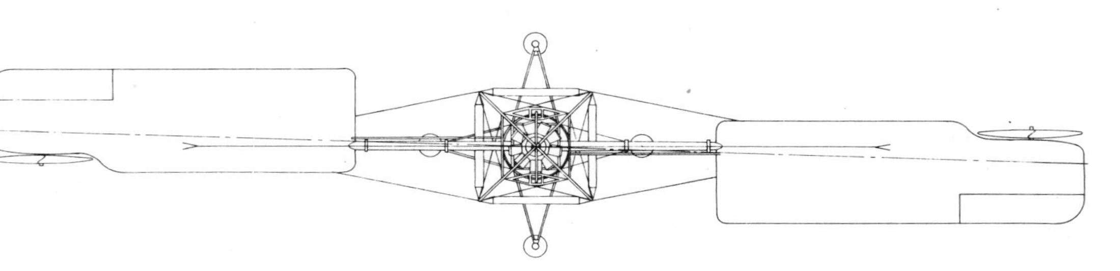

Elevation and plan of Brennan's helicopter.

reserves, and on a few occasions they handled the aircraft. Many teething troubles were experienced and considerable redesign was necessary during the rotor spinning trials, but by 13 January, 1922, we had achieved a lift of 4 ft, and by 14 March the helicopter reached the maximum height of 20 ft possible in that shed. However, there was still a long long road ahead before we could think of tackling outdoor trials, for translation controls still had to be fitted.'

Unfairly, there was growing feeling that Brennan was making a meal of it. Throughout the history of technical development there has always been failure of politicians unversed in technics to understand why a thing would not immediately work. In France and Italy, helicopter development was proceeding; at Göttingen their outstanding young Hungarian aerodynamicist Theodore von Kármán was devising what seemed a workable helicopter; from Spain came further hopeful reports of the windmilling machine made by de la Cierva. Nevertheless, some members of the ARC were unalterably opposed to the idea that helicopters might prove practicable. It was to clarify the possibilities that Air Vice-Marshal Sir Geoffrey Salmond, KCMG, CB, DSO, had proposed a competition open to all comers, though well aware that a niggardly £5,000 could attract only entrants whose machine was already proved, for *de novo* development costs would far outweigh such a sum.

Among changes at the Air Ministry, Sir Geoffrey Salmond was now titled Air Member for Supply and Research (AMSR instead of DGSR), and the new post of Air Member for Personnel (AMP) was established with the appointment of Air Vice-Marshal Oliver Swann, CBE, who as a junior naval officer had been the Navy's first experimenter with seaplanes when at Barrow-in-Furness. In the Department of AMP was a Deputy Director of Personnel (DDP), a Deputy Director of Organization (DDO), and a Director of Medical Services (DMS). The Department of AMSR now included a Director of Research (DOR); a Director of Aircraft Supplies (DAS); a Director of Aeronautical Inspection (DAI); a Director of Equipment (DOE); and there were sub-Departments of Deputy Director of Designs (DDD); Deputy Director of Armament (DDA); and Deputy Director of Instruments (DDI). Above them all reigned Air Marshal Sir Hugh Trenchard, KCB, DSO, the Chief of Air Staff.

It was Air Commodore Brooke-Popham as DOR who emphasized the vital importance of instrument reliability before blind-flying techniques could be evolved, so he proposed that all research into signals, wireless, instruments, navigation, and meteorology should be concentrated at a single station: 'If Great Britain is to maintain the lead it at present holds in the production of aeronautical instruments, and if the Air Ministry is to retain the confidence of the aircraft industry, there is no alternative but to have a well equipped establishment, such as Biggin Hill, and provide a well paid staff to ensure that experiments and research are conducted on the right lines. The development of civil aviation depends largely on the progress made with aerial navigation.' Certainly the value of Biggin Hill as the Wireless Experimental Establishment had been effectively demonstrated far back in 1919 when three American Curtiss flying-boats made the first stage-crossing of the Atlantic, for they carried experimental D/F sets loaned to the US Government by this station.

That hill-top grass aerodrome was too strategically placed to leave solely to technical boffins, for it was a key to defence of London. On 15 March a significant visit was made by Air Vice-Marshal P. W. Game, Salmond's Deputy; Air Commodore 'Stuffy' Dowding; Colonel Fuller the War Office DDT; and Colonel Simon, OC of the School of Anti-Aircraft Artillery and Searchlights at Perham Down. Their plan was to concentrate at Biggin Hill a Home Defence Squadron, a Night Flying Flight, and the A.A. School. Treasury sanction followed for purchase of all land held under DORA around the aerodrome. On 1 July, RAF Weekly Orders stated: 'The Director of Research, Air Ministry, will cease to exercise any responsibility for Biggin Hill aerodrome with effect from 1 July, 1922, and control of this station will be vested in the AOC in that area.' As a result, all experimental instrument and wireless work was transferred to the RAE at Farnborough.

Parachutes also had the attention of that establishment but were still low priority for the RAF, though Ernest Calthrop was vigorously developing his *Guardian Angel* with 28-ft silk canopy. Early in May he gave a private demonstration at Croydon 'with a new type parachute designed to lift a pilot from a machine falling out of control. He now uses a triple parachute – a very small one being first released, which pulls out a slightly larger one, which in turn exerts sufficient pull to release the large man-carrying parachute that finally lifts the pilot from his seat.' A dummy was used, attached to the shroud lines of the parachutes stowed in a container within the fuselage of an Avro 504K piloted by Capt Muir, but though the system functioned effectively it was too clumsy and heavy to attract serious interest of the RAF. There seemed no way of ensuring that the parachute would stream out effectively whatever the attitude of the machine, except by loosely folding and then inserting a circular diaphragm spring across the mouth of each parachute, but this took considerable space. Colonel H. E. S. Holt, inventor of the flare used for night landings, was nearer practical solution with his patented methods (No. 157,472 *et seq*) for rigging-line stowage and hairpin spring pilot parachutes than the indefatigably experimenting Calthrop.

Those responsible for deciding RAF technical equipment were more concerned with risk of fire than the safety afforded by parachutes. To that end the Air Ministry held a competition for safety fuel tanks, each of the 26 entries being tested and judged by the RAE. The onerous requirements were to withstand machine-gun and shell fire and the shock of an aeroplane crash without either bursting or leaking. In course of many months the participants had been reduced to a gas-armoured tank entered by Commander F. L. M. Boothby; a self-sealing rubber-encased tank submitted by Imber Anti-Fire Tanks Ltd; and a somewhat similar self-sealing tank produced by the India-Rubber Gutta-percha & Telegraph Works Co Ltd of Silverton who gained the first prize of £1,400. But designers looked askance at the added weight, and no mandatory requirement followed.

A certain amount of political underplay became evident in June when a contract issued in January for two Short Silver Streak derivatives was summarily cancelled on the score that Geddes had axed Bristol Fighter replacements. With a little prompting, the *Daily Mail* published an article about the Silver Streak, which caused questions to be raised in Parliament, and on 11 July *The Times* published verbatim the debate which followed. From information provided by the Air Ministry's technical department, the Secretary of State misleadingly created the impression that the Silver Streak structure was dangerous, saying it had been flown without load, and in that condition manoeuvrability was good when flown carefully. He stressed that the machine did not fit any specific military specification and was purchased to encourage originality and private enterprise. He spoke of vibration and corrosion, and said that a special sub-committee had been set up to consider such matters, but naturally it would take a long time for the necessary investigation to be made. Yet concurrently there existed a full and favourable report of the destruction test and 200-hour vibration trial made at Farnborough. When Viscount Curzon asked whether Shorts would be allowed to build further machines, the Secretary of State replied that he thought it inadvisable for them to do so.

'Now my reaction to this was immediately to start designing an all-metal hull for a standard F.5 wooden flying-boat,' Oswald Short told me. 'I took these designs to DTD and told him that I could make a watertight hull, but before I could explain about the experiments I had made upon a section of the Silver Streak fuselage for watertightness he said, "It's all very well you saying you can make a watertight hull, but what will happen if it is not watertight?". I replied sharply, "Then I will pay for it.' Finally I obtained an order by signing a letter to that effect, which I asked to be drawn up in the presence of Major Penny and Mr Launchberry, undertaking full responsibility at a risk of £10,000.'

Oswald Short's technical mainstay in backing the reliability of duralumin was C. P. T. Lipscomb, who had returned to the Rochester factory in 1921 after five years' extensive experience of stressing wooden and duralumin airship structures at the Bedford factory during which he had done much metallurgical research. Short now instructed Francis Webber to draught an all-metal version of the Felixstowe F hull incorporating a series of shallow fluted chines on the vee bottom to give a softer ride and prevent panting of the plates. To aid unbiassed assessment of constructional merits the Air Ministry gave a similar contract to S. E. Saunders Ltd for a competitive metal hull with standard F.5 wing structure.

6

The aircraft industry was vigorously continuing to press foreign markets, attaining modest success. In France, Vickers sold the enclosed cabin version of the Viking Mk IV to the French Navy, and now the Dutch Forces in the East Indies purchased ten with high compression Lions and R.A.F. 15 wings. Eventually the Vikings were criticized because of longitudinal sensitivity, for the lever arm, like that of the similar biplane tail on the Vulcan, was too small for the tailplane area. Snags of this nature still were not fully understood, and it certainly was not with knowledgeable misrepresentation that Vickers continued to offer this somewhat unsatisfactory machine. The fact that it was an amphibian had attraction for many countries. One built for the USSR was noteworthy because Sam Saunders had bought back the Vickers shares in his company, so with no further need to trade with a daughter firm, Vickers planked the Russian Viking with special SCT ('securely cemented together') plywood instead of Saunders Consuta. Sales began to mount. A Fleet Spotter Viking was ordered by the US Navy. Two with Rolls-Royce engines were bought by the Argentine's River Plate Aviation Co, followed by an order for four Lion-powered versions with T.64 section folding wings for the Argentine Navy. The Royal Canadian Air Force purchased two Rolls-Royce powered Vikings for forestry patrol, and Laurentide Air Services bought a similar machine for the Canadian lakes. Even Britain, as the country of origin, was showing cautious interest, for the RAF took delivery of two Viking Mk Vs of special specification and intended for development trials in Iraq, where they proved difficult to operate because the hulls often bottomed on the rough desert surface.

Whereas the Swifts for Spain (*illustrated*) had straight wings, those for Japan had marked sweepback to give greater stability.

Overseas maritime interest also favoured the Blackburn Swift. Two were purchased by the US Navy for comparison with their own torpedo-bombers; Japan had two Swifts, which were being evaluated against the Mitsubishi torpedo-carrier designed by Herbert Smith based on his Sopwith Cuckoo; and the Spanish Government bought four Swift Mk IIs.

Like most prototypes of that time, the Swift had encountered stability problems, for its design was largely empirical, based on performance of previous designs. Though Farnborough's wind-tunnel techniques were beginning to be applied to determination of stability characteristics and control hinge moments, there was still a gulf between small-scale wind-tunnel work and full-scale practice. To overcome longitudinal instability the straight wings of the Swift were given a 6 deg sweep-back to bring the C.G. forward. The greater stability then required larger elevators to overcome it, and in turn necessitated diagonal inset balances to give light and easy operation. It was a process of trial and error which took time and could bring hazards to the test pilot through making too big a change. Directional problems were more difficult. The original rudder was ineffective because the big fin held the machine as though on rails, and a still bigger rudder gave only slightly improved manoeuvrability. Martlesham technicians thought they could cure this by halving the fin height and fitting a rudder with unshielded balance; but their efforts merely provided new aerodynamic problems associated with rudder 'tramping' – which was a complex hunting defect. Finally George Petty, the assistant chief designer, contoured a much more pleasing fin and rudder, reversing the original relative proportions, and reverting to an oblique semi-shielded balance, but of greater area.

Of the several companies selling reconditioned machines, Handley Page with his Aircraft Disposal Co was having the greatest success, thanks to the administration of Colonel Darby at the London Office and Major Grant who supervised reconditioning at Waddon adjoining Croydon. Two years' trading had established a splendid reputation for thorough overhaul of airframes and engines, and the company was selling at a good profit, half of which was of direct national benefit, for it was returned to the Treasury. There were many satisfied customers and numerous repeat orders.

Handley Page also achieved success at a competition for commercial aeroplanes held at Brussels on 25 June in conjunction with the first Belgian Aero show. McIntosh and Foot took a W.8b, accompanied by H.P. himself, Savage his manager, and the dapper Cogni to push the prospect of sales. Tests were stiff, reflecting every aspect of operation, including single-engined flying – but the two transport pilots carried off first prize and a promise of an order of six for the Belgian Congo provided a third engine could be fitted to the W.8 as it was imperative to guard against forced landings in the the jungle. Although the W.8bs, and later W.10s with four more passengers, were recorded as 'brutes to fly, with heavy rudder and sluggish ailerons', the pilots of Handley Page Transport were coping, and began to like them well enough, for the twin engines gave greater sense of safety than the single engine of the D.H.18s and 34s.

Nevertheless it was a difficult annual general meeting for Handley Page Ltd that June, with Lieut-Col Barrett-Lennard presiding in his new position as chairman. 'Vast sums,' he explained, 'had been expended by the previous Board in an effort to establish markets for the company's aircraft in India, the States, South Africa, and South America. They had considered it desirable to set up missions on grandiose scale to create rapid demand for that comparatively novel article, the aeroplane; the company also spent a lot in trying to build motor-cars. There is no question that all this was an honest effort to employ the company's works and capital usefully and profitably. It is easy for me now, speaking in 1922, to say it was unwise. Although it has been urged upon me that the rapid post-war cancellation of Government contracts was a hardship, I cannot bring myself to regard it so; compensation was paid, and surely it is unreasonable to expect the taxpayer to go on paying for the construction of unnecessary aeroplanes for the sole benefit of shareholders? As a matter of fact, cancellation of Government contracts has had little to do with the present state of affairs.'

It transpired that the company had debts of £800,000. The £1 preference shares, which investors competed to buy when the company was converted from a private to a public business in 1918, were now worth only 6d. Earnestly Barrett-Lennard assured shareholders that the new Board not only would salvage something for the Bank of Scotland whom they represented, but would put the business on its feet again. S. A. H. Scuffham, last of the Handley Page public relations managers, recorded that: 'Handley Page sat silent and still, as aloof as a great expressionless Buddha throughout these proceedings. Although control had passed in large measure to men with predominantly financial background, the Board knew that only the technical acumen of the company's engineers could get their money back. They were glad to keep the founder of the business – and its presiding genius – as managing director. A long hard climb to solvency had next to be faced.' As a start H.P. was cultivating the friendship of Sir Sefton Brancker, the DCA, and preaching the gospel that separate routes should be allocated to each British airline – undoubtedly believing the most popular should go to the safest airline, which would be the one using twin-engined machines exclusively!

H.P.'s new chief designer, 'Star' Richards, had done remarkably well in completing Volkert's design of the Hanley torpedo-carrier, built to the same specification 3/20 (D of R Type 8) as the Blackburn T.2 Dart, the

Because of its slots the Handley Page Hanley was complex both mechanically and in behaviour compared with the pleasant characteristics of the Blackburn Dart, so it had little pilot appeal. (*Courtesy Air Marshal Sir Ralph Sorley*)

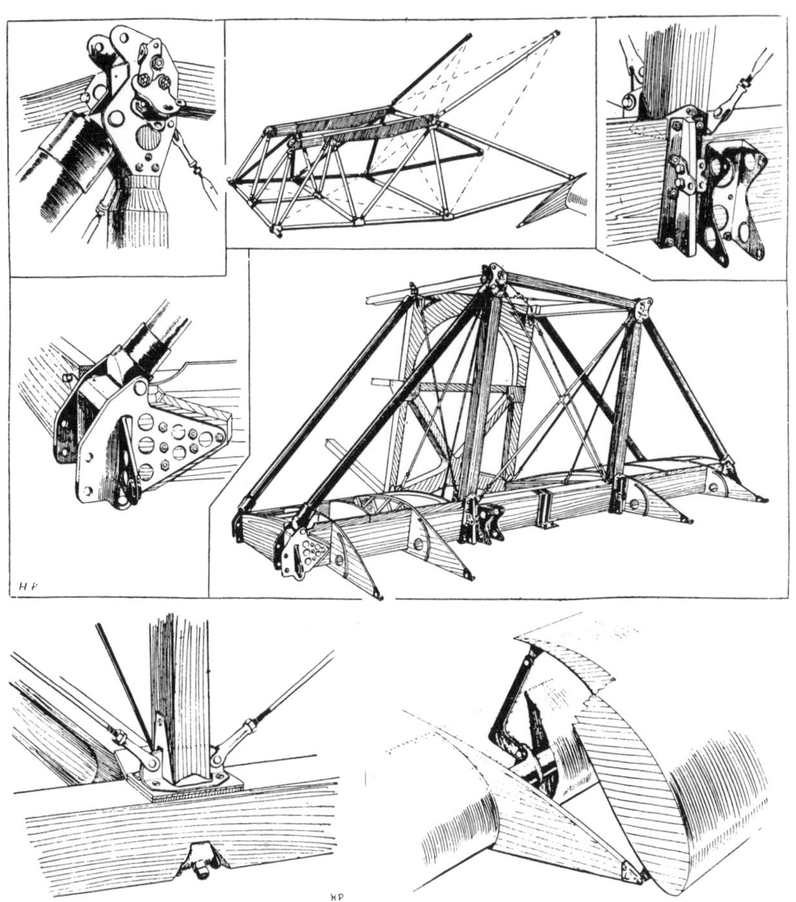

The Hanley was in the die-hard tradition of wood construction with simplest possible metal fittings. *Above*, the tube engine mounting and basic centre fuselage bay with details. *Below*, the crude spacer clip on rear fuselage, and the built-up wood slat and operating link. (*Flight*)

three prototypes of which had finished their Martlesham tests and were with the Development Squadron at Gosport awaiting comparison trials with the Hanley. Known as the 'Heinz' because of the variety of levers with which the pilot operated the full-span slots on top and bottom wings, the H.P.19 Hanley otherwise was a conventional tractor of practically the same dimensions as the Dart, though with higher aspect ratio. Both lower and upper centre-section were braced with splayed struts, giving the impression of an X in front view, the lower struts taking the oleo landing loads. When flown on 3 March by Capt Wilcockson, the undercarriage proved inadequate for landings at the steep incidence necessitated by the open slots, and a more complex but higher drag undercarriage, reminiscent of that of the O/400, was devised by Richards, using sub-axles hinged from a tall pyramid of struts beneath each wing to obtain a steeper ground angle yet present an easier strength problem than a longer version of the original split-axle. Richards personally devised the many ingenious mechanical features, particularly the wing folding which entailed breaking the slot operating mechanism in which 'the ball in the centre-section operating tube has a pin through it, the ends of which engage in the slot in the wing tubes, thus transmitting the torque. When the wings are folded the ball comes out of the socket, and on spreading the wings there is automatic engagement.'

Because of the slots, the loading of the Hanley had been increased to 11 lb/sq ft for the same landing speed of 38 kt as the Dart, which had a wing loading of 9 lb/sq ft and had proved delightfully easy to land on an aircraft-carrier. What had not been obvious to the H.P. technicians was the nature of the big change in trim when the slots were opened. This caused risk if the engine was opened on the approach, for the machine then tended to stand on its tail, despite a hard forward push on the control column, and could result in staggering into the air with ineffective ailerons because of the slow speed, followed by a wing drop and crash. Nor did the pilots like the steep gliding angle produced by the drag, for it necessitated excessive angular rotation to attain the big incidence required for a slotted landing.

The eventual Blackburn Blackburn Mk. II had a raised centre-section, and the trainer version's side-by-side nose cockpit overlapping the sides made it appear as bovine as its 'Bull' soubriquet. (*Courtesy Air Marshal Sir Ralph Sorley*)

Largely because of this the simpler Dart was chosen as the Fleet Air Arm's standard single-seat torpedo-bomber, gaining a contract for twenty-six even before starting comparative trials with the Hanley. It was therefore with considerable hope that Blackburns were pressing on with their bull-nosed Fleet Air Arm reconnaissance biplane to specification 3/21 (D of R Type 7A) based on their Dart experience. Unimaginatively named Blackburn, the prototype, N150, was flown by the company's test pilot, R. W. Kenworthy, early in the summer, and now was at Martlesham; but there was criticism of the stumpy undercarriage and complaints of the draughty and uncomfortable pilot's cockpit, due to wing-suction behind his head – though as all-round view was imperative, a seat in front of the wings was essential.

Its rival, the Avro Bison, with identical pilot location, had already received a development production contract of twelve for an urgent Naval coastal reconnaissance requirement. After modifying the prototype with auxiliary fins each side of the tailplane and fitting a larger rudder, directional control seemed at least temporarily acceptable pending trials with the second prototype which had a long dorsal fin, and wing-gap increased 15 in by lifting the centre-section on short struts to the customary position clear of the fuselage. Soon it was obvious that the Blackburn would have to be similarly altered.

7

Summer! – British pilots visit Le Bourget with four days of racing, demonstrating and socializing organized by the president of the Vieilles Tiges; at Croydon a race meeting where Jimmy James showed off the 'Bamel' re-named Mars I, Raynham disported with his Sopwith Antelope, and Cyril Uwins arrived with a scarlet war-time Bristol monoplane to which Fedden's three-cylinder Lucifer had been fitted, and Sir Sefton Brancker was there to radiate *bonhomie* and award the prizes; with more serious purpose the first Vickers Vulcan made its initial run on the Paris service; Geoffrey de Havilland test-flew Alan Butler's D.H.37 named *Sylvia.* But for the sporting British public even Ascot and Henley paled to

Lined up for the Aerial Derby, Alan Butler's D.H.37 *Sylvia* with passenger cockpit covered to match the Martinsyde F.6 which Fred Raynham had bought for £15. (*Flight*)

insignificance compared with the third annual RAF Pageant on 24 June, though it rained and rained and the enclosures became a sea of umbrellas. The great display maintained its standard spectacular pattern, with demonstrations by such pilots as Flt Lieut P. W. S. Bulman, Flg Off E. R. C. Scholefield, Flt Lieut D. W. Grinnell-Milne; there was Flt Lieut L. W. H. Longton with his crazy flying, and Sqn Ldr Roderic Hill piloting a D.H.10 in battle against two S.E.5as; and among the picked men in the races and formation flying were such pilots as Flg Off H. J. T. Saint, Flg Off J. S. Chick, Flt Lieut A. H. Orlebar, Flg Off C. E. Maitland, Flt Lieut J. M. Robb, Flg Off J. R. 'Pat' King, and Flg Off F. le Poer Trench. For technicians disinterested in ex-war but still active S.E.5as, Avro 504Ks, Bristol Fighters, and Sopwith Snipes, the innovation of a New Type Aircraft Park drew them to study the Blackburn Dart, the Handley Page

That final satisfying bang of the huge set-piece at the RAF Pageant at Hendon. (*Flight*)

Hanley, Parnall Puffin amphibian, a Lion-powered Supermarine Seagull, the Westland Weasel with 380 hp Jupiter and new ailerons, and towering above all, the impressively huge Rolls-Royce Condor powered Avro Aldershot whose very experienced pilot was crusty little Flg Off C. E. 'Tiny' Horrex from Martlesham.

Like many of that generation of prototypes, the Aldershot had had its troubles, as Bert Hinkler discovered when he flew it earlier in the year, for the rudder was impossibly heavy and its effectiveness marginal because of excessive directional stability due to the long dorsal fin. After Avros removed the fin and extended the fuselage by 6 ft, Martlesham considered it sufficiently acceptable to re-power with the 1,000 hp Napier Cub sixteen-cylinder X engine which the Air Ministry had ordered in 1919 at a cost of £60,000 for six development units. To install so powerful an engine, and make so big and heavy a machine, was considered a breathtaking achievement looking far into the future. If this was the trend, then aircraft were going to be expensive and present still greater problems. There was speculation as to whether the smaller but equally heavy D.H.27 would prove a better proposition, but when completed in September its structural weight was greater than expected, and flight trials revealed inconsistencies in handling, unsuspectedly due to flow-interference at the thickened centre-section.

The prototype Avro Aldershot, J6852, made its public debut at the RAF Pageant after considerable modification, and then won the Handicap Race. (*Flight*)

Rival of the Avro Aldershot was the neater looking D.H.27, the Derby, photographed at Stag Lane – where the original wooden offices are revealed above the number on the fuselage.

On 7 August the Aerial Derby was held. Once again it was a disappointment, reported as 'even duller than last year's, and there prevailed an atmosphere of indifference and lack of interest over the whole proceedings that made one yearn for the ordinary Saturday afternoons at Hendon in the good old days'. A new hush-hush Jupiter-powered Bristol monoplane racer with stubby wings, barrel-like fuselage and retractable undercarriage was not ready; Alan Butler's engine in the attractive, functional D.H.37 developed magneto trouble, so he started 10 minutes late but retired after a lap. Leslie Tait-Cox, with an historic B.R.2 Nieuport Sparrowhawk Mars III, retired; and Rex Stocken flying a Martinsyde F.4 landed at Hounslow with engine trouble. Jimmy James won with Folland's Mars I at a speed of 177.85 mph, Rollo A. de Haga Haig in the Bristol Bullet was second and

Barnwell had opposed an out and out racer, but, when he left the Bristol Aeroplane Co, Fedden and Reid collaborated on design of a fearsome barrel-like retractable undercarriage monoplane of 25-ft span.

veteran F. P. Raynham third with his yellow-painted Martinsyde F.6. Larry L. Carter, flying a Lucifer-powered Bristol M.ID monoplane, won the Handicap.

James provided the only exictement by coming in too fast, so that 'the machine ran some considerable distance, getting nearer and nearer a crowd of people standing on the aerodrome. It looked pretty hopeless, but with marvellous skill and judgment he turned the machine carefully to the right at the precise moment – a second sooner and he would have been over, and a second later would have been into the people. Thus he just missed the crowd, and all the damage resulting was a ripped tyre and leaky radiator.' But the win was an eagerly desired triumph for the delighted Gloucestershire team as a step towards military sales. Their sense of eminence was increased when Flt Lieut Bulman aerobatted a Nighthawk 'starting with a zoom such as we have seldom had the pleasure of observing before, he came out with a half loop, and then put up some really beautiful loops, spins, rolls and corkscrew twists, concluding with an example of fast and slow flying past the enclosures'. Both RAF and industrial representatives were beginning to regard him as one whose career should be watched. He was a quietly drawling, faintly smiling little man, known as 'George' because in war days he addressed everybody by that name. Under leadership of Sqn Ldr R. M. Hill at the RAE he had recently concluded with his co-pilots Flg Offs Sainsbury and Gerrard a prolonged investigation of inverted flying manoeuvres.

More important, and holding the attention of the world, was the attempt by Supermarine to win the Schneider contest at Naples with a special Sea King single-seat flying-boat flown by their test pilot H. C. Biard. It was crucial – for if the Italians won, the Trophy would become their permanent property. Their only opponent was Britain because France's C.A.M.S. flying-boats were not completed in time even for the preliminary navigating and mooring tests. Italy had Passaleva flying a 300 hp Macchi M.7 flying-boat, and Zanetti with a 450 hp Savoia S.19 flying-boat. The Macchi should have been disqualified, according to the rules, for in the six-hour mooring test it capsized and had to be righted by her crew, and in the navigability test, split its propeller; but the bulky managing director of Supermarine, Hubert Scott-Paine, known as 'Red-Hot' from the colour of his hair, sportingly declined to lodge a protest.

On 12 August, under a blazing Neopolitan sky, with a mill-pond sea presenting no take-off problems, the contenders in turn thundered round the course. With its boosted Lion giving some 550 hp, the Supermarine, derived from the Sea King Mk II amphibian, with wing area reduced and stripped of its undercarriage akin to the original Sea Lion of the 1919 contest, proved marginal winner, completing the 200 nautical mile triangular course in 1 hr 34 min 51$\frac{3}{5}$ sec during which Biard made 39 turns, averaging 127 kt though capable of 150 kt straight and level. Passaleva was second in 1 hr 36 min 22 sec, and Zanetti third in 1 hr 38 min 45 sec. With the Trophy saved for another day the next contest would be in

The Supermarine Sea Lion II saved the Schneider Trophy for England to contest in later years, but was another classic instance of expediency by converting an unwanted fighter flying-boat. (*Flight*)

England – and it seemed a chance for a seaplane version of the Gloucestershire Mars I rather than hope to make a faster flying-boat.

Maritime international racing could hold the world's Press, but at home it was the Circuit of Britain Race for a Cup presented by HM King George which attracted the enthusiasts with 21 entries ranging from 35 hp single-seaters to twin-engined aircraft – piloted by civilians, RAF officers, and airline captains. The course was too long for special racers of limited fuel capacity, but the entrants represented a typical cross-section of the day. Starting from Croydon at 9 am on Friday 8 September, there were compulsory stops of 1½ hours at Birmingham and Newcastle, with an overnight stay at Glasgow before recommencing next morning with a downward leg having stops at Manchester and Filton, thence to the finish at Croydon. By this means people would have widespread opportunity of seeing the flying and even participating in spirit; at every turning point they crowded to the aerodromes in thousands; not even the RAF Pageants had aroused such interest, for a radio network organized by the Air Ministry enabled Press and public to be constantly and reliably informed of progress.

Describing the finish, *Flight* reported: 'Just before 3.45 pm a speck appeared to the west of the aerodrome, rapidly increasing in size. Through powerful glasses it could be seen that the machine was the D.H.4A, and as no other was in sight it became evident that the race had been won by Capt F. L. Barnard of Instones. With a banked right-hand turn he swept across the finishing line, going a great pace down wind, and after circling the aerodrome, landed and taxied towards the Judges' tent in front of the enclosures, greeted by loud cheers. Meantime another machine was seen to approach, which soon turned out to be Freddie Raynham's yellow Martinsyde F.6. Crossing the finishing line at high speed he swung into wind and alighted, taxi-ing rapidly behind Barnard and coming to a standstill by his wing-tip. While Barnard was being congratulated, Alan Cobham, flying a D.H.9B, came third, and Frank Courtney was fourth with his two-seat Siskin.' At intervals the remaining seven competitors followed, for the rest had been eliminated by mishaps.

The Siskin particularly interested the industry's technicians, for its original had been dismissed as a wartime 'also ran' which had been eliminated by development delay of the Jaguar engine. But this machine was different, for it was not framed in wood but was of all-steel construction, and had wings as described in Patent 210,591 which was subsequently issued in December to Armstrong Siddeley Motors and their advisory expert H. N. Wylie for steel box-spars of thin gauge crinkled steel with a small spanwise tube at its centre to which fittings could be attached, and this was followed by Patent 212,006 showing a method of clamping ribs to the same spar. Lloyd still was primarily responsible for overall design, and particularly associated with methods of securing the sockets of the steel-tube fuselage, though he had been somewhat diverted from aircraft work by months devoted to a ciné apparatus invented by a Mr Bowell which he described as a 'searchlight projector' employing a shutter in the form of an eccentric Maltese Cross. Keen salesmanship by John Siddeley ensured a sequence of 200 of these devices, helping keep the aircraft division going while the big Awana troop-carrying prototypes were being built. Like Volkert of Handley Page, 34-year-old Lloyd had been a student in 1911 at the Northampton Institute where Handley Page had lectured, gaining subsequent experience at Shorts before joining the war-time Royal Aircraft Factory where he was in charge of stressing from 1914 to 1917. Judging by his all-metal Siskin it seemed that the Armstrong Siddeley Co had leapt ahead with small aircraft, for all other fighters were of wood.

Nevertheless, supreme in this type of all-metal space-frame structure was John North at Boulton & Paul. His four-engined Bodmin 'Postal'

Frank Courtney dismounts from the two-seat Siskin having displayed its prowess, though failed to win the Aerial Derby. (*Flight*)

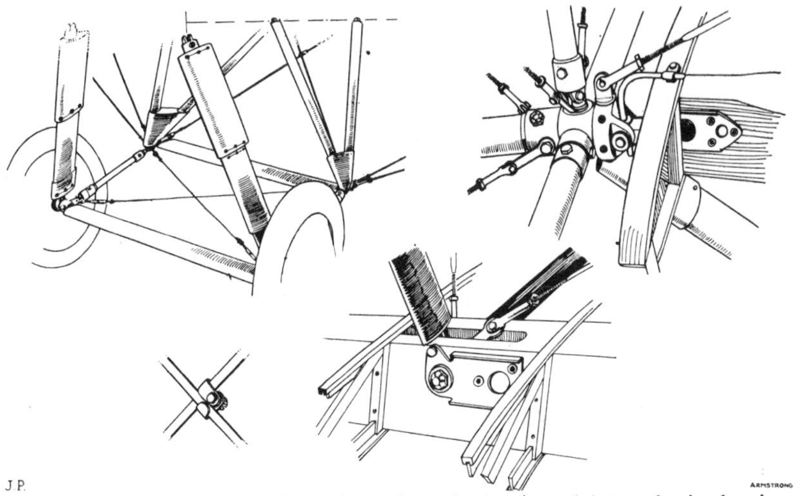

Details of the Siskin long-stroke undercarriage, fuselage/spar joint, and wing-bracing centred within the spar. (*Flight*)

biplane had at last made its initial flight in the hands of Courtney, after much tedious development of the transmission system of the internally located engines. 'I retain vivid memory,' he wrote, 'of Martin, the inspector, busily tending the mechanisms in the engine room of the Bodmin

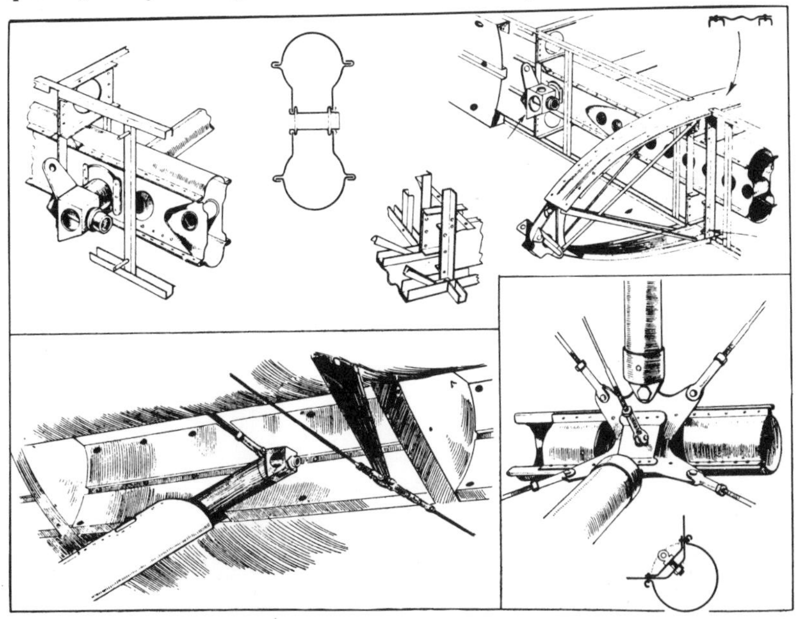

John North at Boulton & Paul quickly took the lead with strip steel construction. Interplane bracing and tailplane struts (*lower left*) were secured to an overhung tube passing through their spars. (*Flight*)

whilst, low on fuel, I was trying to get back to Mousehold aerodrome in a thunder storm.'

Although transmission and cooling difficulties continued to plague John North, he was confident of the eventual success of his costly metal development work, for he had gambled just after the war that if Guy ffisk, the chairman, would transfer the entire aircraft department to Mousehold he would guarantee to lose no more than £5,000 per year in the subsequent six years, after which the firm would be running with full effectiveness. On that basis he agreed that apart from a few hundred as basic pay his emoluments would be $\frac{1}{2}$ per cent on profits of everything designed within the next 14 years. The potential seemed promising, for the RAF had appreciated the Bourges sufficiently for a development contract to be placed for an all-metal version known as the Bolton with span increased from 57 ft to 62 ft 6 in, and the Air Ministry was convinced that elaborate thin steel structures of this nature were particularly suited for British aircraft because of indigenous steel availability – an outlook to which their current administrators had been conditioned largely by the major patentees, Mooney of Gloucesters and Wylie of Armstrongs, while they were serving officers at the Air Ministry's technical department during the war.

One of those with predilection for wooden aircraft was Tom Sopwith – for his war-time experience emphasized how quickly new designs constructed with wood could be built, tested, and abandoned if need be. His chief designer, Tommy Thompson, was tackling a wooden structured, two-seat semi-cantilever parasol monoplane to Air Ministry specification No. 7/22 for Corps reconnaissance duties where downward view was important. Named the Duiker, it bore close resemblance to the high-winger designed by the unconventional Grover C. Loening and L. R. Grumman of New York for the US Air Force, but was intended to have a variable-pitch propeller and Fedden's proposed supercharged Jupiter.

The Hawker Duiker parasol prototype, shown at Martlesham, was not popular, for it was directionally unstable, and on one flight sudden aileron flutter caused a centre-section strut fitting to fail. (*Courtesy Air Marshal Sir Ralph Sorley*)

H. A. Mettam, MA, on graduating at Cambridge that summer, was engaged by Thompson as stressman at £250 per annum. He recollected: 'With blue suit and red beard, Thompson's appearance left no doubt that his rank had been gained with the Royal Navy, and he was a very pleasant person for whom to work. His design staff consisted of very few draughtsmen – memory suggests never more than ten – and just one stressman: myself. There were so few that we left wide open spaces in that long narrow drawing office at Canbury Park Road. Thompson occupied the glass "hutch" which divided the main area into two unequal parts. The chief draughtsman, L. E. Metcalfe, who normally answered the only telephone himself, had the larger part for the aircraft draughtsmen, including F. Cross, R. H. Shaw, Joe Barrett, and young L. P. Pankhurst. I shared the smaller part with Kit Atkinson, who was looking after the last motor-cycle to be designed by the firm, and with Capt A. S. Ellerton, a founder member of the AID, who had been specially engaged to develop the variable-pitch propeller. Another desk was reserved for visiting stressmen from the RAE because it was they who had to check the strength of all new aircraft.

'Of the directors, Sopwith was an infrequent but popular visitor to the D.O. Sigrist did not come often, but when he did he used an effective but irritating technique known as the "Sigrist Principle". At one board he would say: "What is that made of? Duralumin? Why isn't it steel?" When opportunity offered, he would reverse the query and ask why some other part was not duralumin – a material not very well known at that time. I remember Bennett arriving with two 16-gauge dural sheets and calmly crumpling them as if they were paper, explaining they were no use as the only heat treatment man who understood the metal was on sick leave. There was also the episode of a rudder bar for which, as a weight-conscious stressman, I had carefully calculated the size of its vertical torsion tube. Sigrist saw the assembly, stuck the lever in a vice, put his hands on the foot rest, twisted the tube through a most improper angle, and sent it to the office with a message quite as improper. Nor did he spare himself, for he arrived with the workmen, and insisted that staff of all ranks should "clock in" every morning: more than five minutes late any one week brought a warning, and three warnings meant the sack.'

8

Spurred by the novelty of an engineless flight competition held by the Germans in the beautiful Rhön mountains in 1921, and repeated this following August with still more marvellous soaring, the *Daily Mail* offered a prize of £1,000 for the aviator of any nationality making the longest duration gliding flight in England during October, provided it was not less than 30 minutes.

Eleven years earlier, on 24 October, 1911, Orville Wright had shown what the future might hold when he soared his crude biplane glider almost 10 minutes above the sand hills of North Carolina, but now Germany's leadership came as a revelation touched with magic, for Herr Martens with one hour, and Herr Hentzen with two, made the first long duration gliding flights in history with the clean cantilever Hannover sailplane named *Vampyr*. Of the first competition, *Flight* had breathlessly observed: 'If we had been told only a few months ago that it was possible to remain 15 minutes in the air on a machine *sans* motor we would have been mildly sceptical to say the least. How far these experiments in gliding are likely to carry us it might be unwise to prophecy, but there is no doubt they carry us some way along the road to solution of soaring, though this may never be solved satisfactorily, if only because of the impossibility of finding a mechanical substitute for the instinctive methods of birds. The Germans have already accomplished a great deal. They have proved it possible, by taking advantage of gusts and varying wind currents, to remain far longer in the air on a motorless machine than the first aeroplane to fly. He would be a bold prophet who would say definitely and dogmatically that the time will never come when man will be able to remain soaring for hours on end.'

Klemperer's Aachen glider was a trousered low-winger and the one illustrated was used by J. Jeyes at Itford. (*Flight*)

On that same occasion *The Aeroplane* reported that Mr F. Handley Page had been one of those who hastened to view the gliding. 'He confirms the belief that the German study of this branch of aeronautics is a serious scientific undertaking and points out that their leading aeronautical scientists, such as Professor Prandtl of Göttingen, take a deep interest in the tests, and that a number of the machines entered were built by students of technical institutes, under superintendence of their professors. Mr Page witnessed a flight by Professor Klemperer of the Aachen Technical School in a wind of about 25 mph using the Aachen low-wing cantilever monoplane designed by von Kármán. It was launched by a crew of eight, two of whom held the wing-tips, while three, at each end of a long elastic rope round the heel of the skids, walked forward till the rubbers fully extended; then on the pilot's signal the men at the wing-tips let go and the machine slid over the ground, and at the end of two metres was in the air and within the next second was 30 ft up.'

Said C. G. Grey: 'The most successful German machines are thick-winged cantilever monoplanes, generally loaded to some $1\frac{1}{2}$ lb/sq ft, and the distinguishing aerodynamic feature is their slow rate of descent in still air. One showed the wonderful gliding angle of 1 in 19, but as its wing loading was more than usually heavy its speed was fairly high, and despite its fine angle of descent the minimum rate of drop was higher than more lightly loaded craft of inferior L/D ratio. It is believed that in the present state of aeronautical knowledge, with a year's careful experiment, and an expenditure of not more than £400 to £500 on construction, it would become possible to make flights of 10 to 20 miles in favourable conditions.'

What H.P. had more craftily done was to meet Lachmann, with whom he had been corresponding ever since that rival German patent of the slotted wing. Hugh Scuffham, of Handley Page Ltd, later recorded: 'The encounter was promising, and the two engineers got on well together. The Englishman had a folder full of information, and talked slots incessantly. But they still found time to see Berlin and enjoy its night life, in itself a new experience for Lachmann who had just arrived there after finishing at Darmstadt and joining the Opel car works in the absence of any available career in aircraft engineering. Handley Page was certainly an astute man. He proposed to his relatively green companion an arrangement of considerable advantage to both. It was a three year agreement whereby Lachmann, for a salary, would carry out slot experiments in the Göttingen wind-tunnel at the British company's expense, sending regular progress reports to Cricklewood – but with inflation beginning in Germany, it was typical of H.P. that he would only have to make relatively small outlay in sterling!'

In the 1922 competition, the German students and pilots achieved still greater triumphs at the Wasserkuppe – a small mountain of some 3,000 ft in a countryside bare of trees and obstructions. Again Martens and Hentzen made outstanding flights with the 42-ft span Hannover glider they used the previous year. Designed by Dr Madelung, it was a break-through in design, featuring a novel single I-section spar with web of lattice strips, the wide plywood leading edge forming a flattened tube with tremendous stiffness against torsion and needing no rear spar. The fuselage was equally unique with its flattened diamond shape and ingeniously fitted football shock absorbers projecting from the central portion and rendering a conventional wheeled undercarriage needless. But it was as the first practical application of the new principles explaining 'induced' drag that made these German gliders important, proving that the longer the wing span the flatter the gliding angle. It was a lesson not lost on Geoffrey de Havilland, or Capt Sayers of *The Aeroplane*, for they also had been at the Rhön competitions, and both were now designing gliders for the *Daily Mail* competition. The Royal Aero Club had selected Itford on the Sussex hills as the venue, and had received thirty-five entries including a number of oddities such as a pedal driven helicopter.

At this point Britain again met discouragement in the international field of air racing, for in the Coupe Deutsch on 30 September, the Mars I flown by James made disastrous showing. After taking-off on his first lap he was presently seen returning well off course and landed, throwing away the race because his maps, stuck on pieces of plywood suspended from his neck with string, inevitably caught in the slipstream almost throttling him until he broke them free and they whirled away. As one of the onlookers disappointedly said: 'There could be no doubt that James would have stood a very good chance of winning the race. As to the happy-go-lucky spirits which trust to tying maps with string in a machine doing over 200 mph perhaps the less said the better.'

Less predictable was the elimination of France's hero, Sadi Lecointe. He returned with engine trouble, and the propeller stopped as the machine touched down in a series of bounces. 'Sadi appeared to succeed in steadying it, but while still running at great speed it struck a rut, swerved and turned over. A groan went up from the thousands of spectators, and there were cries of "Sadi! Ah, Sadi!" In spite of soldiers with fixed bayonets thought necessary to keep back the French crowd, the multitude broke from the enclosure and streamed at full speed towards the overturned machine. When they lifted the tail, Sadi snaked from the diminutive cockpit and stood up and waved his hands, safe and sound.' The popular Italian pilot Brackpapa also retired with his 700 hp Fiat malfunctioning, and Jean Casale, piloting an I-strutted biplane designed by André Herbemont, retired with similar trouble. It was Sadi's modest compatriot Lasne who won with a pretty little white-painted Nieuport-Delage biplane at an average of almost 180 mph. That the development of racing machines was heading towards a serious and expensive business was rammed home a fortnight later by the Pulitzer race in the USA which was won by a beautifully proportioned special Curtiss I-strutted biplane entered by the Army Air Service and piloted by Lieut R. L. Maugham at a speed of 206 mph over a 160 mile triangular course, racing against other advanced military and naval sponsored prototypes. Though out and out racers, they were built ostensibly as exercises in spearheading future fighter design and their cost therefore was not questioned by the taxpayer.

Stealing a march on the Itford gliding competition, four days before it opened on 16 October, the South Coast Land and Resort Co, promoters of Peacehaven Estate on the South Downs near Newhaven, featured Anthony Fokker, whose name was still ominous to many a Frenchman and Briton, demonstrating the new sport of gliding with his little single-seat biplane before a great crowd of Press and spectators, among whom was Sir Sefton Brancker and bluff Commander Perrin, secretary of the RAeC. The resulting publicity for Peacehaven encompassed the entire British Isles!

When the British gliding competition opened the following Saturday, most competitors were still putting finishing touches to their craft, though several had flown. First away was Raynham, with a 36-ft span monoplane designed by George Handasyde assisted by that same somewhat shy and

Raynham's 36-ft span glider is carried up the hillside at Itford from the competitors' Bessoneaux hangars sheltering below. (*Flight*)

sarcastically defensive young man, Sydney Camm, who had proceeded from storeman to draughtsman at Martinsydes, and in the past year had been drawing the structural details of Handasyde's high-wing passenger monoplane originally intended as an alternative to the similar D.H. monoplane. So hurried had been construction of the glider that Raynham skipped connecting the ailerons to the control column, and instead operated the balance wire directly with his left hand. A short hop showed he could manage, but the rudder was inadequate, and that evening a larger one was built. Gordon England, who had flown the Weiss glider in 1909, found the same snag when he tried a pretty and very small monoplane of his own design and construction. When next day M Barbot cart-wheeled his Dewoitine monoplane through stalling on take-off, it was clear that difficulties could be serious. By Monday seven others had arrived: the Aachen low-winger replica of Klemperer's; the Sayers monoplane similar to the Hannover *Vampyr* but built by the Central Aircraft Co in what proved their last venture; a wire-braced monoplane of somewhat similar configuration built by the famous instructor Merriam, and Newman the works manager of S. E. Saunders; two de Havilland gliders characterized by 50-ft wire-braced wings having the highest aspect ratio of any competitor; a crescent-winged 'Airdisco' monoplane designed by John Bewsher; and two Fokker biplanes. It was Fokker on the bigger of these, with Bewsher as passenger, who showed how soaring should be done. 'Instead of flying out from the hill, as all previous pilots with the exception of Jeyes on the Aachen had done, Fokker hugged the edge of the range, letting his machine proceed crab fashion,' wrote an enthralled visitor. 'It rose and fell as it got into and out of the ascending currents, and for quite a long period appeared standing still. Fokker then turned and commenced the return journey, still at greater altitude than the starting point. Coming back towards Firle Beacon, losing height gradually, he made a right-hand turn into the valley, and alighted smoothly having been in the air seven minutes, which was an extremely fine performance.' Appreciating the lesson, Raynham and Gordon England respectively managed 11 min 23 sec and 4 min 32 sec. Presently Fokker took-off again and achieved 37 min 6 sec amid tremendous enthusiasm. Among scores from the aircraft industry could be seen tall Dick Fairey with Colonel Nicholl; Colonel Ogilvie accompanying Handley Page who was hobbling on crutches; Alliott Roe excited as a schoolboy; Colonel Darby of A.D.C; Howard Wright, doyen of aeroplane designers; and in one of the four canvas Bessoneaux hangars was de Havilland supervising erection of his two gliders. Commiserating with Fokker on the decision of the Chambre des Industries Aéronautiques to ban him from the next Paris Aero Salon in December was Charles G. Grey, that scathing editor of *The Aeroplane*, with quiet Danish-born C. M. Poulson, editor of *Flight*, adding a friendly word.

It was on the last day, Saturday 21 October, that I managed a lucky lift to the gliding site. The thrill! Men, women, children and dogs, were streaming up the hill-side at Firle Beacon where an out of season north-easterly wind was blowing up the steepest slope – and lo! there was Raynham about to be launched in that gossamer-like, translucent-winged creation of Handasyde's. Helpers hooked 'Sandow' bungees to little blocks under the lower longerons, and six men a side stretched it into a V. At a signal from Raynham they struggled forward running, catapulting the glider into the air. With a sibilant whisper it banked left-handed and slid away eastward, slowly dropping. An even smaller monoplane was now ready, and I recognized Gordon England as he climbed into the miniature

That pioneer pre-war pilot-designer, Gordon England, built his 28-ft span miniature glider in a stable in a few weeks. (*Flight*)

A quick camera, at the Itford meeting, catches a most amazing picture as the 50-ft wings of the elegant de Havilland glider simultaneously collapse under the impulse of a 'bungie' launch.

cockpit, cut, like Raynham's, into the wings between the spars. But this time it was disaster. There was a swift rush upward, and almost immediately a wing dropped as though stalled, and the machine dived to the ground with a splintering crump. Men rushed to it. England was dragged out with compound fractured ankle – but the ambulance was nowhere to be found, though it had been on the site the day before. Officials tried to contact the local hospital, but it was an hour before help arrived.

This was not the only crash to follow the Dewoitine. In a hangar was the wreck of the de Havilland glider which had twisted off both wings on launching after modification of ailerons to warping, but luckily it parachuted flatly to the ground without injury to the pilot, E. D. C. Herne of Daimler Airway. Only de Havilland had appreciated the value of extreme aspect ratio in reducing wing drag, but was caught out by the problem of structural rigidity, for despite designing to a factor of 8 in bending, vital aspects of torsion and flutter had not been considered.

Alongside the wreckage was the Aachen, which Jeyes had stalled on launching, the right wing striking the ground and breaking off, turning the glider upside-down; an RAF mechanic sprinted to it, raised the fuselage, and the pilot stepped out unharmed – but his rescuer fell flat on his back, utterly exhausted. Earlier still Merriam had crashed, stalling and cartwheeling the moment he was launched.

While Gordon England was lying on the ground, making jokes to reassure people, I saw Fokker's two-seater carried bodily on an open Chevrolet touring car to the spot where Raynham had landed; soon it was launched, piloted by light-weight Gordon Olley with passenger aboard. Lifting, dropping and rocking, it slowly edged along the hill while Fokker in his car shouted 'Oop igher: oop igher' – though it was all Olley could do

Fokker, the 'Flying Dutchman', was one of the great gliding enthusiasts, both at the Rhön meeting and Itford. His two-seat glider is seen here. (*Flight*)

to stay above the hilltop, but for 45 minutes he struggled to continue. Suddenly the glider dropped and disappeared. There was a gasp from the crowd. We rushed to the edge of the hill – and there was the machine calmly descending to the valley, where it landed by a clump of trees having achieved a world record flight with passenger. But disaster followed. In retrieving the machine, balanced on a plank across the Chevrolet as it climbed the tortuous road up the hill, there was a sudden crack, and the glider toppled backwards and was smashed.

By now Raynham was ready for another try, but waited until the wind was less vicious. Meanwhile a strange, drab little glider was being trundled by several men along the hill ridge from Itford to Firle, and soon was revealed as a tiny 22-ft equal-span tandem monoplane which had been designed by M Peyret, the heavily moustached, bulky man who was helping to push it. Was this one of those freaks of which I had been told – a revival of the Langley dream with thin wings that would probably collapse in similar manner? The wind was strong as ever and extremely gusty, but the Frenchmen were preparing to launch from Firle. The smallest of them, M Maneyrol, climbed into the cockpit behind the forward wing, ducking under a bare central top longeron. With a swish the quaint little tandem shot into the air and began tacking away and back in endless beats, keeping westward of Firle, lifting higher and higher until at a greater altitude than any previous competitor. He was like a bird soaring, confident and supreme, making light of the gale, using the up-current's buoyancy with easy skill.

Raynham, perceiving his own record of 1 hr 53 min might get beaten, was hurriedly launched and turned in the opposite direction to Maneyrol,

The Frenchman, Maneyrol in his compact 22-ft span Peyret tandem monoplane is launched at Itford on his prize-winning flight. (*Flight*)

going down-wind towards Itford, but failed to hold height, and after 8 minutes landed at the foot of the hill.

A heavy-looking craft was being tugged up the hill. Regarded as one of the jokes of the meeting, it comprised a Bristol fuselage bought for 5s, and a Fokker D.VII top wing for another 5s. Simple carpentry had fashioned a rounded nose, and a further 8s 6d for dope completed the composite 'Brokker'. Arriving at Firle, the RAF boys went through the routine of stretched elastic, running mightily to launch the machine. The wheels automatically fell away as it took off – but with a great thump the Brokker dropped to the ground after a record 2 second flight. They tried again. To the amazement of all, the glider, piloted by Sqn Ldr Gray, soared majestically into the air, turned right, and began to beat along the hill between us and the Peyret, the two turning away each time they neared, and the pilots would wave to each other. In the rasping wind as we watched from those Downland heights above the vista of quiet English fields, it seemed a miracle that these two engineless creations were sailing the sky as though this was the way man had always flown.

Will it fly or not? The crowd at Itford gather in expectation to watch the first launch of the RAF's hybrid combination of Fokker wing and Bristol fuselage. (*Flight*)

A cheer went up. Maneyrol had equalled Raynham's duration. The Frenchman waved acknowledgment. Steadily the two gliders continued lap after lap of their hillside beat. Dusk began to shroud the landscape. Presently the gliders were mere silhouettes. Commander Perrin, famous as 'Harold the Hearty', fussed around, arranging cars to light the top of the hill. When the German record of 3 hr 10 min was reached, another tremendous cheer and hooting of motor horns broke the evening quiet. For a further 20 minutes Maneyrol and Gray maintained their soaring, then in turn landed on the summit of the ridge close to their starting point. The Peyret with 3 hr 21 min 7 sec had won the *Daily Mail* prize of £1,000, and the Brokker with 1½ hr gained the Royal Aero Club's £50 prize. Raynham was awarded £50 given by Lieut-Col Ogilvie, and there were lesser prizes for Olley, Gordon England, and Rex Stocken.

Everywhere the lights of towns and villages were patterning the darkness as we descended the hill and made our way to Lewes by car, motor-cycle, bicycle, or on foot. A few saw in this miraculous gliding the beginning of a new sport, though most thought in terms of fitting a small engine to give cheap flying. The little Peyret tandem, with its attractive compactness and simplicity, no longer was a freak but was regarded as the genesis of an ultra-light aeroplane in the belief that novice pilots would find it foolproof because if the front wing stalled then the rear wing would tip the nose, regaining speed immediately. The hopeful proposed an engine of some 12 hp giving 50 mph and an all-in cost of £150 to £200, so that it became the 'poor man's' aeroplane analogous to several light cars then being manufactured. Some even suggested it foreshadowed transports of the future, unaware that the short span and multi-plane interference of this geometry was the antithesis of efficiency because of intrinsic resistance. The real advantage was its small inertia in conjunction with powerful control by clever coupling of the full-span ailerons and elevators using a simple differential of bevel gears at the bottom of the control column so that the ailerons moved in opposite sense to the elevators, and when operated for roll, the elevators moved differentially to augment the power of the ailerons – and that was more applicable to current ideas of fighter manoeuvrability.

Riding the crest of the Itford enthusiasm, *Flight* offered a prize for the best glider design received by the end of November, and the *Daily Telegraph* announced a new prize of £1,000 for the greatest distance covered in a glider during 1923. Meanwhile Gordon Selfridge, with invariable aplomb, got hold of the Peyret and exhibited it free of charge at his Oxford Street store, and then capped the newspapers with an offer of 1,000 guineas for the first flight of 50 miles made by a British pilot in a British-built glider.

9

November fogs were beginning to dislocate air services, yet the London—Manchester route operated by the Daimler Airway was attaining greater regularity than those from England to the Continent. Nevertheless Handley Page was increasingly optimistic over the success of his own service, and changed the W.8b to the W.8c by eliminating the freight compartment and extending the cabin to carry sixteen passengers instead of twelve. Martlesham trials showed the lateral control much improved due to incorporation of a slot between rear spar and balanced aileron though the rudder remained heavy. *Flight* reported: 'It is of interest that tests at Martlesham show that with only one engine running the machine can be flown straight, although losing height steadily, at its all-up weight of 12,500 lb, whereas at 11,500 lb it will actually remain aloft.' But that difference of 1,000 lb represented one third of the passengers or all the fuel. As though to allay fears of a forced landing, *Flight* added: 'Another feature is the low landing speed of 40 to 42 mph, but it appears less owing to the machine's size. Watching it "float" into Croydon one never has the anxious feeling with which one often watches small machines.' Nevertheless its climb at full load was only 370 ft/min.

'Star' Richards had been very busy. Not only had he developed the W.8b but now was modifying the design as a heavy bomber, later named Hyderabad. Currently under his direction, Newell was designing the 'X' – a Lion-powered, high-wing, slotted, strut-braced monoplane bomber, with pilot's cockpit in front of the wing like the D.H.29, but Richards himself was working on the most advanced conception of any design in Britain. Handley Page on his last visit to the United States had sold the idea that with slots for high lift, wings could be small, and the US Navy, in their support of projects for potential fighters, was immensely interested in the compact low-winger which H.P. sketched to illustrate his ideas. Known as the 'S' (H.P.21), three were ordered, allocated numbers A-6402, 03, and 04, with option for a further twenty-seven should they prove successful. Richards put the sketch into practical terms in the form of a cantilever monoplane with tapered wing and circular-section fuselage blending with the round cowl of a B.R.2 rotary, and located the pilot high above the wing with better view than any contemporary aeroplane. He followed conventional wood monocoque structural design, and not only was the fuselage frame covered with thin three-ply but the wings were similarly skinned and the centre-section integrated with the central fuselage portion, the outer wings and rear fuselage being independently bolted to this unit. Typical of Richards' ingenuity was the mechanism controlling the leading-edge slot and full-span trailing-edge flap with incorporated differential for lateral control whatever the position to which it was lowered. All control surfaces were fabric covered, and the tail was adjustable through a large angle. Span was 29 ft 2 in, with an area of only 113.5 sq ft, giving the hitherto unimaginable wing loading of 18 lb/sq ft, yet entailing a landing speed of only 44 mph with slots and flaps at maximum lift. But at the Air Ministry such complication was suspect, and there was expectation that because the Hanley had those problems of trim change, the H.P.21 would be the same.

If this machine proved successful, H.P. thought it might be scaled up as a fast transport. Meanwhile there seemed growing doubt of Croydon's

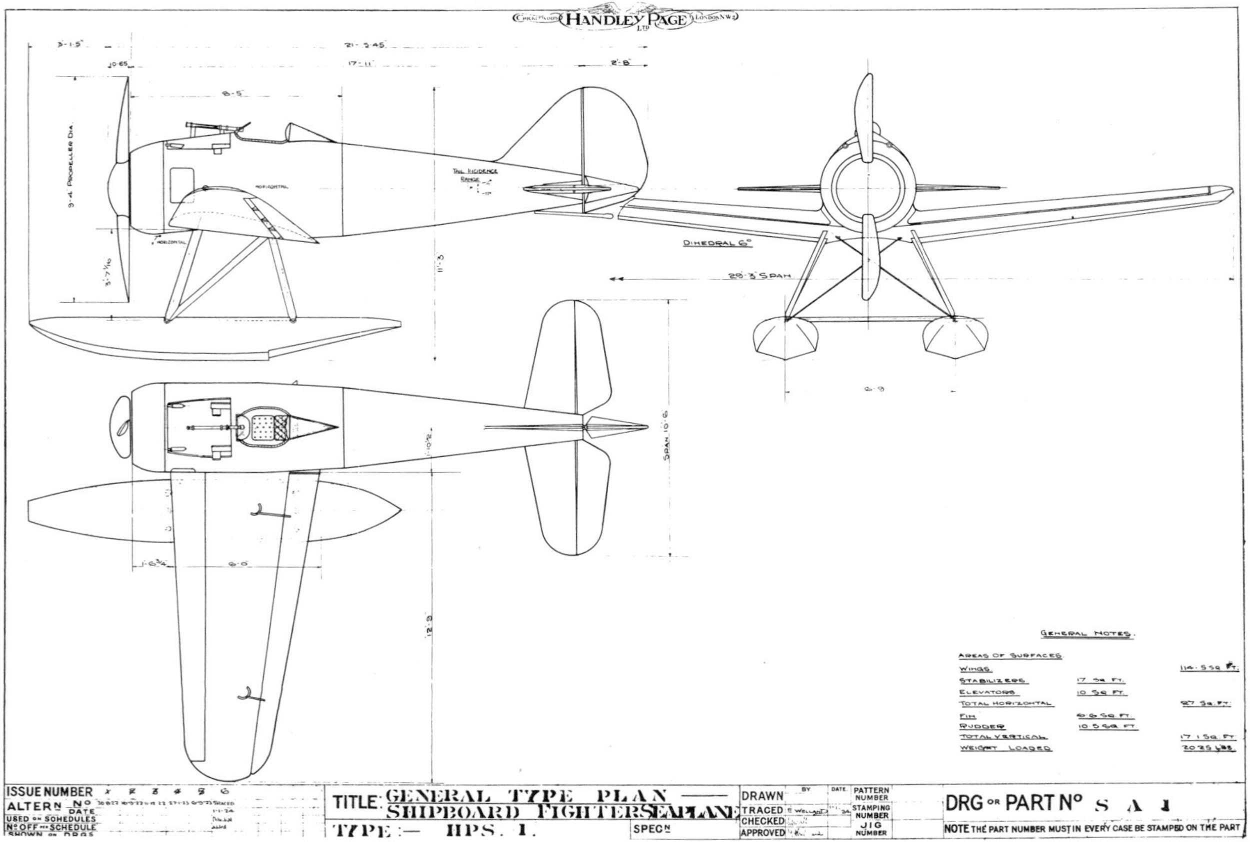
HANDLEY PAGE LTD
CRICKLEWOOD LONDON N.W.2
9'-4" PROPELLER DIA.
DIHEDRAL 6°
29'-3" SPAN
SPAN 10'-6"
TAIL INCIDENCE RANGE
GENERAL NOTES.
AREAS OF SURFACES.
WINGS 114.5 SQ. FT.
STABILIZERS 17 SQ. FT.
ELEVATORS 10 SQ. FT.
TOTAL HORIZONTAL 27 SQ. FT.
FIN 6.6 SQ. FT.
RUDDER 10.5 SQ. FT.
TOTAL VERTICAL 17.1 SQ. FT.
WEIGHT LOADED 2025 LBS.
ISSUE NUMBER
ALTERN NO DATE
USED ON SCHEDULES
NO OFF PER SCHEDULE
SHOWN ON DRGS
TITLE: GENERAL TYPE PLAN — SHIPBOARD FIGHTER SEAPLANE
TYPE:— HPS. 1.
SPECN
DRAWN
TRACED
CHECKED
APPROVED
BY
DATE
PATTERN NUMBER
STAMPING NUMBER
JIG NUMBER
DRG OR PART NO S A 1
NOTE THE PART NUMBER MUST IN EVERY CASE BE STAMPED ON THE PART

Original general arrangement drawing of the Handley Page H.P.21 or HPS.1

suitability as an air terminal, and several firms were asked by the Air Ministry to send machines with full load to try a series of take-offs and landings on the wider space of Wormwood Scrubs Common, where the great airship shed contributed pre-war by the *Daily Mail* still stood. The W.8b accomplished these with ease, but D.H.34 pilots considered the area too small. As an alternative, there was hope that Hendon might be re-opened for it was reported that Grahame-White had at last resolved his difficulties with the Air Ministry. Grahame-White had been in the USA when he was appraised that the Treasury had taken possession of his factory and aerodrome as surety for £200,000 loaned for war-time expansion. Taking the first ship home, he learned from his solicitors that though his was a case for heavy damages against the Treasury, nobody could sue the Crown and his only redress lay in a Petition of Right. By stratagem a writ was therefore presented to the unsuspecting hall porter at the Treasury, and when the case was heard the Judge immediately granted access to the Grahame-White Co so that business could continue. But it proved false hope. The Treasury outmanoeuvred him by a continuous process of postponing the final hearing. Undoubtedly the RAF intended to hang on to that aerodrome, even though its war-time use had only been secured under the Defence of the Realm Act, and at most the Air Ministry considered £250,000 adequate for its purchase – so the position remained stalemate.

There seemed similar prospect for the Air Force with its obsolete aircraft and numerical limitation when Lloyd George resigned Premiership and on 19 October Bonar Law was asked to form a Cabinet. A week later the King dissolved Parliament, and simultaneously the Irish Republicans repudiated the Irish Free State and appointed De Valera as President. Nevertheless, electioneering was quiet compared with the rowdyism which had confirmed Lloyd George as Prime Minister in the post-war fever of January 1919. Polling returned 345 Conservatives, but the eloquent and ruggedly handsome James Ramsay MacDonald, the Labour leader, made a long stride forward with 142 of his party elected, while Liberals cashed in with the balance in a split party respectively loyal to Asquith with 60 seats and Lloyd George with 57. Among those defeated was Winston Churchill, the potential condidate for Dundee. In the new Government, with Bonar Law as Prime Minister, Stanley Baldwin, the stocky Midlander who seemed to typify the middle-class business man, became Chancellor of the Exchequer, and Lord Curzon retained his post of Secretary for Foreign Affairs. On 23 November the King opened Parliament, yet though the pomp and ceremony held its momentary sway, it was the discovery of great treasure by Lord Carnarvon in the tomb of King Tutankhamen near Luxor in Egypt which riveted the excited attention of all: here was the story-book past made real.

But it was towards Paris and its Aero Salon that the attention of the British aircraft industry again turned – primarily to study French designs, for apart from strong engine representation the only British exhibitor was

Handley Page with his slotted wing Hanley torpedo-dropper – though those versed in Air Ministry tactics realized this indicated it had been rejected by the RAF. Other constructors did not feel justified in spending money on a stand, but H.P. had those patented slots to sell to the world. Absence of British participation was sad enough commentary, for the French had seized the moment, and a score of companies exhibited. The only other foreigner was Koolhoven with his interesting Jupiter-powered fast two-seat F.K.31 parasol fighter, which, like the Duiker, seemed influenced by the American Loening. *Flight*, in assessing military progress, commented: 'France is probably ahead of the rest of the world in the matter of supercharged engines, for the Rateau turbo-compressor no longer seems a laboratory experiment, and several machines are designed specially for high altitude work. Combined with a variable-pitch airscrew, the supercharger allows considerable increase in ceiling and speed at great heights. M Louis Breguet has even visualized what appear to be fantastic speeds. One is tempted to ask: What are we doing? Has our Air Ministry sent out specifications for machines to be used at altitudes above 30,000 ft? A hundred aeroplanes capable of reaching 40,000 ft would be more than a match for ten times that number whose ceiling is 30,000 ft, for while the latter were staggering around at that height, the former could manoeuvre with considerable reserve of power and outclimb the others whenever they chose. Do not let us be lulled into a false sense of security by reason that the supercharger is not yet all that it might be.'

Both the Vickers Victoria (*top*) and Virginia prototypes were identical except for the fuselages, and only the lower wing had dihedral, but unexpectedly the Victoria proved 10 per cent faster than the bomber's stately 96 mph.

Perhaps the Air Ministry was lulled by the fact that the RAE was investigating superchargers, for there certainly was no specification for really high altitude aircraft, whether fighters or bombers. Yet England had some potentially very practical new RAF aeroplanes which had just been flown, and others shortly would be ready. Following trials at Vickers, Martlesham was testing the big transport Victoria, which could carry 23 troops and a crew of two, and on 24 November Cockerell made the initial test of the Virginia bomber prototype J6856. Construction of both had been slow, for the RAF had no immediate need to replace their Vimy bomber and Vernon ambulance. Cockerell considered the Virginia's rudders too small but was relieved to find that longitudinally the machine was satisfactory, for the Victoria, which was virtually the same machine except for its bulky fuselage, had been so excessively tail-heavy that sweep-back was proposed if there were production orders. On 11 December, the Virginia, fitted with larger rudders, was delivered to Martlesham for full-load trials. It was soon evident that all the pilots liked it immensely, though there was continuous trouble with vibration from the starboard engine.

By now the private-venture Vickers Vixen two-seater was almost ready. Compared with the bombers and troop carriers the fuselage seemed all too short – a fault habitually repeated by all designers, partly because neutral stability rather than positive stability gave the sensitive manoeuvrability essential for a fighting aeroplane. Vickers hoped this new biplane might replace the Bristol Fighter and D.H.9A, but with the RAF still suffering from the Geddes axe it seemed unlikely there could be more than token orders for another year or so.

CHAPTER IV

PROMETHEUS UNBOUND
1923

'The principal object is to have what is wanted and to have it in time.'

General Sir James Murray (1793)

1

The duralumin age had embraced France, judging by the number of aircraft composed of this material at the Paris Aero Salon. French designers had advanced boldly, primarily attracted to duralumin because it was cheaper than importing the thin steel sheets favoured in England. Because it was thicker, weight for weight, it had the crucial advantage of intrinsic stability; thus a simple method of making spars was by drawing duralumin tube to rectangular section – which was much cheaper than the involved riveted steel used by the British.

The great advocate of duralumin construction had long been Professor Junkers, who deputed his consulting engineer Mierzinsky to write a paper on his behalf for the RAeS which was translated and read to that body early in January by W. J. Stern of the Air Ministry Laboratory, South Kensington. Introducing the subject, Junkers wrote: 'Because of the international situation I interpret this invitation to lecture to have deeper meaning than a token of friendliness towards my person, and see in it an effort at renewing the ties of a genuine humanity which desires to extinguish the sad traces of devastating war by hoisting the flag of peaceful competition. Any hesitancy I had was partly due to the oppression imposed on the German air industry in consequence of that war – a pressure which, although having certain justification as an issue of the war, is felt and resented as an expression of violence overstepping the mark. There was also our instinct of competitive self-preservation, for although we wish to maintain our share in developing aeronautics, we are chained by the clauses of the Peace Treaty to limit our activities with the result that our investigations, constructions and experiments have been accomplished under many a hardship. Nevertheless all scruples have been overwhelmed by an endeavour to bridge the abyss dividing the nations, and this is a task for which scientific intercourse is most highly qualified.'

Junkers's thesis was the interrelation between advanced aerodynamic research on thick-section cantilever wings and his particular method of duralumin-tube girder construction with corrugated metal covering. 'As far back as 1910 I was well aware that the main goal of aeroplane construction must be greatly diminished parasite resistance. The streamlining, or fairing, or covering, must be shaped as a hollow space producing minimum drag with maximum lift. This is the nucleus of my patent. My researches were not based on the usual aerofoil sections, but started as bodies of elementary shape, such as ellipsoids, and it soon became clear when we passed to flatter and flatter ellipsoids of constant perpendicular section relative to the air flow, that the size of the section was not of deciding consequence and thick forms were not only admissible but, within certain limits, superior to thin ones. Thus it became manifest that it is not so much the shape of the suction and compression sides of the wing profile which must be considered but the central line, by which I mean a line equally distant from the upper and lower surface of the wing profile. The camber of this central line, more especially the magnitude of its leading angle and rear angle, are of extraordinary significance in determining the curve of lift and drag.' That was a revealing outlook fundamental to aerofoil design but not yet understood by British designers.

Turning to structural problems, he said: 'The realization of another requirement is facilitated by the thick aerofoil: that of sufficient strength with smallest weight. Characteristic wing stress consists of high bending moments which increase with span; also there is a torsional moment in cambered wings tending to twist it relative to the body, eventually attaining very high values in a dive. The bending moment must be absorbed by tensile and pressure forces. If we consider the wing as a framework, these stresses increase if the structural height is lowered. Although maximum attainable height has been solved readily with the biplane cellule, using an upper and lower wing joined by struts and wires to produce a framework, the air resistance of these connecting pieces is very considerable. My method of overcoming it consists of enveloping the cell in a suitably thick wing section, transforming it into a deep cantilever, the drag of which is distinctly less than a biplane of equal lift, and which also forms a large hollow space for location of crew and loads which would produce drag if exposed.' This was basically what Bruce had done at Westland with the wing structure of the Woyavodsky, though it differed from the multi-spar Junkers wing in having vertical and horizontal spacer struts and wire bracing instead of being boxed with ridged duralumin plate strips riveted to tubular spars.

With jovial malice, Handley Page told the audience that the first Junkers he saw had crashed on landing, breaking the fuselage just behind the K of 'Junkers' painted on its sides, so that in attempting to decipher what the machine was he read the letters 'J-U-N-K'. Pursuing the attack, he questioned the statement that duralumin covering did not deteriorate, for he had examined the Junkers demonstrated in America the previous year and the covering certainly showed signs of corrosion; further, when Junkers seaplanes were supplied to Colombia with duralumin floats, they had to be replaced with wooden ones because of water leakage between the duralumin joints.

He was backed by Major Green, that pioneer exponent of steel construction, who stated: 'In this country we have less faith in duralumin. In fact our constructors were forbidden by the Air Ministry to use it for any parts likely to be highly stressed. As most of an aeroplane is highly stressed, this means we are not using duralumin but prefer high tensile steel.' Both Green and John North (who recently had lectured to the RAeS on *The Case for Steel*) had separately been steadily bringing the design and manufacturing techniques of thin metal members to a reliable standard, and Wilfred Reid of Bristols, accurately assessing the trend, had just engaged Harry Pollard, who had been specialist assistant to North in this work, to carry out similar development at Filton.

First essay at Hamble of the Avro Aldershot prototype re-engined with Napier Cub 'X' engine, flown by Bert Hinkler accompanied by Roy Chadwick.

Patents registered by every firm indicated growing interest in metal construction, but most manufacturers were refraining from any expensive incursion. Thus at Avros, Roy Chadwick was well satisfied with the compromise he had brilliantly executed with the big Aldershot bomber, using a metal-tube framed fuselage as a first tentative step, but with wings and secondary structure of wood. Just before Christmas the long-fuselage prototype J6852, in its new guise with 1,000 hp Cub and double undercarriage, had been demonstrated with splendid publicity before Sir Geoffrey Salmond and Air Ministry officials. Viewed on the ground this 68-ft span machine, its huge four-bladed wooden propeller slowly fanning, was breathtakingly impressive. With Bert Hinkler piloting, it made a majestic take-off in a mere 200 yards, the Cub roaring as 'gently as a sucking dove', as one bardic reporter said. 'I could hardly realize there was 1,000 hp in front of me,' reported Bert, and added for the benefit of the Press: 'The machine is very nice on controls and easy to manage.' Three weeks later it was delivered to the RAE at Farnborough who considered the exceptional stability made it particularly suitable for night flying. Soon it was hinted that a modest production order was likely, both in Cub-powered bomber form and as a passenger/transport/ambulance, powered with Condor III to replace the D.H.10s on the Cairo—Baghdad desert air route. For this second rôle a 22-ft cabin was to be built Vickers fashion with large oval rings and stringers, covered with plywood inside and out so that the enclosed air space gave a measure of insulation against temperature and noise. In the top fairing, ahead of the wings, would be an open cockpit for pilot and navigator, and fuel was in large aerofoil-section tanks on the top wing each side of the centre-section.

The Avro 561, later named Andover, used Condor Aldershot wings, tail and undercarriage, but though the fuselage had the same depth it was re-structured with rounded wooden monocoque cabin amidships, and pilot's cockpit in the nose.

What with the Aldershot, Type 549, the Andover transport Type 561, production of Bisons designated Type 555A in which the top wing was raised above the fuselage, and work vigorously progressing on the 95-ft span twin-engined Type 557 Ava, which would compete with the Armstrong Awana of 10-ft greater span, Roy Chadwick and his staff were very busy. It seemed that Avro must prove so prosperous that Crossley Motors, as major shareholders, were likely to reap rich reward for their support of aircraft. A.V. himself, a happy family man with a bevy of children, remained as fussily inventive as ever. Cliff Horrex, recollecting those days, recorded: 'Chadwick asked me to take a hand in side-tracking him whenever he appeared, diverting his interest to other matters lest his changes and detail modifications began to hold up progress. But A.V. had a very active mind, so when looking at a drawing one was making, his interest might soon wane if it had no bearing on his thoughts at that moment, and one would presently realize he was thinking of something quite different, completely oblivious of his surroundings.' The 504 remained his classic favourite, and he made many modifications of engine and airframe, replacing the elastic shock absorbers of the famous spear-like undercarriage with semi-oleo-cum-compression rubber, adopted tapered ailerons to give even better harmony with rudder and elevators, and installed many different engines, eventuating in the 150 hp Armstrong

Siddeley Lynx seven-cylinder radial. As the Air Ministry proposed to establish flying schools for the Reserve of Air Force officers, A.V. began urging its adoption as a dual-control trainer version, eliminating the skid and radiusing the wheels from a V-undercarriage with oleos taken to the longerons at the wing leading edge.

Reg Parrott, the original Avro designer, had been transferred to Hamble as general manager, with H. Denny as works superintendent, and as assistant had Roy Dobson, a reformed and matured young man after a term at the Manchester factory following his earlier dismissal from Hamble. Because of his City and Guilds training and pioneer design experience, Parrott had that happy ability to proceed with experimental work using the most meagre sketches and instructions, easily taking the wide range of prototypes in his stride. His previous assistant, Harry Broadsmith, had emigrated to Australia when the Avro design team was transferred to Hamble, and formed the Australian Aircraft & Engineering Co Ltd at Mascot, near Sydney, to act as Avro agents, subsequently establishing the first Australian aircraft factory where he built Avro 504Ks from indigenous timbers for the Royal Australian Air Force. However it was proving a precarious livelihood, and although Bert Hinkler had made splendid demonstrations there with the Avro Baby, and eventually sold it, there had been no further orders except for a Triplane sold to Q.A.N.T.A.S. in November 1920.

Blackburns were as busy as Avro with new design and construction, for Major Bumpus not only had modification work on the Dart and bull-nosed Blackburn, but Bob Blackburn decided as a private venture to complete the N.1B flying-boat as a racer for the Schneider contest with Lion engine on the top wing. More important, the Air Ministry, confident that the company had become a major design firm as proved by their naval aircraft, gave a contract to specification 16/22 (D of R Type 9) for a big, Cub-powered, long-distance coastal defence biplane to compete with the twin-engined Avro Ava, and having greater potential than the Cub-powered Aldershot. As a relic of war-time thought these Avro and Blackburn machines were required to have folding wings, though hangars had for some years been standardized with ample width for large aeroplanes.

While the Avro and Blackburn companies were cautiously introducing metal components, Fairey and Parnall preferred wood as the immediate economic material. Their respective designers, Fred Duncanson and Harold Bolas, had been designing single-seat carrier-borne fighters to specification 6/22. It was not difficult to see that the Parnall Plover was by the same hand as the war-time Panther. It was a compact and shapely 29-ft span bi-plane, with full-length combined flaps and ailerons, nicely rounded and faired ply-skinned fuselage of standard longeron and spacer construction, and interchangeable land undercarriage and amphibian twin-float gear.

The equivalent Fairey Flycatcher lacked the speedy appearance and elegance of the Plover, and was strongly influenced by the Pintail, with similar broken-backed type fuselage having equal spacer struts from cock-

Good looking in its day, the Parnall Plover was no great advance, structurally and aerodynamically, on such fighters as the Sopwith Dragonfly at the end of the war. Only six pre-production machines saw Fleet service. (*Courtesy Air Marshal Sir Ralph Sorley*)

pit to sternpost. Nevertheless an appearance of greater sturdiness was provided by thick wooden N-struts, instead of the slender steel of the Plover. A strong long-travel undercarriage, of the type which cushioned so well the IIID and the Pintail-derived Fawn (specification 5/21), gave it a good start in pilot assessment, and handling was better than the Plover's because the aileron/lift-flaps were somewhat lighter to operate as ailerons.

Though Duncanson was in general charge of design, it was Marcel Lobelle who was project draughtsman on the Flycatcher, and it was he who had drawn the Pintail from which the Fawn was derived, complying with general design conventions established by Fairey for the III series.

Hallmark of the Flycatcher in flight was the distinctively upward angled fuselage with wheels hanging far down because of the extended oleos.

The Fairey Flycatcher had a rugged, stocky appearance coupled with easy handling, instilling a feeling of confidence, so pilots liked her.

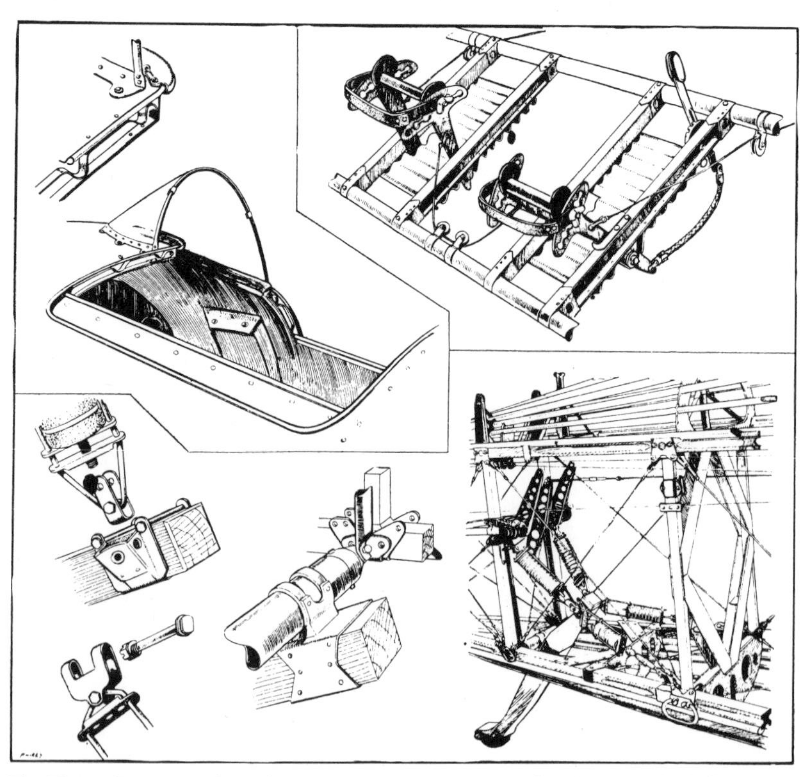

The Flycatcher was typical of Fairey wooden design, but differed little in detail from other manufacturers', though offered a few special selling features such as sliding panels for cockpit access and the novelty of rudder pedals adjustable for leg length. (*Flight*)

In the background is the Fairey Fawn I prototype with short fuselage to specification 5/21, and in the foreground the Fawn II prototype which achieved better stability due to its longer fuselage.

Lobelle was a man of great energy and conviction, who never quite mastered English but rejoiced in many a Rabelaisian story, and was regarded as a genius for clever and simple design, and past master at packing the maximum of men and equipment into the smallest possible envelope. However, Fairey took a hard business view, regarding him less as a genius than a man to whom design responsibility had been delegated.

Meanwhile the endlessly pipe-smoking, gaily sincere Bolas was completing his three-seat Possum triplane, built under the Air Ministry label of 'Postal aircraft' – an inexactitude which disguised expenditure on experimental aircraft, such as the D.H.29 and Westland Woyavodsky, which were outside practical requirements of the RAF but possibly might lead to future developments. The Possum followed the recent favourite formula of fuselage-mounted Lion, with bevel transmission to outboard propellers on the wings – for it was assumed by an unknown, but strong-minded, Air Ministry technician that an engine room for aeroplanes would enable

The Fawn III with 470 hp Lion II was the eventual production type to specification 20/23, little changed from the Fawn II.

minor repairs in flight. He did not seem to realize that the weight penalty of indirect transmission eliminated vital payload, for even in the relatively simple Possum it added at least 1 lb/hp, though the total of 14 lb/hp was ameliorated by light wing loading of 8.15 lb/sq ft.

On 13 May Norman Macmillan made tentative low 'straights' with it at Yate, but the rudder was inadequate for more, and even when enlarged had to be made still bigger. Not until 19 June was it ready for the first flight by Macmillan accompanied by Bolas, but there was considerable transmission vibration which long delayed further trials, and the second Possum was relegated to low priority.

Contrary to its official 'Postal' status the three-seat Parnall Possum had a gun ring in nose and rear cockpits. Transmission was clutchless, and the engine was crank-started from the rear cockpit. (*Courtesy Air Marshal Sir Ralph Sorley*)

To more practical purpose replacement for the Bristol Fighter had been reconsidered, and the order for two all-metal Short two-seat fighters, named Springbok, was reinstated. The prototype, J6974, was taken by road to Martlesham Heath in April 1923, and John Lankester Parker made the initial flight on the 19th when it seemed fairly satisfactory until tests were stopped by the metal skin splitting at the trailing edge due to vibration. The machine was returned to the factory on 23 May to have the skinning removed and the wings converted to normal doped fabric covering, and the second prototype had its wings skinned with thicker duralumin. C. H. Barnes in *Shorts Aircraft since 1900* records: 'From an engineering point of view they required little maintenance, but they were far too ahead of their time; whereas there were plenty of skilled carpenter riggers available to minister to orthodox wood-and-fabric aeroplanes, however fragile, trained sheet-metal tradesmen capable of repairing aluminium-alloy stressed-skinned structures were as rare as icebergs in the Sahara. Furthermore, standard pieces of equipment, such as radio sets, were not designed to go through the limited apertures permissible in a monocoque fuselage, and daily inspection became a job for a boiler-maker rather than a self-respecting rigger, so Service opinion was heavily prejudiced against the Springbok from the start.'

Both machines were accepted respectively in October and November, but there was an intrinsic fault in the relatively short tail-lever arm and small rudder, with the result that its ineffectiveness eventually resulted in J6975 spinning into the ground and killing its pilot. Wind-tunnel tests at the RAE resulted in complete re-design, with smaller chord lower wing, longer fuselage, raised tailplane of greater area, and big fin with tall rudder which had a triangular inset balance. Despite Oswald Short's intolerance of the Air Ministry attitude, six of this developed version were ordered, though later reduced to three.

2

On Tuesday 6 February the third Air Conference began at the Council Chamber of the Guildhall, promoted by the Air Ministry for discussion of problems associated with air transport. Sir Sefton Brancker opened with a lecture on *The Position of Air Transport To-day*, and Commander Burney, MP, promoted the Vickers lighter-than-air theme with *A Self-supporting Airship Service*. There followed *The Progress of Research and Experiment* by Sir Geoffrey Salmond, and Colonel Alec Ogilvie discussed *Gliders and Their Value to Aeronautical Progress*, while Dick Fairey, now chairman of the SBAC, completed the session with a lecture on *Seaplanes*. *Flight* pontificated: 'Opinions differ as to value of such Conferences, some denying they serve any useful purpose, others maintaining they are of greatest importance. We are certain they result in considerable good; not only do the papers give rise to useful discussions, but the Conferences give the Government, through the Air Ministry, an opportunity of stating its policy and "meeting its critics face to face", as General Brancker said in his paper on Tuesday.' Yet the ritual visit to Croydon could have given little hope that there had been real progress in the past year, for there were no new types present, unless the Cub-powered Aldershot could be counted in that category.

The only significant Resolution was proposed by the ailing Holt Thomas intent on maintaining national prestige: 'In view of the necessity to increase rapidity of communication within the Empire, and in view of the progress made by other nations in civil aviation, the Conference calls on the Government to give due and immediate consideration to the foundation of an air mail throughout the Empire.' His hope was that the Government might proceed with his plan for an Imperial air company, for the £600,000 allocated in 1921 for cross-Channel services would last little more than another year. He drew from Brancker assurance that the policy was not only under consideration, but that tenders for three types of airliners would shortly be called for – the first to have maximum economy for European routes; the second for the Indian run with nonstop range of 500 miles against head winds; the third to fly nonstop from London to Malta.

Sir Samuel Hoare, the new Secretary of State for Air, appointed a Civil Air Transport Committee (CAT), chaired by Sir Herbert Hambling of

Parr's Bank, to consider how to proceed. Forty-three-year-old Samuel Hoare, the second Baronet, had been MP for Chelsea since 1910, and proved a most able member of the Conservative team on matters of broad issue, but had no particular knowledge of aircraft. His CAT committee was not unexpectedly composed only of business men, including his brother Oliver, manager of Cox's Bank, Sir Joseph Broodbank of the Port of London Authority, and F. G. L. Bertram as secretary – so it at least was unbiassed about aviation, and recommended that a powerful business company be formed with £1 million capital, of which 50 per cent would be initially subscribed, the Government adding a further £1 million in subsidies over ten years. An accumulative 10 per cent dividend per annum would be paid to ordinary shareholders, any remainder divided equally between them and the Government until the latter had recovered their contribution. As this was virtually the proposal which Holt Thomas had put before the Government, he now brought in financier Szarvasy of the British Foreign and Colonial Corporation, who had rescued the Dunlop Rubber Co Ltd from financial difficulties, to form an Imperial Air Transport company, whose first object would be acquisition of Handley Page Transport, Instone Air Line, Daimler Airway, and the British Marine Air Navigation Co Ltd owned by Hubert Scott-Paine's Supermarine Aviation Works.

Sir Samuel Hoare's next task was to introduce the Air Estimates for 1923–1924. They represented a net increase of over £1 million. 'I am fully prepared to admit,' he told the House, 'the great difference between the political position of France and Great Britain, but even so the disparity in strength is overwhelming. In 1922, only 200 machines both civil and military were built in Great Britain, yet there were 300 civil and 3,000 military constructed in France. Their aircraft industry employs 9,250, but the British only 2,500. While it is inconceivable that we two great allies could ever embark on hostilities with each other, how is it possible to justify that one has an Air Force only a quarter the size of the other? Before members come to an answer as to what the standards of the British Air Force should be, they should be clear as to the responsibilities they wish to impose on it. In the past the Navy and Army had definite Imperial and National responsibilities entrusted to them. Only in the last few months has Home Defence against air attack, and the Independent Air Command of Iraq, been entrusted to the Air Force. Members should be clear as to whether other responsibilities should be given to this new arm, and consequently should consider the cost. In matters of National Defence the question of cost is not a final factor, but it must be taken into account. This year the Navy Estimates are £58 million, the Army Estimates £52 million, and the net Air Force Estimates £12 million. If you decide to apply a One Power standard to the Air without corresponding reductions in the estimates for Army and Navy, it would mean an immediate increase of £5 million, and an eventual increase to keep pace with progress of other great Powers, of £17 million. The House should note this figure of £17 million – for with it our Air Force could be increased fourfold with expenditure little more than double the present cost, though from every point of view you should try to avoid a new lap in the old race of armaments.

'On this account I particularly welcome the comprehensive enquiry that is to be undertaken by the Committee of Imperial Defence into the whole problem of Imperial and National Defence in relation to the three Services. I have been long enough at the Air Ministry to say you cannot isolate the problems of one Service from those of another. Pending a report of this Committee and the decision of the House, we have got to exist and make the most of resources at our disposal, and so long as we cannot have quantity, we must concentrate on quality. The Air Force as it is to-day is a very small Force; it must be a *corps d'élite*, and on that account, judged by expenditure in comparison with size, it costs a considerable sum of money.' He explained that development under consideration for the re-organized airline system would not involve substantial expenditure before the financial year 1924–25, and made it clear that the RAF was on the way towards expansion. Fifteen new squadrons would be formed for Home Defence by April 1925, of which initial expenditure for eight squadrons would arise in the coming financial year. Three squadrons would also be added to the Navy. Significant further expansion was £238,000 for the new Auxiliary Air Force and Reserve.

In the ensuing debate Capt Wedgwood Benn requested assurance that the Admiralty did not intend to depart from the clear and definite policy that the Air Ministry was an independent department. He reminded MPs that during the war the Admirals displayed little interest in the Air, yet now they desired to have their own Naval Air Arm and had become involved in the so-called 'Burney Airship Scheme'. The First Lord of the Admiralty, Leopold Amery, replied that the Admiralty now considered airship reconnaissance might minimize the work of light cruisers, but who was to pay and who control could only be decided by the Committee of Imperial Defence. Promptly Sir Frederick Sykes, speaking for the first time as Unionist representative for Hallam, warned that if the Navy and Army obtained separate tactical units there might be pressure to dispense with development of independent long-range bombing, yet it was on that very strategy that we should concentrate. Lord Hugh Cecil, MP for Oxford University, with decisive urbanity made the telling point that it was difficult to understand a policy of making preparation for defence sufficiently large to be costly but not sufficiently large to be efficient – and with that comment every RAF officer wryly agreed.

Meanwhile the economic scene was worsening. Over 1¼ million were still out of work in Britain, though the great army of unemployed was not very visible. They lurked in back streets, gathered morosely in little groups, their caps pulled down, shoulders hunched in the cold winds of March. A more evident reminder were ex-Service men standing on City and West End kerbsides, wearing their medals, and lethargically attempting to sell toys and trinkets from a tray suspended by string round their shoulders. Yet among the more fortunate there was an air of animation. Young women who were schoolgirls during the war had bobbed their hair and shortened their skirts and were shocking their mothers by using cosmetics: all was emphasis on youth and freedom. Bowler hats and occcasional topper were only for the City; the Homburg soft hat was the mode or straw boater in summer, and young men used lightly starched, downward pointed collars with gay neckwear instead of the tall starched collars worn by their elders. Cigarettes were the vogue, and over cups of tea at Lyons, youth met youth in gay chatter and easily solved the problems of the world. Yet conventionality was everywhere discerned echoing pre-war socialities with its etiquette, political parties, and lionizing celebrities of the arts and sciences. For the poor man, football and betting on horses was his solace for a drab existence.

In the House on 20 March, the bitter-tongued Philip Snowden moved: 'In view of the failure of the capitalist system to adequately utilize and organize natural resources and productive power, or to provide the necessary standard of life for a vast number of the population, legislative effort should be directed to the gradual supersession of the capitalist system.' Bonar Law might shrug that off as mere Socialism, but external relationships were another matter – such as the Chanak crisis when Poincaré almost wrecked Anglo-French relations by withdrawing troops which had been reinforcing the British against the Turks. To Lord Curzon fell the task of restoring British prestige and negotiating a Treaty with Turkey which ended their dangerous front with Russia. But it was the French attitude to Germany that remained the main problem. On 11 January, France and Belgium had begun an Occupation of the Ruhr because Germany was in default over coal deliveries. France was determined to secure reparation payments, for they were vital to her economy – but British policy was reconciliation and restoration of Germany's financial stability. The bitter edge between France and Britain grew more obvious. Resentfully, the Germans adopted passive resistance, refusing to work, and were supported by officials who closed their offices. France set up a military regime to force law and order. In return, the Germans fought them in the streets, wrecked trains, and lynched collaborators. The Mark, once 20 to the £, dropped to 81,000 and continued to fall. Despite British efforts at mediation, the French were determined to continue policing the Ruhr.

Britain fared better with her own enormous war debt to the USA of almost £980 million. Stanley Baldwin and Montagu Norman, Governor of the Bank of England, early in January negotiated with Andrew Mellon, Secretary of the US Treasury, on a basis of repayment spread over 61 years, with 3 per cent interest payable for the first 10 years and $3\frac{1}{2}$ per cent for the remainder. Bitter controversy followed in the British Cabinet, and Bonar Law's consent was reluctant, but, at a price, it helped regain US sympathy with the European scene.

3

His Grace the Duke of Sutherland, Under-Secretary of State for Air, came into the limelight with a prize of £500 for a competition intended to establish the most economical light single-seat British aeroplane. Hope of an aero-equivalent of a sporting motor-cycle costing not much more than £100 was the dream of many a young man, for the Germans were selling sailplanes at £80.

Even Geoffrey de Havilland, not long after the Itford gliding competition, had experimented with a low-powered four-cylinder engine, made for him by the Levis Motor-cycle Co, which he fitted to his remaining glider – but he found it an impracticable solution. However, W. O. Manning at the English Electric Co of Preston, had been designing a fascinating 'motor-glider' despite preoccupation with a new biplane flying-boat characterized by a steeply dihedral lower wing to give lateral stability on the lines of the pre-war flying-boat built by Saunders to Percy Beadle's design. With an all-up weight of only 360 lbs including pilot, Manning's 36-ft span cantilever monoplane, the Wren, was a miracle of light construction in which he daringly eliminated most metal fittings, joining the spruce longerons and spacer struts with $\frac{1}{4}$-in ply gussets, cross-braced with piano-wire looped through an eyeletted hole in the ply, and similarly simplified the wings. Powered by a tiny two-cylinder 400 cc A.B.C. motor-cycle engine, nominally $3\frac{1}{2}$ hp but developing about 7 hp at 4,500 rpm, the Wren was tested by Fairey's friend, Sqn Ldr Maurice Wright, MA, at Avenham Park, Preston, but aileron application caused alarming wing-twist. After adding torsional stiffening the Wren was tried from Lytham sands and performed remarkably well, for though the power loading was the highest ever at 51.5 lb/hp,

Virtually a powered glider, the dainty English Electric Wren was a miracle of lightness with cantilever wings weighing 85 lb and fuselage 39 lb. The bare engine weighed 33 lb, and the propeller 1 lb 11 oz.

Wright found he could fly throttled to 4 or 5 hp and despite marginal power attained 50 mph. Climb was hardly spectacular, so though the Wren could be landed almost anywhere, climbing out from a small field surrounded by hedges and trees would be impossible, nor was it practicable to circle out as control was insufficiently firm at such low speeds.

At Bristol, Roy Fedden had been considering with growing interest the possibility of ultra-light flying, and had designed a small, flat-twin air-cooled engine of 1,070 cc capacity, named the Cherub, rated at 18 hp at 2,500 rpm and weighing only 85 lb, but at the proposed price of £75 profit was not likely to be noticeable. W. S. 'Bill' Shackleton, a fair-haired young Australian designer, had recently joined the Air Navigation & Engineering Co of Addlestone owned by Norbert Chereau, war-time builder of Blériots and Spads, for whom Shackleton designed a clean, somewhat Germanic, semi-cantilever high-wing monoplane of 3-ft less span than the Wren, intended for the Cherub. An untapered, high aspect ratio wing was at eye level in front of the pilot, but his forward view was improved by scalloping the decking with tumble-home contour. A patented feature was the novel triangular-framed spar with apex and lower corner fillets of spruce and sides of ply, though it would have been a sounder structure if inverted but more difficult to mount. The 13 ft 6-in fuselage had the customary light spruce framework, and also was covered with thin ply. Even the wheels were of wood with laminated rim, five wood spokes plywood covered, and a hardwood hub bushed with phospher-bronze. With more power than the Wren, Shackleton could afford a more robust and therefore heavier machine weighing 460 lb, with power loading of 23 lb/hp and a top speed of 78 mph, but to meet competition requirements an inverted Blackburne vee two-cylinder of 320 cc less than the Cherub would have to be substituted.

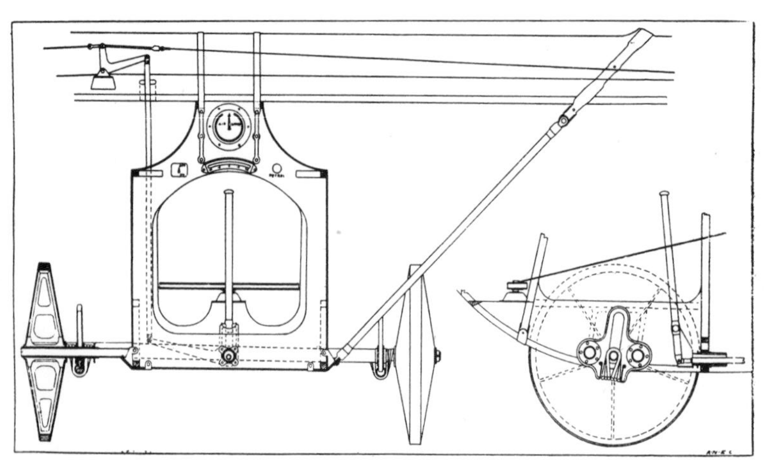

Original drawing of cross-section at pilot's cockpit of the A.N.E.C. single-seat light aeroplane shows simple undercarriage and ingenious wooden wheels.

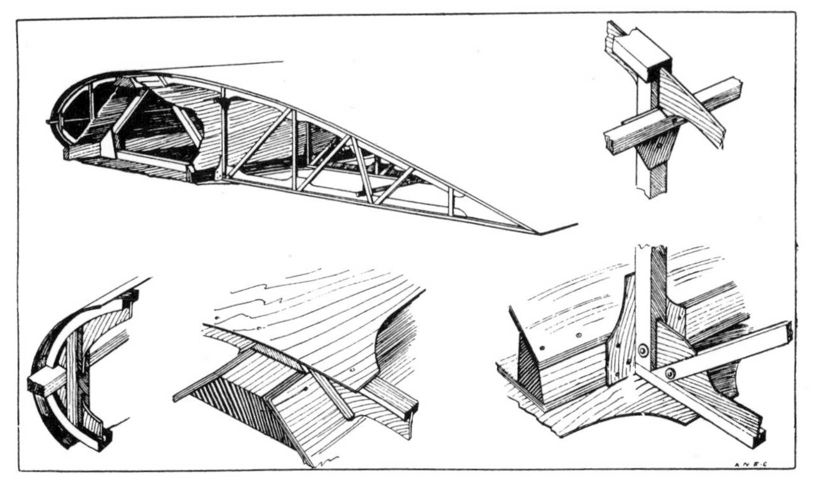

An unconventional triangular semi-cantilever spar for the A.N.E.C. enabled simple attachment and was excellent for combined bending and torsional strength. (*Flight*)

First flights of the Wren and the A.N.E.C. in April were promising enough to incline other designers towards light single-seaters. Solving such technical problems was an inexpensive gamble appealing to their sporting instincts, particularly as every major firm except Martinsyde had Government orders which promised mild assurance for the future. Ideas became reality in the middle of April when the *Daily Mail* announced a further prize of £1,000 open to any nationality for the longest motor-glider flight of not less than 50 miles over a 15-mile triangular course using an engine not greater than 750 cc capacity. Inevitably the Duke of Sutherland's contest now became known as the *Daily Mail* Motor-Glider Competition.

Concurrently the Air Council endeavoured to stimulate a very different development by offering prizes amounting to £50,000 for a helicopter, or equivalent machine, capable of successful flights to 2,000 ft carrying pilot, fuel for one hour, and 150 lb of military load, top speed not less than 60 mph and ability to hover, descend, and land without damage in winds not exceeding 20 mph. Crucially it must make 'a vertical flight from a position of rest and be manoeuvred while in the air over a given ground point as directed by the judging committee, and must descend vertically from a height of not less than 500 ft *without using engine*, and alight without damage within a confined circular area having a radius of 100 ft.' There was no rush for entries, and Brennan's helicopter was ineligible because he was working as a Government servant.

Presently questions began to be asked in Parliament as to the cost of Brennan's efforts. The Under-Secretary replied: 'Work in connection with a Brennan helicopter, which has previously been carried out under the Ministry of Munitions, was transferred to the RAE in June 1919 from which date until 8 July, 1923, the expenditure, inclusive of salaries, wages, materials and costs, has amounted to approximately £41,000. It is difficult with complicated experiments to know how long they will take, but I think on the whole, in view of the point at which these experiments have arrived, that it would be a pity to end them now.'

The industry took a very different view. The Council of the Royal Aeronautical Society made representations to the Air Ministry, pointing out that there was much greater need for research along lines offering better promise of sound aeroplane and engine development.

Major Green had been investigating the possibility of helicopters for John Siddeley, but decided against them on the score of excessive horse-power, grave difficulty in making bearings and working joints structurally safe, and absence of any workable theory of rotary wing stability. 'We have been informed at the last two Air Conferences,' he said, 'that experiments on full-scale have been made, but we have not been supplied with any details of the results obtained. In all seriousness, a balloon or airship is far more promising. It seems unlikely that helicopters will ever afford a useful means of getting from place to place.' Even Dr Watts, the great exponent of propeller design, agreed with Green, and said that if a helicopter ever becomes possible it would only be for low altitude work. But it was cost that was the real obstacle, for as Geoffrey de Havilland said: 'No private concern could entertain such expenditure in the present condition of the industry', and with official backing for the Brennan in mind, added: 'It is difficult to avoid the conclusion that the whole experiment is a waste of money.'

Handley Page, with ever enquiring interest in any new development, was equally cautious, but in commenting on the possibilities of helicopters could not forbear a crack at the proposed Imperial company: 'Those who might have had doubt as to what were the objects of the company, or on what its funds could be expended, would at least find in helicopter development some object sufficiently large on which the capital could go!' Nevertheless he thought it dogmatic to say that the helicopter, though not so efficient as an aeroplane, would not be in use 20 years hence. In fact recently he had been in Madrid and saw a very interesting semi-helicopter. 'This consisted of an ordinary aeroplane fuselage at the centre of gravity of which was fixed an inclined vertical shaft, the angle of inclination being a few degrees back from the vertical', he explained. 'On this inclined shaft was what appeared to be a four-bladed propeller, between 15 ft and 20 ft in diameter, the blades consisting of fabric-covered planes of approximately 1-ft chord, the cross-section being ordinary aeroplane aerofoils. Instead of this horizontal propeller being driven positively by the engine, it was free to rotate. The fuselage was fitted with an ordinary engine propeller, and when the machine commenced to run along the ground to take-off like an ordinary aeroplane, the blades of the horizontal propeller started to rotate, and owing to the forward component of the vertical reaction on the blades, they rotated in a way which seemed at first contrary to what one would expect, namely, the leading edge of the aerofoil moved forwards. Owing to the increased speed obtained by the velocity of rotation, a higher lift was available than if the planes were stationary. To equalize lift on both sides, a very ingenious device was incorporated, for the blades were hinged at their interior end, approximately 2 ft from the axis of rotation, and held in position by rubber shock absorber cords. In consequence, on the side in which the velocity of rotation and translation were the same, the blade was allowed an upward movement, thereby diminishing the effective angle of attack. The reverse took place on the other side, so that the angle of attack was increased and lift correspondingly increased. By this means the lifts on the two sides were equalized. Although somewhat complicated, this machine had actually flown, and seemed to alight very slowly. Whether such a device was really good and worth the complication is open to question.'

It was not open to question. This was Cierva's fundamental rotating-wing invention of articulation which was to open the door to successful design. After many minor crashes and setbacks, the persevering Juan de la Cierva y Codorniu had proved the vital value of his invention, and secured British Patent No. 196,594 granted on 18 April, 1922, which stated: 'The rotating wings are mounted on a base plate which revolves in bearings around a shaft supported by a tubular pyramidal structure. The wings are stayed by bracing wires and fixed to the base plate by hinges which permit movement of the wings in the direction approximately in a plane passing through the shaft.'

4

For some time Bonar Law's health had been failing: like Holt Thomas, he had incipient throat cancer which was weakening his voice, so Baldwin had been deputizing in the Commons, and Curzon presiding over the Cabinet. On 20 May Law submitted his resignation to the King. Whit Monday intervened, and on the following day, to Curzon's surprise, Baldwin was invited to form a Government. Regarded as a typical blunt Englishman of common sense and good will, Baldwin seemed a popular choice. He appointed Curzon as Foreign Secretary, and Neville Chamberlain became Minister of Health with the task of handling a new Housing Act by which it was hoped to encourage extensive building of low-priced houses by private enterprise. The keynote of Baldwin's policy was Protective Tariff against imports, whereby he hoped to reduce unemployment. On that count he was prepared to go to the Country, for it contradicted the pledge given a year earlier by Bonar Law that there would be no fundamental change in the fiscal arrangements of Great Britain.

But it was the ultra-light aeroplane which was holding the attention of flying enthusiasts, spurred by a lecture on *Low Power Flying*, given by Sqn Ldr Maurice Wright to the Royal Aeronautical Society's more practical rival, the Institution of Aeronautical Engineers. Since the war he had been the Air Ministry's chief technical test pilot, visiting seaplane and flying-boat firms to try their prototypes, and thus met Manning in connection with the Phoenix company. It was because that company had no regular pilot that he tested the Wren in his spare time for the fun of it.

Commenting that at Itford the wind was so strong that a standard Avro with stationary propeller could have soared like the Peyret, he said of the Wren: 'The low wing loading of $2\frac{1}{2}$ lb/sq ft called for larger controlling surfaces than expected to retain course in rough weather; in any case such machines will be more sensitive to air disturbances than those more heavily loaded. If higher wing loading reduces the amount of controlling required it pays to adopt it at the expense of slight increase in power to fly level. On the other hand the Wren's wing loading is not as low as some machines which used to fly successfully in pre-war days, such as the 50 hp Boxkite.

'Perhaps it is a pity that the engineless glider has not remained a little longer so that aerodynamic problems could be investigated more fully. When once these machines take power-driven form it is tempting to boost engine size and efficiency to obtain improved performance, and to neglect the aerodynamic side. With careful development it may be possible to build a single-seater of 12 hp, carrying a load of 230 lb, which would have a top speed of 80 mph at 500 ft, cruise at 60 mph at 100 mpg for a range of 360 miles, land at 28 mph, and have a ceiling of 8,000 to 10,000 ft. To fulfil public demand for a utility aeroplane it must be able to take-off and alight in a field of average size and have sufficient climbing gradient to clear obstacles such as high hedges on the boundary. That will not necessitate a high rate of climb; ground speed will be low and consequently the ratio of height gained to distance covered may prove better than many existing commercial machines.'

Roy Chadwick agreed there should not be less than 12 hp if the machine

Whatever the expedients of design it was impossible to build a sufficiently strong ultra-light aeroplane weighing less than the Wren. Roy Chadwick was nearest with his biplane Avro 558 which was 80 lb heavier largely because of a more powerful engine. (*Flight*)

was to be robustly constructed. 'I have gone into the subject rather carefully,' he said, 'and reached the conclusion that the light biplane offers possibilities. You reduce overall size, and the narrower wing chord should give better elevator control.' At Hamble he was producing an unmistakably Avro conception despite its I-struts for the wings and the wheels half buried in the fuselage. With 30-ft span it was only 6 ft smaller than the 504, although the chord was a mere 3 ft and the laden weight 480 lb, giving a wing loading of 2.89 lb/sq ft and power loading of 26.7 lb/hp.

A.V. himself thought a monoplane might do better, and sketched a shoulder-wing machine, with tapered wing of 36-ft span slightly resembling his pre-war monoplane, but instead of enclosing the pilot, gave him an open cockpit between the deep wing spars. Both machines had a fuselage of four light spruce longerons with sloping spacer struts attached with ply gussets, forming a Warren girder. Though designed for identical all-up weight, the monoplane wing loading had to be increased to 3.4 lb/sq ft, so this necessitated a 696 cc Blackburne engine of 24 hp, at a power loading of 23.5 lb/hp, instead of the 500 cc 18 hp Douglas fitted to the biplane.

Sticking to the biplane formula was the Viget of 25-ft span, designed as an after-hours project by some of the Vickers design team with the approval of Rex Pierson, but unmistakably it bore characteristic design signs of its *alma mater*, with aspects reminiscent of the Vulcan single-engined airliner. More conventional than the Avro, the Vickers light-plane had normal strutted and braced wings, but the fuselage of longeron and spacer strut construction was somewhat expensively braced with swaged rods for which Vickers rivalled Bruntons as specialist producers. A 750 cc Douglas engine drove a large slow-revving and therefore efficient propeller by chain and sprocket with $2\frac{1}{2}$: 1 reduction, and with a wing loading just under 3 lb/sq ft it was expected there would be brisk climb. Much more of a gamble, using a light two-cylinder 750 cc two-stroke air-cooled engine specially designed by a young amateur, Sir John Carden, was the 18-ft span Gloucestershire Gannet, for its designer, Henry Folland, had not understood the necessity of light span loading, and had also penalized it with a wing loading of 4.46 lb/sq ft.

At Short Bros an interesting monoplane meeting Maurice Wright's specification was ready for testing, powered with a Blackburne 700 cc engine which had neat chain drives to an outboard propeller on each wing. Named the Gull because of the superficial resemblance of its beautifully shaped circular-section monocoque fuselage with wings having each outer half swept heavily back at the leading edge, it had been originally designed as a glider by Oscar Gnosspelius – that lesser known pioneer of charm and explorative mind, who at last had abandoned dreams of a man-powered ornithopter. The almost cantilever wing had four spars, and behind the second was 'a substantially vertical backwardly facing step extending transversally to the line of flight', as patented in January 1922 by A. E. Short, H. O. Short, A. Gouge, and O. T. Gnosspelius, as the result of experimental work with the latter's 'aerodynamic pendulum' devised as a

In flight the Gnosspelius Gull looked like its namesake but sounded like a bluebottle. At an empty weight of 360 lb, it was 30 lb lighter than the Viget, heaviest of British entrants for the Lympne contest, and climb was barely 200 ft/min.

substitute for a wind-tunnel. Wing loading was 3.4 lb/sq ft and power loading 23.8 lb/hp.

Construction was paid for by Gnosspelius as it was not officially a Short design. Lankester Parker tested it at Lympne on 26 May, where it had been taken by lorry. Take-off was a mere 80 yards in no wind, but because of the close proximity of the cockpit to the ground, he misjudged his landing, flattening out a little too high and, instead of skimming to the ground at small incidence as intended because of the negligible fuselage clearance, he stalled and dropped with sufficient jolt to break his seat mounting. During tea-time it was repaired, and 'Lanky' made a second flight, climbing to 2,100 ft in 20 minutes, achieving 65 mph full out, and a stall of 30 mph. For a few days the Gull remained at Lympne. Larger wheels were fitted, and further testing was sufficiently promising for Oswald Short to decide there might be sales, so he ordered construction of a further Gull, and advertized delivery in nine weeks so that a purchaser could enter for the Motor-Glider contest.

George Handasyde, with the help of Sydney Camm, had designed a shoulder-wing monoplane somewhat similar in conception to the Avro, but

Intimidating in appearance, the tiny RAE Hurricane had a tapered, small-span, thick wing, and a triangular fuselage giving a very small cockpit even for light-weight George Bulman. (*Flight*)

with parallel-chord wings except for slight leading-edge taper at the tips, and 6-ft less span, powered by the slightly larger 750 cc Douglas flat twin, and the Air Navigation Co were well ahead with its construction.

A very different shoulder-wing monoplane was being built as a spare-time job by the RAE Aero Club at Farnborough as a miniature racer with narrow triangular fuselage and a mere 23-ft span, 80 sq ft area, and wing loading of 6 lb/sq ft. Named the Hurricane, it was supposed to attain 100 mph with only a 600 cc Douglas. Like the somewhat disappointing Zephyr pusher biplane of 29-ft span the Club had built on the lines of the F.E.8 war-time fighter, the Hurricane was designed by S. Childs, who many years later, in conjunction with C. F. Caunter, wrote the first official history of the Royal Aircraft Establishment.

The RAE's heavy Zephyr tail-boom biplane. After cooling problems as a cowled pusher its 500 cc Douglas engine with 2 : 1 chain drive was fitted to the RAE Hurricane. (*Flight*)

As a short cut to motor-gliders, Handley Page agreed to fit an engine to the Itford C.W.S. glider which Capt W. H. Sayers had designed in nineteen hours and the Central Aircraft Co built in nineteen days, but it was soon clear that major re-design was necessary. Sayers drew a new machine of similar shape but with more rigid wings, mounting an A.B.C. engine on a streamlined pylon fitted to the fuselage ahead of the pilot in the manner of the Wren. During construction H.P. seemed far from impressed, and when light-weight Gordon Olley from his airline failed to make it take-off amid the tufty herbage of Cricklewood aerodrome he instructed Harold Boultbee, that early draughtsman of Bristols, to fit a 500 cc Douglas and to move the wheels further forward to prevent the nose pressing firmly on the ground under the high engine thrust – but take-off still was difficult.

'This reminds me of the early days,' said H.P. with a broad grin as he watched two lines of workmen and draughtsmen trying to launch the machine with elastic cord. 'Lots of enthusiasm, but no flying.'

Boultbee made a swift design for a new machine, using the same simple jigs for an identical tail and rear fuselage, but sloping the forward top longerons more gradually, mounted the A.B.C. conventionally in the

Handley Page's own version of an ultra-light had very small span and was fully slotted and flapped, but it was too late for the Lympne competition and never flew. (*Flight*)

extreme nose driving a 3-ft propeller, lowered the cantilever wing, and totally enclosed the pilot with a plywood cover except for a viewing space each side. On one thing H.P. was adamant – a slot in each aileron to give better lateral control. But the little engine proved a beast to start, and when H.P. tried, it back-fired and broke one of his fingers. Even though Olley managed to fly after a bungee launch, H.P. remained dissatisfied and thought he could do better with a new design incorporating full front slot, slotted ailerons, and slotted flap, with expectation of low landing speed despite the minute area of 60 sq ft and heavy loading of 8 lb/sq ft. A miniature was drawn with similar diamond fuselage but the top longerons so steeply sloped that they joined the lower longerons at the pilot's feet, above which a tall and narrow streamlined pylon carried a 700 cc Blackburne with thrust-line concentric with the 20-ft wing which was mounted semi-parasol fashion so that the cockpit could be located beneath the front spar. If it worked, here was the genesis of a compact and attractive machine – but would that little engine and small span afford reasonable climb? It would still be weeks before details of the complex mechanism in that slim wing could be draughted by Richards, agreed and issued to the shops; time was at a premium.

The round-up of British entries was completed by two very different low-wing monoplanes respectively designed by de Havilland and Bolas. D.H. could not help drawing an attractive shape; Bolas was less successful at pretty lines, but always was ingenious. Of the two, the D.H.53, presently named Humming Bird, was ready to fly before the Parnall Pixie. Both the D.H. and Parnall were semi-cantilevers, braced with 4-ft compression struts to the fuselage, and spanned 30 ft 1 in and 28 ft 6 in respectively, though the Pixie also had spare racing wings of only 18 ft. Inevitably D.H. followed his usual wing shape with curving raked tips and tail with typically sweeping rudder – but Bolas used a markedly forward swept rear spar for the length of the ailerons, resulting in an almost triangular outer wing portion; tailplane and fin were obliquely triangular, and the rudder, hinged with angled axis above the unbroken elevator, was as typically small as his earlier Panther and Puffin. The de Havilland rudder had twice the area and a long fuselage lever arm deliberately greater than D.H. would have given a faster aeroplane. In accordance with recent de Havilland practice, the D.H.53 had its attractive little fuselage covered with millimetre ply; but Bolas for greater lightness used spruce Warren-girder bracing and covered the structure with fabric. At a glance the D.H.53 had greater pilot appeal, and with its conventionally-sprung undercarriage, compared with the cantilevered steel-tube legs and flexible axle of the Pixie, seemed the more practical.

Gordon Olley, in the cockpit of the cotton-covered Sayers-Handley Page, found even the increased power of a Douglas insufficient as the high thrust-line battled with the elevator at take-off. (*Flight*)

5

With spring and early summer came the round of England's gala occasions, but the season of air racing opened sadly with tragedy on 23 June in the Grosvenor Challenge Cup race of 400 miles for aircraft of less than 150 hp. Ten entered: Hinkler flew the Avro Baby, Longton a Sopwith Gnu, Perry a Lucifer-powered A.D.C. 504K, Rex Stocken a Le Rhône 504K, Raynham a Clerget 504K, Hamersley another Lucifer 504K, Dr E. D. Whitehead Reid an S.E.5 he had converted to 80 hp Renault, Sqn Ldr Robinson a Boulton & Paul P.9, Cyril Uwins with the Lucifer Bristol Taxiplane, and Major E. L. Foot, affectionately known as 'Feet', flew the pretty, crescent-winged, Lucifer-powered Bristol M.1D monoplane. On the

This 1917 production Bristol monoplane was bought from the Disposal Board in 1919, hopefully reconditioned for re-sale, but found no buyers so in 1922 was converted to a scarlet-painted flying test-bed for the 100 hp three-cylinder Lucifer. (*Flight*)

last lap the wing of Foot's machine failed, and he crashed fatally near Chertsey. Weeks later it was established there had been a fatigue failure of a landing wire, probably due to the rough running of the three-cylinder engine, and the wing had collapsed under down load in a bump. News of the crash reached spectators at Croydon before Longton arrived as winner, with Raynham second and Hinkler third, so greetings were subdued, for Foot had been very popular.

At the end of June came the RAF Pageant, with the King and Queen in the Royal Box, accompanied by Queen Alexandra, the newly married Duke and Duchess of York, Empress Marie of Russia, Princess Beatrice,

Pomp and Circumstance! At the 1923 RAF Pageant Queen Mary, ever regal, walks with King George V to the Royal Box, accompanied by Air Vice-Marshal John Higgins. (*Flight*)

Because of its unprecedented power, the Napier Cub in the massive Avro Aldershot attracted awed comment when displayed at the RAF Pageant. (*Flight*)

Princess Victoria, the Grand Duchess Xenia, and the Crown Prince of Sweden; in attendance were Air Vice-Marshal John Higgins, who was chairman of the Pageant committee, Sir Samuel and Lady Hoare, the Duke and Duchess of Sutherland, Air Chief Marshal Sir Hugh and Lady Trenchard, Earl Beatty and the Lords of the Admiralty, Major-Gen Sir Frederick Sykes, Major-Gen Sir Sefton Brancker, Members of both Houses of Parliament, and Naval, Military and Air Attachés of foreign governments. The seal of social approbation, like that of Ascot, had been

No less the centre of attraction at Hendon was the little Wren, which floated into the air like thistledown after help by a man pushing at each wing-tip to overcome the resistance of the tufty grass. (*Flight*)

conferred. True the bright sky of morning turned to overcast threatening rain, but it was a monumental day of flying, beating even the glories of the previous years, and the great event of a fly-past of new aircraft was considered unique. In single file they taxied past: the Blackburn Dart, the Cub-powered Avro Aldershot, Westland Weasel and Walrus, the new Blackburn Blackburn four-seat fleet-spotter, Avro Bison, Supermarine Seagull, the fascinating Fairey Flycatcher followed by the rival Parnall Plover, the Handley Page Hanley, Fairey Fawn, new Gloucester Grebe, and finally the ethereal little Wren. Of new aircraft, only the Armstrong Siddeley steel-framed Siskin III and Awana, and the Vickers Vanguard and Victoria were missing. The low ceiling added to the thrill of brilliant aerobatic displays. The Nighthawk battled with the Napier-powered Bourges. Longton exquisitely whirled the Flycatcher. Three Vickers Vernons landed troops to succour the garrison of the set piece. Formations swept the skies. Yet the mainstay still were the S.E.5as, Snipes, D.H.9As, Bristol Fighters, and Vimys.

Though Bonar Law had appeared willing to cut the RAF, any precipitous move had been overruled by the appointment of an investigating sub-committee chaired by Lord Salisbury, Lord President of the Council. In June the committee's report was published, and the Prime Minister on the 26th announced that in addition to meeting essential air power requirements of Navy, Army, Indian and Overseas commitments, a Home Defence Air Force would be provided of sufficient strength to protect against air attack by the strongest air force within striking distance of this country. It would be organized in part on a permanent military basis, and in part a territorial reserve, but so arranged that sufficient strength would immediately be available for defence, and fullest possible use was to be made of civilian labour and facilities.

In the first instance the Home Defence Force should consist of 52 squadrons, to be created with minimum delay – in effect adding 34 squadrons to the authorized strength of the RAF – and the Secretary of State for Air had been instructed to take preliminary steps to implement this decision. Subsequent expansion was to be allowed for, but would have to be re-examined in the light of the air strength of foreign Powers at that time. Nevertheless, in confirmation with British obligations under the Covenant of the League of Nations, HM Government would gladly co-operate with other governments in limiting the strength of air armaments on lines similar to the Treaty of Washington as applied to the Navy. Sir Samuel Hoare said it was difficult to give an exact estimate of expenditure, but thought it would not exceed £500,000 this year, and the average, including capital and maintenance, would not be more than £5,500,000.

Here at last was hope for the manufacturers who had fought to survive on starvation contracts, but who, at the instigation of Trenchard and his far-seeing staff, had managed to hold their technical personnel together with design contracts which yielded no profit but at least paid their salaries. Nevertheless the theory postulated by Lloyd George and enunciated by Trenchard at the end of the war remained the tenet, whereby it was believed that threat of renewed European war would be detectable ten years before the event and give ample time to achieve operational war strength of fighters and bombers. Thus it precluded heavy expenditure on armaments which might lead to a preponderance of obsolete aircraft, and instead ensured that the medium for production was available and capable of expansion, yet kept within limits which Britain could reasonably afford while rebuilding her world trade and stabilizing home economics.

To those prepared to accept military commitment the Air Ministry announced that as the RAF Reserve was being increased for Home Defence, ex-Service pilots would be admitted and 'gentlemen qualified as civilian pilots who had not previously held commission in the RFC, RNAS, or RAF'. Subject to completing the necessary annual training, officers would be paid a retainer of £30 per annum.

Air racing next held the stage with the King's Cup circuit of Great Britain on 13 and 14 July. Of particular interest among the seventeen entries were a Supermarine Sea Eagle amphibian piloted by Henri Biard; the angular single-seat Armstrong Siddeley Siskin, G-EBEU, flown by Frank Courtney; and the new and pretty single-seat Gloucestershire Grebe which was obviously a re-designed, modernized and streamlined S.E.5 powered with a 325 hp Jaguar radial, and was flown by their new test pilot Larry Carter, for Jimmy James was *non persona* after the fiasco of last year's Coupe Deutsch and had joined Smith's Instruments as a salesman.

The Grebe was certainly an ingenious development of the S.E.5-derived Nieuport Nighthawk, on which David Longden, the Gloucestershire managing director, had bravely gambled in buying all the war-time stock, and one of these fuselages had been adapted with all the skill Folland had shown in making the 'Bamel', but the wings were new, and subject of Patent 225,257: 'The top plane of a biplane is a thick high-lift section, and the bottom plane is of normal section. The outer ends of the top plane are tapered in thickness for about one-fifth of the span, and the thickness is also reduced towards the centre in the region of the propeller slipstream, both giving slightly reduced resistance. Ailerons are provided in which the chord increases towards the wing-tips for better control.' To test the idea, the Sparrowhawk demonstrator with B.R.2 engine, G-EAYN, had been successfully tried with these wings and was known as the Grouse, resulting in an Air Ministry contract for the Jaguar-powered Grebe version.

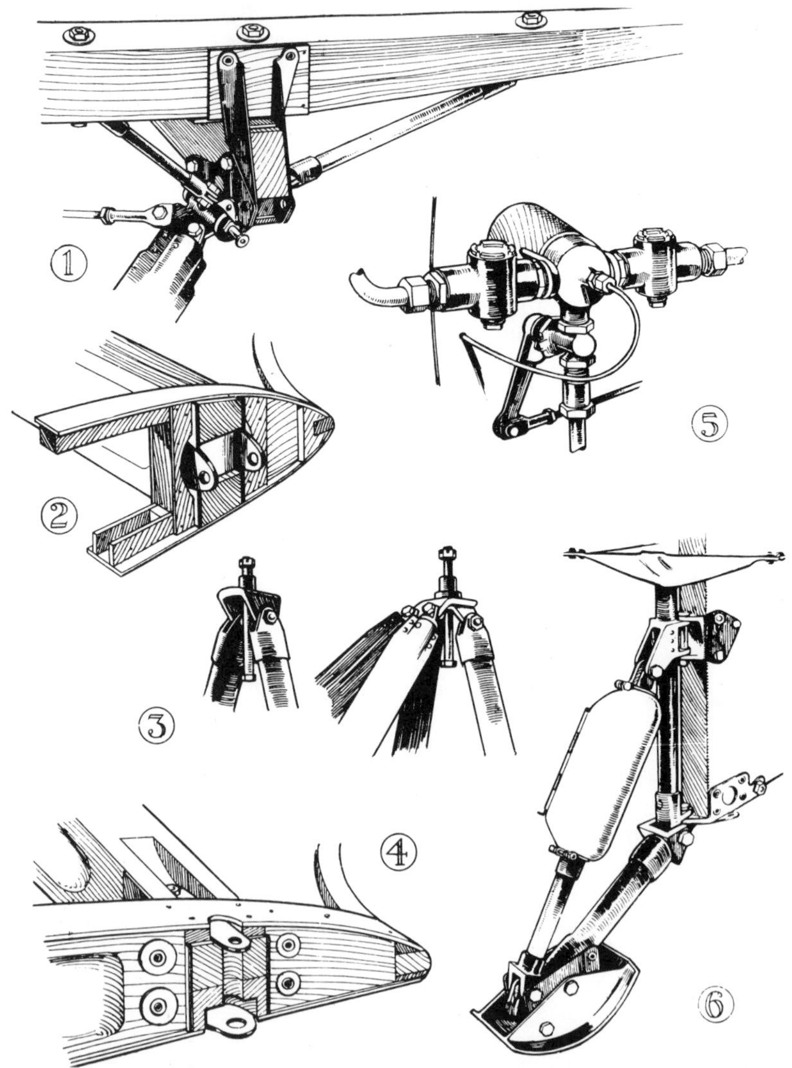

Typical wood and metal work of the day. 1, the Gloster Grebe undercarriage/spar attachment to longeron was in mid-bay. 2, wing root fitting at that point. 3, cabane struts had vertical bolts connecting the fittings, 4, of the top wing halves. 5, combined oil and fuel cock with fuel equalizer box. 6, the S.E.5 type tailskid. (*Flight*)

Prettiest development of the S.E.5 was the Gloster Grebe, design details of which remained almost unchanged from the Nieuport Nighthawk. Handling characteristics were regarded as delightful. (*Air Marshal Sir Ralph Sorley*)

Except for a ponderous Vickers Vulcan the rest were well-known stand-bys, but de Havillands had fitted a Napier Lion to the D.H.9 flown by Alan Cobham and entered by that irrepressible comedian, George Robey, whose rival, Harry Tate, not to be outdone, entered a D.H.9C piloted by Hubert Broad, Cobham's new and slightly-built assistant.

Start and finish at Hendon were tame affairs. Only a few hundred spectators turned up, mainly zealous aviation enthusiasts, though there were no other big attractions to keep the crowds away. 'Just as it was nearing 4.30 p.m. the purr of a Jaguar was heard, and in came Courtney, fairly high, but descending to the necessary 500 ft as he crossed the line to the accompaniment of cheers. At this moment another came in view, and, with a squeak of delight from George Robey, Cobham thundered past – obviously a popular arrival judging by the applause. Photographers were hard at it when Broad with Harry Tate's 9C arrived and was accorded a

Though the Vickers Vulcan was hardly a racing machine even with a Lion engine, as shown, G-EBFC came seventh in the 1923 King's Cup. Its sister, G-EBLB, was the ninth and last of the type to be built. (*Courtesy Air Marshal Sir Ralph Sorley*)

The stumpy little Rainbow racer was a typical relic of Sopwith war-time design technique inspired by Harry Hawker in the days of Herbert Smith.

great reception. Perhaps the most delighted man was John Siddeley, who received the King's Cup from the Duke of Sutherland, his lordship felicitously commenting that this day had seen the triumph of the air-cooled engine. That seemed a little hard on Fedden, who hoped that in the subsequent Aerial Derby on August Bank Holiday, the Jupiter-powered 21-ft span racing biplane Thompson had completed to Sopwith's ideas might prove faster than the famous but much modified ex-'Bamel' now called the 'Gloster'.* Though the Sopwith-Hawker Rainbow seemed a reconstruction of the earlier ill-fated Sopwith Schneider of 1919, with the same wing chord and length of 19 ft 3 in identical with Tabloid and Pup, the Gloster had a foot less span, was cleaner, more powerful, and faster, and Larry Carter thundered round the triangular course at an average speed of 192.4 mph, easily beating Walter Longton's 164.5 mph with the Sopwith. At the prize-giving, Larry from sheer modesty disappeared, and David Longden received the Cup from Lieut-Col Francis McClean.

Success now attended the de Havilland company early in August with their new D.H.50 biplane which followed the D.H.37. It was a carefully considered venture to replace the D.H.9C, representing a scaled-down D.H.18B with four-seat cabin between engine and pilot, and had a 230 hp Siddeley Puma with nose radiator. As one technical journalist reported: 'Looking at the new D.H.50 in a casual way, one might be forgiven for failing to detect anything remarkable in its design. Resembling in general appearance other de Havilland machines that have become famous, the D.H.50 looks, and is, a clean, straight-forward piece of work, but certainly not remarkable for originality. It is not until one begins to go into detail, and looks at figures of weight, useful load, and so on, that the real merits become apparent. When it is stated that the D.H.50, although possessing all modern features such as oleo undercarriage, damped tail skid, gravity petrol feed, carried four passengers yet weighs some 100 lb less than the

* Subsequently this led to confusion over the company name of Gloucestershire, so their aircraft tended to be prefixed with the name Gloucester and, eventually, Gloster.

D.H.9 from which it has descended, one begins to get an idea of the progress made. Even so, it seems possible that it will not be the economy, performance, nor general excellence of design that will make this type popular in years to come: most likely it will be the delightful manner in which the D.H.50 responds to her controls not merely at full speed and cruising, but at or near the stall. Here is an instance where a machine has come out right from the first. Any designer knows this occasionally happens, and that one can never be quite certain whether it will or will not happen, but in this case there is not the slightest doubt that the machine is exactly right, and one congratulates the designers on their achievement.'

That the D.H.50 was a great success was emphasized when Alan Cobham took it to the Gothenburg International Aeronautical Exhibition a few days after its first flight. Britain was well represented with the Avro Aldershot, Fairey IIID, Handley Page Hanley, Vickers Viking, a Bristol Fighter fitted with Jupiter radial, the Armstrong Siddeley Siskin, the new Gloucestershire Grouse and Grebe; engine manufacturers exhibited the Jaguar, Lynx, Jupiter, Lucifer, Lion, Cub, Eagle IX and Condor. Said a partisan observer: 'There is nothing in the whole show to compare with the modern British machines.' However, France was represented by Breguet, Caudron, Farman, Hanriot, and Lioré et Oliver, and Germany, despite restrictions, showed three Junkers monoplanes, two Dornier flying-boats, and small wood-constructed aircraft built by two other companies.

The Civil Aircraft Competition was the magnet, and to assess the winner the Swedes devised a complicated formula based on arrival time and distance, daily return flights via Malmö to Copenhagen for consumption tests, ability to carry load, and high speed in relation to all-up weight. It needed a mathematician to ensure that the rules were properly applied, so Charles Walker went with Alan Cobham, together with Admiral Mark Kerr, their overseas representative.

Said Cobham later: 'For those who have not flown the D.H.50 let me tell you it is a pleasure in store. I think it is the most delightful machine I have ever handled. It is the last word in pilot's comfort, and even in blinding rain-storms, goggles are unnecessary and perfect view is obtained. She is 10 mph faster than the D.H.9, and cruises on half throttle at 100 mph. The competitive flying between Gothenburg and Copenhagen and back started on 7 August and continued five days. Machines were carefully weighed, loads properly checked, petrol consumption measured, and flight time correctly taken. The objective was the maximum load carried at greatest speed on smallest fuel consumption – fuel consumption being the basis for the horsepower used. There were also points for punctuality, construction, cabin comfort, etc. After a hard struggle, in which the Junkers were my chief competitors, we succeeded in getting 999 out of a possible 1,000, one point being lost because we were 90 seconds late in starting one morning. Great credit is due to Dr Malmo, who arranged the contest. We won first prize, 15,000 kronen (£800) and the gold cup presented by the Crown Prince of Sweden.'

The cash prize was retained by de Havillands, but Cobham vociferously insisted on retaining the gold cup. Undoubtedly he deserved it, for to the fatigue of flying, with a start at 7.30 a.m. and return to the hotel at 7 p.m., were added the rigours of a banquet every night with visits to the local dancing hall, resulting in less than four hours sleep before beginning again next dawn. In *D.H.*, Martin Sharp adds a slant typifying the salesman's attitude of those days: 'One of the points which Walker managed to get across the judges concerned structural durability. This was because his keenest competitor appeared to be a self-confident young German named Hermann Goering who was flying a Junkers monoplane. Walker examined its corrugated aluminium skin. "No thin perishable material whatever in the D.H.50," argued Walker with all and sundry, "robust wood members and good metal fishplates throughout. Proved in all climates, and any carpenter can repair it." ' His dogged, learned attitude seemed to clinch matters.

Although refusing to encourage complex and expensive British experiments, the Air Ministry were prepared to evaluate outstanding foreign designs such as the advanced all-metal Junkers-J 10, seen here with enclosed rear cockpit for transport duties. (*Courtesy Air Marshal Sir Ralph Sorley*)

The next great event following the Gothenburg Exhibition would be the Schneider International Seaplane contest at Cowes on 28 September. Blackburn's war-time N.1B flying-boat hull had long been carefully stored at Brough, and in its new form would have a sesquiplane V-strutted wing structure, with a Napier Lion on the top surface of the upper wing. However, the Blackburn staff were pressed for time to get it designed because Bumpus was engrossed not only with design and construction of the 88-ft span Cubaroo, but was making preliminary studies which seemed to favour a high-wing monoplane for a three-seat deck-landing reconnaissance replacement for the Blackburn and Bison fleet-spotters. In any case any chance with an unproved racer was insignificant, for Supermarines had further cleaned up their Sea Lion and fitted a special Lion engine, and Hawkers entered the Sopwith-Hawker racer on floats – though within days it was written-off while still a landplane, for its experimental spinner came adrift while Longton was flying, smashing the propeller, and in the forced landing in a small field the little machine turned over, luckily without injuring the pilot. Against the British competitors in any case there was formidable threat from the entry of three from France, three from the USA, and two from Italy.

Although goaded in Parliament by Capt Wedgwood Benn to publish the report of the National and Imperial Defence Committee in full, the Cabinet uneasily kept it secret, but released the recommendations of a special sub-committee, consisting of Lord Balfour, Lord Peel and Lord Weir, delegated to consider 'the relations of the Navy and the Air Force as regards control of Fleet air-work.' In general terms they clarified that the Air Ministry raised, trained, and maintained the Fleet Air Arm, which at sea was placed under operational control of the Admiralty who were responsible for design, construction and maintenance of the aircraft carriers. Fleet Air Arm duties were aerial reconnaissance, naval gunnery spotting, bombing and fighting. RAF squadrons might similarly operate over the sea from shore bases in co-operation with the Navy, but *liaison* between the two Services was by a special organization termed 'Coastal Area' under general control of the Air Ministry, and commanded by a senior RAF officer communicating directly with the C-in-C Atlantic Fleet. For big operations the Naval C-in-C would notify requirements to the AOC Coastal Area, but minor operations were controlled by the Area Commanders at Leuchars or Lee-on-Solent. On board aircraft carriers, Air Units were under command of the Naval C-in-C, who gave direct orders for flying to the Captain of the aircraft-carrier, who in turn passed them to the senior RAF officer on board. Aircraft design and research to meet naval requirements would be conducted by the Air Ministry after consultation with the Admiralty, and the former would draw up specifications and call for designs from the Trade. New uses of aircraft, possibly involving new types, could be proposed by either Service. There would also be a 'Joint Technical Committee on Aviation Arrangments on HM Ships', composed of representatives of Admiralty and Air Ministry to examine technical problems and suitability of aircraft intended for carriers.

Such broad principles hardly merited the deliberations of a special committee, for other decisions merely clarified obscurities and differences of opinion on such matters as responsibility for discipline when an RAF officer left the ship where he was under command of the Captain, and went for training to an aerodrome controlled by an RAF officer. As a sop to sentiment, it was recommended that 'the uniform of a Naval flying man, who except for his period of training was under the Admiralty, should be distinguished from the flying men under the Air Force by some differentiating badge or mark. This would be the outward and visible sign that he still remains a member of the Service which he originally joined.'

The Government considered it impracticable to supersede separated ministerial heads of the three fighting Services by subordinating them to a Minister of Defence, nor was the alternative plan of amalgamating their departments practicable. Nevertheless co-ordination by the Committee of Imperial Defence was regarded as insufficient, so it was ruled to be consultative but not executive, the power of initiative resting with Government departments and the Prime Minister. For the time being there would be a lull in the Admiralty's struggle for absolute control of its Air Service.

It was not naval matters but the great maritime contest of the Schneider Trophy which was currently attracting sporting and technical attention. Sam Saunders had placed his East Cowes works at the disposal of the contestants, and early in September the Americans arrived with an obsolete 300 hp Wright-powered Curtiss Navy twin-float practice biplane, two Curtiss Navy biplane racing seaplanes with 450 hp Curtiss D.12, and a big Navy Wright I-strutted biplane seaplane powered with an awe-inspiring, high-compression, 700 hp vee twelve Wright – but high hopes for its success were almost immediately dashed when the three-bladed metal propeller failed during a practice flight in the vicinity of Selsey Bill, punctured the floats, and the machine turned upside-down on alighting. A few days later four French machines arrived – two of which were C.A.M.S. 33bis flying-boats with twin 360 hp Hispano-Suiza in tandem, followed by a somewhat similar Latham flying-boat with twin 400 hp Lorraine-Dietrich engines in tandem. A Blanchard seaplane which had a French-built Jupiter was withdrawn. The Italians had cancelled as their big engines were insufficiently developed, but like the French, their expectation was that brute power would win the day, for the less powerful Navy Curtiss, though a beautifully clean float seaplane, was regarded as only a fine weather machine and the Sea Lion flying-boat too slow.

It was because of its suitability for rough seas that Supermarine hoped their dated but pepped-up entrant would make a showing. At least it was ready. Not so the Blackburn Pellet, for its completion was a month late, and when launched from the Brough slipway the strong ebb swept it

For the 1923 Schneider Trophy race, the 1922 Supermarine Sea Lion II winner was revamped as the Sea Lion III with greater power, more seaworthy hull, and larger wings to cope with a 400-lb weight increase, but it was too obsolete to win.

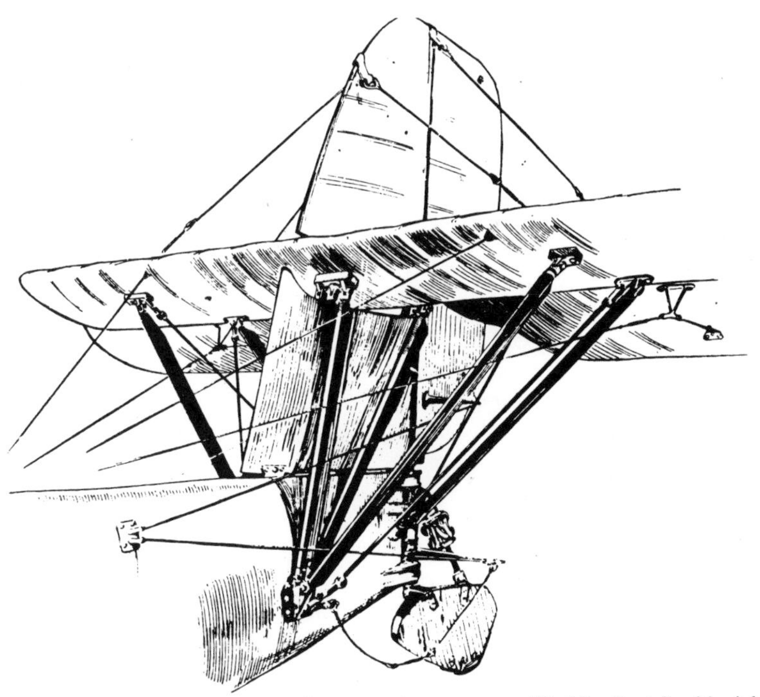

Drag of the Sea Lion III was excessive everywhere, as exemplified by the tail with eight struts and many wires and cables. The original water-rudder tailskid of its amphibious grandfather, the Sea King I, was retained. (*Flight*)

sideways, submerging the starboard wing-tip float, and the machine slowly turned turtle. They managed to haul it out without great damage; the engine was stripped by Napiers and reassembled, and by working night and day the airframe was taken apart, dried, repaired, re-built, and larger wing floats fitted. Lacking time to fly it, the Pellet was railed to Southampton and assembled at Fairey's Hamble Works. Only at dawn on the day before the race was assembly completed and the machine launched. With Blackburn's test pilot R. W. Kenworthy at the controls, the machine lifted a cascade of spray in taxi-ing and shipped considerable water, but as the sea was calm a take-off was made. A strong backward pull on the controls had to be maintained for level flight, but before investigation could be extended the cooling water boiled, and it was necessary immediately to alight south of Calshot seaplane base, whence the Pellet was presently rescued and towed to Cowes by one of Saunders's motor-boats. Working through the night, the Blackburn and Saunders teams by-passed the wing surface-radiators and fitted a cylindrical Lamblin radiator borrowed from Folland, and the wooden propeller was replaced by a metal one loaned by the Americans.

On the day before the race, 27 September, under ideal conditions, competitors, led by the Sea Lion III, undertook taxi-ing and take-off tests. 'The scene at Cowes was one to gladden the eye,' wrote an eye-witness.

'The sun was shining, the sea calm, and the hum of aero-engines filled the air. A Supermarine Seagull amphibian circled overhead, and farther out a twin-float Fairey IIID floated around majestically. A yellow biplane with rotary engine became a Parnall Panther, but despite his ancient engine the pilot was enjoying himself at quite low altitude, presumably relying on his flotation bags in case of trouble.

The Blackburn Pellet beginning to porpoise in a battle between high thrust-line and hydrodynamic instability. The elevator is hard up: two seconds later it was down.

'At about 11.30 the Blackburn Pellet was finished, and officials were notified that Kenworthy would attempt navigability and water-tightness tests. The engine was started, and it taxied down the Medina to the starting line. Getting into position, Kenworthy opened out his Lion. The bows rose at considerable angle, and the machine began to lift, sinking back and rising again in a form of "porpoising", but did not give the appearance of pitching to any considerable extent, though hitting the sea with a series of resounding smacks. In a few seconds the Pellet began swinging to starboard; the wing-tip touched, and dramatically the machine turned on its nose and sank. For what seemed a long time there was no sign of the pilot, but suddenly he appeared, bobbing up like a cork, and was picked up by one of the many motor-launches which sped to the crash.'

In the calm morning sunlight next day, the sea was silky blue, and the Hampshire shore green and clear, and buildings could be discerned eight miles away at Southsea. In a line the seven contestants pulled lightly at their moorings in the Solent, the big twin-engined Latham being nearest inshore. In the anchorage beyond were several clipper-bowed steam yachts with tall yellow-painted funnels, and smaller craft were anchored nearer Egypt Point. Time-keepers and officials crowded the Victoria Pier. The Promenade was thronged with thousands of spectators. Even the British Navy arrived in the shape of a P-boat to add their mite to the US *Pittsburgh* and the French Navy's *Verdun*. The wind began to stir enough to extend the White Ensign on the castle-like building of the Royal Yacht Squadron, the national flags flying on the pier head, and to ripple the string of bunting along the Esplanade. The smooth sea of yesterday broke into ripples which would make it easier for the pilots to distinguish the surface when skimming low across the water.

The course was a shallow triangle, with starting line between two mark boats at the mouth of the Medina, with an initial leg of $18\frac{1}{2}$ nautical miles to a white cross on the ground near the white windmill at Selsey, then a steeply banked left-hand turn to Southsea pier and back to Cowes, repeated four more times to give a total of 186 nautical miles.

At 11 o'clock the Americans started, followed fifteen minutes later by the Supermarine, and by the French after another fifteen minutes. Even while the Sea Lion waited for the starting signal, a small dark speck appeared in the direction of Southsea, the second turning point, and then another. It meant that the first of the Americans had been flying at over 170 mph. Soon the number '4' on the rudder showed this was Rittenhouse. Twenty seconds behind came No. 3, piloted by Irvine. Meanwhile Biard took-off. When the signal was given for the three French flying-boats to start only No. 9, the C.A.M.S. piloted by Hurel, crossed the starting line, one having fouled a buoy and the other suffering engine trouble. Almost immediately the Americans passed on their second circuit, and Biard on his first, though it was clear he was far slower. The C.A.M.S. was entirely

American development of racers at first followed the pattern of British compromise, but soon led to Army and Navy biplanes with wing radiators and metal propellers. (*Flight*)

outclassed, and after the first lap was forced down with engine trouble near Selsey Bill. The two Curtiss Navy racers and the Sea Lion continued lapping and on the fifth and last the American winner, Rittenhouse, achieved over 181 mph, averaging 177.3 mph for the course, with the second American at 173.46, and Biard a long way behind at 151.16.

At the Royal Aero Club banquet that night at the Cowes premises of the Royal London Yacht Club, Sir Sefton Brancker, eye-glass twinkling under the electric lights, told his audience: 'Our glorious defeat is likely to do us good; moreover had we won despite our great handicap, the Admiralty would have said that they were right in not assisting, and that everything was splendid. They now have something to worry about – though there are plenty of millionaires who now have the chance of being patriotic enough to provide funds to bring back the Cup from America next year.'

Of that, Scott-Paine commented: 'We lost to the better machines and men, but I need to apologize to Capt Biard because we did not give him a good enough machine. As to the engine, Mr Vane of Napiers gave us a unit that would have gone on for ever. The Supermarine was in fact the reserve machine, for the Hawker-Sopwith was our hope, so we did not do too badly – at least we finished! But it is no good cursing the Admiralty for lack of support. It is the Treasury which shut the door against expenditure on this race. Money is the real necessity to enable us to win. My firm will, however, do its best within our limited means to bring back the Cup next year.'

7

To design a racing seaplane which could beat the Americans needed most cautious consideration because the cost could only be financed from profits, yet production orders gave no hope of a big return, and many firms would even have difficulty in paying a dividend on investors' capital. Construction of miniatures such as motor-gliders seemed the limit to which it was reasonable to go unless there was strong likelihood that by building a private-venture prototype a production order would follow.

Sopwith and Fairey were hopeful enough to investigate possibilities, for it would be a splendid advertisement to win the Schneider – but even design work they had in hand was stretching resources too far to attempt more. The difficulty was further emphasized when the latest Navy Curtiss R2C racer, powered with a 488 hp Curtiss, won the US Pulitzer Trophy Race at 243.68 mph, photographs revealing a superbly clean machine, far more elegant than the Gloster racers, with monocoque fuselage and 22-ft span I-strutted biplane wings with flush surface radiators.

At Hawkers, the Duiker had proved a disappointment when flown by Fred Raynham in July. Perhaps he had not been very clear about its faults, but it was directionally unstable, and there was longitudinal instability when gliding. No attempt was made to correct this, for it was thought that the big cut-out in the centre-section was the culprit – but with the pilot located at its centre, it was impossible to close the hole for this was his only upward view and means of access. Then Martlesham found wing flutter and the machine was abandoned. Nevertheless the Air Ministry had confidence enough to give Hawkers a further prototype order – this time for a night fighter to specification 25/22. Named the Woodcock, it developed into a rather clumsy two-bay single-seater of 34 ft 8 in span, though with an interesting continuation of the general appearance of Sopwith war-time fighters. The first engine fitted was an Armstrong Siddeley Jaguar, now proved by a 100-hour test. After the initial flight, Raynham said the rudder seemed completely ineffective and that the machine was unsafe to spin; it was also heavy laterally, for it had excessively wide ailerons integrated with Fairey-like flaps, and the control system shook badly in a form of flutter at high speeds. Modification, and even major re-design, was needed. It seems this led to dissatisfaction of Thompson by Sigrist, and the Hawker designer abruptly left, going to Sam

Thompson's two-bay Hawker Woodcock night fighter lacked the neatness of Folland's designs, but its big wings were initially thought essential for low landing speed.

Saunders at Cowes, where he reopened the aircraft design department which had been non-operative since the disaster of the Kittiwake flying-boat. There he was joined by Henry Knowler as assistant chief designer. Recollecting those days, Knowler told me: 'There was a considerable gap between Percy Beadle leaving S. E. Saunders and old man Saunders deciding to re-start in flying-boats. The initial team in 1923 consisted of Bertie Thompson, who came from the Air Ministry via Hawkers, David Nicolson, and myself. My task was to design the Valkyrie, for which we obtained an Air Ministry contract on the strength of my knowledge and experience as assistant to W. O. Manning on design of the Cork and Kingston flying-boats while I was working at the aviation department of English Electric. The Valentia flying-boat was a Vickers machine which was designed in my time at Weybridge when I was with Rex Pierson, but its Consuta hull was designed by George Porter, the almost unknown Saunders motor-boat designer who later did design work for Uffa Fox on yachts. While I was engaged on the Valkyrie, Thompson undertook a small civil machine called the Medina, but it happened that the Medina proved a dead loss and the Valkyrie not too bad – so Tommy left, and I stayed on as chief designer.'

Hawkers now re-engaged George Carter, the ex-chief draughtsman of Sopwith Aviation Co, as chief designer. He had been out of the industry almost three years, and had let his house to the Hawker chief draughtsman, L. E. Metcalfe, who told him of Thompson's resignation and transfer to Saunders. To the Hawker directors, Carter seemed the obvious successor, for his years of experience of Sopwith and Sigrist design policy would be invaluable in giving time-continuity for their efforts to break into the peace-time aircraft industry with military designs. Carter had not been there long when Fred Raynham asked him to take on a withdrawn but hard-swearing, hook-nosed, tall 30-year-old named S. J. Camm, who said he had learned draughting during the war at Martinsyde Ltd, and latterly had been George Handasyde's only draughtsman, but he did not reveal that he was author of a little book on aircraft design, nor his modest renown as spare-time mechanic for Raynham's personal Martinsyde F.6 tourer.

Many RAF pilots favoured the Hawker Hedgehog because of easy handling characteristics and absence of 'plumbing', but only a few were ordered. (*Flight*)

Carter's first job was to re-design the Woodcock with single-bay wings and a Jupiter installation, eliminating the landing flaps, increasing the distance between the spars, and fitting narrow-chord ailerons. Concurrently a scaled-up version to specification 37/23 was designed as a private venture to comply with Fleet Air requirements for a three-seat reconnaissance machine. Compared with the 32 ft 6 in span of the Woodcock, the new Hedgehog, which had folding wings, had a span of 40 ft $\frac{1}{2}$ in, and the landing flaps were reinstated. The three-seat arrangement was compact, with the pilot directly under the front spar where he had remarkably good forward and upward view. The gunner was in line with the trailing edge, well clear of the cut away centre-section, and the navigator had a cockpit between the two. The machine was regarded as a more easily handled replacement for the not entirely satisfactory Bison and the cumbersome Blackburn, coupled with the advantageous simplification of an air-cooled engine.

Another newcomer was David Hollis Williams, who in recollecting those days said: 'I started with Hawkers working under George Cornwall who was chief technician; Carter was chief designer; Sydney Camm had just joined as a senior draughtsman. One day Freddie Sigrist came into the drawing office, and went to Carter's glass cubicle on its raised platform in the middle of the office, had a conversation, and went out. Carter came down the office and said, "Anybody know the sharp end from the blunt end of a boat?" Nobody seemed to volunteer so I put my hand up and told him I knew a little. He said, "Here's a job for you. Sigrist wants the handicap of his racing yacht reduced." About a week later Sigrist came into the office, and after a few moments Carter came out and said, "Who did I give that job to?" I raised my hand again. Sigrist pushed Carter away and put a stool against my desk, and we got down to this business of reducing his handicap, and it proved quite successful. Then he asked if I would like to crew for him – so at week-ends I started sailing with him and got to know him quite well. He was a lean, dark, almost Spanish-looking type of medium height, and was an absolute whirlwind of energy and activity. He was virtually works director – ran the shops, but took no part in designing these days except to criticize. Everything was "bloody awful and couldn't be made", and inevitably became a more practical job through re-designing it.'

Of that period Herbert Mettam told me: 'Although remaining grateful to Thompson for giving me my first stressing job, I left the firm before he did. I therefore did not get to know his two immediate successors until several years later. Both George Carter and Sydney Camm were experienced men whose advent greatly strengthened the Hawker design staff. Camm initially served under Carter, but was soon given a free hand.

'I had joined Westland as head of the technical office on 1 July, and took over from J. D. Williams. His assistant was John Digby, who had joined Westland on coming down from Cambridge in 1922 and continued as my assistant for several years. John and I were friends and contemporaries at Cambridge, and were both coached by H. A. Webb, one-time mathematician of Farnborough. W. M. 'Bill' Widgery joined from the Handley Page wind-tunnel, at which firm he had earlier been an apprentice, just in time to take over my lodgings when I got married on 29 September. By the time I had joined, all the stressing and aerodynamics of the Woyavodsky Dreadnought had been settled, except for very minor details. Arthur Davenport recently had been promoted chief designer, and Tom Reilly was chief draughtsman – but it was R. A. Bruce who was the omnipotent head of all, whether design, construction, or business.

'Perhaps it is of interest that I was paid £250 per annum by Hawkers, but promoted to £360 by Westland, and that was £110 more than Williams had got before he left for a job as chief draughtsman at the

Aircraft Disposal Co where he was paid £360, but went on to become chief draughtsman with Boulton & Paul. For my old job at Hawkers it also happened that George Cornwall got £360, though there was a short interval when Henry Davies was the stressman. As a sidelight on the contemporary scene at Kingston, we usually lunched at Bentalls, now a much enlarged department store with branches at other towns, and their charge for soup, meat, sweet and coffee was only 1s 6d – moreover a request to the three ladies of the orchestra for any item of their extensive repertoire would result in it being played without further reminder on our next reappearance.'

Far ahead of contemporary fighter conceptions was the Shipboard Fighter HPS.I (or H.P.21), interchangeably a seaplane or landplane. The monocoque fuselage was lightly framed and thinly skinned. Floats were built but not tried.

That September the H.P.21 prototype was completed and flown by Fred Raynham, but the rudder proved inadequate and heavy, and torque reaction of the rotary engine put the wing down badly at take-off, which complicated the lateral control use of the combined aileron-flap. At the stall there was a big change in trim, and considerable tail shake which seemed insurmountable as it was not realized that this was a wake effect from the wing/fuselage junction. In view of the complexity H.P. put the entire test programme in the hands of Colonel Ogilvie, who already had served him well in setting Handley Page Transport on its feet. Eventually, with a balanced rectangular rudder, the machine was sent to Martlesham for corroborative assessment, but before the RAF pilots had opportunity of handling this advanced looking fighter it was crashed by Raynham through breakage of the control column at its root, necessitating an

Almost ready for first test flight: the Handley Page H.P.21 receives final inspection in the erecting shop. By later standards the root fillet was inadequate – for nothing was known of interference.

attempt to land by bending down and pulling at the control column socket, touching down blind, but miraculously without injuring himself.

Richards' Hyderabad military version of the W.8b had far better luck at Martlesham, and was so much liked that an initial batch was ordered for No. 99 Night Bombing Squadron of the extended RAF. Currently he was working on the W.9 as a three-engined version of the W.8 both in commercial and military rôles.

With future pilot requirements in mind, the Air Ministry had invited tenders for the establishment of Flying Schools, and five were accepted:– The Bristol School at Filton operated by the Bristol Aeroplane Co, using a two-seat version of their Lucifer-powered Taxiplane for elementary training and Jupiter-powered Bristol Fighters for advanced training;

With altered rudder and smaller fin, the H.P.21 was flown to Martlesham. Wind-tunnel tests had predicted the need for big tail adjustment to trim between flaps-and-slots closed or open.

Derived from the Bristol Taxiplane by using a similar but narrower two-seat fuselage, G-EBFZ was the first of the Bristol Flying School trainers (P.T.M.). They were firm on controls but with heavy rudder and vibro-massage from the Lucifer.

Armstrong Siddeley Motors Ltd would operate a school at Whitley near Coventry, with Renault-Avros for preliminary training and D.H.9s for advanced until replaced by a dual-control version of the almost completed Armstrong Siddeley Wolf; Wm Beardmore & Co would establish a school at Renfrew using Avros and D.H.9s; de Havilland Aircraft Co Ltd would operate a school from Stag Lane with the D.H.6 and Renault-Avro for preliminary training and D.H.9s for advanced flying; and at Brough it was agreed that the Blackburn Aeroplane Co Ltd would shortly establish twin-engined training with reconditioned Kangaroos fitted with dual control in tandem cockpits, and later would institute a seaplane school equipped with Dart floatplanes. At each school it was intended that eighty pilots should take refresher courses. As yet there was no intention of training *ab initio* pilots.

With an impressive feeling of power from its de-rated 290 hp Jupiter, the modified Bristol Fighter Trainer, Type 89, with Frise ailerons and balanced big rudder, was a delight to fly.

Entries for the great Motor-Glider Competition had to be in by 1 October. There were twenty-three British machines, though five were not ready in time to compete, and there were also four foreign entries – two French Peyret shoulder-wing monoplanes which had proved markedly successful at an earlier *Avionette* meeting at Vauville in August and two Belgian Poncelet mid-wing cantilever monoplanes.

During the week-end preceding the competition on 8 October, the grassy airfield of Lympne, on table-land high above the marshes spreading to the sea, was busy with competitors carrying out the pre-trials mandatory transport test. Wings had to be folded or dismantled within limited time and the machine man-handled along a mile course which included roadway, and then pass through a 10-ft gateway to show the practicability of getting from a field. The D.H.53s were in immediate trouble, for the tailplanes were wider than the gate, so Walker proposed the simple expedient of sawing off the tips.

Because of its efficiency and novel design the A.N.E.C. was bought by the Air Ministry for further evaluation, but it was the war-time Kitten which had been the first ultra-light to wear RAF roundels. (*Courtesy Air Marshal Sir Ralph Sorley*)

On Monday, fifteen of the British were ready on a calm and misty morning with low cloud, ideal for the principal test of fuel consumption around the 12½-mile triangular course extending to the edge of the South Downs. Luck of the draw fell to Longton, who took off with Wren No. 4 at 7 a.m., skimming low into the distance. In turn the majority of competitors followed, but because the wind gradually freshened they did not all attempt the economy test, though Jimmy James, with A.N.E.C. No.17 achieved 87.5 mpg, and Longton on his second attempt covered 85.9 mpg.

Additional prizes had been offered by the Abdulla Co of £500 for the highest speed over two laps without restriction on fuel consumption, and £150 each by the Society of Motor Manufacturers and Society of British Motor-Cycle Manufacturers for the greatest number of circuits *in toto* by a British machine and pilot providing 400 miles minimum was covered and no alteration made to aeroplane or engine. Sir Charles Wakefield gave a prize of £200 for the greatest altitude, and the Duke of Sutherland and members of the RAeC added £100 for a landing competition. With such encouragement to fly whatever the weather, Major Hemming with D.H.53 *Sylvia II* achieved 300 miles on the first day, and continued to add laps on the next, but the overloaded little Douglas engine broke its crankshaft –

enabling Bert Hinkler to take the lead with the Avro 560 monoplane by completing 575 miles that evening. Meanwhile several were out of the competition: the Douglas of the Viget broke a rocker-arm causing Cockerell to land some six miles from the aerodrome, and he then wheeled the machine back with folded wings along the road to Lympne. The experimental Carden engine of the Gloucestershire Gannet prevented it flying at all, and the 400 cc A.B.C. of the Sayers-Handley Page was often impossible to start or ran so erratically that it could not fly long enough to compete. The Douglas of the Hurricane also broke a rocker-arm, but was safely forced-landed by Bulman, and Piercey had to forced-land his A.N.E.C. because of a sooted plug. Then A.N.E.C. No.18 cracked a cylinder, but James and Piercey continued in turn with No.17 in the lap contest, and were only 50 miles behind the Avro. Poor and gusty weather on Tuesday and Wednesday caused most competitors to remain grounded, but Hinkler continued circuit after circuit, carrying all before him.

Thursday saw Longton take-off the buff-coloured Wren in an attempt to beat the A.N.E.C. fuel consumption, but he only equalled it, though on an earlier calm day at Brooklands he had achieved 127 mpg. Meanwhile Bolas had the big wings taken from the Pixie, as there was no chance of winning with fuel consumption, and fitted the special 18-ft span wings – which enabled Norman Macmillan to achieve an average of 76.1 mph in the speed tests, but James on the A.N.E.C. was only 2 mph less despite wings suited for quick take-off and climb rather than speed.

Capt Norman Macmillan, MC, AFC, author and outstanding pilot, about to fly the Parnall Pixie and win the speed prize at Lympne. In the following year he became chief test pilot of Fairey Aviation Ltd.

A gale blew throughout Friday, making flying with these lightly loaded aeroplanes too dangerous to attempt, but the rain and the wind enabled competitors to attend to their engines, decarbonize cylinders, clean sparking plugs, and grind-in valves, or change carburettor jets in the hunt for economy. The machines themselves did not require much attention, though some competitors doped fabric strips along hinge-lines in attempts to improve control. Shortly before sunset a faint streak of light in the west promised a better morrow, and the clouds gradually cleared as evening drew on.

Conventional, except for folding wings and dural fittings, but typically Vickers in appearance, the Viget, like the Avro biplane, had a top speed less than 60 mph. (*Flight*)

With Saturday came clear skies and tremendous visibility, though a fairly strong wind. All the aviation personalities seemed gathered for this last day; nor were the hordes of schoolboys missing, and hundreds of the general public watched from the aerodrome enclosure, the fields around, and even the main road running past the south side despite efforts of the police to keep it clear. There was an air of anticipation, an excitement at seeing all these little aeroplanes and their fabulous pilots – for this meeting had all the atmosphere of earlier days when flying was a novelty, for here was the same struggle to perform with inadequate and unreliable low-power engines, though piloting skill had improved so enormously that it made accomplishment much safer.

Long before the crowds arrived, the Viget piloted by Cockerell took the air shortly after 7 a.m., lifting like a rocket, and was followed by Hemming on the D.H.53, Sqn Ldr Wright on No. 3 Wren, Hamersley on the Avro 558 biplane, Hubert Broad on D.H.53 No. 8, Piercey on the A.N.E.C., and Longton on No. 4 Wren. Soon so many were in the air that it was difficult to follow progress. Several attempted the altitude flight, including Raynham, Piercey, and Stocken in the Gull. Then Alexis Maneyrol

Prior to Hamersley's climb in the Motor-Glider Competition, A. V. Roe himself (*left*) supervises adjustment of the engine of the Avro 558. (*Flight*)

brought out No. 15 Peyret monoplane, climbing for the altitude prize, and was shortly followed by Capt Simonet with Poncelet No. 21. Meanwhile Bert Hinkler recommenced lapping with the Avro monoplane, ultimately completing 80 laps with a total distance of 1,000 miles, easily winning the two prizes of £150.

Capt Broad added spice to the meeting on gliding down with the D.H.53 Humming Bird after abandoning an altitude attempt, for at 400 ft he began a series of Immelmann turns and made two loops. Though other light-planes seemed so fragile he convinced everyone that the de Havilland conception was certainly robust. There was more excitement when Olley attempted to take-off the Sayers-Handley Page with a 'bungee' start, for it

Much more efficient and practical than the Avro biplane was Roe's conception of a light cantilever monoplane, the 560, though its biconvex section gave a sudden stall.

wobbled and flickered, just managing to cross the starting line, and then dropped with a bump, and it was a relief when Olley opened the lid and stepped out with a grin. Baron de Lettenhove was even more terrifying with No. 16 Poncelet which still had not got airborne when only 30 or 40 yards from the railings. Just as the pilot ruddered away, the right wing lifted high in the air, and with considerable drift the machine sank to the ground, the left wheel buckled, the left wing-tip touched, and there followed a spectacular cart-wheel, throwing the machine with a thud onto its back.

By this time the celebrities were gathering: Sir Samuel Hoare arrived with Lady Maud, and made a tour of inspection with the Duke and Duchess of Sutherland and Lord Hugh Cecil. About noon the Duke of York arrived, and was conducted around the line of machines, pilots, and constructors by General Festing, Sir Geoffrey Salmond, Sir Sefton Brancker, and General Bagnall-Wild.

Maneyrol came gliding down and landed, his barograph showing 10,000 ft, giving a corrected altitude of 9,400 ft. Hamersley followed with the Avro biplane having reached 13,000 ft, and Piercy on the A.N.E.C., hot on his tail, after attaining 13,600 ft. Maneyrol immediately set off on a

Before it flew, several remarked on the dangerously oblique angle of the Peyret's lift struts. (*Flight*)

second attempt, and was followed by Hamersley and Piercey whose previous climbs had been stopped by frozen carburettors.

An hour later Maneyrol was seen at great height, slowly descending, presently circling the aerodrome at a few hundred feet. As he approached to land, suddenly, at only 100 ft, the light tubular bracing struts failed under down load, the wings collapsed, and the machine dived to the ground, killing the pilot instantly. There was horrified silence; then a rush of useless helpers. The Duke of York sent his equerry to convey sympathy to M Peyret and other French entrants, and there were messages from Sir Samuel Hoare, the Duke of Sutherland, and officials of the Air Ministry.

Presently, on a sadder note, the competitions continued. Broad tried to change the atmosphere by again aerobatting the Humming Bird, and towards evening Longton brought out No. 4 Wren and to everyone's astonishment gave a thrilling performance of his crazy flying speciality,

Arthur Longmore, pioneer CFS instructor in 1912, newly promoted Air Commodore from the Army Staff College at Camberley, has verbal instruction from 26-year-old Capt Hubert Broad on handling the delightful D.H.53. (*Flight*)

weaving the machine around in a series of side-slips and skids and complex undulations. *Flight* reported: 'Altogether the Wren is one of the most amazing aeroplanes ever constructed, and absolutely the only criticism any one could level against it is that for cross-country flying it is a little slow. As against that must be set the fact that it is as near fool proof as any aeroplane we have ever seen. The proportioning of fin surface, dihedral, control surfaces, etc, seem as near perfection as one is ever likely to attain, and the aerodynamic efficiency is extraordinarily good in view of the performance attained with an engine of but 400 cc capacity.'

Pleased at the success of the A.N.E.C. at Lympne – 'Bill' Shackleton its designer, Mrs Shackleton, Dr Hope, and the two pilots Jimmy James and Maurice Piercey. (*Flight*)

An hour before the competition officially closed Piercey landed from his altitude flight on the A.N.E.C., so numbed with cold and impeded by poor view from his cockpit that he overshot and nearly hit the fence, but triumphant at achieving a corrected height of 14,400 ft, having beaten Hamersley's ultimate 13,850 ft on the Avro biplane, and thus winning the Wakefield prize of £200. The Duke of Sutherland's prize of £500 and the *Daily Mail* £1,000 for the greatest distance flown on a gallon of petrol was divided between Flt Lieut Longton on the Wren and Jimmy James on the A.N.E.C. for their tie of 87.5 mpg. The Abdulla £500 for greatest speed was Norman Macmillan's, and since the gusty conditions prevented a landing competition, the £100 prize was given to Hamersley for his fine altitude attempt.

Flt Lieut Walter Longton, known fondly as 'Scruffie', was a skilful mechanic and brilliant pilot, equally at home with twin bombers or the little Wren. (*Flight*)

So encouraging had been the development of little aeroplanes specially for this competition that the Duke of Sutherland stated he would offer another prize next year for a two-seater contest, and Sir Samuel Hoare intimated the probability of an Air Ministry prize for machines with an engine capacity of some 1,500 cc. But the Lympne meeting had been a turning point as well. As Terence Boughton wrote in his *Story of the British Light Aeroplane*: 'As at Itford, the cheerful, unselfish spirit of the early days of aviation seemed to have returned. Here were many of the great figures of the old flying years before 1914, some of whom were destined to be greater still in time to come, gathered together in not too serious competition with one another and with the less celebrated to make the aeroplane, so recently a weapon of war, into a vehicle for the use and enjoyment of ordinary man. It was a meeting of enthusiasts, of men who were in aviation not for the material rewards it offered, but simply because its unique blend of subtle theory and practical engineering, of mental agility and physical skill, of eccentricity and adventure, made it the one life they really cared about.'

On the wave of enthusiasm aroused by the competition, a race meeting for light aircraft was held a fortnight later at Hendon. The interval gave de Havilland time to replace the Douglas of the D.H.53 with an inverted version of the Blackburne which had proved the most successful engine at Lympne. Hubert Broad, with No. 8, opened the meeting with a magnificent display of aerobatics which confirmed the confidence with which this pretty little monoplane could be handled. He was followed by Jimmy James on No. 17 A.N.E.C. monoplane and Stanley Cockerell with the Viget 'soaring heavenwards, lark-like, without appreciably changing his position over a particular spot on the aerodrome'. Three eliminating heats of a handicap race followed for a £50 prize given by Sir Charles Wakefield, each respectively won by Hamersley on the Avro biplane after Hinkler's monoplane with special small wings had dropped out, Broad on the D.H.53 and James with the A.N.E.C., with Broad winning the final race. Finally there was a speed contest for another £50 prize from Wakefield which was won by Macmillan on the small-span Parnall Pixie II. Reported one hopeful enthusiast: 'If Saturday's meeting is any guide, we can see next summer providing a number of successful and enjoyable week-end light plane meetings at Hendon, and as the Underground will have a station just outside the aerodrome, a popular revival of the pre-war flying meetings is not at all unlikely.'

9

In October Bristol's earlier famous designer, Capt Frank Barnwell, returned from Australia, where he had found it difficult to work in harmony with Sqn Ldr Wackett. Wilfred Reid, appreciating the position from Barnwell's letters, had negotiated a contract to join Canadian Vickers Ltd – so the way was open for Barnwell to rejoin his old company.

As he had predicted, the two-seat mono/biplane Bullfinch was not good enough to replace the Bristol F.2B Fighter, particularly as its structural weight had increased due to change of mind by the RAE on allowable stresses for thin steel. Reid had already attempted several proposals to meet specification 3/21 for a Napier-powered machine, but such an engine inevitably was opposed by Fedden. Under pressure from Sir Henry White-Smith, the Air Council agreed to a radial, but required it to be supercharged, and defined new requirements to specification 3/22. Although the alternative of using alcohol fuel and higher compression was discussed, development was concentrated on the Orion version of the Jupiter with RAE exhaust-driven supercharger.

The Bristol directors decided to build a private venture biplane pending official contracts, and the Bloodhound, as it was named, was initiated as a

The Short Springbok I spun into the ground due to rudder blanketing, so after wind-tunnel investigation a revised version with longer fuselage, deep rudder, and smaller-chord lower wing was built, but the open centre-section was retained.

Built on familiar de Havilland lines, and almost the same size as the D.H.9A, but lighter, the D.H.42 Dormouse was pretty and relatively cheap, but not good enough. (*Flight*)

In an endeavour to secure better performance of the D.H.42A Dingo variant, a Bristol Jupiter replaced the Armstrong Siddeley Jaguar, but the gain was only 2 mph, and ceiling increased from 16,000 ft to 17,500 ft. (*Hawker Siddeley*)

civil aircraft registered G-EBGG on 3 May, 1923. Flown by Norman Macmillan, it was found that the fin and diagonally-hinged rudder were inadequate. Much bigger surfaces were fitted and the sweep-back was increased. In this form an RAF order for three was received in June for competitive trials against the de Havilland Dormouse, Short Springbok, Vickers Vixen, and the Armstrong Siddeley Wolf as contenders for the Duiker specification.

Longitudinal stability was still unsatisfactory and Barnwell now largely solved matters by increasing the dihedral, lengthening the fuselage, and tilting up the engine thrust-line. In this guise, with wood wings and steel-tube framed fuselage, the prototype was flown to Martlesham three months later, where it attained 230 mph and a ceiling of 18,000 ft. However the Dormouse was preferred, largely because its wood construction was deemed a better production job, and to improve on its slightly inferior performance a version known as the D.H.42A Dingo, powered with Jupiter III, was ordered for further investigation. Virtually these two de Havilland machines were little more than up-dated D.H.9As, with modern engine and lighter tare and full-load weight, for Geoffrey de Havilland always seemed to design the same machine in slightly varying form, but with the same artistic lines.

By comparison the Vickers Vixen was a squat, heavily aggressive looking two-bay biplane, suggesting a break-through in design, though in fact its performance was little different from the others, nor was it likely to offer easier production or maintenance. Nevertheless when Broome took it to Martlesham as a private venture, the RAF pilots seemed impressed, particularly with its possibilities in the bomber rôle. There was always the difficulty when a number of RAF pilots flew a new type of aeroplane that their assessments were by no means unanimous when performances were all nearly the same. Sometimes preferences depended on personal foibles, but in most cases the firm's pilots had ironed out the worst faults of each new prototype. Often Martlesham recommendations were eventually swayed by matters of access and servicing and ease of arming, other things being more or less equal as far as flight handling went.

Less obvious signs and portents had bearing on the future. It was even indicative that two Daimler D.H.34s flying from Hamburg to London in the teeth of a gale had to alight twice for fresh supplies of petrol, and arrived at Croydon only a few minutes before dusk. Though it was hoped regular night flying services might soon be initiated, there was still no effective organization, and the absence of special instruments necessitated visual contact on every flight. For airship protagonists there was the encouragement of a flight of 119 hours made by the French-owned ex-Zeppelin. The USA had gone one better by completing construction of their first rigid, the ZR-1, which made its first trial flight on 4 September.

Of interest as a far echo of the crashed Tarrant Triplane was that the man who designed it had got the Americans to build a six-engined, 120-ft span version which successfully flew on 22 August at Wilbur Wright

Field, but the disposition of engines was much more sensible than the Tarrant's. France similarly was investigating four-engined aircraft as possible airliners and bombers, but Britain was content with the inexpensive method of awaiting results of other people's efforts.

More significant was a series of US tests on the vulnerability of capital ships to aerial bombing. The obsolete US battleships *Virginia* and *New Jersey* were sunk respectively in twenty-six and seven minutes by a twin-engined Martin bomber powered with supercharged 400 hp Liberty engines; but British armament specialists were more interested that the bomber carried five Lewis guns in accord with prevailing American opinion on the importance of defence. In Britain the Admirals still insisted that the Navy was the basic defence because for ocean use aircraft were too limited by short radius of action. Sir Percy Scott, who had been responsible for war-time defence of London, saw more clearly, and boldly stated that the best place for battleships in war was at home because the floating forces of the nations were so menaced by air attacks from above and torpedo attacks from below that they would be unable to carry war from one continent to another. Bitterly he criticized the Government for spending millions on new docks at Singapore, and advised the Dominions to contribute nothing. 'The Government should be spending their money on modern weapons of defence,' he said, 'instead of wasting it by supporting antique ideas originating from men who are either ignorant or who have not taken the trouble to analyse the basis of their convictions.'

The comments were well timed to attract the attention of the Dominion Premiers attending an Imperial Economic Conference in London, during which Sir Samuel Hoare outlined the financial difficulties of establishing Imperial air routes, though disclosed that he had set aside money for construction of an experimental civil aeroplane capable of the long-distance journeys necessitated for such routes. 'The designs are now almost ready, and we hope to be able to put in hand the building of a machine of that kind at once.' Hope was the operative word. To show members of the Conference that Britain was aeronautically determined, the Air Ministry arranged a great demonstration at Hendon with between fifty and sixty military and civil aeroplanes of every available type from the impressive Condor-powered Avro Aldershot to the light aircraft of the Lympne competitions. Meanwhile, Baldwin's policy of Tariff Protection was being manipulated to bring Austen Chamberlain and Lord Birkenhead back into the Government; but though Baldwin agreed, he found it impossible to face opposition of some of his lesser ministers and decided to put the issue before the Country – so Parliament was dissolved on 16 November.

Here was Labour's chance, as well as re-union for the Liberals. Lloyd George portrayed himself a Free-Trader and offered to collaborate with Asquith. One of their supporters was Winston Churchill. It proved a landslide for the Conservatives who held only 258 seats compared with the 346 before the Dissolution, Labour providing the main opposition with 191, and the Liberals electing 158 with principal gains in Greater London.

Baldwin was under immediate pressure to resign, but decided to face Parliament when it met in the New Year.

December saw de Havillands achieve an unexpected contract. The RAF had been impressed by Broad's handling of the D.H.53, so twelve were ordered for communication and practice flying with a permitted military payload of one Service hat and one raincoat! But in the hands of Alan Cobham it achieved even greater triumph and further proof of the tremendous stamina and push this pilot had revealed in his many outstanding taxi-flights for de Havillands around Europe. As C. Martin Sharp recorded in his book *D.H.*: 'St Barbe decided that the D.H.53 should appear at the Brussels Aero Exhibition held in December, and Cobham was to go with it. "What's its range?" Cobham asked. "Don't worry," said St Barbe, "you can read all about that in the train." . . . "Why in the train? Won't it fly?" . . . "Do you mean you'd fly it there?" exclaimed St Barbe.

'Cobham was prepared to fly the thing across, but it needed bigger tankage – two gallons more! Hagg had a small tank made to fit behind the pilot's head, but it was not high enough for gravity feed, so he suggested a rubber tube for Cobham to blow down to give pressure feed as there was no pump small enough. When Cobham got to Lympne the 100-gallon bowser trundled up to refuel him for the Channel crossing. "Two gallons, please," said Cobham, trying to maintain his dignity. He took off and headed for Cap Gris Nez. "This is the maddest thing I have ever done," he thought. "No wonder they named it the Humming Bird – this little propeller is revving at 3,000 rpm! One magneto! And then an ordinary motor-cycle engine, revving as it never revved before!" He went through a rain storm, reflecting that the buzzing wooden propeller had not even a metal edge. By the Grace of God he got to the other side and flew up the coast with a following wind. Needing the pressure feed he blew down the rubber pipe. A little later the engine started running roughly, then kept cutting out, coming on, cutting out and coming on again. It was due to dribble in the carburettor resulting from the apoplectic efforts at blowing. Eventually he got there, and a big song was made about the fact that the trip had cost 7s 6d – no allowance for nerves. In the Exhibition the placard on the de Havilland stand described the flight achieved on Saturday 8 December as "Le plus long cross-country vol d'une aviette de 6 cv effectué jusqu'à ce jour" – but it took the wit of C. C. Walker to put the matter in to proper perspective. "Cobham," he said, "did that flight mainly by force of character." '

Even though the return was made ignominiously by train and ship because of snow and bitterly cold head-winds, that flight of Cobham's in itself marked the great advance since the Wrights first flew their pioneering power machine on 17 December, 1903, twenty years earlier, for twelve seconds. At that time the *Auto Motor Journal* recorded: 'A cynic has observed that the joys of successful prophecy are among the most unalloyed given to mankind. At present we are enjoying these delights to the full. Four months ago we referred to the brilliant experiments of the brothers Wright in North Carolina, which we have recorded and illustrated from time to time in the pages of this journal. On that occasion we commented "They are gradually increasing the lifting power of their aeroplanes, and any day we may hear of their putting on a motor and propeller, and actually accomplishing free, independent flight. The Empire of the Air is still to be conquered but we have certainly got a further glance into the promised land than ever before."

'This prophecy now appears to have been completely fulfilled. The brothers Wright have attached to their aeroplane a motor-driven propeller, and, utilizing their enormous and unique experience of how to manipulate the aeroplane under the varying stresses and strains to which it is subjected, by what Mr Chanute had graphically described as "the whirling billows of air" that constitute the wind, their machine has successfully accomplished a free, independent flight.'

And now, with an engine of almost the same power, a pretty little monoplane had revealed the great progress in structure and aerodynamics by flying nonstop 150 miles from Lympne to the Belgian capital in a flying time of four hours – but not only that: speed of aircraft had risen from some 25 mph to 266 mph recently recorded by the Americans, and load-carrying capacity had increased from one man to several tons. Already there were indications that the sky was limitless.

CHAPTER V
IMPERIAL OUTLOOK
1924

'For the present we may content ourselves with thinking of the advantage of sailing through the air at about double the pace of the fastest steamer, skimming whole Continents, and without dust, confusion, or examination of baggage at the frontiers.'

Daily News on Blériot (26 July, 1909)

1

New Year's day saw aviation insurance placed on a sound footing by the formation of the British Aviation Insurance Group, combining the Union Insurance Society of Canton and the White Cross Aviation Insurance Association – sole survivors of the many firms whose underwriters had lost heavily since the war in dabbling with aircraft despite premiums of 30 per cent. The new monopoly company, by discriminating, could well nigh control civil aviation with even greater effectiveness than the Air Navigation Consolidation Order which had just been issued defining strict licensing requirements for aircraft and personnel – but in the hands of that great figure Capt A. G. 'Lamps' Lamplugh, the group's principal surveyor, all could be assured of a scrupulously fair deal when it came to vital insurance cover.

But the makers of law were themselves in trouble. On 6 January Parliament met. The *Daily Mail* led the clamour for a Coalition as it was clear that the country rejected Protection, and it could only be a matter of time before Baldwin would be defeated if he attempted to continue as Prime Minister. On 17 January J. R. Clynes moved an amendment to the King's Speech 'that the Cabinet has not the confidence of the House', and Asquith sealed its doom by stating he would vote for this. Four days later the Government was defeated by 328 votes to 256. Baldwin resigned, and Ramsay MacDonald took office. Never before had there been such sheer inexperience as that of the new Labour ministers. Looming over them was the massive unemployment of one million. The City of London expressed utmost concern, and in the RAF there was disturbing belief that Labour would tamper with the promised modern equipment instead of scrapping

A replacement for the Bristol Fighter was the aim of many designers, but it was almost a matter of luck to devise one with the same appeal. Although the Armstrong Siddeley Wolf had superficial similarity its handling was poor. (*Courtesy Air Marshal Sir Ralph Sorley*)

the war-time collection of Snipes as first-line fighters, the D.H.9As struggling with tropical conditions, and overloaded Bristol Fighters spotting for the army.

Many of the new types were nearing completion, and two were tested in January. John Siddeley had recently bought the ex-RAF aerodrome at Whitley, and here Courtney flew the Armstrong Whitworth (formerly Armstrong Siddeley) Wolf on the 19th, and found it satisfactory though slower than the Bristol Fighter it was originally intended to replace. A week later at Brooklands, the revised Hawker Woodcock was flown by Raynham and showed considerable improvement in handling and negligible flutter, but when tested later for climb in bitter weather the Jupiter's exposed valve gear proved vulnerable to icing, cutting out the engine. Similar trouble had been previously experienced with the Jupiter-powered Westland Weasel, and to overcome it, Farnborough had fitted cylinder 'helmets' after wind-tunnel investigation. A similar scheme was therefore

In an attempt to reduce the drag of radial cylinders, and in the hope of improving cooling of the rear fins, experiments were made with ducted helmets on several aircraft such as the single-bay Hawker Woodcock. (*Courtesy H. F. King*)

The Westland 'Postal' Dreadnought being assembled in the Vimy Shed, with D.H.9As in the foreground undergoing reconditioning. It was not realized that the cabin flare into the wing might cause adverse flow.

applied to the Woodcock, and partial 'helmets' were fitted to the Hedgehog prototype which Raynham flight-tested soon after the Woodcock, finding it even nicer though subsequently longitudinal instability was revealed with aft C.G., and a larger tailplane with balanced elevators had to be fitted.

Most advanced looking of all new aeroplanes was the Westland Woyavodsky Dreadnought, now structurally complete and ready for fabric covering. Its massive cantilever wings seemed to fill the entire far end of the erecting shed known as the Vimy Shop – a building which had the biggest unencumbered width of any factory of that time, though financial stringency dictated that, as earlier recorded, it still had a gravel floor, and the fitting shop was merely a line of benches along its walls, together with four lathes and several drills. Ahead of the Dreadnought three reconditioned D.H.9As could be erected side by side, ready to push onto the aerodrome. Wings, ribs, and propellers were made in a separate building alongside the wing-covering and dope shop, nearby a well-equipped woodmill. The rest of the Petter factory, including the war-time erecting shop, had reverted to manufacture of industrial semi-diesel engines.

Westland, Bristol, Handley Page, and Hawker had each been lucky in securing a prototype contract for a medium single-engined day bomber to specification 26/23, capable of carrying a 1,500-lb bomb load. The Air Staff considered that the Condor-powered Avro Aldershot, of which fifteen had been ordered, was too big and unwieldy. A machine of some 10-ft less span than its 68 ft, and a laden weight of some 8,000 lb instead of 11,000 lb seemed to fit Air Staff requirements better, and should give far greater performance using the same 650 hp Condor. Within the fixed limits of a specification, inevitably the solutions of different designers have broad similarity in appearance though differing greatly in detail. Of the four, the Bristol Berkeley, Type 90, differed chiefly in locating the pilot ahead of the three-bay biplane wings, whereas the three-bay Handley Page Handcross, designed by Richards, had the pilot conventionally behind the centre-section. The Hawker Horsley and Westland Yeovil were cleaner two-bay biplanes, with pilot and gunner positioned similarly to the Handcross, giving better intercommunication than the Berkeley. Undoubtedly the Bristol design was influenced by the return of Clifford Tinson to Filton after his association with Chadwick in designing the Avro Aldershot, for both machines had a low aspect ratio fin and rudder and comparable fuselage structure of steel-tubes and swaged rod bracing, though the Berkeley wings were framed in composite steel and duralumin as the result of Pollard's high-tensile steel development.

Of immediate urgency in expanding the RAF to 52 squadrons was replacement of the many squadrons of Snipes with fighters having modern reliable engines and better performance, and much depended on the pretty Jaguar-engined Gloucester Grebe, which not only had the patented H.L.B.1 and H.L.B.2 single-bay upper and lower wings but an oleo undercarriage. It was the prototype which had been flown by Larry Carter in the 1923 King's Cup race and then displayed in military guise and colours at the RAF Pageant that June.

Hugh Burroughes recapitulated: 'After G.A.C. had been put on the Air

The Gloster Grebe quickly became the favourite fighter, but also could carry four 20-lb bombs under each wing. (*Flight*)

Ministry approval list for design of single-seater fighters, the company was quick to move by producing a private-venture variant of the Nighthawk, first as the Grouse I with the B.R.2, representing the last of the rotary engines, later with the Armstrong Siddeley 185 hp Lynx, and finally as the Grebe with Jaguar IIIA built to the current Air Ministry specification for a day and night fighter, and 132 were bought for squadron service. It may have been only slightly faster than its competitors, but its outstanding quality was manoeuvrability. It had what we still called "harmonized" controls, and it was a great favourite with the squadrons on that score. So far as G.A.C. was concerned, the Grebe contracts were the first physical evidence of the new policy to provide the RAF with new squadrons. You can get an idea of the difficult conditions in the aircraft industry in 1923/4 from the fact that under our contract with the Air Ministry we had to sub-contract large numbers of wings and tail units to such firms as de Havilland, Avro, and Hawker, who were very short of work. By the time the Grebes were in full production, Bristols had forged ahead with their Jupiter radial. At a number of meetings at Cheltenham, Mr A. H. R. Fedden pushed the case for his engine with his usual forcefulness, and we were induced to modify a Grebe to try one. With the extra 100 hp it soon proved a worthwhile development, and, renamed the Gamecock, 90 were eventually ordered.

'At that time, the trend towards all-metal construction was growing fast, so the next fighter aircraft was the all-metal Armstrong Whitworth Siskin which, despite marginal inferiority in speed and range, was regarded as a safer aeroplane. There was ample evidence that all aircraft manufacturers were actively seeking means of improving their respective competitive positions. Conversion from the wooden construction of the 1914/18 War to composite wood and metal structures and later to all-metal structures was being pursued. This did not commend itself to all our senior executives, some of whom thought it premature, but I succeeded, with the support of David Longden, in pursuading my colleagues to agree that I could buy a half interest myself in the Steel Wing Co, a firm which had been dedicated to all-metal construction since 1919. I subsequently sold those shares to the G.A.C. at the same price I paid for them, and this

A Hucks starter was still standard equipment for starting engines, as shown with the Armstrong Whitworth Siskin II at Martlesham. (*Courtesy Air Marshal Sir Ralph Sorley*)

Rare picture of a Hucks starter being used to start the rear port engine of the Bristol Pullman.

enabled the G.A.C. to secure valuable sub-contracts for construction of the all-metal Siskins which succeeded the Gamecock as the standard RAF day and night fighter.'

The Siskin III had taken five years to evolve from the original RAE design for an S.E.5a replacement, becoming an entirely different design characterized by an almost sesquiplane arrangement with big span and chord top wing and small lower wing which served chiefly as a lower member to box the structure with V-struts and flying wires. An R.A.F. 15 aerofoil section was used, increased by 10 per cent in depth. Although conventionally fabric covered, the Siskin was the first 'all-steel' structured aeroplane to go into production if that distinction is denied the R.E.5, which under supervision of the same chief engineer, Major Green, had a fuselage similarly built with steel-tubes and swaged rod bracing, though the Siskin had 'crinkly' corrugated steel box spars and lattice U-section metal ribs instead of the steel-tube spars and wooden ribs of the R.E. At Martlesham the Siskin's awkward angular appearance and slow but easy manoeuvring earned the comment 'uncouth but with a heart of gold', for it was the easiest single-seater to fly that had yet been introduced. In the air it was unmistakable, with dominant top wing, and oleo legs extended like a long-legged bird. Production deliveries from Armstrong Whitworth to the RAF were about to begin, and a batch of Mk. Vs had been ordered by Roumania.

2

On sale at the beginning of January, price 6d from HMSO, was the long awaited 'Air Ministry Agreement made with the British, Foreign and Colonial Corporation Ltd, providing for the formation of a heavier-than-air air transport company to be called The Imperial Air Transport Co Ltd.' Dubbed the 'Million Pound Monopoly Company' by the Press, it was unavoidably monopolistic to conform with the recommendation made in March 1923 by the Hambling Committee whose task had been to consider the best way of encouraging commercial aviation with subsidies.

By this Agreement, the company had to guarantee subscription of half the shares of the £1 million operating company which would purchase, on approved terms, Handley Page Transport Ltd, Instone Air Line Ltd, Daimler Airway, and British Marine Air Navigation Company Ltd, or in default must establish an air transport service to operate equivalent services. During the ten years subsidy a yearly average minimum of one million miles must be completed, for which payments would be made in diminishing instalments from £137,000 for the first four years dropping to £32,000 for the tenth year giving a total of £1 million. 'No pilot is to be employed regularly on the Air Service without passing a course of instruction on the types of commercial aeroplanes he is to pilot; and all pilots should be such persons as shall be enrolled in the Air Force Reserve or the Auxiliary Air Force. Members of the technical and administrative ground personnel shall be British subjects or subjects of the British Dominions, Colonies, Protectorates, or Mandated Territories in which they are employed; and 75 per cent shall be enrolled in the Air Force Reserve or the Auxiliary Air Force Reserve or in the Dominion or Colonial Reserve or Auxiliary Force.'

The four acquired airlines would have representation on the Board, and two Government directors would be appointed by the President of the Air Council. Officers of the Air Ministry were to have full access to technical and operational information. The new airline was debarred from shares in any aircraft or aero-engine construction firm, but equipment must be of British design and manufacture, and when required, experimental types constructed for the Air Council were to be tried under operational conditions. Other clauses provided for proper equipment of the company's aircraft, transmission of meteorological information by wireless, and continuance of the aerodrome facilities owned by the Government.

At least at holiday time the British public were increasingly supporting flights to the Continent, but could hardly conceive the hazards to which they were subjected. Frank Courtney, probably the most widely experienced of all pilots at that time, recorded what it was like: 'Imagine a trip, in very bad weather, from Paris to London, and see what the pilot of a modern passenger aeroplane does, and why he does it. At Le Bourget the clouds are at 800 ft; rain or drizzle is coming down on and off; visibility is 2,000 yards; a 20 mph wind is blowing from the west, and is stronger higher up. I will have to take a chance on the weather in the Channel, for Lympne aerodrome, high on its hill, also reports very low cloud, but the crossing seems passable.

'While taxi-ing out to take-off, I am getting wet and uncomfortable because of the open cockpit. Once airborne, though still comparatively close to the ground we are just below cloud. I start on the usual compass course with a mentally-calculated allowance for drift, which I shall correct on a known landmark 10 miles hence; but after 6 miles I find there are tree-covered hills, shrouded in wisps of cloud. It is now obvious that a compass course is impossible, so I turn left and pick up the main road from Paris to Boulogne. I follow it for some time, but the ground is gradually rising to the hills south of Beauvais, and the clouds become slightly lower, owing to the wind on the hills. Visibility has decreased to some 500 yards, a distance I am covering in 10 seconds, so I fly about 150 ft off the ground, which feels a mere 50 ft, and am compelled to stick to this road as completely as a motor-car, for if I lose sight of it I am to all intents lost. Flying so low in such bad visibility keeps me so busy looking for obstructions that the compass is doubly useless, for I cannot fly straight long enough for it to settle to correct heading, and in any case I am literally unable to take my eyes from looking ahead.

Outstanding among aircraft on the London–Paris run was the D.H.34, but it was regarded with doubt because of high landing speed if forced down in a typically small British field. (*Courtesy Air Marshal Sir Ralph Sorley*)

'After some minutes of this uncomfortable flying we pass the hills, and with lower ground comes comparative peace for a while, but I must stick to that road, for the ground gradually slopes up and up towards Poix. I get lower and lower until I cannot pass; I am right on the tree tops, and wisps of cloud increasingly blind me. I must turn back. It is unbelievably tricky to turn a heavily loaded machine under such circumstances. I am tempted to leave the road and take a short cut to lower ground near Amiens, but I must not as the low clouds are drifting in patches, and I may find myself in a position where I cannot go forward nor find my way back. So I turn back along the road until I reach the branch road I want and follow its detour to Amiens, thence along low ground by the Somme to the coast. The next part is easy, for though the clouds are low on the sand-dunes, I can fly skimming those flats. I intend to leave the coast at Boulogne, but the high waves show that the wind in the Channel is much stronger than over the land, so to lessen risk of drifting off-course in such bad visibility, I follow the coast of Cap Gris Nez, and then start across the Channel, flying at 150 to 200 ft, with 500 yards visibility and patches of drizzle. It is easy flying, though disconcerting that a patch of nil visibility might occur any moment, but at last white cliffs loom ahead, with cloud all over them. I am not sure whether they are Dover or Folkestone cliffs, but it does not matter, for I follow them round until I can cross the low ground of Dungeness. A straight line to Croydon is impossible, as there are too many cloud-covered hills to pass, so I select a known low-ground course, and in continuing bad visibility, keeping a steady look-out, I traverse a considerable detour and possibly arrive at Croydon.

'Ground services are hardly incommoded on the sort of day described, whereas the aeroplane has a difficult journey which it is only just able to accomplish; with weather even a little worse it could not have done so. Moreover, completion of the journey has been entirely dependent on what risks the pilot was prepared to take, and there is the added fact that on such a flight, which is fairly common in bad weather, the avoidance of collision with a machine coming in the other direction is frequently a matter of luck. Imagine what it would be like at night under circumstances even partially as bad. One sees then that there is little hope for regular commercial night flying; lighting the air routes, and other assistances appear to me useless, for the great problem is to fly and navigate in weather when these lights cannot be seen. So long as airlines depend on ground aids for visual flying, aeroplanes will be an undependable method of travel. I have looked at these matters from all points of view, and it seems absolutely inevitable that we must find a method of flying and navigating in no visibility at all; we must be able to fly quite blind.

'To a small extent, instruments called turn indicators, of various patterns, have been fitted to commercial machines for use inside clouds. But flying by turn indicator is essentially different from visual flying, and must be assiduously practised before a pilot can hope to fly blind for prolonged periods. For sundry reasons, these indicators are all inaccurate in some degree, and so are almost all aircraft instruments, but experience shows it is possible to keep a good course inside clouds if the pilot has plenty of practice. He cannot get this in satisfactory manner if he has the primary duty of taking a load of passengers somewhere. One's first efforts with a turn indicator generally result in a course which would considerably worry passengers. Flying by instruments for, say 200 miles, is a new form of flying, so I strongly contend that all commercial pilots should be given plenty of practice in this blind flying when not engaged in their ordinary flying.'

3

Yet another blow to airship hopes was the delayed disclosure that on 27 December Italian fishermen at Cap St Marco, Sicily, had found the body of the commander of the *Dixmude*, the ex-L.72 Zeppelin acquired by the French as war reparations. French warships rushed to the spot and dragging operations located some of the wreckage, revealing that the airship had been destroyed by fire. In the course of the next few weeks a Commission of Enquiry confirmed that lightning had been the cause.

Concurrently the Duke of Sutherland had just returned from a visit to the USA, and at a Press interview said that American naval and military aviation was about equal with ours, though their aeroplanes were faster – but in commercial aviation Britain was far ahead except for the US Air Mail, which, at the cost of many pilots' lives, flew through all weathers and at night, using a chain of beacons set at 25-mile intervals across the Continent. However, when it came to airships America was indisputably ahead. Their new ZR-1 airship, *Shenandoah*, would shortly set out on a Polar expedition, and was using helium to eliminate fire risk, though it was a very expensive gas and had to be transported from Texas.

Almost immediately the *Shenandoah* encountered near disaster, for on 16 January, after four days on the mooring mast at Lakehurst, New Jersey, she was wrenched from her anchorage by a violent gale, and rapidly disappeared into the rain-swept darkness. Fortunately Capt Heinen, the German Zeppelin pilot in charge, eventually got her under control and, although the bow was smashed in, succeeded in bringing her safely back in the early hours of the following morning. Meanwhile the British Government, blind to these risks, was examining the scheme prepared by Commander Burney on behalf of Vickers Ltd for an Imperial Airship Service, using rigids to be designed and constructed by Vickers or an associate company.

Coinciding with the loss of the *Dixmude* was the death at 91 of Alexandre Gustave Eiffel, the great French engineer whose interest in meteorology led him to investigate aerodynamic problems described in his *La Résistance de l'Air*, which helped to set the empiricism of early designers on a basis of facts. Eiffel made no charge when testing models of new designs, whether aerofoil sections, fuselages, or complete machines – only stipulating that results should be published in his works so that knowledge might be available to all. Yet to the world it is the dominating tower he built for the Paris Exhibition of 1889 which represents his fame, rather than the aerodynamics which are his true memorial.

England similarly lost one of her great aeronautical pioneers with the passing on 2 January of Colonel James L. B. Templer at the age of 78. Back in the 1870s, as a captain in the King's Royal Rifles, he pioneered man-lifting kites, developed ballooning, and eventually became the first Superintendent of the Army Balloon Factory at Farnborough and was largely responsible for the first British Army airship, *Nulli Secundus*. The Royal Aircraft Establishment was the direct outcome of this almost forgotten man's great work.

The United States, with their National Advisory Committee for Aeronautics (NACA), had set up the equivalent of the British and French research centres. Dick Fairey, like the Duke of Sutherland, was much impressed with the progress the Americans had made, particularly the recent rapid development by the Curtiss company of clean high-speed single-seaters, with light powerful engines, and metal propellers made from solid duralumin forgings machined to correct aerofoil section and bent to the appropriate pitch – as devised by S. Albert Reed, an engineer who on

Nose and structure of the Napier Lion II powered production Fairey Fawn III at Hayes, with a Flycatcher fuselage to the right.

retirement turned to research at his private laboratory. In describing contemporary events, the later chairman and managing director of Fairey Aviation Ltd, G. W. Hall, FRAeS, recorded: 'The Fawn was a turning point in the history of the company in so far that Fairey was very unhappy with the Government's specified general purpose aircraft, for he felt it was asked to do too much in too many varied duties, and in consequence would not be a satisfactory Service aeroplane. With this still in mind he saw the 1923 Schneider Trophy Contest and was much impressed and stimulated by the Curtiss D.12 racers. Afterwards we find him working long hours on his own drawing board at home (and playing endless games of patience), eventuating in an approach to the Government to support his new ideas for a two-seat bomber faster than any existing single-seat fighter. His proposals were rejected – but Fairey had seen a vision and had no intention of being diverted. He raised all the money he could, virtually mortgaging the company, and sailed to America. He came back with rights for the Curtiss D.12 engine, the Reed propeller, wing surface radiators, and various design features. In greatest secrecy, on a private venture basis, he set to work with his team to produce a new aircraft.' This was the genesis of the Fox.

Capt Norman Macmillan, who joined the company in January as sole test pilot, adds to the picture: 'Fairey saw that the American success was primarily due to a clean engine of small frontal area mounted in a well-streamlined seaplane. He decided there was every reason to show Britain what could be done by similar methods in the design of military aircraft. By 1924 the company was firmly established as a post-war aircraft manufacturing organization, despite the difficult early post-war years, and now it received Air Ministry approval for all stress calculations. At the end of the war the firm had about £6,000 in the kitty, and by early 1924 this had risen to about £23,000. With that sum available for private enterprise, Fairey went to the United States, and purchased those licences. Colonel Nicholl had turned from flying to administration as deputy managing director, and Major T. M. Barlow, chief technical officer at A & AEE, Martlesham Heath, came as chief engineer very shortly afterwards.'

Thirty-year-old Tom Barlow had won a scholarship to Birmingham University before he was 17, took an Engineering honours degree, and subsequently an MSc as a research scholar on electrical transformers.

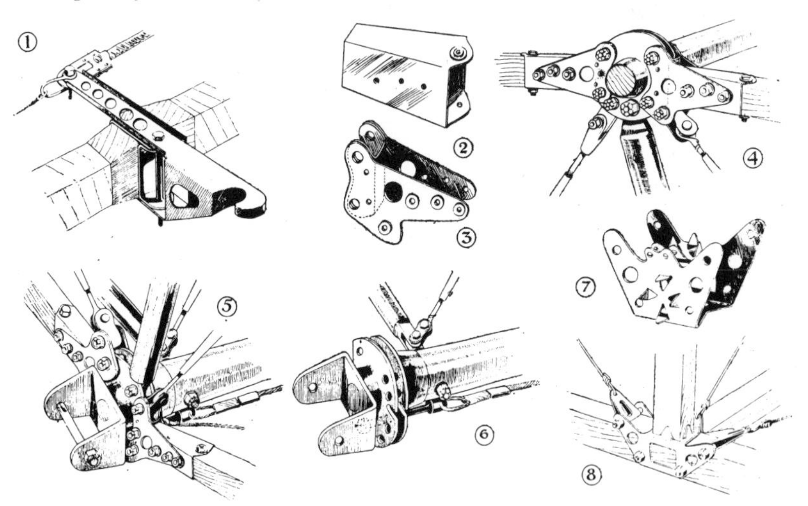

Details of Fairey design: 1, lock holding folded wings to fuselage. 2 and 3, longeron fittings of assembly 4. 5, spool assembly on fuselage at wing spar attachment. 6, spool before fitting longerons, etc. 7 and 8, typical rear fuselage strut fitting on longeron. (*Flight*)

From apprentice at Siemens Dynamo Works he became assistant engineer, and then engineer at the Factory department of the Home Office, but joined the RNAS in 1915 for engineering and aircraft technical duties, becoming CTO at Martlesham in 1918 and meeting Dick Fairey during trials of the Pintail. Friendship developed, for the aircraft manufacturer was impressed with Barlow's balanced and knowledgeable outlook.

The idea of a scheme for 'approved design firms' was only just beginning. Performance checking was done by Capt R. N. 'Loopy' Liptrot and his small Air Ministry staff in Kingsway, while structural integrity was dealt with by Frank Cowlin as head of Airworthiness at Farnborough. As one stressman of that time recorded: 'The budding aircraft engineer had few books of reference. Bairstow's *Aerodynamics*, Pippard and Pritchard's *Aircraft Structures*, and the first slender volume of *A.P. 970* together with a few R & Ms constituted his library, apart from standard engineering textbooks. At this stage there were no special instruments, no strain gauging, no electronics – a hand-held spring balance was the usual instrumentation by which a pilot brought confirmation that control was too heavy, and a stopwatch measured the phases of instability. Because aircraft were simpler we worked faster than now, using comparatively small teams. Eleven months was a fair time from inception of design to first flight. There were usually two or three prototypes going through, the peak loads occurring at three to four month intervals, so one could keep trying new ideas without waiting several years wondering whether the right decisions had been made.

'Because prototypes were relatively cheap it was usual for the Ministry to give contracts for the same requirement to several rival firms. Because of this competitive situation and final selection from the actual hardware, Government control during design and construction was practically non-existent except for the AID, though there was an initial design conference to determine general principles, then a "mock-up" conference, and a final conference. Decisions were taken solely by the firm, and this led to speedy work. Meanwhile the world's wind-tunnels were pouring out information, new materials and better engines were coming to hand, and a vast field of development suddenly began to open. Basic problems have always been the same; knowledge, processes, materials and tools were more primitive, but pressures generated by competition were as real then as now. Other firms' prototypes always seemed to get to Martlesham first, with the result that heavy overtime disrupted holidays, and wifely complaints were the rule.'

Private enterprise still had free rein. Fairey's was an example of cool judgment in building a machine at his own expense to prove that the official specification was outmoded. There were others who ventured in a different way, such as the young ex-RAF pilot D. H. Williams, BA, not yet widely known as Hollis Williams, who was making his mark at Hawker Engineering. He recorded: 'Hawkers had a new contract to recondition D.H.9As, and a shop was being cleared to make room. In one corner, among a pile of

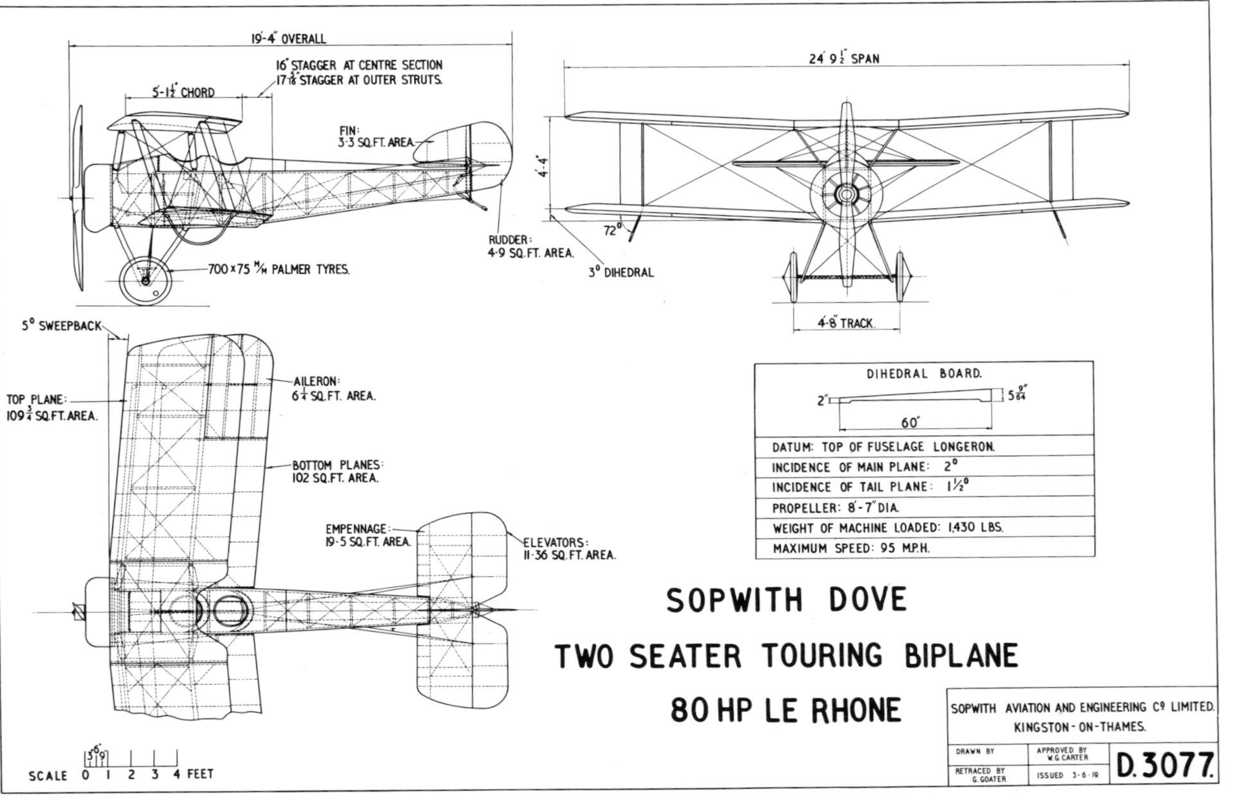

Original Sopwith Dove general arrangement 'approved' by Carter as chief draughtsman.

obsolete components, were six partly finished uncovered airframes intended as conversions from Pups to Doves for a joy-riding company which failed to find the money. Coley, the scrap-metal merchant in Kingston, was going to have them as junk. Lowe-Wylde was one of the Hawker draughtsmen, and suggested we acquire one of the sets. As I had no manufacturing apprenticeship, I thought that if I could reconstruct an aircraft it would at least give experience of the special trades – fabric doping, tin bashing, and so on. We were uncertain whether it was best to try to buy from Hawkers or wait until the airframes were in Coley's yard and possibly damaged, but sensed a favourable moment to ask Sigrist. He said: "What do you want an aeroplane for?" I replied, "To put it together and fly it." Sigrist said, "More bloody fool you! . . . Anything you can put on a lorry is £5 to you." Lowe-Wylde and I sorted out the best airframe; found wings, wires and struts, but couldn't find the tailplane. Luckily there was a Snipe tail, so this was loaded on the lorry, and the lot removed to a small lock-up garage in Kingston.

'For another £5 we acquired an 80 hp Le Rhône from Coley, but found some of its pistons cracked, so went to him in high indignation. He said, "What do you expect for £5? An airworthy engine?" In no uncertain terms we told him. He wouldn't take the engine back, but pointed to a case and said, "That engine has never been installed. Just as received from the makers with 30 minutes test run in the log book. £5." We paid. My pay was only £5 a week, and it would have been impossible to rebuild the machine but for the help received. Coley went out of his way to be helpful, and Dallas, who was just starting a business with a 1-ton truck, provided the transport and very decently never remembered to put in a bill.

'Presently Lowe-Wylde left, but we had covered the wings and fuselage with surplus fabric bought from Bentalls of Kingston, and doped it after heating the garage at great risk to neighbouring cars by using open coke fires. The AID, at that time responsible for civil as well as military inspections, were called in. Promptly they scrapped the wing covering, and for good measure scrapped the wing spars as they had dried out. I felt beaten, so appealed to Sigrist, who said, "Send them in. Hawkers will replace the spars and cover and dope for £30." That was a nominal price by any standard, but there would still be a long story to unfold before that little machine ended its days as the Sopwith Pup at the Shuttleworth Trust at Old Warden, Bedfordshire.'

4

Reaction of the new Government to rearmament was awaited with interest, but for the moment its members seemed bemused with the turn of fortune which had brought a clerk to Premiership, an engine driver to the Colonial Office, a foundry labourer to the Home Office, and a millhand as Lord Privy Seal. However it was to tall and commanding 49-year-old Brig-Gen C. B. Thomson, who joined the Labour Party in 1919, that the Secretaryship of State for Air was given, and shortly after appointment he

Air Vice-Marshal Sir John Maitland Salmond (*left*), AOC British Forces in Iraq – a man in whom the RAF had absolute confidence. A. W. Martyn was the original managing director of the Gloucestershire Aircraft Co Ltd, and Hugh Burroughs (*right*) became a key figure in its activities.

was elevated to the peerage under the significant title of Baron Thomson of Cardington. His had been a distinguished military diplomatic career, though with no aeronautical experience, but as one paper commented: 'General Thomson's past record is of such nature that at least there should be great promise of his taking a very strong line in our Air Council.' As Under-Secretary for Air, MacDonald appointed William Leach, MP. Yet despite the avowed pacifism of the Labour Government, there were signs that rebuilding the RAF would continue. On 3 January the Air Ministry had announced that 400 officers were required for flying duty under a new short service commission scheme in which, as C. G. Grey caustically said: 'Young men of 18 to 22 at the beginning of their adult lives would be taught to live like officers and gentlemen for five years, and then cast on the world at 23 to 27 with a small gratuity and no visible means of support.' A flood of applications followed from young men whose overwhelming ambition was to become a pilot.

A new Home Defence Area, on a larger scale than the existing Inland Area, was planned by the RAF. *The Times* of 5 February reported: 'Air Marshal Sir John Salmond, KCB, CMG, CVO, DSO, Air Officer Commanding Iraq, is coming home in April, and Air Vice-Marshal J. F. A. Higgins, CB, DSO, AFC, Air Officer Commanding Inland Area, is leaving this month for Iraq to take over Sir John's Command. It is generally understood that Sir John Salmond, after a period of leave, will assume command of Home Defence. The new Commander of Inland Area, who begins his duties in March, is Air Commodore T. I. Webb-Bowen, CB, CMG. Another appointment which reflects the increasing importance of the RAF in the defence of the Empire is that of Air Commodore E. L. Gerrard, CMG, DSO, to the Palestine Command from 1 April. Air Commodore C. L. Lambe, CB, CMG,

DSO, will take over Command at Halton, the large aircraft mechanics training centre of the RAF, at the end of the month. The last appointment, that of Air Commodore F. C. Halahan, CMG, CBE, DSO, MVO, to the Department of the Air Member for Supply and Research (AMSR) is the most interesting of all in view of the announcement that the post of Scientific Development Director is to be a military appointment.'

The Air Ministry had announced that: 'The rapid development of aeronautical science and the increased requirements of the RAF have now made necessary a re-allocation of the responsibilities of the Director of Research. As from April next the control of one side of the work will be transferred to a Director of Scientific Research (DSR), while the other will be assigned to a Director of Technical Development (DTD). The appointment of Director of Research will then lapse. The DSR will be a civilian having high scientific qualifications and, if possible, a knowledge of aeronautical research, and the DTD will be an experienced senior officer of the RAF; both will serve under the Air Member for Supply and Research.'

To goad the Government into expression of policy the former Minister of Air, Sir Samuel Hoare, on 19 February moved: 'That this House, while earnestly desiring further limitation of armaments so far as is consistent with the safety and integrity of the Empire, affirms the principle laid down by the late Government and accepted by the Imperial Conference that Great Britain must maintain a Home Defence Air Force of sufficient strength to give adequate protection against air attack by the strongest air force within striking distance of her shores.'

He said that when he took office in October 1922, at the most critical time of the Chanak crisis, there were only 20 first-line aeroplanes available for Home Defence, but now benefiting from action by the late Government, we had about 80, or 100 if army co-operation aircraft were added. 'M René Fonck, the great French war pilot, who is a Member of the Chamber of Deputies, calculated that a force of 500 aeroplanes could in the space of a single night obliterate a city as big as Paris. Currently France has about 1,000 first-line aeroplanes of which 600 are for the French Independent Striking Force, and 400 for duty with the French Army. That is a striking disparity of 10 to 1 compared with Britain'. As a first stage the late Conservative Cabinet had agreed to increase the RAF for Home Defence to 600 first-line machines; he trusted the Labour Government would honour that, but hastened to add there was no suggestion of hostility towards our friends and allies or that a breach between the two countries was likely, and he was just as apprehensive as any member of the House at seeing a new armament race starting.

While admitting the case had been put with temperance, Mr Leach, replying for the Government, said he declined to be alarmed by the picture of disparity between ourselves and France. It was his party's legacy – not its responsibility. However he had been asked what the Government was going to do with the expansion scheme? He would say in similar explicit terms that there was no change in the policy of the Government for the time being. As to orders for new types of aircraft to meet requirements of the expanded Force, there were regular and definite stages, and sufficient machines were being ordered to equip squadrons formed during the coming year. Pending delivery of such machines, some of the squadrons would temporarily only have trainers and the present type of Service machines, but time would remedy that. In regard to civil aviation, agreement for a single Imperial Transport Company was a *fait accompli*, and would be properly fulfilled. The Government was anxious to foster civil aviation, and would take whatever measures were open. Similarly airships would be explored, encouraged and fostered in every proper way that was available. As to the part of the Resolution calling for a Home Defence to match the strongest air force in striking distance – was that a practical proposition? If we had 50 to 1 against the strongest we would not be adequately protected in the opinion of some, and if we were ringed with defence aeroplanes from Humber to Thames, and around the south to Cornwall, there would still be some saying this was inadequate protection. The only impregnable defence that he could see was a changed international atmosphere. If we continue to put fear at the helm and folly at the prow we should steer straight for the next war.

Inevitably these sentiments drew upon the Under-Secretary's head a fierce attack from those who saw the inevitability of war and necessity of preparedness. Nevertheless it was now clear that Labour intended to complete the expansion programme. That was confirmed in the Lords, following a Resolution moved by Lord Londonderry in identical terms with Sir Samuel Hoare's, and provided opportunity for Lord Thomson to make his first official statement on Air policy in a speech which created favourable impression, during which he said that for the Labour Government the flower of idealism was rooted in common sense. They therefore reserved the right, while carrying out the expansion scheme, to confer with other nations to try to find a method of all-round reduction in armaments.

Implementing the Labour Government's undertaking were the Air Estimates published on 7 March which revealed an increase of £2½ million, though the Army was cut by £7 million and the Navy by £2 million – yet the Air allocation was a mere £14½ million compared with £45 million for the Army and £55½ million for the Navy. The Secretary of State for Air explained: 'The largest individual increase is for technical equipment and research. Reconditioning of existing machines and engines is being continued so far as it is judged economical and compatible with efficiency; but it is proposed to equip new squadrons (including additions for co-operation with the Navy) with new type machines. In this connection it is of utmost importance that there should be no relaxation in experiment and research, and increased provision has been made for this purpose. No provision is made for development of airships as that is under active consideration, and the decision will be communicated to Parliament in due course, any necessary provision of funds being made by Supplementary Estimate.'

The Labour Party intended to complete the expansion programme but a year later the RAF still depended on Sopwith Snipes, and production Vickers Victorias had only just been ordered.

The Air Estimates were regarded by the de Havilland Board as cause for optimism, and, on learning of their staid Mr Turner's intention of relinquishing the chairmanship, appointed their keen shareholder-director, Alan Butler, in his place.

5

On 17 March four of America's big single-engined Douglas bombers left California, flying westward on the first stage of yet another attempt at encircling the world. With the same thoroughness that the US Navy organized the 1919 stage-by-stage flight across the Atlantic, whereas Alcock and Brown had flown it with minimum official help, so now the US Army Air Service had gone to tremendous detail in planning the flight, despatching ground crews to every programmed landing point. In the following week a British team comprising Sqn Ldr A. S. C. MacLaren as navigator; Flg Off J. Plenderleith as pilot, and Sgt Andrews as engineer, began an attempt eastward. Their send-off was indicative of the whole

Vickers Vulture II, G-EBHO, used by MacLaren, had increased tumble-home, widening the planing bottom to improve take-off, and Vulture I, G-EBGO, was shipped to Tokyo as a stand-by for him.

venture: no shouting, no fuss – just a few of those directly concerned watched the heavily laden Lion-powered Vickers Vulture, a derivative of the Viking amphibian, slide down the slipway at Calshot and head towards Southampton on what seemed an interminably long run before it managed to lift, then circle back and disappear in the distance. A sarcastic onlooker suggested it had been designed to taxi round the world rather than fly, and when Brancker asked C. G. Grey why he had not been present, the great editor retorted that the take-off of one amphibian was much like another. 'Believe me Charles,' said Brancker. 'This one was not!' However the attempt was not as casual as it appeared, for not only had Vickers devoted much time in organizing, but Napiers had sent spare engines to Tokyo and Toronto, and Shell-Mex laid down special supplies of petrol and oil throughout the route. Late in the afternoon of the start came news that the Vulture had encountered fog the other side of the Channel, and had alighted near Le Havre. Meanwhile the Americans were waiting at Seattle to start the Pacific crossing.

On 31 March formation of the new airline, now named Imperial Airways Ltd, was announced. The Board comprised Sir Eric Geddes, chairman; Lord Invernairn; Sir George Beharrell; Lieut-Col Barrett-Lennard; Sir Samuel Instone; Mr Hubert Scott-Paine; Lieut-Col Frank Searle, managing director; and Sir Herbert Hambling and Major J. W. Hills as Government directors. Major G. Woods Humphery was appointed manager, and Mr S. Dismore of Handley Page Transport became secretary.

Immediately came news of strikes and disagreements. Only a few return flights of British aircraft abroad were made in the next two days. An announcement from Sir Eric Geddes stated: 'Until last night all pilots were still in the employment of the four existing companies, and the new Imperial Company has as yet made no appointments. The directors decided that in the case of pilots who pass the rigorous medical tests, terms should be offered which in the form of an annual retainer and flying-time pay will ensure an annual income of £750 to £850 according to seniority. For this they will be required to fly an average of two hours a day. The records of existing companies show that the average earnings of pilots during the past 20 months has been in the neighbourhood of £680 a year.' This was unacceptable.

To safeguard themselves the pilots formed an association named the Federation of Pilots, with Capt F. L. Barnard as secretary; the ground personnel similarly formed a Federation of British Aircraft Workers – both organizations being registered as trade unions. The aid of MPs was invoked. Meanwhile all British flying ceased, and foreign air services cashed in. Imperial Airways issued a notice inviting applications for pilots, scheduling an even less attractive graduated income scheme of £755 rising to £855 after five years. The pilots were furious, and refused to join the new company, claiming that under the old *régime* a Handley Page pilot received £915, an Instone pilot £857, and Daimler pilots averaged £1,000 a year. They also strongly opposed the appointment of Woods Humphery. Almost a month passed before reaching a settlement. Imperial Airways, as the new hope for civil aviation, had got off to a poor start.

To the British public the peril of a monopoly company and the battle of pilots were only of passing interest. Personally I was much more concerned with preventing abandonment of the aeronautical course at the Northampton, where I had been studying general engineering for eighteen months, mixing with the last of the student ex-Service men whose mature views were stimulating compared with the new input of ex-schoolboys whom lecturers were at pains to address as 'Mister' in order to encourage maturity. My year had attracted more than sixty youngsters whose specialization interest ranged widely, but not one proposed taking the aeronautical course. I lobbied a number, and at last succeeded in getting three, including one of the ablest, Reggie Stafford (later the penultimate Handley Page designer and final Technical director), to believe that aeronautics might offer a career. Ours was a 'sandwich' course in which six months in mid-second and third years were spent at appropriate firms to gain practical experience. Stafford and I went to Handley Page Ltd at Cricklewood. To our surprise we were even paid 25s per week – Stafford at a lathe, and I as assistant in the wind-tunnel where R. Reynolds, ex-RAF pilot, had succeeded Geoffrey Hill as aerodynamicist, and G. C. D. Russell, keen ex-apprentice and now a slick and very capable young man, was his assistant.

H.P.'s sanctum just within the factory entrance, the drawing office, assembly shop, metal-working and machine tool section, dope room, store, and wind-tunnel were all under a single long roof: the rest of the factory had been let or sold. Volkert had returned from Japan, like the rest of the Sempill mission, following a catastrophic earthquake. It was only a step

Following Fokker's lead, the Handley Page W.8 design was revised as a three-engined airliner. Here the pale blue prototype W.8e, O-BAHG, is having the first engine-run at Cricklewood. (*Flight*)

from the D.O. into the almost empty 200-yard erecting shop. The orange-winged 1913 Handley Page monoplane hung under the adjacent roof girders, emphasizing the great constructional advance in eleven years. Facing the clattering roof-high doors to the aerodrome were a deserted Hanley and two pale-blue W.8es bearing Belgian registration letters. Built for Sabena as a result of the Brussels trials in 1922, they were powered with a 360 hp Rolls-Royce Eagle IX in the nose and a wing-mounted 240 hp Siddeley Puma each side. The prototype, O-BAHG, had just flown its acceptance trials in the hands of Capt Wilcockson, but there had been criticism by the airline of vibration and they were temporarily grounded, though urgently required for regular service between the capitals of the Congo and Katanga, a 1,200-mile journey of 45 days by surface transport, yet could be achieved with stage-by-stage flight in only two days.

Too late for the Dreadnought, Westland built the most up-to-date wind-tunnel possible, but placed it in the open wing shop subject to vagaries of flow.

Behind these two big biplanes were benches and perhaps fifty men. Partitions, made with O/400 wings on their leading edge, hid dope shop, stores and, far at the end, the entrance to the wind-tunnel tucked behind an enclosure full of the stage properties of Martin Hearne, a friend of H.P.'s, well known as a professional comedian and amateur pilot. The crash of a cable-linked weight informed the wind-tunnel staff that the outer door was being opened, warning that H.P. might be visiting to study progress. I learned that the previous wind-tunnel assistant, W. 'Bill' Widgery, who had worked with Hill on early slots, had recently joined Westland to design and operate a new type of venturi-shaped wind-tunnel which was expected to be more accurate than the standard National Physical Laboratory (NPL) design such as used by Handley Page. At that juncture my aerodynamic knowledge was merely that of an enthusiast gleaned from text-books – for the aeronautical lectures at the Northampton would not begin until the Autumn term. Under the tuition of Reynolds and Russell – aided by model-maker Tom Miles, who at slack moments would play his violin after ascertaining that H.P. was away on business! His polished mahogany models revealed the fascinating story of experiments that H.P. had conducted, including the pinion tip which led to lateral slotting instead of longitudinal. Even more fascinating was one of the 'S' fighters built for the US Navy but now dumped on the scrap heap. There I could sit in the cockpit and dream of flying, but lift coefficients, drag coefficients, hinge moments, pitching moments had become my jargon; filed results were my reference; and drawings of many proposed but never built designs revealed that manufacturers might design ten aeroplanes for every one constructed as a prototype.

Came a day when I was hailed to the great H.P.'s office. He seemed to fill the spartan little room, leaning back in an arm chair, one leg on the desk to rest a painful knee, hands clasped across his bulging waistcoat. Big brown eyes bored into me. 'Wawl!' he nasally drawled. 'I hope you don't think there's a fortune in building aeroplanes?' He grinned. 'What made you take up aeronautics?'

'I just like aeroplanes.'

There was a pause. He nodded. 'That's as good a reason as any. Enthusiasm is everything. It's a gamble whether there are any material rewards. What are we paying you?'

I told him. 'Oh! I thought I'd agreed a quid with the labour manager. Wawl – I shall expect my money's worth!' And with that, he nodded me away; yet throughout the long years he always remembered that I had been at the Northampton where once he lectured in the evenings on aerodynamics and the art of aeroplane design.

Money was also having its say at Supermarine, for C. G. Grey recorded in an unpublished MS: 'Scott-Paine and James Bird had a tremendous row. Bird said heatedly, "What will you take to get out of this?" Paine named some fantastic sum – I have a notion it was £200,000 – and Bird slapped his hand on the table and said, "Done with you," and Scott-Paine was out. With the money thus gained, backed by the sea-faring experience he had as a youngster with early naval motor-boats and augmented by his years of

It took a year after RAF tests to gain an initial contract for eight folding-wing Supermarine Seagulls which were delivered in 1923, but pilots found twin-float seaplanes more practicable. (*Courtesy Air Marshal Sir Ralph Sorley*)

First of Mitchell's multi-engine designs was the Supermarine Swan to Air Ministry specification 21/22. Outboard of the engines the wings folded forward, though the reduction in stowage width was not worth the complication and weight.

craftsmanship with Supermarine flying-boats, Scott-Paine formed the Power Boat Company at Hythe.'

Concurrently R. J. Mitchell acquired his earliest assistant with a degree, a graduate from the Northampton, Alan N. Clifton, BSC, who became responsible for strength and performance calculations which previously the designer himself had done. Of Mitchell, Clifton later recorded: 'He was a most likeable man, modest and even shy, but full of fun, and the life and soul of a party. He played tennis and golf and snooker, and learned to fly when the Hampshire Aeroplane Club was formed. Capable of intense concentration, he did not like to be interrupted when thinking hard. Sir Henry Royce said he had the ideal temperament for a designer – "Slow to decide and quick to act."' Following his success with the single-engined tractor amphibian Seagulls of the previous year, of which a number had been built both for the British and Australian Air Force, and as the pusher Scarab for Spain, Mitchell had designed a scaled-up twin-Eagle IX version with similar hull having cruiser bow and flared chines like a rim around the bottom. It was built to Air Ministry contract, named Swan, and had dual possibilities of naval reconnaissance or with cabin and hull adapted for passenger carrying. Its centre-section as far as the wing-fold joint, where the amphibian oleo leg was attached, had bold Warren bracing with engine close each side high in the middle of the V to be clear of waves and to reduce turning effect with one propeller stopped. Beneath the central inverted V-struts was the pilot's cockpit, superimposed on the hull in a turret-like structure. Characteristic were triple fins and rudders above the monoplane tail. When Henri Biard flew it in March he was quickly convinced that the Swan was a great advance on the long-winged F.5 flying-boats used by the RAF. Later, the Swan N175, re-engined with two 450 hp Napier Lion IIBs, the undercarriage removed, and registered G-EBJY, went in August to the MAEE Felixstowe for evaluation, showing such promise that Air Ministry specification R.18/24 for a military version was issued soon afterwards.

Mitchell was also busy with his special interest of a replacement racer for the out-dated Sea Lion, contemplating an idealized floatplane, with various arrangements of shaft-drive in order to get a smoothly pointed nose, and weighing the merits of biplane versus monoplane. He was also sketching possibilities for the new Light Aeroplane Competition, the rules of which had just been announced by the Air Council, with prizes totalling £3,000. Requirements were simple. The aeroplane must be a two-seater with dual control and have an air speed indicator visible from both seats. The engine must not exceed 1,100 cc. Aeroplane, engine, and magneto must be entirely designed and constructed in the British Empire, and entrant and pilot must be British subjects. The disposable load, exclusive of fuel, must be not less than 340 lb, and a Certificate of Airworthiness in normal category except for engine was initially mandatory, but eventually the Air Ministry decided that a Certificate of Exemption was permissible at the profitable cost of £12 10s each.

There would be eliminating tests of dismantling, housing and re-erecting, and separate flights to demonstrate controllability from either cockpit. Marks would then be awarded for high-speed and low-speed tests, take-off distance to clear a 25-ft marker, and distance to land over a 6-ft fence. For these last two, the Duke of Sutherland gave a prize of £500 and Capt C. B. Wilson, MC, added a prize of £100. The Duke expressed his intention of purchasing a two-seater evolved from those competing, though not necessarily of the winner's design.

Triumphant production line of D.H.53s for the RAF.

Despite the helpful Air Ministry contract for twelve D.H.53s, Geoffrey de Havilland was far from convinced of the practicality of ultra-light aircraft, and was backed by Cobham after his cross-country experience with the D.H.53. When the competition was discussed by the Technical Committee of the SBAC, Charles Walker argued it was not worth while putting design effort into so useless a category, for it would require the same number of man-hours as a bigger aeroplane, such as the 90 hp R.A.F.-engined D.H.51 two-seater they were currently building. He found no supporters, and the Committee decided to endorse the competition rules.

Nevertheless four other manufacturers were of similar mind and decided not to enter. Thus, hard-headed John Siddeley considered there was no possibility of a light aeroplane market, so staked his money on military ventures; Boulton & Paul directors were sceptical after the ill success of their 80 hp two-seat P.9 and had earmarked every penny for metal construction; Folland was too busy with developing the Grebe which was achieving big money for so new a firm; Handley Page saw no point in wasting more money on ultra-lights and had Boultbee designing a side-by-side high-wing cabin monoplane of 80 hp.

Of those who took a purely sporting view, Chadwick induced A. V. to agree that last year's light single-seat biplane had greater possibilities scaled-up as a two-seater than the more successful Avro monoplane; Shackleton had already designed a two-seat version of the A.N.E.C. single-seat monoplane, but almost immediately joined Beardmore to resuscitate the aircraft department, where he proposed an almost identical machine to get things going while negotiating a licence to build the big metal-skinned Rohrbach monoplane. Robert Blackburn settled for a side-by-side biplane designed by A. C. Thornton. Barnwell at Bristol was intent on a cantilever low-wing monoplane; Flt Lieut Nick Comper, an ex-Airco apprentice later on the staff of the NPL, then a Cambridge undergraduate, and now a lecturer in charge of the Cranwell engineering laboratory, was designing a side-by-side biplane; at Hawker Engineering, Camm had been given the drawings of the old Tabloid, and Carter told him to produce a modern version with half the power and half the weight while he got on with the Hedgehog and Horsley; Harold Bolas, like Barnwell, favoured a monoplane based on his 1923 entry but had the bright idea of making it a convertible mono-biplane. Oswald Short remained intent on publicizing duralumin monocoque; spurred by success in designing a miniature metal flying-boat hull for a little twin-engined monoplane ordered by Lebbaeus Hordern, to whom in 1920 he had supplied the F.3 air yacht conversion and in 1921 a Shrimp seaplane, he decided on a low-winger with similarly constructed fuselage. At the rival marine firm of Supermarine, Mitchell was producing a conventional lightly-loaded biplane with ply-covered fuselage. Vickers also pinned hope on a biplane, with a scaled-up version of the Viget to be made by Avro at Hamble. Westland proposed to test full scale the rival merits of a very small biplane argued by Robert Bruce against an advanced looking parasol monoplane with lozenge-tapered wings advocated by Arthur Davenport.

Only Geoffrey de Havilland seemed convinced that there was a potential market equivalent to private cars. Although the B.E.-like D.H.51 was

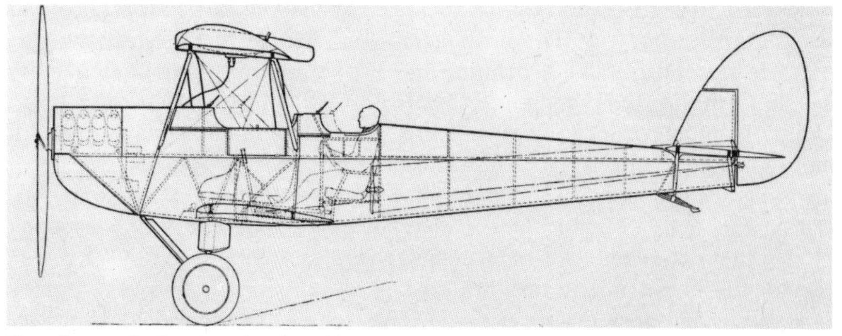

Manufacturer's drawing showing side elevation of the de Havilland D.H.51, with arrangement of cockpits, seats and controls.

Lineage was unmistakable between Geoffrey de Havilland's B.E.2a and the D.H.51 a decade later.

nearing the final stages of construction, he was certain that the ideal solution lay somewhere between that machine and the D.H.53. What he needed was an engine of about 60 hp weighing not more than 350 lb – then a two-seat design could follow. He tackled his friend Frank Halford of B.H.P. fame, who, with his sole assistant John Brodie, had been working for the Aircraft Disposal Co refining the 90 hp R.A.F.1A so that it could develop 140 hp. D.H. proposed that Halford should take four of the eight cylinders and appropriate parts of this 'Airdisco', as it was named, and fit them to a new crankcase and shortened crankshaft. Halford was busy uprating the Puma to 335 hp for A.D.C. and also designing a 1½ litre engine for Aston Martin cars. Reluctant at first, he eventually accepted D.H.'s idea, but the terms of his engagement held him to A.D.C. Since de Havillands could not afford the cost of developing a prototype engine, D.H. now had to argue with the A.D.C. directors that the project should be undertaken at their expense as it would help find a market for their huge war-time surplus stocks of cylinders and pistons.

'For the past six weeks the eyes of the world in general have been centred upon the rival efforts of the British and American aviators in their respective flights round the world, during which time the Americans disposed of some 3,000, and the British 5,000 miles of the total distance to be covered,' reported *Flight*. 'Then during the week-end, like the whirlwind that accompanied it, comes the Frenchman, Lieut Pelletier Doisy, who – without any preparation to speak of and flying a standard Breguet military machine – covers about 3,000 miles of much the same route as the British in not more than six days.' His aim was Tokyo.

MacLaren and his team had experienced many difficulties even to attain Karachi, then, after overhauling the engine, set off on 26 April for Tokyo, but failure of the reduction gear caused a forced landing after 380 miles. On 3 May, Lieut Pelletier Doisy, also having overhauled his engine, chased after him, but at Calcutta found the wing fabric so badly deteriorated by the strong sunlight that it had to be repaired. Meanwhile two Portuguese military pilots, Capt Brito Paia and Lieut Sarmento, and their mechanic, in an attempted flight with a Breguet from Lisbon to China, reached Karachi the day after Pelletier Doisy left. On the other side of the world the Americans arrived at the Aleutian Islands, where they awaited 'supplies' –

Marking a break-away from convention was the all-metal, I-strutted, Breguet reconnaissance two-seater biplane of the French Armée de l'Air and other air forces.

which indicated they also had engine difficulties. Less publicized was a flight around Australia begun on 6 April by two officers of the Royal Australian Air Force, Wg Cdr Goble, OBE, DSO, DFC, and Flt Lieut Ivor McIntyre, using a Fairey IIID, who completed their 8,500-miles flight on 19 May despite similar delays due to engine trouble.

Hazards of long-distance flying remained primarily those of sudden engine failure or bad weather and the strain of fatigue. Such flights might be into the unknown of diverse countries and oceans, but not into the uncertainties of aerodynamics in the manner that befell Stuart Keep of Westland on 9 May. That week the Woyavodsky Dreadnought, silver painted except for grey cabin and cowling, was at last ready for flight. It was Britain's current ultimate in construction and aerodynamics. Like Germany's Junkers, it was a pointer to the future and was expected to establish a new generation of monoplane fighters and transports.

Adjustment and testing of its Lion engine had occupied a couple of days, and on the 8th the machine was tugged onto the grassy little aerodrome so that Keep could carry out taxi-ing trials accompanied by his managing director the elderly but very vigorous Robert Bruce. All seemed well. Several fast runs were made with tail lifted, and then the machine was brought in for final inspection.

In the calm sunshine of the next afternoon the Dreadnought was again pushed out, and for an hour mechanics and inspectors fussed around it. Keep, wearing battered helmet and oil-stained leather coat, strolled from the office block accompanied by Bruce and Davenport, with Mettam, Digby, and other seniors following. From the D.O. and workshops men and women, discreetly in the background, came to watch. Keep climbed aboard. The engine started, and presently he taxied to the far end of the aerodrome, beyond which the church tower and buildings of Yeovil loomed distantly. Turning, he made a run along the crown of the aerodrome; returned and made a faster one, the machine lifting after 200 yards, but almost immediately was wheeled down with throttled engine. A further run seemed entirely satisfactory.

To the British, the Westland Dreadnought represented a greater step aerodynamically and structurally than anything previously attempted. Its crash re-affirmed the biplane traditionalists.

The watchers were silent, intent on this novel monoplane as it faced the light breeze once more. The whirling propeller blurred. The roar of the engine echoed across the field, and the machine began to move. It looked magnificent. The wheels left the ground. For a moment it seemed Keep was holding it down, skimming a yard or two clear – then the nose began to rise into far too steep a climb as he drew level with the Vimy shed. With engine full out, the great wings seemed impotent. Tensely Bruce glanced at Davenport. The right wing began to drop as the machine suddenly sank, and with a crash and cloud of dust, hit the ground with nose and wing-tip, tearing off the undercarriage, pushed the engine into the cockpit, slid 50 yards, and stopped. Neither Westland nor any other manufacturer in those days had ambulance or fire engine. There was still only the small trolley of Keep's earlier crash, though now it carried six fire extinguishers – and the factory nurse, Sister Thomas, standing by with a box of bandages. Said Mettam: 'She began to run to the machine, which was 200 yards away, and Widgery and I formed up on either side to help her along until a car caught up. She undoubtedly saved Keep's life by applying tourniquets to his smashed legs before he was got out, and presently the ambulance arrived from Yeovil Hospital.

'When we gathered in the works canteen afterwards, Reilly, the chief draughtsman, sensibly but somewhat primly said there should be no discussion of the crash. The explanation of that disaster which has remained in my mind, though I do not know if it was officially accepted, was that there had been an uncontrollable stall due to lack of aerodynamic knowledge, for it was generally believed that scale effect on a thick wing-section would cause the centre of pressure to move back at the stall, yet later work showed the exact opposite and this would have been accentuated with engine on.'

Details of the accident were kept quiet. Keep's broken legs were amputated. No journalist made headline news that this was a secret machine of very advanced design. They were far more interested in the 'death-ray' invented by Grindell Matthews who, it was said, claimed it the future weapon of war, lethal to troops and capable of bringing down aircraft. There were questions in the House. The Secretary of State for Air said: 'The Service Departments have been placed in a difficult position in dealing with Mr Grindell Matthews, partly because of the vigorous Press campaign conducted on his behalf and partly because this is not the first occasion on which this inventor has put forward schemes in which extravagant claims have been made.'

Lieut-Cdr Kenworthy disclosed that Grindell Matthews had been paid £25,000 by the Admiralty for an invention to direct vessels by wireless, and nothing had come of it. In February 1924, the Air Ministry had offered the inventor opportunity of demonstrating his ray apparatus, but nothing transpired until he was interviewed by Air Vice-Marshal Sir Geoffrey Salmond on 24 May. Two days later Grindell Matthews demonstrated to representatives of the three Services that from a distance of 15 yards in his laboratory he could light an Osglim electric lamp and stop a small motor-cycle engine. The experts were suspicious. He was asked to stop a small petrol engine provided by the Government, and if successful he would be paid an immediate £1,000 provided he allowed the Government 14 days to consider further financial arrangements for development of the invention. But Grindell Matthews left the country, telling journalists that to wait meant dropping a French offer, so he had refused to give the Government a further test.

On 25 May Mr Leach, the Under-Secretary of State for Air, speaking on the RAF Estimates, summarized the last few years of aerial endeavour as moribund and development of airships as disappointing. 'After the war, when the general slump began, plant, airships and material, were offered free to anybody who would have gone on with the scheme of airship development. Not an offer was forthcoming. Efforts were also made to enlist the interest of the Dominions, but these came to nought. So for three or four years nothing had been done except that the Research Department had been accumulating knowledge. However, throughout these years Commander Burney, the Member for Uxbridge, has displayed considerable faith in lighter-than-air travel, and kept alive the flame which would otherwise have been extinguished.' He did not reveal that Vickers had lit this flame, but proceeded to criticize the so-called Burney Airship Scheme which his Government had recently rejected. Nevertheless: 'The Government proposes to begin research and experiments at Cardington. We will first recondition one of the existing ships for research, and at an early date will proceed with building a new airship of 5,000,000 cu ft capacity. We will construct the necessary intermediate and terminal bases overseas to enable two airships – one built by the Government and the other by contractors – to fly with safety between here and India. The contract for the second airship is ready for signature as soon as the House agrees. The total cost is £350,000. The whole scheme involves a three-year programme and gross expenditure of £1,350,000. The question of technical staff and of expert research workers has been fully explored, and the Ministry is satisfied that we are properly and fully equipped. We are looking for successful achievement in opening up a new era in Dominion relationship; with airships, communications between our blood relations overseas can be put on a closer basis than ever before.'

Dryly Sir Samuel Hoare said: 'To-day the hon gentleman has come before us, not only as a full blooded Imperialist, but as a full blooded militarist as well. The scheme of our late Government was a commercial scheme. The sole original contribution of the present Government to the development of British aviation is the building by direct labour of a military airship that the late militarist Conservative Government would never have dreamed of proposing. The Under-Secretary has given a travesty of the late Government's scheme, for we proposed a loan, without interest, of £400,000 a year for seven years, repayable from profits, and at the end we expected six large airships operating a commercial service between England and India. The present Government's scheme is likely to be much more expensive, for it has no ultimate object in view, and at the end of three years would be little further forward. There are few Air Ministry officials who know anything about airships, so a great airship department must be created, and a big Government organization set up for a single airship – and, what appears no less dangerous, a new construction organization at Cardington. What does the Government intend when the experimental period is over?'

Mr Leach replied: 'As to where we are going, what the nature of our policy, what is visualized years ahead – I could not tell yet. The Government is looking at a more or less dead industry which we want to put on its feet if possible. The airship industry has a chequered past; there is no certainty about its future security. The six airships of the Burney scheme would have taken seven years to provide, whereas the three under the Government scheme – two new and one reconditioned – should be provided inside three years.'

Vickers, in June of the previous year, had sent Pratt with Commander Burney to Germany to negotiate an agreement between the Zeppelin firm and the proposed British airship company whereby the former would exchange 40 per cent of its shares for 25 per cent in the latter, and would supply drawings of a 5,000,000 cu ft airship of 3,600 hp with dynamic lift of 15/20 tons and gross lift of 150 tons, giving a disposable load of 75 tons for 150 passengers at a speed of 80 mph and cruising range of 2,100 miles. On that basis Vickers made proposals to the Government for a £4 million company, towards which they would subscribe £200,000. One facet of their report stated: 'Mr Ricardo, the engineer no doubt remembered as responsible for tank designs, has perfected a carburettor to use petrol and hydrogen combined which will allow the useful load to be increased by 40 per cent and accordingly increase the ship's earning capacity.'

Sir Barnes Wallis, who as B. N. Wallis had rejoined Vickers in 1922 when Commander Craven gave him a job on the sales side at £650/annum, told me that Major Scott the airship commander, Major Colmore, at that time a Lieut Cdr R.N, and Colonel Vincent C. 'Dopey' Richmond, had earlier approached Burney with a view to participating in the proposed company. 'When Burney told me of this, I refused to join in, for I considered that Richmond and Colmore knew nothing about actual design. I therefore schematically designed a suitable airship at home, showed it to Burney, who was sufficiently convinced to go ahead and persuaded Vickers to purchase the empty Howden shed and proceed on their own. The Airship Guarantee Co was formed 29 November, 1923, the directors being Sir A. Trevor Dawson, Sir Vincent Caillard, Mr A. Cartwright, Commander Craven, and Lieut-Cdr C. D. Burney. H. B. Pratt who had been general manager in charge of Vickers war-time airship construction was no longer interested, and I, as his ex-chief designer, was appointed chief engineer.' To that company the Government now contracted the second airship. An advance of £150,000 would be paid on 22 October towards the estimated cost of £300,000 together with £50,000 towards expenditure on plant.

Simultaneously work began at Cardington on the Government airship, led by Colonel Richmond as chief engineer, who, after extensive wartime experience of airship maintenance, had made a study of Zeppelins in Germany. He sub-contracted detail design and construction to North at Boulton & Paul, though assembly would be at the Royal Airship Works at Cardington. Major Colmore, who recently had succeeded Grp Capt

Peregrine Fellowes as Deputy Director Airships, was the key who would link detailed knowledge of design and progress of both the privately built R.100 and the Government-built R.101, though the respective designers, largely at Burney's instigation, were kept competitively in the dark as to each other's progress and methods. Undoubtedly Colmore, an ex-naval officer and pre-war ship builder, had expert knowledge, for he had commanded the Mullion airship base, and was the Air Ministry's airship constructional overseer from the time of R.9.

Among those joining Wallis was Nevil Shute Norway, stressman and novice author. He had been on Walker's staff at de Havillands since leaving Oxford, but found promotion could not be rapid because 'the company was staffed by seniors who were all young men.' In *Slide Rule* he records: 'I joined the Airship Guarantee Co as chief calculator. I knew nothing of airships, but they had three consultants who taught me the fundamentals of my job. Professor Bairstow was our authority on aerodynamics, Professor Pippard on structures, and Mr J. E. Temple was the most practical and useful of them all because he had been chief calculator for Wallis on the R.80 built by Vickers at the conclusion of the war. My job was to get a staff of calculators to do the work on R.100, translating the theories of the consultants into forces and stresses in each member of the ship and so provide the draughtsmen with the size for each girder and wire.

'I spent many hours reading old reports and records to find what had previously been done in airship calculations, and when I came on the report of the R.38 accident enquiry I sat stunned, unable to believe the words I was reading. I had come from the hard commercial school of de Havillands where competence was the key to survival, and a disaster might have meant the end of the company and unemployment for everyone concerned. It was inexpressibly shocking to find that before building the vast and costly structure of R.38, the civil servants concerned made no attempt to calculate the aerodynamic forces acting on the ship, and I remember going to one of my chiefs with the report and asking if this could possibly be true. Not only did he confirm it, but he pointed out that no one had been sacked over it nor even suffered censure. Indeed, he said, the same team of men had been entrusted with the construction of another airship, the R.101, which was to be built by the Air Ministry in competition with our own ship, the R.100.'

7

Though the Labour Government had endorsed the previous Conservative Government's policy of increasing the strength of the RAF, and when implementing airship construction knew of their wartime possibility as naval scouts, Ramsay MacDonald was bent on becoming peacemaker of the world. On assuming office, he sent a friendly warning to Poincaré, backed a month later with a sterner note, against France's determination to subordinate Germany. To the same end on 9 April a committee of British

The second D.H.50 prototype about to take-off with spring-loaded flaps automatically deflected. (*Flight*)

and Americans, headed by General Charles C. Dawes, issued a Report, written by Sir Josiah Stamp, recommending stabilization of German currency on a gold basis, and that Germany should make progressive annual reparation payments rising from 1,000 million marks to 2,500 million in the fifth year, and the methods of raising it were described in detail. With France's May elections Poincaré fell, and Edouard Herriot, leader of the Radicals, became Premier, and like Britain, accepted the Dawes plan, visiting MacDonald at Chequers for discussion at the end of June.

Endeavour to secure closer communication and contact between one country and another was increasing. The necessity of exports to buttress home requirements was obvious, and firm after firm in the British aircraft industry was seizing every opportunity to demonstrate and sell its products, though successive British governments and ministries were far from helpful. Sir Sefton Brancker was an exception to the run of officials, and, accompanied by the Air Minister, he was flown by Hubert Broad in a D.H.50, readily loaned by de Havillands, to the Prague Aero Exhibition on

Sir Sefton Brancker, Lord Thomson and his P.A., pose for the Press before embarking on a D.H.50. (*Flight*)

30 May, accomplishing the 600 miles in 6½ hours. Theirs was the second prototype, G-EBFO, fitted with another of Hagg's ingenious devices in the form of full-span flaps (Patent 219,388) which automatically deflected with decreasing air speed, but could be over-ridden manually. Tested by D.H. himself, the flaps streamlined with the wing-section above 70 mph, but at 45 mph moved to full angle, reducing landing speed by 7 to 8 mph.

Of the Prague Exhibition, *Flight* reported: 'That one Czechoslovakian firm should show no less than 11 machines, of eight different types, affords proof of the determination and enterprise of a nation at last liberated from centuries of suppression and augurs well for the future of the Republic under the wise and firm guidance of its first President, Professor Masaryk.' British enterprise was at least emphasized by a communal stand, boldly placarded and emblazoned with a Union Jack shield surmounted by gilded eagle wings. On display were a Lynx Avro, a Blackburn Dart, the Rolls-Royce Condor, a Bristol Jupiter, and a Napier Cub. Afterwards Brancker said: 'It was a very good thing that this country decided to exhibit, and I think the £12,000 which it is estimated our show will cost, of which half will be paid by the Air Ministry, is money well spent.' On Sunday morning 1 June, Brancker and Thomson left Prague in the D.H.50 at 7.30, stopped at Cologne for lunch, and arrived in London in time for dinner.

To help Broad promote sales of the D.H.50 at Prague, Alan Butler, chairman of the company, had flown there in 5½ hours nonstop with his D.H.37 *Sylvia*. Nevertheless, it was disappointing to the firm that they were unable to exhibit the new two-seat D.H.51, powered with the same air-cooled 80 hp eight-cylinder vee R.A.F. 1A which had powered the B.E.2c. Despite the engine's official ancestry the Air Ministry had refused to accept it as airworthy, so to conform with their requirement for dual ignition instead of single, de Havilland fitted Remy coil ignition in addition to the standard magneto, only to find that a 10-hour type test would be required, or flight of similar duration around Stag Lane aerodrome in lieu – a ruling which irritated the company for it would cost £100 or so, and they had no proprietary interest in the engine itself. By that time it was too late for Prague, but in any event, neat though the D.H.51 might be, it was not one of Geoffrey de Havilland's wisest gambles, for few would want to be bothered with the complexity of two-bay rigging or of housing a 37-ft span machine, however pleasant it was to fly – but at least it could be used in the newly opened de Havilland Flying School.

The *London Gazette* of 10 June published despatches from Air Vice-Marshal Sir John Salmond, written before he left Iraq, which told of air co-operation in Kurdistan against the pro-Turk intrigues of Sheik Mahmoud. 'Using two Military columns in conjunction with Nos. 1, 6, 30, and 55 Squadrons, these operations were carried out in difficult hill country over little known routes, of which even the most recent maps were of small value. Throughout, I was impressed by the many and particular advantages which the informed use of air power had given me for this kind of warfare, and I venture to suggest that this experience foreshadows important developments in the conduct of small wars.' By personally flying he found it possible to survey in the course of a 550-mile reconnaissance the situation on all sectors. Not only had aircraft been used to blockade insurgents in their mountain retreats, but his Vernons had dropped food and equipment to the troops, and evacuated 200 dysentery cases to Baghdad 200 miles distant, whereas previously they would have been transported by donkeys on a six-day journey of prolonged suffering. To sceptics who had criticized proposals to hand control of Iraq to the RAF, Salmond's despatches were convincing proof of the value of aircraft.

Backbone of the RAF for General Purpose duties, whether formating at Hendon RAF Display or operating in Kurdistan, was the D.H.9A. (*Flight*)

The fifth RAF Pageant, on 28 June, provided spectacular emphasis in a day of 'perfect weather, perfect organization, and perfect flying.' Some 80,000 crowded the enclosures, and thousands used vantage points outside. There were the usual magnificent displays of squadron flying, and a French escadrille of five Nieuport-Delage *avions de chasse* performed graceful individual aerobatics. C. M. Poulsen described the Wing drill of two D.H.9A bomber squadrons as 'the finest flying display ever seen, both from the spectacular and technical point of view'. Their eighteen aircraft took-off in massed formation, separated into two squadrons, and in a manner never seen before passed headlong through each other in what was described as a 'heart-stopping evolution'. But these were antique machines, like the 504s, Snipes, and Bristol Fighters participating in their turn.

Once again the prototypes of the fly-past proved the major index of progress. In single file they taxied past the Royal Enclosure, and took-off with loud and varying roar. First came the diminutive D.H.53 and the Parnall Pixie, mechanics steadying their wing-tips, the Blackburne Tomtit engines crackling like motor-cycles. A Handley Page three-engined airliner lumbered after them in magnificent contrast, followed by the Condor-powered Avro Andover air ambulance, its large square windows giving the

Several ultra-lights such as the Parnall Pixie (*illustrated*) were purchased for evaluation by the RAF, but all except the D.H.53 were dismissed as toys. (*Courtesy Air Marshal Sir Ralph Sorley*)

Yet another version of the W.8 formula was the Handley Page Lion-powered Hyderabad bomber which in full war-paint looked intimidating compared with its civil brothers.

From the Vixen, Rex Pierson developed the workmanlike Vickers Venture, of which six were built, but Martlesham turned the type down because of longitudinal instability, and preference for the Armstrong Whitworth Atlas.

impression of an airliner. The Parnall Possum triplane followed, the whirr of its transmission gears distinguishable, and then came the Lion-powered Vickers Virginia and the Handley Page Hyderabad heavy bombers. No less interesting were the medium-sized machines: the Lion-powered Vickers Venture army co-operation two-seat biplane, its rounded nose cowl glittering; the Jupiter-powered de Havilland Dormouse two-seat fighter-reconnaissance biplane which was clearly a slicked-up re-design of the D.H.9; the spectacular Jupiter-powered Bristol Bullfinch single-seat fighter monoplane with olive fuselage instead of silver like the others; the Hawker Woodcock single-seat fighter with helmeted Jupiter engine looking somewhat antiquated compared with the dainty Jaguar-powered Gloucester Grebe II. I had also hoped to see the Bristol Bloodhound in new guise with longer fuselage, uptilted thrust-line, and bigger and balanced tail surfaces, but it had suddenly been classified 'Secret'. The Blackburn Cubaroo was also missing, though it had been flown by Flt Lieut Bulman in mid-June.

Final thrill of the RAF Display was always the set piece of great ingenuity, massed flying, and tremendous explosions. (*Flight*)

With the final set-piece, the RAF staged the most splendid spectacle ever seen at a Service display, for they had made two huge replica steamships, representing a cargo vessel and an armed enemy merchant cruiser, which rode serenely on the green grass sea at the far side of the aerodrome. Smoke was seen issuing from the funnels, crews moved on the decks, and on the bridge of the British ship signallers were flag-wagging. A Supermarine Seagull amphibian appeared high overhead and was fired at by the cruiser which was moving with evil intent towards the cargo ship and sent a pinnace to seize its papers and scuttle it. But the Seagull had radioed for fighters, and along came three Fairey Flycatchers which dived at the cruiser with machine-guns stuttering and silenced her. Five Blackburn Darts came thundering low, discharged their torpedoes, and a few moments after, with a tremendous boom and awe-inspiring flash, a great rent appeared in the enemy's bows and she exploded in a mass of dense smoke and fragments. 'Cor!' said a youth by my side. There was a stir of chatter and movement. The fifth and best Display was over.

8

Little more than a fortnight later, on 16 July, three of the four US Army's round-the-world big Douglas DWC biplanes arrived at Croydon from Le Bourget, led by Capt R. H. McIntosh who had joined them with his Handley Page W.8b for the Channel crossing. Unlike the disregarded arrival of the American flying-boats after crossing the Atlantic in 1919, the RAF, Air Ministry, and RAeC had distinguished representatives to welcome them, including Sir Sefton Brancker, Air Vice-Marshal Sir Geoffrey Salmond, Air Commodores Longcroft, Webb-Bowen, and Dowding. First to land was the *Boston*, piloted by Lieut Lowell Smith as leader, followed by bald-headed Lieut Eric Nelson with *City of Chicago*, and then the *New Orleans* flown by Lieut Leigh Wade. Next morning they flew to Brough where floats were fitted for the Atlantic flight via Iceland and Labrador. Pilots and navigator returned to London by train for a banquet at the Savoy given by the RAeC. Replying to felicitous speeches by Lieut-Col Francis McClean, Lord Thomson, and Air Marshal Sir Hugh Trenchard, the American Ambassador said: 'Round the world in 80 days, or nearly 20,000 miles in 264 flying hours, with 10,000 miles yet to do! Who a generation ago would have believed such wonderful accomplishment possible? So marvellous is the age in which we live that in a short time we shall be talking to all parts of the world by wireless telephone. Whatever America may achieve in scientific progress she will feel herself associated with the great British Empire. We are of the same race, with the same ideals and aspirations, the same form of government, and the same

Crew of the United States Army world flight stand in front of the Douglas World Cruiser *Boston*, but their leader, Eric Nelson, is camera-shy to the right. (*Flight*)

hopes for advancement of our great civilization. England is facing her problems, and stands before the world as a nation which kept her promises and outlived the wreck of war. We have had our triumphs, but I desire to congratulate that brave airman Sqn Ldr MacLaren, now struggling in the wilds of North Japan. The prayers and good wishes of the American people have followed his splendid flight, and all hope it will end successfully, and we believe it will.'

But it did not. MacLaren and his crew had more than their share of bad weather, engine trouble, and accidental damage, and after leaving Tokyo were held up in the Kuriles by dense fog, rain, and gales, but managed to reach their supply depot ship *Thiepval* at Kamchatka, only to encounter more fog on the next lap until forced to within 50 feet of the water when 30 or 40 miles short of Attu, their destination. Describing what followed, MacLaren wrote: 'As we neared the coast, the fog got lower and lower. We were only a few feet above the water when suddenly black cliffs loomed, and only a quick swerve saved us from crashing into them. By this time we were pretty nervy, so decided to alight in the open sea. Plenderleith brought us down perfectly, but just after touching, a large wave caught one wing-tip float, smashed it, and buried the end of the wing under water. This made the machine swing suddenly round, and the sea smashed the starboard float and wing-tip. We hastily put on life belts. The fog was too thick to see more than 50 yards in any direction. Our only hope was that my navigation had been correct, which placed us just south of Bering Islands. It was essential to keep the machine on the move as this was the only method of keeping the wings out of water. Plenderleith taxied slowly northward, while Broome and I ran up and down the wings trying to keep them balanced horizontally. After three hours of anxiety, half a mile to the north, we saw a coast line. Reaching the shore, we anchored just off the surf, jumped into icy water, and gained the beach. That was the end. The machine could not be repaired. The *Thiepval* arrived next morning, and salved the remains. We would not have missed the adventure for worlds: we did our best, but failed.'

Meanwhile the Americans resumed their flight on 30 July, having decided that British weather was worse than anything they had experienced. Flying through fog they arrived at Kirkwall for refuelling, and on 2 August started the 600 miles to Iceland. Soon they encountered thick fog and lost sight of each other. Lieut Smith and Wade returned to Kirkwall, but Lieut Nelson sighted the US destroyer *Billingsley* 300 miles from Iceland, and reached his destination at 4 p.m. The other two made another attempt next day. Smith achieved his goal after flying through fog and rain, but Lieut Wade had engine trouble and damaged his machine in alighting on the rough sea off the Faroes, but was towed by the *Billingsley* to the US Cruiser *Richmond*, where in hoisting it aboard the gear broke and the machine was wrecked.

Meanwhile Argentina with Major Pedro Zanni flying a Napier-powered Fokker C.IV, and Italy with Signor A. Locatelli piloting a twin-Rolls

Dornier Wal entered the lists, but the former crashed at Hanoi, and the latter, forced down with engine trouble en route to Greenland, had to abandon his flying-boat on being rescued by the USS *Richmond.*

While spectacular world flights were achieving headlines, acknowledgment was made to 30-year-old Alan Cobham with award of the Britannia Trophy for the many splendid flights made during 1923 as chief pilot of the de Havilland Aeroplane Hire Service. Thanks to his tremendous initiative that business had made an outstanding success of its unsubsidized flying.

It was further triumph for Cobham when the King's Cup race for landplanes and seaplanes was flown on 12 August in a great circuit from Martlesham Heath northward to Leith and Dumbarton and then south to Falmouth where they turned east to the finishing point at Lee-on-Solent. There were no compulsory stops, and competitors could alight where they liked any time during the 950-mile course.

Though scarcely daylight at the starting time of 5.30 a.m., several thousand aviation enthusiasts arrived by car, cycle, and foot to see the ten entrants take-off at minute intervals. Courtney, on a V-strutted Siskin III with large auxiliary fuel tanks each side of the centre-section, was first away, climbing steadily. A second Siskin followed, flown by Flt Lieut H. W. G. Jones. Next came H. J. Payn with the civil registered Vickers Vixen G-EBIP, and then Alan Butler with his handsome gold-winged D.H.37 *Sylvia.* Raynham's old yellow-painted Martinsyde F.6, piloted by J. King, streaked away, and was followed by two D.H.50s, the first piloted by F. L. Barnard, winner of the original King's Cup, and the second by Alan Cobham who was regarded as hot favourite. Two Supermarine Seagull amphibians, respectively piloted by Henri Biard and the Master of

Alan J. Cobham after winning the King's Cup with his D.H.50.

Sempill, were last to depart, the final entrant, a Fairey IIID seaplane flown by Macmillan having started from Felixstowe.

The early morning haze settled into low cloud and poor visibility. Many spectators made for the finishing point, Lee-on-Solent, 160 miles away. Hampered not only by poor weather but by mishaps, competitors were having a difficult time. A propeller spinner failure forced Courtney to land at Brough. King in landing at Newcastle for petrol knocked off his undercarriage. Biard narrowly escaped disaster when his propeller, just behind his head, worked loose and flew off, but he landed safely in a field near Blaydon. The Master of Sempill stopped at Renfrew to mend a broken drift wire, and at Anglesey and Padstow for petrol, so was virtually out of the race. Cobham in landing at Ayr for petrol and oil almost came to grief by running into a haycock, so did Alan Butler, and Barnard actually collided with a haycock there and smashed his propeller.

It was Jones on the Siskin who flashed past the finishing line 1 hour 45 minutes after mid-day, followed by Butler at 1 hour 59. When Macmillan appeared seven minutes later there were frantic signals because he landed without crossing the line, and had to take-off again to do it. Fourth was Payn on the Vixen; fifth came Cobham who recently had married a delightful actress. The slide-rules of the RAeC handicappers, those Farnborough men, Capts Goodman Crouch and Dancy, quickly confirmed that Cobham was winner with Macmillan second and Butler third. Not until four hours later did the missing Sempill arrive, amidst sighs of relief.

That month an important announcement from the Air Ministry stated that the Air Council was so greatly impressed with the possibilities afforded by development of light aeroplanes that, in addition to prizes for two-seaters at the Lympne competition in September, it had been decided to assist financially for two years the establishment of ten light aeroplane clubs provided their constitution had been approved. Each club would have an initial grant to buy approved light aeroplanes, but would be required to contribute an equivalent amount and to insure against loss or damage of equipment. The club would be responsible for management and maintenance. Periodic inspection would be undertaken by the Air Ministry, who would make an additional grant for each member qualifying for a private licence on the club aircraft. Putting the scheme into operation was delegated to the RAeC.

Eliminating tests for the ultra-light two-seaters would begin on 28 September. C. G. Grey dryly prophesied that whenever there was a competition, work on the entries only started a fortnight before the event, so the Lympne machines would have their finishing touches while the pilot was on his first lap. He was right, for though by midsummer several entries were structurally complete but awaiting their engines, only the Bristol Brownie cantilever low-winger had a margin in hand because its engine was built by the same firm, so it received the first of the Cherubs – the sole small aero-engine to have passed the Air Ministry type test of 25-hours endurance running, including 10 hours nonstop. Rated at 24 bhp at 2,500

When the Bristol Brownie G-EBJK first flew, piloted by Cyril Uwins, and then by its designer Frank Barnwell, it had slight aileron flutter at the stall until the control system was modified. Like Bolas earlier, Barnwell erroneously thought a small rudder prevented accidental spinning.

rpm it actually developed 32·66 bhp at 3,200 rpm. The rival Anzani, Blackburne, and A.B.C. Scorpion had not even been presented for test yet, though nine machines had Cherubs, four had Blackburnes, three had Anzanis, and one had the A.B.C.

Cyril Uwins tested the Brownie on 6 August, and then handed it to its designer, Frank Barnwell, most erratic of pilots. He survived and decided it was worth building a second machine with identical wire-braced, steel-tube framed fuselage, but with a wing having steel spars and ribs instead of wood.

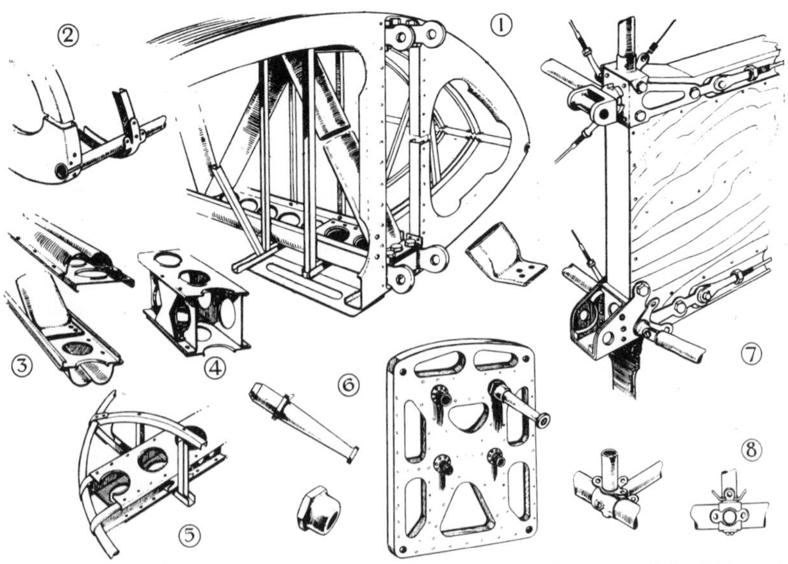

Wire-braced steel-tube fuselage frame and steel-structured wings were Bristol Brownie features. 1, spar lower boom and Warren web-bracing. Details at 2, 3, 4 and 5. 6, forward bulkhead and machined tubular engine support. 7, fuselage wing root and intercostal wooden box spar. 8, fuselage fittings. (*Flight*)

The two-seat A.N.E.C. and Beardmore *Wee Bee* were near sisters designed by Bill Shackleton after his successful essay with the unique single-seat A.N.E.C. the previous year. (*Flight*)

Not until September did others fly – the Beardmore shoulder-wing monoplane with the second Cherub, and Camm's pretty little Scorpion-powered Hawker Cygnet with which he had managed to achieve an empty weight of a mere 270 lb – reflecting his earlier genius at designing model aeroplanes. By mid-September the Pixie low-winger, Cranwell C.L.2, Supermarine Sparrow, and the A.N.E.C. had also struggled into the air, and though the Avro Avis had flown with a Blackburne engine, it had to be delivered by road to Lympne through lack of its Cherub, and the Vickers Vagabond built by Avro was not even flown at its manufacturers. The Short Satellite's attractive metal monocoque fuselage proved to be penalized by excess weight, for though flown solo by Lankester Parker on 16 September it failed to get airborne when tried with a passenger on the 19th. With less than ten days to go, others followed, but some, as Grey predicted, were being finished at the opening of the competition, and already

When the engine was cowled and a fine-pitched propeller was fitted, the Short Satellite just managed to get airborne with two up, but was better after a 40 hp A.B.C. Scorpion was substituted.

Wind-tunnel prediction of the lifting power of the flaps-cum-ailerons of the 22 ft 9 in span Westland Woodpigeon was too optimistic, so its stablemate G-EBIY was fitted with 27-ft wings but was still underpowered.

several found that performance with two aboard was marginal. The miniature Westland biplane, the Woodpigeon favoured by Bruce, proved impossible to take-off until the rough airfield grass was mown smooth as a tennis court, and the distinctive Widgeon parasol of Davenport's was handicapped by the same engine faults as the Supermarine. Longer wings were made for a second Woodpigeon.

The Air Ministry backing of flying clubs spurred Geoffrey de Havilland to redesign the D.H.51 with single-bay wings of 32-ft span and fit the patented automatic flaps. Though nearer the target, it still seemed too cumbersome and with too complex an engine and was only repeating the unsuccessful attempt at civil sales of the Boulton & Paul P.9 five years earlier. But at last the Aircraft Disposal Co accepted Geoffrey de Havilland's arguments and agreed to manufacture a four-cylinder engine to Halford's design, utilizing the 140 hp Airdisco cylinders, pistons, and details. The way was open. D.H. drew a scaled-down version of the D.H.51A intermediate in weight with the new generation of questionable ultra-lights. Designated D.H.60, it had a span 3 ft less than the D.H.51A, and was intended to have half the weight yet carry two at the same cruising speed of 80 mph so that it could battle against head winds. Constructional simplicity, and therefore minimum cost, was to be the key note.

Concurrently interest briefly turned to the forthcoming Schneider contest due to begin at Baltimore in the USA on 14 October. Convinced by G.A.C.'s long racing record, the Air Ministry had tardily ordered a Gloster racing machine, for unless something was quickly done the Americans would have a fly-over and retain the Trophy for all time. The 'Bamel' had therefore been fitted with floats as a trainer, and a contract was received for a scaled-down re-design. Built in a mere two months, and powered by a direct-drive Napier Lion, the Gloster II was sent to

Felixstowe in mid-September with little time to test, dismantle, and freight it to the USA. Flown by Hubert Broad on the 17th: 'The performance of the machine was very promising and at the conclusion of the flight he brought it down on the water beautifully. It had no sooner alighted, however, than the undercarriage or floats gave way and the machine began to sink. A fast motor-boat put out from the RAF station and made an exciting rescue of Broad, but a few moments later the machine sank, and so far all efforts to salve it have failed.' The one and only British machine was out.

Built to the 'Bamel' formula but with thinner cabane, smaller wings, and ply-covered fuselage, the Gloster II, shown taking off at Felixstowe, was a quick expedient that cost only £3,000.

Though the Italians had challenged, they failed to send a machine. Sportingly, the American Navy decided not to fly the course rather than claim an empty win as their right. Said C. G. Grey: 'This act of self-abnegation on the part of the Americans is one of the finest examples of true sportsmanship in the history of international competition. One of the conditions of the Schneider Contest is that permanent possession of the Trophy can only be gained if a country wins it three times within five years. If the Americans had chosen to claim a walk-over, they would have then registered two successive wins and if they won it next year the Trophy would belong permanently to the National Aeronautic Association of the USA.' The RAeC cabled to express warmest appreciation of this sporting action, and implied that England would be fully represented in the 1925 race. If a machine was to be ready in time it was imperative to start preparations without delay, but it took the Air Ministry some weeks to decide that they would sponsor construction of a new Gloucestershire biplane seaplane racer and the Supermarine monoplane which Mitchell had been evolving. Additionally G.A.C. received a development contract for a landplane version of the Gloster II so that variations of metal propellers and different cooling systems could be evaluated. Of vital importance was a contract for Napiers to make a 6 : 1 compression-ratio, direct-drive Lion which it was hoped might develop 700 hp. Further engines of this type and the aircraft would be loaned to the two aeroplane firms for their pilots to fly in the Contest.

Re-engined with Jupiters, the Handley Page Hyderabad was re-named Hinaidi, and in production had an angular fin and rectangular rudder. The eventual series was restructured in metal.

9

My session of practical work at Handley Pages had finished by the time the Lympne competition opened, and I was visiting my home in Suffolk before returning to College for the aeronautical course. It was only when I went to nearby Martlesham and saw a Hinaidi with its original W.8 type fin and rudder that I realized I had drawn too short a fuselage for the model of it with experimental triangular fin and rectangular rudder with backset hinge which we tried in the H.P. wind-tunnel. I wrote to Reynolds, fearing that in full scale there might be serious difficulties. He replied with a teasing letter that the model fin and rudder showed greatly improved characteristics, and results would be even better with the actual longer fuselage. 'Research must be tempered with common sense,' he said. 'And in any case the old man has decided to use your tail.'

C. M. Poulsen, the editor of *Flight*, describing contemporary events at Lympne, said: 'Started in gloom, figuratively speaking, the Air Ministry competitions for two-seater light planes may be said to have finished in

The Parnall Pixie IIIA was an ingenious biplane conversion of the Lympne light monoplane by mounting a smaller unbraced top wing on N-struts, but gave no overall advantage. (*Courtesy Philip Jarrett*)

The Parnall Pixie III Lympne light monoplane. (*Flight*)

sunshine. The eliminating trials, which everybody had regarded as child's play, proved far more difficult to pass than expected, and only a relatively small proportion succeeded during the Saturday and Sunday set aside for that purpose. Of the 19 machines entered, only eight were admitted to the competitions when these started on Monday morning. Even this small number had, by Tuesday evening, been reduced to six by almost simultaneous failure of the two Parnall Pixie IIIA biplanes – so there remained in the competition only No. 1, the Bristol Brownie; No. 3, the Cranwell biplane: No. 4, the Beardmore Wee Bee; No. 5, the second Westland Woodpigeon biplane; and Nos. 14 and 15, the Sopwith-Hawker Cygnet 1 and Cygnet 2. What was the cause of all these failures? The answer is that the capabilities of the 1,100 cc engines were overestimated by those who drew up the rules. It takes very much longer to develop an aero-engine than develop a new type of aeroplane, so it is not surprising that most of the engines failed to stand up to the strain of the competition.' However, engines which had been refractory became tractable, and during Thursday, Friday and Saturday morning feverish activity reigned, dispelling the gloom of the first few days, and in the Grosvenor Cup race on Saturday afternoon, no less than nine completed the 100-mile course at full speed.

Model making as an early hobby led Camm always to think in terms of minimum weight, and his pretty little Hawker Cygnet was the quintessence of this expression. (*Hawker Siddeley*)

'It might be thought that as far as the aircraft industry is concerned', reflected Poulsen, 'the net result of the competition is that the Air Ministry has caused it to spend something like £30,000 to win £3,000. While on the face of things this is so, much of the value was lost because so many machines were prevented from showing what they could do. However the Air Ministry propose to try out at Martlesham the various competing aircraft.' Instead, a few days after the competition, a statement was issued regretting that none of the types could receive official recommendation, and clubs were warned that delay was inevitable.

Nevertheless, the widely varying Lympne designs were of great technical interest. Bristols had useful if alarming competition experience in testing the all-metal cantilever wing of their second machine, G-EBJL, for it exhibited aileron flutter, with the wings warping first one way and then the other until it died down when speed was reduced. Parnalls confirmed the practicability of a convertible arrangement by superposing a top wing to gain take-off marks. Hawkers proved that strict weight control paid dividends, as exemplified by fuselage longerons spindled to X-section, in conjunction with light box spars and lattice ribs. The Short Satellite emphasized Oswald's conviction that shapely and strong fuselages could be made fairly simply using duralumin monocoque. And although the Westland Widgeon piled up due to a down-current greater than its climb, Bruce and Davenport had at least been able to determine it was more efficient than the Woodpigeon which had been one of the few to last out the competition. Like the Westland machines, the Vagabond, Avis, and Sparrow had full-span combined ailerons and flaps, and the resulting complications revealed the difficulty of attaining reasonable lateral control.

Despite hope that the competition would attract amateur-built aircraft, the sole example was the side-by-side C.L.A.2 biplane built by the apprentices who constituted the Cranwell Light Aeroplane Club. It handled well, but was underpowered with a Cherub so its top speed was only 60 mph, and it was facetiously alleged that on the slow-run test it was 2 mph faster

By contrast with the Woodpigeon the Westland Widgeon had the novelty of a parasol monoplane with double tapered wing, but construction of both followed conventional practice and ensured a much sturdier machine than the Cygnet.

Power of the Thrush engine was insufficient for the Widgeon's relatively high wing loading. The full-span ailerons deflectable as flaps were based on the patented flaps of the war-time Westland N.1B seaplane.

A novel feature of the Vickers Vagabond, possibly suggested by the Armstrong Whitworth Ape, was hinging the rear fuselage behind the aft cockpit so that it could be slanted by handwheel control to alter the relative incidence between wings and tail. (*Vickers*)

There was a noticeable family resemblance between the pre-war I-strutted racing Avro 511 and the slightly lighter Avro 562 Avis (*illustrated*) which managed to fly with half the power.

The 34 ft 4 in span Sparrow had full-length flaps on the wider-chord top wing only, and was designed by the Supermarine 'stress merchant' R. Dickson, but became eliminated from the Lympne contest by repeated engine failure.

than its average round the high-speed course. C. G. Grey was quick to commend the effort, saying: 'If the Air Ministry is really sincere in its desire to encourage light aeroplane clubs it will go out of its way to give a little extra encouragement to those who are ready to follow the Cranwell example.'

As in the previous year, Lympne provided the greatest excitement on the last day. Winning the competition would largely be swayed by marks awarded for speed range, but these could not be counted until ten laps of the high-speed course had been completed in trouble-free sequence. So disastrous had been delays due to repairs and replacements outside the permitted schedule that by Saturday only little Piercey on the Beardmore Wee Bee and Uwins on the Bristol Brownie had succeeded in completing those laps. On his earliest speed attempt Piercey averaged 70.1 mph with engine prudently throttled, but Uwins had only achieved 65.1 mph though hoped to increase his total by improving his take-off and landing score in further attempts. Piercey thus seemed the potential winner unless Raynham, who had attained 82 mph with the Cygnet, could remain airborne for ten laps instead of the five which had been terminated by engine failure.

The Sparrow rebuilt as the parasol Sparrow II for the 1926 Lympne trials. (*Flight*)

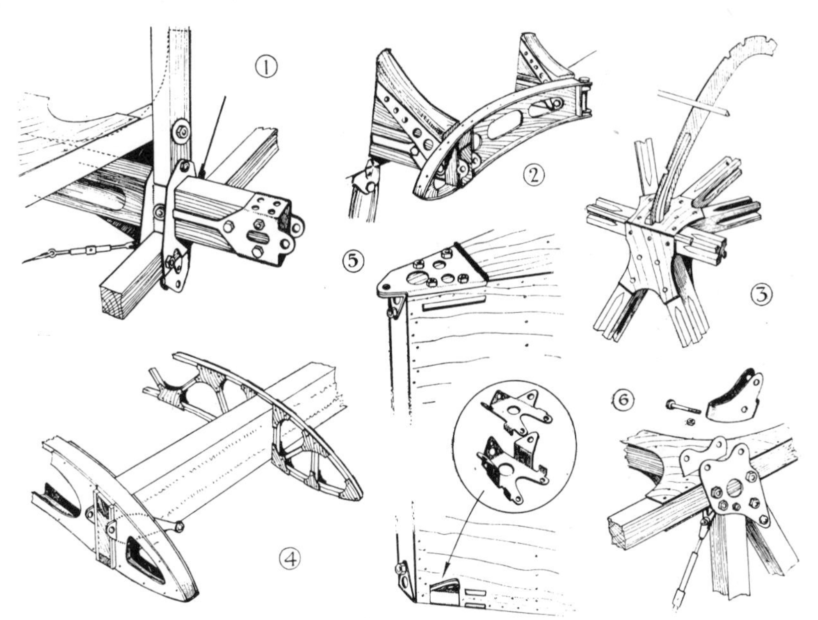

Cygnet details show Camm's technique: 1 and 2, fuselage root spar construction. 3, spindled fuselage members. 4, box spar and simple spruce rib. 5, sternpost fittings. 6, solid fuselage members at centre-section struts. (*Flight*)

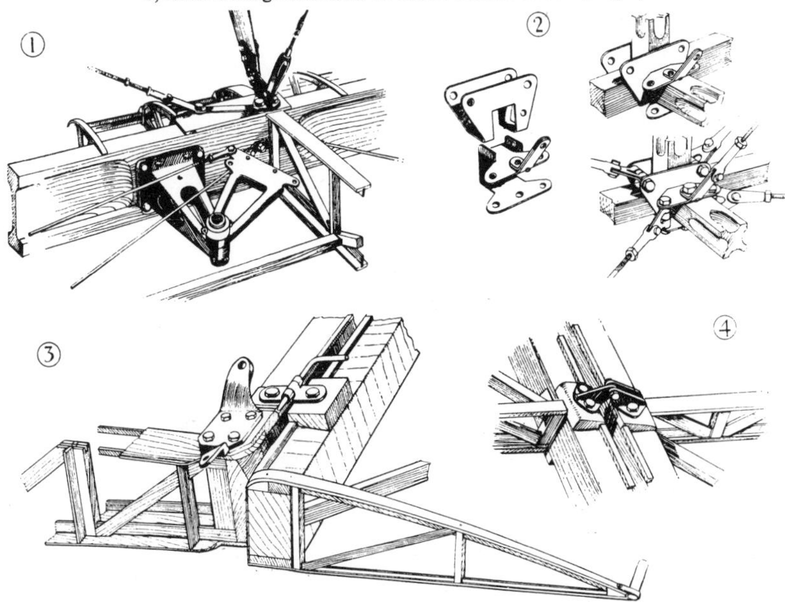

Construction of the Vagabond was conventionally straight-forward. 1, ailerons had a differential crank. 2, fuselage strut clips of mild steel. 3, a root flap could be released to give folding clearance for the lower wing. 4, aileron hinge. (*Flight*)

Cramped: cockpit, little man; M. W. Piercey in the *Wee Bee 1*. (*Flight*)

Realizing the main prize might slip from his grasp, Piercey took-off with the Wee Bee and began lapping at almost 80 mph, but after the fifth he did not reappear – the big end of his engine had failed only two miles from the finish. Raynham was equally unfortunate. He continued lapping, but as the Cygnet approached the South Hill turning point, with only 15 miles to go, it suddenly swung away with a broken rocker-arm, and landed in a field.

So the prize of £2,000 was Piercey's on the strength of his earlier speed run, for though the Cranwell C.L.A.2 was attempting to complete four hours' lapping, it could not win because of its slowness. Uwins with the Brownie was second and also won the Duke of Sutherland's prize for best

Comper's series of amateur-built light aircraft such as the C.L.A.2 were sporting ventures and always caused interest, but did not prove particularly practical.

Side-by-side seating was an attractive companionable feature of the Blackburn Bluebird, but most professional pilots criticized lack of upward and lateral view when banking. (*Flight*)

take-off and landing marks. Comper and his Cranwell, more fortunate than all others with the way his engine kept going, had lapped the course 61 times at an average of 42 mph, and for this gallant effort he and his little band of RAF helpers received the Motor Traders prize of £300. Nor were the efforts of unlucky Raynham with the Cygnet entirely lost, for he was given the second prize of £100 presented by Capt Wilson for the take-off and landing competition.

When Major John Buchanan, of the Air Ministry Technical Department, gave a paper before the RAeS on lessons of the competition, it was clear that designers and pilots considered a minimum of 40 hp would be required for two-seaters. Said Fred Sigrist: 'We have been trying to produce too efficient a machine. Light construction means expensive construction. A larger engine is wanted, enabling heavier and cheaper construction. A larger engine could be produced as cheaply as the small engine, and the question of fuel consumption is immaterial. If a man ran a machine and kept careful tally of expenditure, the petrol would be so small a percentage that it was not worth bothering about.'

That had also been Robert Blackburn's view. His attractive side-by-side Bluebird had been finished some days before the Lympne competition, despite pressure of production for the Dart at the Olympia Works, but persistent trouble with the 1,100 cc Blackburne three-cylinder radial prevented any flying until October, when it was tested by A. G. Loton, the CFI of the Brough RAF Reserve School. The little engine was clearly insufficient for reasonable climb, and, despite the machine's many attractive points, its success depended on the engine manufacturer producing the more powerful radial he had promised. In every respect the Bluebird preceded the same line of thought as Geoffrey de Havilland with his D.H.60 except that the sociable seating was a better selling point. Nobody at Blackburns had been aware of the secret little Airdisco four-cylinder Cirrus which would soon be ready for bench tests at Croydon, nor was it in D.H.'s interest to reveal not only that he had inspired its conception but had almost completed design of the little biplane with tandem seating which he intended naming the Moth, powered with this engine.

Not light aeroplanes, but scandal over Ramsay MacDonald had been titillating the public. He had just returned from the League of Nations at Geneva, when a reporter discovered the Premier had been given a Daimler and 30,000 shares of £1 in the biscuit firm of McVitie & Price, by Alexander Grant, one of its directors, who subsequently had received a baronetcy. It was not mentioned that Grant had also endowed the National Library of Scotland, nor that he had been selected for honours by the previous Government.

Almost as intriguing was withdrawal of a prosecution against J. R. Campbell, acting editor of *The Workers Weekly*, who had been arrested under the Incitement to Mutiny Act of 1797 for an article urging soldiers not to fire on fellow workers. Conservatives tabled a Motion of Censure, and Liberals demanded a Select Committee of Enquiry. When debated on 8 October, Sir Patrick Hastings, the Attorney General, made it clear that though the Prime Minister and Home Secretary had discussed the unwisdom of the prosecution, the final decision had been his alone. But MacDonald rashly said: 'If this House passes either the resolution or the amendment we go.' And go it was, for the vote was carried by 364 votes to 198. Parliament was dissolved next day, and arrangements made for a general election on 29 October.

The country already was concerned by Ramsay MacDonald's sympathy for Russia and his proposed 'Bolshevist loan' as *The Times* called it. Four days before the poll, came the bombshell. *The Times* headlined: 'Soviet plot. Red propaganda in Britain. Revolution urged by Zinoviev. Foreign Office bombshell'. The crux was a letter copy, apparently signed by the President of the Communist International in Moscow, addressed to the Central Committee of the British Communist Party urging the proletariat to insurrection in the struggle for ratification of the Russian Treaty. Protest from the Foreign Office to the Russian Chargé d'Affaires in London firmly convinced the public that the letter was authentic, though the original was never produced.

The result was catastrophic for Labour who lost 40 seats. Conservatives returned 419 members, gaining 161; but the Liberals dropped to third party with 116 seats lost and only 40 members returned. Asquith was out, but Winston Churchill was in again.

Baldwin formed his Government on 7 November by reappointing most of his previous Cabinet. Austen Chamberlain became Foreign Secretary, and Neville Chamberlain was Minister of Health. Winston Churchill was made Chancellor of the Exchequer; Joynson-Hicks, Home Secretary; Amery, Colonial Secretary; and Sir Samuel Hoare again became Secretary of State for Air. Of what had been inherited on Air affairs, a commentator said: 'The regime of Lord Thomson and Mr Leach certainly had no ill consequences, even if nothing startling has been accomplished. Perhaps Lord Thomson, who became personally popular everywhere, will go down in history as the Air Minister who definitely revived airships. Although his Government did not see fit to adopt the Burney scheme, they did succeed in making a real start with a definite airship policy. The great thing is that airships have been resurrected.' But was it?

In support of that attitude, the first public act of Sir Samuel Hoare was a visit to Cardington Airship Station on 18 November, accompanied by his Under-Secretary, Sir Philip Sassoon, and Air Vice-Marshal Sir Geoffrey Salmond the AMSR, together with Grp Capt Peregrine Fellowes, Major Colmore, Colonel Richmond, and Major Scott.

Although the importance of aeroplane research and development was in no way diminished, airships had come into the foreground of Parliamentary interest. The R.33 was being reconditioned and considerable electrical equipment was fitted to record air pressures and stresses during violent manoeuvres. Similarly R.36 was to be put in order and the nose strengthened for a proposed experimental trip to India to test the feasibility of regular services. As to the new 'giant' airships, information was released with particular emphasis on the Government-sponsored R.101 and its pioneering of steel girders in place of duralumin. Appropriate to official prestige it was bigger than its rival, for its length was 720 ft with maximum diameter of 130 ft, and the R.100 was 695 ft with diameter of 132 ft, though both were of identical 5,000,000 cu ft capacity.

At the Lord Mayor's Banquet Sir Samuel Hoare in explaining air policy said: 'Last year I was able to announce that the vital gap in defence was to be filled and a force of 52 Home Defence Squadrons would gradually be created. This year I am happy to say that, thanks to the foundations laid 18 months ago and to the continuity of policy adopted by my distinguished successor, Lord Thomson, such substantial progress has been made that by the end of the financial year 18 of these squadrons will have been formed. As to airships, if we can reduce the time of the journey between London and Bombay by 10 days, and between London and Melbourne by 20 days, the benefit to the Empire cannot be over estimated. From the viewpoint of defence, better air communication will help solve economically many urgent questions in the Near East.'

Empire air communications certainly appealed to Sir Sefton Brancker, who on 10 November began a stage-by-stage flight in the second prototype D.H.50, G-EBFO, piloted by the redoubtable Alan Cobham, with A. B. Elliott as his mechanic, in order to attend a conference with the Indian Government in January concerning the big airship scheme.

Although this, and many other long-distance and world flights, marked great progress in man's endeavour since the war, it was evident that British conception of aeroplane design had advanced but little in that time – and this was emphasized when John Kenworthy, chief designer of the Aircraft Disposal Co which now owned the assets of the liquidated Martinsyde business, fitted a 380 hp Jaguar to one of the stock Martinsyde F.4s with the result that its handling and performance was better than the RAF's latest fighters, the Siskin and Grebe, for it climbed 10,000 ft in 4 min 50

Six years had passed since the Great War ended, yet the re-engined war-time Martinsyde could more than hold its own with contemporary fighters. (*Courtesy Air Marshal Sir Ralph Sorley*)

sec, and had a top speed approaching 160 mph. Throughout the industry, current advance in engineering design was rarely more than cautious replacement of wooden structural members with equivalents of similar dimension and shape in light gauge high-tensile steel, instead of departing boldly from the classic war-time biplane and the standard structural arrangements which had given satisfactory results with timber. Only the elegant monocoque fuselages of Shorts showed equivalent advance to the metal machines of Germany and France, though British designers were extraordinarily successful in achieving aeroplanes with better handling qualities than their Continental rivals.

At the Grand Palais in the Champs Elysées, where on 5 December the doors opened to the ninth International Aero Exhibition, duralumin remained the favourite with French designers. No aeroplane in Britain could equal the cleanness of the Bernard (S.I.M.B.) racer, the Nieuport-Delage sesquiplan, or the I-strutted Breguet XIX sesquiplan made famous by Pelletier Doisy's flight to China, and which, with the exception of the covering on the wings and rear portion of the fuselage, was entirely metal.

The sole British aeroplane was a Siskin V shown in skeleton with steel-tube fuselage frame and wooden wings having conventional front and rear interplane struts with larger chord lower wing than the V-strutted '1½ wing' of the RAF Siskin, and as there was no lower fin it also had a different rudder. Shown with it were a Jaguar engine and a seven-cylinder Lynx. Nearby was the Aircraft Disposal's stand exhibiting a Siddeley Puma, B.R.2, and Wolseley Viper, whilst on an equally small stand was the Bristol display of the little Cherub, the three-cylinder Lucifer, and the big Jupiter which France, Italy, and Czechoslovakia were licensed to build.

'It is only necessary to make a cursory mental comparison between the variety of exhibits at Paris and the lack of variety among British aircraft to realize that the attitude of the technical section of the Air Ministry has limited British experimental work to a dangerously circumscribed field,' reported *The Aeroplane*'s technical editor. 'The Air Ministry, excellent though it is, is a government department and has an inevitable tendency characteristic of such institutions to play for safety, to discourage enterprises of which the result cannot be foreseen, and is an entirely unsuitable body for the control of an experimental undertaking which is essentially of a highly adventurous nature, calling for precisely those qualities least in favour in official circles. What is wanted in this country is the production of a reasonable quantity of experimental aircraft of types classed as freaks by the majority of our more self-satisfied "experts". It is particularly important that these freaks should be built under unfettered control of people who believe in their virtue, and that their design should not be marred by the well meant efforts of conservative minded technicians to bring them into partial conformity with their own orthodox ideas.'

There were exceptions to this rule, such as the Supermarine racer on which Mitchell was engaged, but the difficulties of the Westland Dreadnought, the geared drives of Bodmin and Possum, the Duiker parasol monoplane, and the Brennan helicopter only served to strengthen opposition to novelty, though limited support might continue. Thus there was 30-year-old Capt Geoffrey Hill, MA, who, after his aerodynamic work on slots with Handley Page, was awarded a Studentship by the Royal Commission for the 1851 Exhibition, and had worked in the seclusion of a Sussex cottage on ideas which might render aeroplanes impervious to lack of control when stalled.

Inadvertent stalling had become the *cause célèbre* of aviation. Some fifty lives a year were being lost in the RAF through spinning. It had become one of the principal objects of research by the RAE, who were particularly engrossed with the Handley Page slot in several complicated variations which rendered adoption unlikely through impracticability. Meanwhile Farnborough pilots were making many spins with different types of aeroplane to assess the factors involved. Hill's analysis dovetailed into a tailless monoplane design having a 31-degree swept-back wing with 6-degree wash-out twist, rudders underneath, and floating tip 'controllers' of T.P.3

The structure of Geoffrey Hill's tailless glider was conventional and simple despite sweep-back and twist, and weight-saving was the criterion of achievement.

Hill aboard the finished glider on the downs at Little Hollands, Sussex.

section which remained substantially normal to the air stream, and thus unstalled whatever the attitude of the wing, retaining powerful longitudinal and lateral control. Airscrew 4 aerofoil section was used with slightly reflexed trailing edge and no dihedral. Variable gear between stick and controllers gave low ratio for high speed and higher for landing, and both rudders could be swung outward simultaneously as an air brake, doubling the resistance.

For a year Hill had been building it as a glider, helped with out-of-pocket expenses from the Aeronautical Research Committee after successful model tests, and the Air Ministry had agreed to loan a Cherub for its eventual metamorphosis into a two-seat light aeroplane.

'As a glider, it was completed in December 1924, and tried in a remote spot in the South Downs where, just 12 years before, my brother Roderic and I had flown a biplane glider we designed and built,' recorded Geoffrey Hill. 'My current expedition was carried out under great difficulty owing to the time of the year, which was exceptionally stormy. The ancient tent in which the glider was housed needed constant attention to keep it standing at all. For this reason gliding was cut down to the minimum. The first was on 13 December. An elastic rope was used for launching, and against a 12 to 15 mph wind the glider got off in about 10 yards.' He made four glides, the longest being 300 yards at a measured gliding angle of 1 in 11 despite the open skid framework on which he sat beneath the centre-section. Control seemed effective. 'The success of the trials rapidly bore fruit. Further assistance was offered by the Air Ministry who authorized loan of a draughtsman and fitter from the RAE to expedite matters, and I was able to get such parts as steel sockets and fittings made at the Factory, where from the beginning I had advice on every imaginable subject. I think many people have no conception of the vast storehouse of information, so freely and willingly given, behind the letters "R.A.E.".'

In the equally speculative world of industry a hopeful indication of success was the fourth annual general meeting of the de Havilland Aircraft Co Ltd held at Stag Lane on Monday 22 December, which, in the absence of Alan Butler who was abroad, was presided over by A. E. Turner who had become secretary. A very satisfactory year's trading enabled a dividend of 10 per cent for the fourth year in succession, with £5,000 transferred to reserves. Orders had been received for the D.H.50 from Imperial Airways, three companies in Australia, the Czechoslovak Government, and Northern Airlines; others were being negotiated. 'At the present time, more than half our business is with the Air Ministry,' said Turner. 'The remainder comes largely from civil aviation in all parts of the world in which we have been at pains to establish a connection. Although we hope that Air Ministry business will get still larger, we firmly believe there is an immense field for civil aviation. We are therefore devoting considerable time and energy to furthering this branch of our enterprise.' As capital was required for expansion of the premises and provision of new plant and equipment, shareholders were invited to subscribe to a further 25,000 shares of £1.

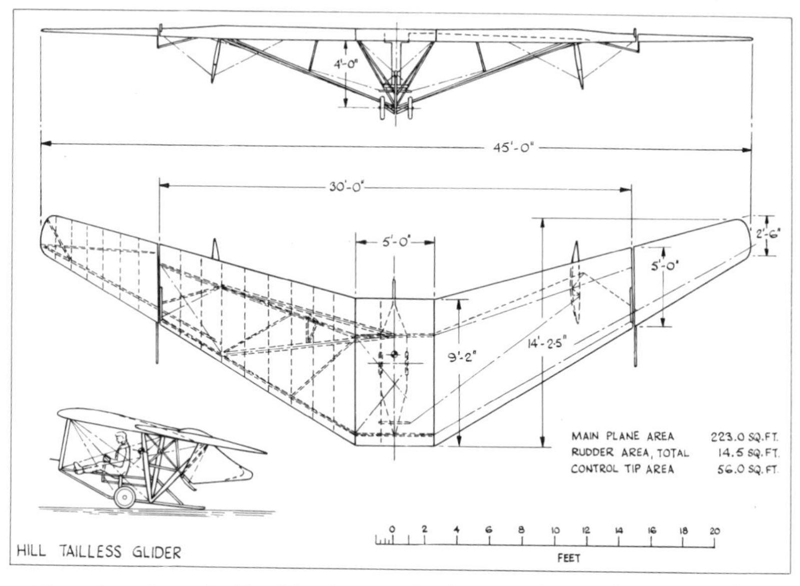

HILL TAILLESS GLIDER

Two days later de Havillands were in the news in tragic manner, for on Christmas Eve, a D.H.34, piloted by Capt D. A. Stewart and carrying seven passengers, crashed shortly after taking-off from Croydon, and all were killed. After long investigation a Court of Enquiry come to the conclusion that: 'The flight of the aircraft for about half the time was normal. Thereafter some defect developed, but whether in engine or installation, or otherwise, there is nothing to show. The dive to earth was due to loss of control combined with a stall, which occurred whilst the pilot was endeavouring to make a forced landing in circumstances of great difficulty.' The crash caused much controversy, reviving criticism of the high stalling speed of the D.H.34, for a similar accident had occurred the previous year when another of this type stalled on the approach during a forced landing at Ivinghoe Beacon. As a result of that crash Capt Stewart's machine was one which had been rebuilt with wings of greater area, by adding 18 in to the wing-tip overhang and 6 in to the chord, the greater area reducing landing speed by 7 mph. Said the *Morning Post*: 'What the report of the Croydon crash shows is that aviation is attended by the gravest dangers whose very nature is unknown and which therefore cannot be averted.' Even *The Times* warned: 'If passengers are to be carried safely in the air, if the public is to be spared suffering such as that which the relatives and friends of those who lost their lives in this disaster had to endure, no financial consideration must be allowed to stand in the way of experiment and research by which, as far as is humanly possible, the risk of accidents may be reduced to a minimum.'

CHAPTER VI

INNOVATION AND RESUSCITATION 1925

'I say that if this instrument with a helix is well made, that is to say of flaxen linen of which one has closed the pores with starch, and is turned at great speed, the said helix is able to make a screw in the air and to climb high.'

Leonardo da Vinci (1452–1519)

1

'People are already beginning to forget the brief though intense war of 1914–18,' wrote C. G. Grey in the New Year. 'A new generation is arising which only remembers it as something heard of as children. University Graduates were leaving their preparatory schools when the Armistice was signed. Even the younger officers of the RAF were schoolboys when the war was at its most critical point. To-day our people are better fed and clothed and housed than ever before the war. They work shorter hours, have more time for amusement and better forms of amusement – even the Americanized kinema, frivolous revue, and glittering dance palace are improvements on the old beer-swilling music-hall and reeking pub of 25 years ago. British credit in the world's financial centres stands higher than ever it did. Our strikes are no more numerous or important in proportion to work done than before the war. The proportion of unemployed is no greater than in any period of moderately bad trade before the war. And the social unrest is only an indication of groping for a higher and better standard of living.'

Grey's interest was history and ethnology as well as the progress of aeronautics. He was not alone in believing that in another ten years Britain might again be at war, and that this time it would involve the world. Speaking of his recent visit to the USA, he said: 'Among the people who really make things move in the US Army and Navy Air Services one found that almost all regarded an alliance as essential and natural for Great Britain and America. Some spoke of it as a Nordic alliance, to include the Dutch, Scandinavians, Northern French, Belgians, Northern Italians, and the true Germans as distinct from square head Prussians. As enemies of that alliance they class the Russians, East Prussians, Turks, Tartars and Japanese. US Services regard war with Japan as a certainty. They are preparing for it as thoroughly as they can with the small amount of money they can extract from Congress. Japan owns by Mandate most of the Pacific islands covering the route of American shipping to the Philippines, Hawaii, and Malay Archipelago, and is endeavouring to get a trade stronghold on China to exclude American and European business. Japan has captured Korea and has established Army, Navy and Air bases there. In fact Japan is openly manoeuvring to make the Pacific Ocean a Yellow Lake. And America is not going to stand for it.

'The position is like that of 1914 when Germany hoped to keep us out of the war while she settled France in a short six month campaign so that she could give all her attention to smashing Russia. Whereas in 1914 we were on the wrong side, and ought to have held France quiet by diplomacy while we helped Germany to smash Russia, we now have the opportunity of doing the right thing by allying ourselves with our kinsmen in America and not with our natural hereditary enemies as we did when we allied ourselves with Russia in 1914. At least we shall enter the next war knowing the military virtues of our American and German allies and the vices of our Russian and Japanese enemies.'

A few weeks later the *Chicago Tribune* published a despatch from its London correspondent outlining a more divergent orientation of British political ideas for an Anglo-Franco-German pact. A secret memorandum had been circulated by Austen Chamberlain to the Cabinet, ambassadors, and various officials stating: 'The position of Europe is full of uncertainty and incalculable danger. The only possible British policy is to consult British interests, leaving the League of Nations to develop at some future time into an instrument for international peace. Aerodynamics have transformed the Channel from a barrier of great military importance into a ditch devoid of military value. It is therefore a matter of supreme importance to prevent any Continental Power from obtaining control over the whole coastline. If Germany invades France or Belgium a situation will be created in which Britain would be open to attack by air; therefore it is of supreme moment to guarantee France and Belgium against any possibility of such invasion. The policy of isolation in European affairs may be right for the USA, which is safe behind two oceans, but for Britain it would mean proclaiming impotence and invite attack. The situation of France and Belgium must be considered from the viewpoint of British security alone, but whatever has to be done must be done publicly, for it was the secrecy of the pre-war Franco-British arrangements that allowed Germany to embark upon war. If in 1914 Germany had known that the British Empire would assist France against any invasion or infringement of Belgian neutrality there would have been no war.

'When Great Britain and the United States refused to ratify the recent Security Pact they created insecurity in Europe, and France was forced into a policy of adventurous expediency. Instead of being able to concentrate on economic reconstruction, which would have allowed her to pay her debts, she had to organize Europe into an armed camp to oppose the rise of a spirit of revenge in Germany against which she regarded herself as safeguarded only by the temporary expedient of the Rhineland occupation. Germany was not destroyed by the war. She is recuperating. To destroy the feeling of revenge in her, she must be brought into the comity of European Nations and made to co-operate as a member of the League of Nations. This cannot be done without the goodwill of a pacified France by guaranteeing that her territory will not be invaded with impunity. Russia hangs like a shadow on the horizon. She has ceased to be European, and none can say what she does represent, but it is clear that European salvation must be worked out without waiting till that great country finds herself.'

Neither British aircraft constructors, nor any others in Europe and the USA, were particularly influenced by such considerations. Primarily they were men whose overwhelming interest had always been aeroplanes and the fascinating problems associated with flying – but necessarily they had also become vigorous business men, or would not have survived. The extent of their profits dictated their ability to venture, but they were prepared to gamble on prospects. Thus the combined affirmation by Conservative and Labour governments that the RAF would expand from 18 to 52 squadrons for Home Defence led several manufacturers to extend their working facilities. With the establishment of the Armstrong Whitworth Flying School at Whitley with facilities for testing experimental aircraft, such as the secret new two-seat Army Co-operation biplane, the Atlas, John Siddeley decided to transfer his company's aircraft production shops, managed by Sam Hiscocks of war-time Airco fame, to an up-to-date factory which was forthwith built on this aerodrome, but the design team remained at the original little office at the 'Deasy' Parkside Works of the automobile and engine section of Armstrong Siddeley at Coventry. An enthusiastic journalist described the Siddeley car produced there as 'silent and inscrutable as the Sphinx' – which so delighted fierce little Siddeley that he adopted the Sphinx as the firm's trademark, sending an artist to London to copy the sculpture of it on the Thames Embankment to get the details right.

Similarly A. V. Roe & Co, while retaining Hamble as the experimental section dominated by the effervescent A. V. himself, had recently purchased 163 acres at Woodford, Cheshire, for a new aerodrome, vacating the old one at Alexandra Park, Manchester. Woodford was outside the smoke ring, with three times the area of the old field. Three hangars of over 100-ft span were removed from the old site and reassembled at Woodford as erection shops for Manchester-built production aircraft.

Already the Aldershot III had proved too cumbersome for the quick adaptability of rôle the RAF preferred, but its excellent stability enabled short-term use as a night bomber, operating from Bircham Newton, Norfolk, despite a general feeling that single-engined machines should not

Pilots reported that the Avro 504N was the best trainer in the world: sensitive enough to show faults, no vices – forgiving, and instantly recovering from spins, but primitive, and draughty. (*Flight*)

be used for this purpose – so eventually they would be replaced by the twin-engined Handley Page Hyderabads now being built. All new Avro designs would spring from Hamble, where A.V.'s old enthusiasm, the 504, had been successfully developed as the 180 hp Lynx powered 504N, with long-travel oleo undercarriage modelled on the Siskin's, and devoid of the old spear-like skid. Martlesham trials resulted in its adoption as the RAF's new standard trainer, and large-scale production began at Manchester. Presently there were orders from Belgium, Brazil, Chile, Denmark, Greece, Peru, Siam, South Africa, and Sweden. For Alliott Roe it was a quiet and happy triumph that his 1913 design had been resuscitated for another splendid lease of life, and the Central Flying School at Upavon was no less pleased. Though Avro profits were keyed to an antique aeroplane, the company was not resting on its laurels but, like all manufacturers, was offering various design proposals fated to be turned down by the pundits of the Air Ministry – such as an advanced looking Condor-

Designed for night bombing, but eventually built for long-range coastal defence, the impressive 95-ft span, Condor-powered Avro Ava had a retractable prone gun position in addition to gunners in nose and abaft the wings. The second metal-structured Ava had square wing-tips and 2-ft greater span. (*Courtesy Air Marshal Sir Ralph Sorley*)

powered two-seat fighter monoplane with thick-section elliptical wings, a similar Napier-powered version, and an all-metal single-seat fighter on which Chadwick was currently employed in the intervals of supervising 'mods' for Bisons and his splendid but highly secret coastal defender, the great twin Condor-engined Ava which had been completed the previous year and now was at Martlesham.

At Stag Lane similar extension was under way. *The Aeroplane* reported: 'The works of the de Havilland Aircraft Co Ltd are swelling visibly. On the occasion of a recent visit this expandment was, owing to the recent unpleasantly wet weather, made all the more noticeable because the plodding back and forth of myriads of workmen had transformed the normally dry and tidy sleeper-and-cinder tracks, that lead to the works and interconnect the various shops, into paths of thick glutinous mud. The greatest alterations are at the erecting shops and a large two-storied shed

The 68-ft span de Havilland D.H.54 Highclere was test flown by Broad on 18 June, but not until early 1926 did it reach Martlesham, where this photograph was taken. (*Courtesy Air Marshal Sir Ralph Sorley*)

facing it previously used for aircraft storage, for the whole space separating the two buildings is being roofed in, which will practically double the area of the total works. The old two-storied Flight shed is now entirely woodworking, the ground floor being a sawmill and the upper is well equipped with woodworking machinery and a gluing shop. The old drawing office in a long wooden hut opposite the general offices has been moved to a new two-storied building on the site of the old wood mill, and, though not yet complete, will eventually hold the general offices on the ground floor – and the whole top floor, adequately lighted by large windows and skylights, will contain the drawing office. In the erecting shops are the inevitable batches of D.H.9As being reconditioned, the D.H.54 single Condor-powered airliner is fast assuming shape; a D.H.51 recently purchased by General Weir is being overhauled prior to delivery; there is also the all-steel Dormouse, and parts of the D.H.60, known as the Moth, which is being produced as *the* two-seat light aeroplane and fitted with the 60 hp air cooled four-cylinder Cirrus, the general arrangement of which is due to Capt de Havilland.'

Nor was it only the factory which was making progress. Alan Cobham, with the DCA aboard, had reached Calcutta, but there Brancker was prostrated with fever, and it was decided that after his conference he would proceed to Rangoon by sea while Cobham continued by air. In the interval Cobham made demonstration flights and, *en route* to Darjeeling, flew towards the impressive Kanchanjanga, climbing to the D.H.50's ceiling of 17,000 ft from which there was a splendid view of Everest. A reporter ventured the opinion that: 'With a properly equipped modern aeroplane of higher power than the D.H.50 it should be possible to land stores and personnel quite near the summit and so save the long march which has been necessary hitherto before the final assault proper on the mountain becomes possible.' On 8 February Cobham and Brancker began the return journey home.

Emphasizing the strengthening of air defence, came an Air Ministry announcement: 'For various reasons the main part of the Home Defence Air Force must necessarily be established on a regular basis, but it has been decided that part of the new organization should consist of non-regular units. These will be divided into two classes: Special Reserve squadrons and Auxiliary Air Force squadrons, together forming a quarter of the total strength as at present authorized, seven being Special Reserve and six being Auxiliary.'

The full strength would eventually be 360 officers and 2,400 airmen, but a start was to be made with two or three squadrons in the spring, others in subsequent years forming proportionately to the regular units of the RAF. Hendon was to be headquarters for two Auxiliary and one Special Reserve squadron, and Scotland would have an Auxiliary squadron at both Renfrew and Turnhouse. All would be equipped as bombers. Special Reserve units would be numbered from 500 onward and the Auxiliaries from 600. The latter would be given a territorial designation – thus No. 600 City of London (Bombing) Squadron. The wise men, headed by Trenchard, had got their way. With Brancker's tape-cutting powers it was hoped that civil aviation would similarly become unshakably established, and that British Empire air routes would set the pattern for world communication.

2

Since the war the Royal Commission of Awards to Inventors had considered an enormous number of claims. The Aircraft Manufacturing Co and George Holt Thomas swept the board with more than £200,000, and a further £6,500 was received from the US Government. The Sopwith Aviation & Engineering Co Ltd were given £100,000. G. Constantinesco and W. Haddon received a significant £70,000 for their vital C.C. synchronized gun actuating mechanism compared with £5,500 to Capt V. V. Dibovsky for his interrupter gear. The value of the Bristol Fighter was emphasized with an award of £50,000, and that of the Avro 504K with £40,000.

These were regarded as comparatively big sums, for the Treasury with Churchill at the helm had adopted a policy of 'deflation' and hoped soon to return to the gold standard with sterling at pre-war level. Unemployment had dropped slightly below the million figure, and it seemed economic conditions were improving. Neville Chamberlain was taking cautious steps towards the Welfare State with a Pensions Act and a Rating Act.

As an expression of pride in national progress and to demonstrate the possibilities of this country's territorial possessions, the Government-sponsored British Empire Exhibition at Wembley, which had been opened in 1924, was continued in 1925, drawing thousands from all over the world to its cultural pavilions, huge amusement park, and spectacular cowboy rodeos. For his part in organizing the splendid engineering section, Ernest Petter, proprietor with his twin brother Percy of Westland Works, was knighted. Tall, white haired at 40, the two dignified brothers were tight with money, and it took all Bruce's tact and power of argument to convince them that expenditure on the aircraft side was necessary, for their chief interest was the engine works of Petters Ltd. Said one of the departmental managers: 'The whole secret of Westland's early days lay in the happy family spirit under the leadership of R. A. Bruce as managing director. We were all free to approach him, austere though he could be, and he listened to our viewpoints and considered them with fairness. When he was at variance with Sir Ernest Petter or P. W. Petter we all felt R.A.B. was right, and if he had decided to form his own company practically all of us would have kept with him.'

Sir Ernest lived at Ipswich, where he directly managed Vickers-Petters who manufactured larger industrial oil engines than those produced at Yeovil, and consequently he was less often seen than Percy Petter, the Yeovil director, but his rare visits always occasioned a furore. Undoubtedly he was a great character, but with the adroitness of an impresario used his connection with Vickers to counter Bruce on matters of aircraft policy, though in fact his contact with Vickers aviation department was negligible. However there was a *rapport* because Westland had been the only post-war outside business to build Vimys, and as a symbol an up-ended Vimy fuselage could be seen as a flagstaff at the entrance to the Yeovil Works.

Vickers remained busy with experimental variations of their bombers, transports, and amphibians. A new version of the Viking, the Mk VII later named Vanellus, had been ordered by the Air Ministry as a three-seat fleet spotter with pilot in front, a Scarff gun position behind him, and another abaft the mainplanes. Instead of the previous forward-hinging wings, they folded conventionally back and this necessitated a monoplane tail with big single central fin, but as it was expedient to lower the machine with spread wings on the 50-ft wide lift of the *Eagle* aircraft carrier, span was reduced to 46 ft. The Vanellus was chosen for comparative trials with the rival Supermarine Seagull III in the spring, but was completing manufacturer's tests in the hands of bulky Flt Lieut E. R. C. 'Tiny' Scholefield who had

RAF experience in Iraq with two Vickers Viking Mk Vs (*as illustrated*) revealed sufficient advantages in amphibian design, compared with pure flying-boats, to warrant ordering a cleaned-up single-bay version – the Mk VII or Vanellus.

been appointed chief test pilot in midsummer 1924 on leaving the Farnborough team.

As with the Viking types, the Vixen had only moderate success, though six were ordered for two-seat armed reconnaissance rôle and named Venture, but, despite favourable Martlesham report, Vickers were informed that the design was too large to be acceptable, the view inadequate, longitudinal stability insufficient, and landing run too long. Disappointment was somewhat alleviated by an order for twelve from the Chilean Government who had been very satisfied with four Vixen Is, known as Valparaiso, supplied the previous year. A Vixen Mk III with larger wings and greater capacity also failed to interest the Air Ministry, nor did the prototype G-EBIP when fitted with floats, for the MAEE at Felixstowe found it showed no superiority over the standard Fairey floatplanes. Undaunted, the prototype Vixen II, G-EBEC, had its fuselage again lengthened, larger wings, rudder and elevators fitted, and a 650 hp Rolls-Royce ungeared Condor III was installed, subsequently replaced by a geared high-compression version – in which form 151 mph was attained

Destined for continuing development, the Vickers Vixen with Condor engine was turned down because the engine was deemed too heavy, powerful, and expensive for a machine of this class.

and a ceiling of 27,500 ft. At least Pierson's determined development of the Vixen had resulted in an invitation to tender, with Blackburn and Parnall as rivals, for a somewhat similar machine to specification 5A/24 for a trainer float-seaplane, the Vendace, powered with a 275 hp Rolls-Royce Falcon III, and the D.O. was busy designing it.

To further strengthen the ambitious Vickers programme of making all sizes of aircraft, Acland authorized Pierson to proceed with the Vespa – a private venture rival to the Bristol Bloodhound, Short Springbok, and Armstrong Whitworth Ape contestants for the Bristol Fighter replacement. Currently a new version of the Springbok with larger fin and rudder and greater-chord top wing than lower was being built, and the new Vickers was somewhat similar, but with even greater resemblance to the war-time Parnall Zeppelin fighter which had been designed by Camden Pratt, now Pierson's project designer.

A unique variation of the Vickers Virginia theme was the Mk. VIII with Condor III engines and 'fighting tops' for all-round defence. This was the much modified 1922 prototype J6856 with which several arrangements were tried. It had fixed streamlined access ladders, and no fins.

Production hope was keyed to the Virginia bomber, of which there had been six variants since the prototype flew in November 1922, the only profitable production having been twenty-two of the Mk V, which had, like the prototype, dihedral on the lower wing only, and twenty-five Mk VIs with equal dihedral to facilitate wing-folding. From the squadrons there began serious criticism of deficient longitudinal stability – for the operational C.G. had moved fractionally further aft. Investigation was therefore in hand to decide whether the instability could be cured by giving the wings slight sweep-back.

Even slower to attain orders was the Victoria, for the prototypes had been under assessment at Brooklands, Farnborough and Martlesham for 2½ years, and only now was an 'Instruction to Proceed' (ITP) issued as the preliminary to contracts for sixteen at a cost of £9,000 each.

Associated with Vickers as consultant was Michel Wibault who, as Avions Wibault of Billancourt in Paris, had taken over the old R.E.P.

Production Vickers Victoria IIIs had neat nacelles for the Lions raised above the lower wing which carried the radiator beneath. Top wing as well as lower had dihedral. (*Ministry of Defence*)

company, and since the war had become a leading exponent of metal construction, using fabricated light alloy components of simple shape which did not need the costly processes of the Boulton & Paul type of thin strip-steel manufacture; moreover he eliminated fabric covering, adopting a finely corrugated skin for the wings and slightly wider spaced corrugations for the fuselage. Its simplicity and practicability appealed to Pierson. He engaged the celebrated, idiosyncratic Harris Booth, who gladly cut free from the Alula wing project, to investigate metal construction, and an order was placed with Avions Wibault for one of their single-seat, strut-braced parasol monoplanes as this would take Vickers a long step forward at far less expense than initiating and developing metal constructional methods of their own.

Handley Page regarded the Vickers progress with large aircraft as an intrusion into his domain. Was not 'Handley Page' defined in the Oxford Dictionary as a synonym for a large aeroplane? It seems probable he had offended Trenchard when the latter expressed doubt that civil aviation was of value to any country and not only wasted time but money which should be spent on additional squadrons. This provoked H.P. to say: 'That is an extraordinary point of view. I am one of those who believe the military side of aviation should only be in the background of civil aviation.' Since then he had remained in the background as far as military contracts had gone for twin-engined military bombers until a recent order for an initial squadron of Hyderabads, though the prototype had first flown in 1923.

By now Handley Page Ltd had overcome its debts, helped by the sale of Handley Page Transport to Imperial Airways for £17,000 in cash and £34,000 shares of £1. Barrett-Lennard had intimated that as the creditors were satisfied he would resign as managing director at the next annual general meeting.

Meredith, the works manager, had recently left, his ideas having clashed with H.P.'s. 'There was no personal rancour,' explained their public relations manager. 'He founded his own firm of material suppliers and did a good deal of business with his former chief.' Nothing was said about his leaving. Major Savage suddenly appeared in his place and that was all. Savage had become technical manager of Handley Page Transport when Woods Humphery left; now that Imperial Airways had been formed he was free to join the parent concern. Under him, as works superintendent, was 'the man who was never wrong' – Meadows, who 'had a reputation for a wealth of picturesque language, a rough manner with his work people, and being difficult to get on with.' Nevertheless he kept his team on its toes, firmly aided by sandy-haired MacRostie the shop's masterful foreman.

3

Several prototypes of note made their first flights in the early part of the year. On 3 January Norman Macmillan flew Fairey's American-influenced Fox two-seat day bomber, based on the Fawn specification of Air Staff requirements, but built, as Fairey said, 'from the designer's point of view without any restrictions other than those directed by commonsense.' With carefully faired fuselage, pointed nose, and single-bay wings which had N-struts with junctions faired into the wing with *papier-mâché*, it presented an idealized picture compared with the clumsier two-seaters of that day. Said Macmillan: 'I flew the prototype at Hendon aerodrome because Northolt airfield was a bog that winter. Streamlining had been carried to excess by using two surface radiators on the lower side of the upper wing roots, and these were insufficient to prevent boiling within a few minutes. Nevertheless, in a few brief flights of eight minutes duration, all that the cooling permitted, I was able to assure Fairey that the Fox was a winner.

With single-bay wings, *papier-mâché* strut fairings, pointed nose, and stringer-rounded fuselage, the Fairey Fox prototype with imported Curtiss engine seemed a big advance on contemporary two-seaters though its engineering design was conventional.

The system was then altered to a retractable honeycomb below the fuselage allied with a surface radiator on the centre-section, and this kept the temperature within controllable limits. A maximum of 156 mph was attained – no less than 50 mph faster than the officially approved Fawn. This was not regarded with undisguised pleasure by the officials whose duties were concerned with Air Ministry specifications, and who had dealt with the Fawn but had nothing to do with the Fox. In the circumstances, it was no easy matter to sell the Fox to the Air Ministry, despite its performance, which exceeded that of then current single-seat fighters.'

Weightier, more powerful, and with better performance than the Grebe, the Gloster Gamecock was an inevitable development, but some of the pleasant handling quality was lost.

Although an experimental all-metal version of the Grebe fitted with a Napier Lion had been built for the Air Ministry, and was named Gorcock, Gloucestershire's Gamecock was the latest favoured prototype fighter. Initially tested by Larry Carter, it was flown early in February to Martlesham Heath. Originally ordered in August of the previous year as Grebe II, its chief difference from Grebe I was replacement of the Jaguar by a Jupiter IV which, in the absence of a supercharger, had a pilot-controlled variable timing gear in conjunction with high-compression pistons giving 6·3 : 1 ratio (regarded as high compression). By closing the inlet valve late near ground level, compression was delayed well up the stroke; but by gradually advancing the timing it became normal at the predetermined altitude for maximum power, which fell off thereafter. Making the best of a difficult system, Bristol stated that: 'Increasing the engine performance at altitude is far preferable to an oversize high-compression throttled engine because certain definite advantages, such as a rising torque curve, short compression and long expansion stroke, are secured with no chance of pre-ignition and detonation.

Fedden had been working on this development for the past nine months, using double epicyclic timing gear – but Martlesham pilots were not enamoured. Fully advanced, the inlet opened 9 deg early and retarded 20 deg late. On the first attempted climb the engine ran irregularly at 15,000 ft 'due to faulty carburation consequent on low air temperature'. Modification was introduced to heat the air intake from the lower exhaust pipes, but carburation remained defective, so helmets were fitted to the cylinders and exhaust heat fed through the air intake. 'It is difficult to disentangle the effects of the various modifications on rpm and climb,' reported Martlesham. 'The difficulties increased because on several climbs the variable timing slipped back from 5,000 ft to 4,000 ft. This may well account for the reduced ceiling in spite of the increased rate of climb at lower altitudes. Under the circumstances it is impossible to fix a definite ceiling and rate of climb for this aircraft owing to continual changes of timing, intake heating, and airscrews combined with a short time allotted for the test.'

The pilots were puzzled over the Gamecock's stability, which they reported as good, and commented that 'a feature of the aircraft is the uniformity of feel of the lateral and longitudinal controls' – but there was a disconcerting 'sinking' when flattening from a steep dive, and a similar effect was noted on very tight turns. Comparative flights were therefore made against a Siskin and the A.D.C. Jaguar Martinsyde. Rate of turn was slightly greater than the Siskin's, though the Gamecock was as good in changing from right to left-hand vertical turns, and was superior to the Martinsyde in radius of turn, manoeuvrability, and climb, though not in speed. Again the worry was reported: 'In flattening out from a steep dive of 150 mph considerable pull is required on the stick, and the Gamecock does not follow a true path but sinks bodily. This is not due to lack of elevator control, which is probably sufficiently powerful to break the machine if used too heavily. It is thought that an elevator of modified size or shape would make for tighter turning circle and also overcome the fault, but endeavour should be made not to upset the present equality of controls.' There was a feeling that the sinking was 'probably an apparent defect, perhaps due to down-wash on the elevators, which are set considerably lower than those on the Siskin and Martinsyde.' Such comments were usually accepted without question, and much needless alteration might follow, for Martlesham pilots were the Air Ministry's chosen arbiters as to whether each aeroplane met operational requirements, but their undoubted flair for flying sometimes obscured lack of scientific knowledge and analysis. Thus it is probable that because of heavier wing loading than the Grebe, the pilots were encountering 'g' stalling at a different stick force compared with the Siskin and much more lightly-laden Martinsyde. It also is possible that they did not spin the Gamecock, for in squadron use the difficulty of recovery, and flutter, soon became apparent, and the accident rate presently proved high.

Hugh Burroughes, recalling the Gamecock, said: 'Ninety were ordered by the Air Ministry in 1925. The wing structure was subsequently redesigned, but, as we found later, at some risk, for the top wing was increased in span by a margin which proved too great. The excess overhang led to flutter, and this, together with spinning problems arising from the short fuselage, imposed a period of modification delays which allowed our competitors to catch up and brought to an end our short ascendency in the S.S. Day and Night Fighter field, and in terms of fighter aircraft it set us back five critical years. However, with the object of widening the field for G.A.C., I made contact with Dr Hele-Shaw and Mr T. D. Beacham who were working on their variable-pitch airscrew, and secured an exclusive licence. We went through the usual process of making a few prototype propellers as private ventures, and fitted one on a Grebe and another on a Gamecock. Both showed only marginal improvement over the fixed Falcon wooden propellers, but proved that the functioning was reliable. At that time engine companies were not interested because they thought the extra thrust needed at take-off could be provided by stepping up power at some increase in weight, though less than that entailed by the variable-pitch gear. None of the authorities were really interested. In fact we were premature.'

George Bulman, of the RAE Aerodynamic Flight, who had come into prominence at the RAF Displays and by flying the RAE Hurricane racer, was currently seconded to Hawker Engineering to test Carter's wooden-framed Horsley bomber and subsequently the metal-structured Heron fighter designed under jurisdiction of Carter by Sydney Camm following his success with the little Cygnet. Raynham, their former pilot, had decided to throw in his lot with Ronald Kemp, who had been a pioneer Farnborough pilot, and Colonel C. H. D. Ryder, in their Aerial Survey Co Ltd which had secured a contract for photographic survey of Sarawak and North Borneo following success with an Irrawaddy Delta survey the previous year. Their rival was the Aircraft Operating Co, of which Alan Butler was chairman and Major Hemming the managing director, and they

Like all prototypes, the Hawker Horsley required several major changes to make it satisfactory for Service use, and its earliest guise with fuselage side radiators was soon changed. Illustrated is the much later target-towing version.

were air surveying in British Guiana. Recently Major Mayo, OBE, MA, had joined them, following dissolution of Ogilvie & Partners of which he was a leading member, and he also became consultant to Imperial Airways. Colonel Ogilvie had gone on a world tour prior to specializing on I.C. engines, and Dr H. C. Watts, M Inst CE, became resident consulting engineer to the newly formed Metal Propellers Ltd of which H. Leitner was technical director. The fourth member of Ogilvie & Partners, Colonel W. A. Bristow, MIEE, remained an independent consultant with particular knowledge of fuels.

There was still a distinct Sopwith fighter ancestry in the appearance of the Hawker Heron displayed in this typical view taken by *Flight*'s famous photographer, John Yoxall.

With orders placed in the next few months for Grebe, Siskin, Gamecock, and the first of the Woodcock about to be delivered, there was little chance that the Hawker Heron would also be standardized despite its 156 mph and commended manoeuvrability – but it has significance as a display of Camm's conception of elegant shape and earliest metal structural design as an experiment in substituting drawn square-section mild steel-tube with radiused corners in lieu of conventional wooden longerons and struts, and wing spars of Accles & Pollock drawn three-bulb-section steel-tube with wood ribs as demanded by Sigrist. The engine mounting structure with ball and socket joints was patented by F. I. Bennett, the engineer-secretary of the company.

Said Camm, who was regarded as difficult, though I knew him better in his enthusiastic and friendly moods: 'I am one of those lucky individuals who has been able to convert a boyhood hobby into a paid profession, as I commenced making model aeroplanes when a schoolboy, and followed with formation of a model aeroplane club which held weekly flying meetings in Windsor Great Park. One or two biplane gliders were built, and finally a full-size biplane with 20 hp two-stroke was projected but never completed. Prior to 1914 I used to watch with enormous interest the flying displays at Brooklands and Hendon, and then in the middle of that year I got a job with Martinsydes at Brooklands. It was a period which gave me a great deal of practical experience, and during this time I was fortunate in being able to inspect almost weekly the captured enemy aircraft stored at the Agricultural Hall, Islington. It was towards the end of my time with Martinsydes that I was given a position in the drawing office, and late in 1921 joined Handasyde, and during this period we produced the glider for the Itford Hill competition.'

With knowledge of new types recently flown, and others that would be ready shortly, C. G. Grey wrote critically: 'At the moment our Air Power for war on land is only just beginning to be developed, and our war in Iraq and on the Afghan Frontier is being done by aircraft of 1917 vintage. But we are progressing. The RAF in Iraq are getting the big Vickers troop carriers with Rolls-Royce or Napier engines – which, incidentally, they turn into bombers in their own workshops, contrary to the intention of the Air Ministry's technical experts responsible for the original specification. And the single-seater squadrons, mounted on antiquated Sopwith Snipes with clumsy rotary engines, are each allowed one sample Gloucester Grebe or an Armstrong Siskin to see how they stand up to the climate and sand. Thus in due course squadrons on foreign service will be re-equipped with machines of design which by then will be not more than six or seven years old.

'But the most important progress is in the new Home Defence area. Here we are acquiring entirely new squadrons equipped with entirely new types, such as Fairey Fawns with Napier engines, Vickers Virginias with Napiers, and Avro bombers with Rolls-Royce Condors, as well as Grebes, Siskins and so forth. It is true they represent their designers' ideas of years and years ago, but if all the firms were given a free hand to produce what they liked as experimental machines then in 12 months they would jump five years ahead of the official programme envisaged by Air Ministry experts who are not only a curse to the Trade, but a menace to the safety of the flying personnel of the RAF. They play for the safety of their own jobs all the time instead of working for the safety of the pilots. They have no initiative themselves, and they crush all initiative in the Trade because they are afraid to sanction or accept any innovation in design or construction lest the failure of the experiment should be blamed on them.'

In that judgment there was a strong element of truth. On the one hand there were the industry's designers utterly immersed in the swiftly advancing tide of aeronautical knowledge, always one move ahead in understanding the lessons of their previous design: on the other were men who preferred the safety and pension of the Civil Service, and who rarely had practical experience of design and research, whatever their theoretical training, yet had become administrators in the position of judges endeavouring to assess technical merits but subservient to the Treasury requirement that public money must be spent cautiously.

4

On 19 February Sir Samuel Hoare issued the Air Estimates for 1925–26. Parliament was asked to authorize £15,513,000 compared with £14,720,000 the previous year. Additionally the Colonial Office would provide £2,744,100 from Iraq and £372,600 from Palestine and Transjordania. There was also a vote of £1,320,000 to come from the Admiralty under a vote for the Fleet Air Arm.

Said Sir Samuel: 'The present strength of the Air Force, apart from training units and establishments, is the equivalent of 54 squadrons. Of these, $43\frac{2}{3}$ are organized as such, and there are 21 flights averaging 6 machines (half the strength of a normal squadron) for the Fleet Air Arm. Of the established squadrons $25\frac{1}{3}$ are stationed at Home, 8 in Iraq, 6 in India, and $4\frac{1}{3}$ in Egypt and Palestine. The number of completely formed Regular squadrons for Home Defence is 18. During 1925–26 the number of squadrons formed will be: Regular 2; Special Reserve 1; Auxiliary Air Force 4.'

The Times warned: 'It certainly cannot be said that the increase in the Estimates errs on the side of extravagance. Only two regular units are to be added to the small force of eighteen completely formed Regular squadrons available for Home Defence. At that rate it would take until 1936 to raise the strength to forty squadrons, and by no means too many for the defensive work which may some day be demanded of it.' Few realized the extent of wastage. Replying to a question by Sir Frederick Sykes, the Under-Secretary for Air said: 'During the 12 months ending October 31 last, 339 aircraft were written off charge owing to accidents, and 81 for general deterioration. In each case the average age was about five years and the average flying life about 130 hours.'

At least the Trade was encouraged by the vote for technical equipment, experiment and research, for there was a net increase of £763,000 'mainly due to the new squadrons being formed for Home Defence and to the larger orders being given for aircraft of modern types. It is to be remembered that, as with all types of fighting equipment, there is a marked tendency for aircraft to advance in power and complexity and consequently in cost.' Of the much criticized RAE, Hoare reiterated the wartime statement: 'It is the considered policy of the Air Ministry to reserve this Establishment for experiment and research, and not to employ it on the normal work of production for the Air Force.'

When the House debated the Estimates he evidently had no illusions about another European war and the necessity of Imperial defence. He said there had never been such unanimity among leaders of all parties on the question of defence. The Government was trebling the combatant strength of the Air Force compared with when he previously came in to office. It would take four or five years to complete the first stage of the programme because in peacetime greater expenditure would dislocate the normal life of the country. By the end of 1926 half the 52 squadrons needed to complete that stage would have been formed, but even then we would still have numerical inferiority, compared with the greatest European Air Power, in proportion of one to three.

The strong Labour Opposition party was expected to be critical, but Philip Snowden, in opening the Debate, took the mild line that the former defence function of the Navy had been superseded by the RAF, so there should be cuts in the Senior Service, but George Lansbury and Clynes preached impossible pacifism, and Mr Saklatvala followed suit for six and a half columns of Hansard, referring to universities and public schools as 'these Nobs of Society'. However J. H. Thomas supported the Air Estimates and said that Saklatvala's speech was either spoken in ignorance or deliberately intended for foreign consumption as his sentiments did not represent any material force in this country. Communist James Maxton, of histrionic appearance, darkly said that his constituents on the Clyde were not going to vote a penny for armaments until the working classes had something worth defending.

The only important outcome of the Debate was revelation by Sir Samuel Hoare that the Air Ministry had decided to issue to the RAF a type of 'free fall' parachute called the Irvin* which was standard equipment in the USA. Two months later *The US Army Air Service News Letter* reported: 'Grp Capt M. G. Christie who is Air Attaché of the British Embassy, Flt Lieut Soden and F/O Lacey of the British Air Service, successfully accomplished live jumps from de Havilland planes at a height of 2,000 ft over Chanute Field recently. Capt Christie had but an hour or so instruction before his jump. Lieut Soden and F/O Lacey spent a week at the Air Service Technical School taking intensive training in parachutes.' Thereafter the job of the two latter was to tour British squadrons with a Bristol Fighter, subsequently replaced by a Fairey Fawn, fitted with a side ladder from which pilots were induced to try a parachute descent.

Choice of the American parachute led to strong protest by Colonel Holt whose 'autochute' of early post-war years had similar functioning and quick-attachment pack. With either, the pilot had to pull a rip-cord which opened its container-pack and released a spring-loaded pilot parachute that drew out the main canopy, though Holt interposed an intermediate parachute to reduce shock of opening. Using a Holt autochute, Cpl Dobbs, who made the original test descents, was the only British man who had parachuted from a spinning aeroplane, yet the Air Ministry refused to take over development – but the Irvin required no new manufacturing facilities and could be bought in production quantities for only £40, though to private persons it was £90 because of import duty and trans-shipment. There was no one to push the case for the long established but bulky *Guardian Angel* as Calthrop was suffering illness that soon led to his death.

* There is some confusion about the correct title. When Leslie Irvin's British company was registered a 'g' was added to the name in error.

5

On Sunday afternoon 22 February, de Havilland's dainty little Moth – wings and tail clear-doped and shining, the fuselage still in grey undercoat – was pushed onto the muddy Stag Lane aerodrome, and Geoffrey de Havilland, in leather coat and helmet, climbed aboard. The propeller was swung and the engine warmed. Waving away the chocks he taxied out, and turning into wind took-off in a mere 100 yards. Those who had seen his R.E.1 of 1913 could have mistaken it for the same aeroplane. Steadily he climbed, banked and banked again, checked the stall, and came gliding in. Taxi-ing to the group of senior de Havillanders, he beckoned to Broad, who climbed onto the wing for a hasty word, and then, opening the side door in the fairing, stepped into the front cockpit. This time de Havilland made a more extended flight, and Broad tried the dual controls. They returned jubilant. Without doubt this was a winner.

Ancestor of the Moth was the 1913 R.E.1, which with a doubtful 70 hp achieved 78 mph and climbed 600 ft/min, but the slightly smaller Moth could cruise at that speed.

On 2 March, the Moth was exhibited to the Press, and Broad took many representatives for a short flight. All were enthusiastic. Typical of their reports was: 'Even in the present waterlogged state of the aerodrome the Moth takes off quickly, and climbs really fast. She obviously handles extremely easily, performs all the usual stunts in an exemplary manner, and lands very slowly. She is undoubtedly a thorough practical training machine and should certainly meet the needs of the light aeroplane clubs, and that of the private owner.'

Not long afterwards, the RAeS Students' Committee, of which Stuart Scott-Hall was chairman and I a member, arranged a visit to de Havillands, and flights in the Moth were the highlight. By then its fuselage had been painted black and bore the registration G-EBKT. To cope with

Registered G-EBKT, with fuselage painted black and wings clear doped, the prototype de Havilland Moth became the centre of interest for every potential flying club as the first really practical dual-control light aeroplane. (*Flight*)

the soggy aerodrome, larger wheels had been fitted, and several propellers tested for best compromise between climb and full-out revs. After the long-winged, Renault-powered Avro 548 on which I had been given brief dual by Colonel Henderson at Brooklands, the Moth seemed a tiny machine, with easy, natural response, yet adequate stability to fly hands-off, though a slightly tiring foot pressure was necessary to hold directional trim, for the rudder as yet had no balance. I was told that production machines would cost £900 – a sum beyond all hope in the kind of job I

A hinged coaming and careful proportioning of gap and longeron height allowed easy access to the Moth's front cockpit. Every detail showed the impress of a pilot-designer. (*Flight*)

might get next year, for I had noted that 'scientists with a degree or equivalent, willing to observe on experimental flights' were offered no more than £4 10s per week by the RAE.

The students' section of the RAeS was lively enough, and the Society's lectures were excellent, despite criticism of being too scientific rather than practical, but finances were in a parlous way, for expenditure had exceeded income during the past three years by £279, £316, and £359. Hopefully the Council approached the Air Ministry for a grant-in-aid, but though they proved sympathetic, the Treasury was not. Colonel Lockwood Marsh had thereupon tendered his resignation as secretary, but luckily Laurence Pritchard, that ex-Admiralty mathematician who edited the Society's *Journal*, took on the job without pay, supporting himself by writing thrillers. An appeal was launched at the Industry to subscribe to an Endowment Fund of at least £10,000, and negotiations began to induce the Institution of Aeronautical Engineers to amalgamate with the RAeS. The inevitable sharp-edged comment came from C. G. Grey: 'With relatively few exceptions the Society for the past few years has served no useful purpose except advertizing the personnel of the official research departments, who set up a pseudo highbrow atmosphere which discourages many aeronautical engineers from attending or offering papers of a more practically interesting and less obscure nature.'

Dr Adolph Rohrbach, late war-time aeroplane designer of the Zeppelin company, added emergency sails to his unique metal-skinned Ro.II Eagle-powered flying-boat built by the Rohrbach Metal Aeroplane Co, A/S, of Copenhagen. (*Flight*)

Some lectures had been the medium of commercial publicity, such as Handley Page on *The Slotted Wing*; *Radial Engines* by Fedden; Professor Junkers on *Metal Construction*. Recently the latter's compatriot, Dr Rohrbach, had lectured on *Large All Metal Seaplanes*, arguing the case for heavy wing loading appropriate to his design of high aspect ratio, rectangular cantilever wings. Despite a fascinating film showing constructional detail, ground handling, launching, outstanding manoeuvrability when taxi-ing and turning, and good behaviour in a rough sea, the lecture was received with polite scepticism. Typically *Flight* said: 'We do not think that the marked superiority of the monoplane over the biplane in

Manning's Lion-powered English Electric Ayr flying-boat was an attempt at eliminating the fallibility of wing-tip floats, but never flew when tried on the Ribble estuary.

As soon as the Ayr began to accelerate the starboard sponson completely submerged.

such large machines has yet been conclusively established, and we believe the future will show room for both. But in the meantime it is satisfactory to know that a machine incorporating so many highly interesting features, as does the Rohrbach, will now be constructed in this country, and thus the British Air Ministry will have an opportunity of actually comparing performance behaviour with other types.' This explained the registration on 26 January of Light-Metal Aircraft Co Ltd – a private company with £1,000 capital, of which the first directors were A. Rohrbach and E. Lerp of Berlin, A. E. L. Chorlton of London and J. G. Girdwood of Glasgow, each remunerated at a princely £50 per annum, with an additional £25 for the chairman. The stated purpose was to establish a licence agreement between the Rohrbach company and William Beardmore & Co Ltd, of whom Alan Chorlton was the aircraft manager – for Beardmores duly received a contract for two flying-boats of the type on which the lecture had centred, and for a still larger landplane to be built on the same system, all powered with British engines. To speed matters, the two flying-boats would be built by Rohrbach at their Copenhagen works. Meanwhile Bill Shackleton was designing, to the order of the Latvian Government, a two-seat biplane fighter, the W.B.XXVI, remarkable for absence of bracing wires. Other projects were a single-seater parasol monoplane fighter, and a two-seat, low-wing reconnaissance seaplane.

Though the Air Ministry had shown untypical initiative in ordering the Rohrbach flying-boats, their interest was more closely focused on conventional biplane flying-boats, exemplified by the Kingston I development of the Phoenix P.5, the Short metal-hulled Cromarty derivative S.2 which had Felixstowe F.5 wings, and the even more practical Supermarine Southampton military version of the Swan having 10-ft greater span.

The small Phoenix Ayr flying-boat had just been launched, but when Marcus Manton, the company's recently appointed test pilot, attempted fast taxi-ing, it rolled far to the right, and the starboard steeply-dihedralled sponson-wing submerged under a cascade of water. Take-off was impossible. Because of the low root position of the sponson-wing, not even tip floats would be able to hold it clear. Hope for survival of the English Electric aviation department therefore depended on the Kingston, but in late spring of the previous year the prototype, with Major Brackley at the controls, had hit submerged wreckage, badly holing the bottom, so that the hull filled with water, tipped vertically on its nose, and the machine was only salvaged with difficulty. Since then Marcus Manton had completed successful trials with N9709, the first of five semi-production wooden-hulled Kingstons, finding that in comparison with the P.5 the increased tail lever arm, and further forward C.G. due to repositioning of the Lion engine well forward, resulted in greatly improved stability. But ill luck still attended, for on 28 April it sank at Felixstowe on official trials when the engines pulled from their mountings immediately after becoming airborne, and the wing structure collapsed. The pilot, Flt Lieut D. V. Carnegie, and the presiding genius of the NPL tank, G. S. Baker, escaped with slight

In its day the metal hull structure of the English Electric Kingston seemed very advanced but too complex and costly compared with wood.

The Kingston II metal hull was flown with the Kingston I wood wings, tail, and steel-tube superstructure of N9712, in itself a relic of the war-time Phoenix Cork.

A similar exercise was the Short S.2 which had a 44-ft dural hull, based on Cockle experience, fitted to a war-time Felixstowe F-5 superstructure. (*Short Bros.*)

injuries. In view of the promising Short S.2 metal hull, Major Penny, the Air Ministry flying-boat specialist, decided that English Electric should build a duralumin hull for the last but one of the five Kingstons, N9712, retaining the wooden wing structure. Supermarine also were contracted to design a metal hull for the Southampton.

The metal-hulled Short F.5, N177, had its initial flight by Lankester Parker on 5 January, and he found that the multi-chine fluted vee bottom produced much less spray than the standard wooden F.5 hull, but airborne the machine was a little nose-heavy. Why it was not flown again until two months later is not clear, but on 14 March Parker delivered it to Felixstowe where every effort was made to prove hull susceptibility to damage and corrosion; yet the 'Tin Five' was triumphant, its 16-gauge planing bottom even withstanding a stall onto the water from 30 ft without springing a seam. As a result Shorts were invited to tender for an all-metal twin Condor-powered flying-boat to an existing specification, 13/24. Oswald Short told me: 'After sending in the Tender I waited some weeks before telephoning to ask if I was going to get an order. Two different officials said they knew nothing about it. I was advised to ring DDT himself and ask if the Blackburn Co had got an order for a metal flying-boat. I did so, and to this direct question he could only answer "Yes". On hearing that, I called in my works manager W. P. Kemp and told him to order all materials for building the Singapore, as we called it, without a moment's delay.

'I then telephoned Sir Geoffrey Salmond and said, "You know that we have lost the order for the Singapore?". He replied "Yes, Short, I am very sorry that you have." I replied, "Well, I am going to build it if I bankrupt the firm." He answered cheerily, "Good fellow. I will come and see you tomorrow morning."

'Next day I showed him the plans, and he immediately asked, "Why have I not seen these?". He said he liked the appearance of the machine and asked how much I wanted in that financial year if he gave an order. I replied that with such an order no doubt the bank would grant a larger overdraft, though ours was already £50,000 and the bank were pressing for payment. He would not hear of borrowing more and said he would give the order, but in that financial year the Air Council had only £10,000 left and the balance would be paid next year.'

In the previous year Oswald Short had decided to make a hull model testing tank by excavating and lining the chalk cliff behind No. 3 Shop. Referring to it, Lankester Parker said: 'High speed travel through or on water raises many problems not easy of solution. It is important to keep the drag as small as possible, and vital to have a degree of stability at all speeds. One form of instability is a high frequency "chatter" which no structure, or occupants, could stand for long; another is a fore-and-aft oscillation known as "porpoising"; there can also be failure to keep straight. The only tank available to Shorts was the Froude at the National Physical Laboratory, but Teddington was a long journey from Rochester, and the tank was much occupied with ship models. Oswald therefore decided to have his own so that tests could be made without delay and almost continuously. Financial limitations necessitated a relatively small one – at first 250 ft, later extended to 350 ft – and most of the equipment and measuring devices were designed by the experimental department at Rochester, then under the control of Oscar Gnosspelius and Arthur Gouge. I have no hesitation in saying that the pre-eminent position which Great Britain achieved in flying-boat design in later years was a direct result of Oswald's foresight and courageous decision.'

Gouge, studying at evening classes in Chatham Technical School, had graduated with an engineering BSc against all difficulty. Squat, sturdy, and with a wall-eye, he hardly looked the scientist, but his was a fundamentally logical mind. Assisting him was a junior, Jack Lower, who had trained at Chatham Dockyard. The tiny Cockle flying-boat with two 32 hp engines

Trying to get off the Medway, flown by light-weight Lankester Parker, the 36-ft span Short Cockle heads into wind, down-tide to obtain every advantage, but not until geared Cherub IIs were fitted could take-off be certain.

The Supermarine Southampton's elegant boat-built mahogany hull was the accepted Linton Hope type of construction and though drawing 3 ft of water it lifted easily. (*Grp Capt G. Livock*)

and hollow vee bottom was the first model they tested, and it became the first metal hull to be built. Completed without wings in April 1924, a 24-hour flotation test revealed only minute leakage, but the assembled machine was so underpowered when tried in October and November that Parker could only coax it from the water for one brief flight. In January 1925 the original tiny fin and rudder were replaced with units twice the size to control swinging if one engine failed. As Lebbaeus Hordern refused delivery because of inadequate performance it was agreed the Cockle would be sent to Felixstowe for further investigation.

Tests in the new tank with Webber's original Singapore design with fluted planing bottom failed to equal those in the NPL, so Gouge made an equivalent model incorporating a similar bottom to the Cockle, and found its resistance then agreed with the original NPL results. Reduction of beam gave further improvement and almost eliminated porpoising. At that Oswald Short told Webber to redesign the hull without flutes. Resentful at being overruled he resigned and presently joined John Lloyd's staff at Armstrong Whitworth. A few days later Gouge was astounded to receive a letter from Oswald Short appointing him chief designer.

Meanwhile it was the beautiful new Supermarine Southampton flying-boat which was receiving unstinted approbation from RAF pilots. The prototype, N218, to specification R.18/24, was significant of the trust the Air Ministry had in R. J. Mitchell, for the unusual course had been taken of ordering six 'off the board'. Construction had begun in August 1924, and the prototype made its first flight in the hands of Biard on 14 March, 1925. No modification was required, and the Southampton was delivered to the MAEE at Felixstowe on the following day.

Distinctive with patented up-swept adjustable incidence tail and triple rudders, white-painted struts and glowing varnished wooden hull, its compactness and obvious robustness, together with excellent controllability aided by large Frise ailerons, made it a tremendous advance on the existing soggy wartime-designed F.5s in standard use. Squadron mechanics appreciated the facility of its uncowled engine installation and simple gravity feed from tanks located under the top wing to prevent fire hazard. The only criticism was that had the propellers been still higher the Southampton

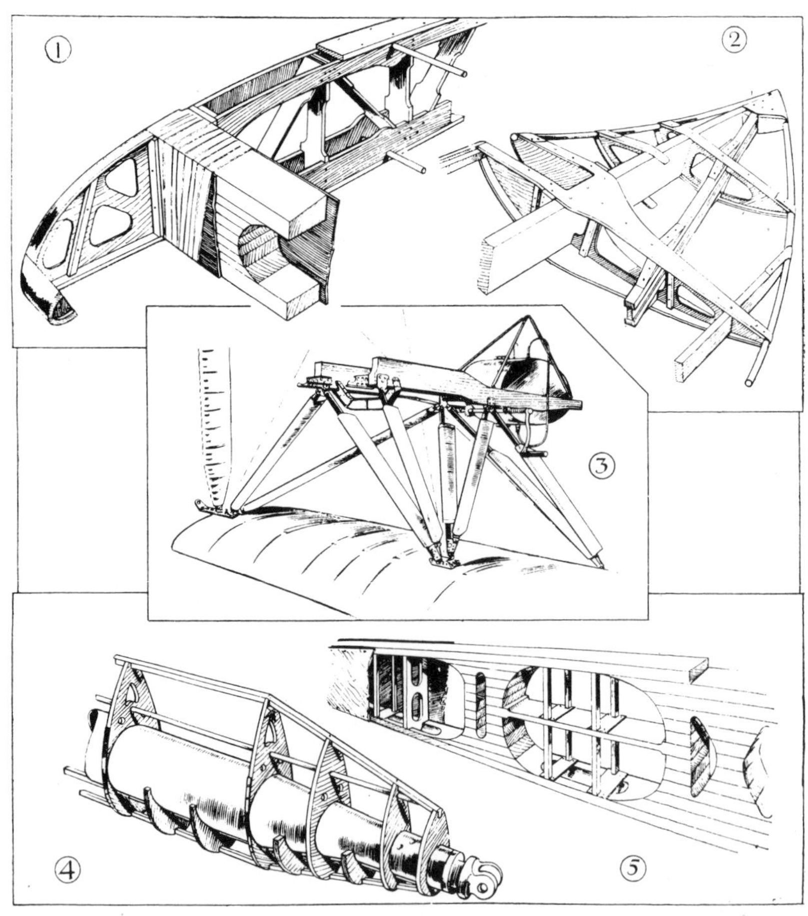

Wing details of the Southampton were standard practice with 1, wood box spar; 2, inset aileron on solid spar; 3, engine mounting, but diagonal interplane struts not shown; 4, faired tubular wing struts similar to Kingston; 5, laminated built-up tail spar. (*Flight*)

could have operated in even rougher seas, for the flared bow ran cleanly. The Linton Hope hull was double planked, with inner skin of cedar and outer of mahogany with varnished fabric between, and the double bottom had ten watertight compartments, each with draining plugs and vents, and the two steps could be readily replaced if damaged. The steel-tube interplane struts were fabric covered over wood formers, and the wing box spars had spruce flanges and ply sides, with laminated distance pieces at load points. Pilot and co-pilot sat in tandem in front of the wings with splendid view and freedom from the cascade of spray. A gunner was located in the extreme nose, and another abaft the wings with gun location each side of the hull ensured aggressive defence.

'The next design R.J. got out was never built because of doubts about the shaft drive,' said Alan Clifton after retirement as the last Supermarine chief designer, 'but by 1925 Mitchell had evolved a float monoplane for the Schneider Trophy which caused a sensation. It was an exceptionally clean design, with a central skeleton of steel-tubing which included daring cantilevered float struts and attachment of the wooden cantilever wings. Forward of this was bolted the mounting for the Napier Lion, and aft it joined a wooden monocoque rear fuselage. Thin Lamblin radiators were mounted on the underside of the wings. This revolutionary design received the order to go ahead on 18 March.'

6

On 17 March, Brancker's D.H.50 piloted by Alan Cobham completed its epic 17,000 mile flight, landing at Croydon from Paris at 1.45, to the cheers of a welcoming crowd. Six days later, at a Royal Aero Club banquet in their honour, Tommy Sopwith, chairman of the SBAC, revealed that the Treasury had refused to authorize payment for the flight, so C. R. Fairey had guaranteed to raise the money. 'When the proposition was put to the industry not a single firm wavered. Not only did all come in with their whack but others joined. De Havilland paid more because his machine was being used, and A.D.C. paid more because it was their engine. The Anglo-Persian Oil Co and the Wakefield Castrol people also helped.'

Sir Sefton – described by an evening paper as 'a dapper little man with an eye-glass' – replying to the Duke of Sutherland's toast of the Guests, said: 'I take no responsibility for the performance except for wangling a very small sum from the depressed and somewhat scandalized financial department. I think we have proved that flying is not a stunt. We were not blazing a trail, but following the footsteps of Sir Geoffrey Salmond who did the flight five years ago, but we are the only people who have come back by air as well. All the credit is due to the de Havilland machine, the Siddeley Puma engine, and the pilot and mechanic. The flight was easy with such a crew. We got in and out of many difficult aerodromes, and Elliott spent many unpleasant hours working on the machine. I myself merely sucked in fresh air all day and banquets all night and delivered hot air in many languages.'

From Croydon *The Aeroplane* reporter wrote: 'It is very satisfactory to see how every one recognizes the wonderful work of young Mr Elliott. He has had little limelight, but a very large proportion of the success was due to his intimate knowledge of machine and engine and to his special genius of adapting himself to varying conditions. He seems one of those fortunate people who are good at every thing they take up. One has no hesitation in predicting a big future for Mr Elliott in aviation.'

Others were no less certain in predicting a big future for the tireless Mr Cobham who had carried out a great feat of pilotage, but his reward from de Havillands was dinner and a visit to the London Hippodrome. Within a few days he was off again with the D.H.50, accompanied by Elliott, on a secret journey to Lisbon for the *Daily Mirror* to pick up photographs of the Prince of Wales's tour of South Africa. Cashing in on Cobham's increasing fame, the de Havilland company sent him off on 29 May with the Moth prototype G-EBKT on a publicity flight to Zürich. To give greater range an additional tank was fitted in the front cockpit. Leaving Croydon just before 5 a.m. he reached Zürich at 11 a.m., refuelled in 45 minutes and arrived back at Croydon at 5.30 p.m. Newspapers shouted his praise, though one journal acidly observed: 'One wishes that the daily papers would boom other good flights the same as they do those of Mr Cobham, and remember that in 1920 Bert Hinkler with the 35 hp Avro Baby flew non-stop 700 miles from Croydon to Turin, and the following year flew from Sydney to Bundaberg in Australia, a distance of 800 miles nonstop. Many people object to publicity, but Mr Cobham does not. His valuable work will always be recognized and at the same time his equally valuable leg will always be pulled.' But as Cobham had said before the flight: 'It is to get the public airminded' – and steadily this was happening.

However, newspapers at the beginning of April intimated that Colonel Frank Searle, managing director of Imperial Airways, had resigned. The *Evening News* commented: 'Searle's aim has always been safety first and then high efficiency, but non-aeronautical members of the Board, notably the chairman, Sir Eric Geddes, are more concerned with attracting revenue and cutting down expense to show that the business is a commercial success.' Since Imperial's formation a year ago some 800,000 miles had been flown, regarded as colossal with so few aircraft, but it had made a considerable loss and this caused a split in the Board which led Geddes to ask Searle to resign as 'the Board considered it their duty to the shareholders'.

Said Searle after leaving: 'The Board consists of very successful men who know the danger of trusting the unknown, but none of them has successfully operated any new form of transport. My experience has in fact been unique, but they must find it difficult to reach a correct conclusion when they hear so many conflicting opinions on the same subject. Such men want a lot of convincing, and in such company my rhetoric may have left a lot to be desired, but I at least have the satisfaction of knowing that what has been accomplised has placed the company in a very strong position for its second year of operation.' His close and friendly contact with Brancker had ensured that the efforts of the Civil Aviation department were directed towards initiation of more economical types of aircraft. Handley Page would soon have the triple Jaguar-powered W.9 Hampstead ready for test, and it was rumoured that de Havillands and Armstrong Whitworths were building three-engined airliners with steel-tube fuselages for the spring of next year. They were the D.H.66 Hercules and Argosy, representing latest design techniques and good performance with economic passenger cost per mile.

Avros had just tested a civil transport adaptation of the Condor-powered Andover Type 561 Ambulance/Troop carrier and were about to hand it over to Imperial Airways as G-EBKW for cross-Channel proving flights. Its more specifically designed equivalent, the D.H.54, of almost identical dimensions and performance but with slightly greater payload though smaller cabin of 15 ft compared with the Avro's 22 ft, was in final stage of construction, and represented a scaled-up version of the D.H.34 but with the slower landing speed demanded by airline pilots. As equivalent in size and performance of these two land machines, Fairey had built to Air Ministry order the twin-float seaplane Fremantle – but, compared with the floats which Shorts were tank-testing for the Gloucester and Supermarine Schneider racers, its wooden pontoons seemed archaic though they contained fuel for a nonstop flight of 1,100 miles. Taking a leaf from de Havilland practice, the cabin portion of the fuselage was planked with mahogany and the rear fuselage with plywood.

The twin floats seemed remarkably small for so large a machine as the 650 hp Condor-powered Fairey Fremantle, and it was subsequently used with land undercarriage as a flying laboratory for RAE 'wireless' tests.

Even at this stage it seemed improbable that the single-engined de Havilland or Avro would be selected in addition to the new three-engined machines; nor had the single-engined Fremantle much likelihood of production compared with the Supermarine Swan which Major Brackley was investigating on behalf of Imperial Airways. However, the Fremantle was regarded by the Air Ministry as a possible for a round-the-world flight, piloted by an RAF officer and with Capt Freddie Tymms, of the

Department of Civil Aviation, as navigator. To that end additional fuel could be carried to give a range of 1,700 miles, and the cabin was arranged with berths for a crew of five.

Such matters had not even passing interest to the public, for the world had far greater problems. The policy of Baldwin was causing the Bank of England to maintain high interest rates to prevent exchange of gold, and this hit the development of industry and trade, and tended towards cuts in wages. The unemployed were still well above a million, and the dole only offered starvation subsistence. Miners particularly were hard hit, yet the motor trade was booming. Car manufacturers had shown courageous imagination in building up their industry, and whereas in the 1921 slump Morris had been building only 3,000 cars a year, his policy of slashing prices had reduced the Cowley two-seater from £465 to £198 and led to demand for 20,000 cars a year despite strong competition from Sir Herbert Austin. With equal initiative Austin had risen from the slump with a miniature car – the Austin Seven, which drastically undercut Morris by selling at £165 for those who could manage its cramped accommodation. Both cars opened an era of mass-production and 'motoring for the million' as Austin termed it. Those who could not afford cars or motor-cycles were happy with a plenitude of trams and buses, and long-distance motor-coaches competed with profitable trains. Now that even the outskirts of towns could be reached by public transport, the urban sprawl was extending. Baldwin was moved to prophesy: 'It is no exaggeration to say that in 50 years, at the rate so called improvements are being made, the destruction of the beauty and charm with which our ancestors enhanced their towns and villages will be complete.'

With no sense of oppression, even those in poor employment went their way. Surely it was an expanding world? There were *thé dansants*, films, revues, Bridge, and that young Mr Noël Coward with shattering plays of social commentary, Edgar Wallace thrillers, uproarious farces at the Aldwych theatre – and everywhere one heard songs Binnie Hale was singing in that enormous success *No, No, Nanette.* Home radio on a crystal set from 2LO at Marconi House in the Strand, Jazz, the Charleston, skirts up to the knee, soft hats instead of bowlers and toppers, and even the sartorial extravagance of Oxford 'bags' with legs as wide as skirts: these were the days, thought the young.

There were also dedicated young men at universities intent on pursuing a career. To those at Cambridge, Air Chief Marshal Sir Hugh Trenchard delivered an address on 29 April in the Union Society's Debating hall, under the auspices of the University Aeronautical Society, presided over by Flt Lieut E. L. Howard-Williams, MC. In his deep booming voice, Sir Hugh explained the general organization of the RAF and the problems of defence. Wastage of aircraft in peace was about 30 per cent per annum, while in war it was 80 per cent per month; so the side which could keep re-equipping with pilots and machines would probably win. Nevertheless there was no need to locate squadrons all over the Empire; as long as there were operational facilities, units could be very mobile – 'still more so when the great aircraft carriers of the future, airships, come into being'. To some, that belief was questionable, for only twelve days earlier the fallibility of airships had again been demonstrated when the R.33 – now G-FAAG – broke from her mooring mast at Pulham where she had been left with a skeleton crew because conditions were unsuitable for experimental work. The nose tore completely away, but luckily the top fabric folded over the gaping hole and the ship was steerable, though a 30 kt wind drifted her to Holland. Only when it slackened could headway be made to Pulham, which was reached the following day at 2 p.m., and the ground crew managed to get her into the airship shed. But Sir Hugh pursued his theme: 'The great problems of the world are economic, and if the Air Force cannot conduct defence of the Empire as cheaply or even more cheaply, and as efficiently as the older Services have done for hundreds of

Luckily R.33's first officer, Flt Lieut Ralph Booth, was aboard when the airship broke away. Within five minutes all engines had been started, and a dramatic flight followed which became worldwide news. All the local inhabitants rushed to the air station to help with her berthing. (*George Swain*)

years, then there is no justification for it. But if the trade routes of the world are studied it will be seen how close they run to the shore. Submarines brought us to verge of disaster in the last war by attacking our ships near our ports. Merchant ships under air convoy in narrow waters will be far less vulnerable, though no doubt the Navy has a great part to play. Yet the air not only lends itself to protect Empire trade but equally for attack on any one else's trade.'

To the interest of all, he announced that the Air Ministry was anxious to start Air Force units at Oxford and Cambridge, and at other universities later. 'Your squadron during term time must be kept going by instruction on engines, rigging, wireless etc, with possible flying as observers at Duxford or other Air Force stations, which with further flying during the long vacation, will, I trust, be the means of stimulating interest in the air throughout the University. I hope this unit will be formed in October, and feel sure I may count on the backing of undergraduates to make the movement a success and that they will do all in their power to impress on the nation as a whole the value of this arm.'

Construction of three experimental Yeovil bombers in the foreground intermingled with rebuilds and erection of D.H.9As in the main section of Westland known as the Vimy shop.

That May, interim examinations having concluded, it was arranged that, with another student, my second summer of practical experience should be with Petters' Westland Aircraft Works at Yeovil. We found a dreaming countryside and small market town dependent more on gloving than oil engines and aircraft. Unlike Handley Page Ltd no pay was offered. We were sent to the foreman of the combined metalworking and assembly department known as the Vimy shop. Despite the dusty gravel floor the building seemed airy and light because of its considerable height. Two Yeovil bombers predominated, their fuselages not yet covered, though the first was almost ready for flight. Alongside was a rebuilt D.H.9A looking like new, and another awaiting its wings. There was nothing for us to do. 'Better have a look round for the time being,' said the foreman. We

The illustration is typical of the wingless wrecks sent to manufacturers for 'reconditioning' into fully equipped, new D.H.9As.

investigated wood mill, wing shop, wind-tunnel, and checked over the girls who were covering and doping the wings. Next day I visited the recently appointed chief draughtsman, the ever energetic 'Johnnie' Johnston, who of old had been the liaison between de Havillands and Westland during design and production of the D.H.9A. A long narrow room, with large windows on one side, housed the twenty draughtsmen, with offices at the far end for Arthur Davenport and the two stressmen, Mettam and Digby. The managing director, Robert Bruce, had a more splendid room in the Petter office block 200 yards away; adjacent to him was a room for Stuart Keep, their heavy-weight ex-test pilot, who now struggled around, using sticks to help balance on two Desoutter metal legs six inches shorter than of old, but he made light of his injuries from the Dreadnought crash, and was now technical supervisor, with W. G. Gibson, another ex-RFC man, under him as works superintendent.

The workmen were very friendly, even though they regarded us as 'two young toffs from London', and for several weeks we had the nominal task of helping strip the miscellany of tatty wings, fuselages and tails dumped

To all intents, except for the contract and inspection record, this was a brand-new dual D.H.9A outside the little Westland factory. Administration offices to left.

by the RAF under the misleading description of 'D.H.9As for reconditioning'. Clearly they were an excuse for replacement by virtually a new aeroplane incorporating any serviceable parts after stripping and heat treatment. We spent hours and hours undoing and throwing away little nuts and bolts and piling up the fittings.

Major Laurence Openshaw, MA, was the new pilot. During the war he was one of Harry Busteed's team at Grain and Eastchurch, but after leaving the RAF he became a mining executive in Italy where he participated in the Grand Prix with a Fiat racing car painted the official British racing green, and which now was garaged in the same aerodrome hangar which contained the great stripped wing of the Dreadnought. In the next two hangars we found the Westland light aeroplanes and the prototype of the smaller Limousines. In that machine, and the three-cylinder Thrush-powered Widgeon, Openshaw presently took me flying. Meanwhile we began to design and build a 'run-and-jump' biplane glider.

The Westland Condor-powered Yeovil prototype, with big wooden propeller, ready at Yeovil for its first flight. Courtney stands in front of the port wheel.

Came the day when the Yeovil bomber was ready after many engine runs. Undoubtedly it was impressive. The Condor engine was started by an experimental light-weight Bristol gas-starter engine stowed on the floor of the fuselage. The great blades of the 14ft 6 in propeller majestically swept round, geared to only 900 rpm at cruising power. But Openshaw, despite recent practice at the Bristol Flying School as a Reserve officer, was considered by the Air Ministry to have been too long from test work to be trusted with the big bomber, and Frank Courtney was employed, at the magnificent fee of £100, to make the first flight. It was done with care and Courtney showmanship. This tall, dark-haired man, with aquiline nose and *pince-nez* glasses, was almost as impressive as the aeroplane, and everyone seemed to hang on his words when he landed 30 minutes later. It seemed the machine was satisfactory and could go to Andover where Openshaw would fly it. In the afternoon the Westland pilot took his place in the gunner's cockpit behind Courtney, and we watched the big biplane smoothly take-off, turn and head away over the town and fade into the distance. They had no parachutes.

A year later, the third Yeovil bomber, with modified rudder, awaits test with simplified undercarriage, Leitner-Watts propeller, and non-lifting fuel tanks. The last Westbury can be seen beyond the port wing.

Some days later Openshaw told me he needed an observer to pay out the suspended static head on its long trailing tube. Would I like to act as his observer at Andover? Somewhat embarrassed he added: 'I'm afraid they're too mean to insure you, and only cover me for two thousand, so you'll have to sign a blood-chit.' Insurance or not, here was the chance of flying in a bomber exemplifying the last word in techniques – and of course I went. Even the journey to Andover held its thrills, for it was in Openshaw's racing car. Through dusty Wiltshire roads of a summer's day, we raced at what to me was the unbelievable speed of 70 mph, gales thudding at my head and the big engine roaring. The countryside lay quiet and still, tractors unknown, slow plodding horses pulling an occasional mower or haywain; maybe we met two cars in 60 miles.

At Brooklands, Cricklewood, and Filton, the Yeovil bomber's rivals were being tested. Though I had seen the mahogany wind-tunnel model of the Handcross when at Handley Page Ltd, I had gained only a glimpse of the prototype behind its protective screens, and it was far from complete when I left, yet was the first of the four flown. Just before Christmas Hubert Broad made the initial test, but found it disappointing, for controls were heavy and sluggish, and performance less than calculated. There were the usual changes of propeller and alterations to control surfaces before the Handcross was flown to Martlesham many months later.

Even the neat-looking Horsley, with wing geometry alleged to be scaled from the Cygnet, proved disappointing when flown by George Bulman in the summer, for it was longitudinally unstable, and the cooling system, with radiators each side of the fuselage just ahead of the pilot, was useless. Installing a single radiator below the nose solved the cooling problem, but the instability was intrinsic in the modified Göttingen thick aerofoil, chosen because of its high lift coefficient and to enclose the fuel tanks within its contour, for it had greater destabilizing C.P. travel than usual,

To the same specification as the Westland Yeovil, Hawker Horsley, and Bristol Berkeley was the clumsier Handley Page Handcross which lacked slots in view of handling difficulties with the Hanley.

which also necessitated an abnormally far back rear spar to attain stressing strength in torsion. The radiator re-positioning, by bringing the C.G. further forward, improved matters but ultimately sweep-back had to be incorporated.

Although the Bristol Berkeley mock-up, as designed by Reid, had been inspected as far back as 4 January, 1924, and the Air Ministry had urged construction of a prototype with view to evaluation that summer, J7403 was not erected until November. Further delay ensued on discovering that the longerons were below strength and must be replaced, so it did not fly until 5 March, 1925, but nevertheless was the first to get to Martlesham where it arrived on 22 May.

When Reid resigned, Barnwell modified the Bristol Berkeley design with longer fuselage and taller rudder of de Havilland profile. Because of the nose cockpit the Berkeley afforded the best view, but pilots preferred the conventional aft location of other contenders. (*Courtesy Air Marshal Sir Ralph Sorley*)

Pre-production Hawker Horsley (*above*) and its scale-model Cygnet progenitor flown by George Bulman in similar attitude around the Brooklands hangars. (*Flight*)

Early Atlas Army Co-operation biplanes had no slots, but tended to drop a wing when landing. Between 1926 and 1933, production totalled 450 including trainers. (*Courtesy Air Marshal Sir Ralph Sorley*)

One other aeroplane of great importance was flown that May – the two-seat, single-bay Armstrong Whitworth Atlas designed as replacement for the well-loved Army co-operation Bristol Fighter. Although the same size, it looked smaller and stockier, and its all-up weight was more than 4,000 lb compared with 3,250 of the Mk III Fighter. Powered with a geared Armstrong Siddeley Jaguar, the prototype J8675 achieved over 140 mph, but Courtney found that it stalled sharply, and the ailerons were heavy.

7

On 3 May, France's great aeronautical pioneer, Clément Ader, died at the age of 84. His engineering ingenuity was enshrined in a mass of patents – electrics, telephones, petrol engines, steam engines, air-cushioned vehicles, aircraft – and as his memorial there is his bat-like, twin-engined *Avion* in the Musée des Arts et Métiers, Paris. Contention has always surrounded its flight because de Freycinet, the Minister of War, had placed a seal of secrecy upon the inventor for some years. Nevertheless Ader's wing is correctly swept forward with wash-in to give longitudinal stability, and the three-axes control was intended to give independent downward bending of the outboard trailing edge, analogous, but not identical, to the Wrights' differential wing warping, as well as longitudinal control in combination. That his machine was not developed, despite favourable reports by General Mensier who witnessed the tentative initial test, was due to a new Minister of War who utterly disbelieved the practicability of flying. Disheartened, and unable to afford further enormous expense, Ader burnt his earlier machines and all his records.

While Britain saluted Ader, another French pioneer, Robert Esnault-Pelterie, was causing annoyance and worry. He had sued his compatriot constructors for infringement of stick control patents during the recent war, obtaining damages of some £120,000; encouraged, he now instituted action in the French court against the Aircraft Manufacturing Co Ltd, Beardmore & Sons Ltd, Grahame-White Co Ltd, A. V. Roe & Co Ltd, and the Bristol Aeroplane Co Ltd, claiming £1 million damages.

The R.E.P. patents reveal a different arrangement from the customary 'joy-stick' giving longitudinal and lateral control application singly or in any combined relationship, but in any case was pre-dated by a patented stick and wheel combined control used by Alliott Roe for his first aeroplane. A.V. was astounded at being sued, and handed the letter from the Frenchman, and a copy of his own patent, to the Air Ministry, observing: 'There is not sufficient subject matter to constitute invention in view of my prior claim.' The SBAC also took up the cudgels against the Air Ministry, pointing out that in any case the aircraft had been sold and delivered to the British Government in this country regardless of where they were used, and thus the purchaser was responsible for taking them to France where the infringement was alleged. At that, the Government agreed to fight the case and Esnault-Pelterie was non-suited, thus freeing British constructors from liability – though it seems that an *ex-gratia* payment was made, for A.V. was told by the Air Ministry that to fight the case would have cost £40,000.

In the middle of May the second Gloster II racer, replacing the one crashed the previous year but powered with the latest Napier Lion up-rated to 700 hp at 2,700 rpm, was flown as a landplane by Larry Carter at Cranwell, as this was the only aerodrome large and smooth enough to make flying so fast an aeroplane reasonably safe; the intention was to make comparative tests of several metal propellers, radiators, and so on. Three weeks later the G.A.C. announced that the machine had crashed, and Carter had fractured his skull and broken a leg. 'The pilot was about to carry out a speed test, but was forced to make an emergency landing at 200 mph, which carried away the undercarriage, the machine sliding on its fuselage for 150 yards. Fortunately it did not turn over because the propeller, which was vertical at the moment of impact, bent under the fuselage

Final clean-up of the 'Bamel' design was the Gloster II landplane with ply-covered fuselage, knife-edge cabane and Lamblin radiators on the undercarriage legs. Despite flutter, the tail did not break when the aircraft crashed.

and acted as a slide. The machine had been flown successfully by Mr Carter three times before the accident.' Tail flutter had developed while flying at 40 ft above the speed course, but as the cause and cure seemed apparent, work on the Gloster IIIA and Supermarine S.4 seaplanes was pressed forward, and Hubert Broad and Bert Hinkler were selected to fly them.

Coincidentally, Napiers had the highest profit of their history, earning £237,542 for 1924–25. It was almost embarrassing. Montague Napier, their dark and bearded chief, advised the managing director, H. T. Vane, that rather than increase the dividend it should be at the same rate as before, but tax-free – then their workers might not realize that addition of tax resulted in greater percentage profit! He was concerned that the Cub, which he had presumed would replace the Lion as their main production, seemed to have only a limited future. Writing to his lawyer co-director, Henry Cooke, he suggested they had better experiment with air-cooled engines, pointing out that Armstrong Siddeley and Bristol had a five-year lead with air-cooled radials, but as no one had made a modern air-cooled vee engine 'we might for some time keep this line to ourselves.' Soon he had doubts, and wrote to Vane: 'The question arises whether we are right in making the new eight-cylinder, but the situation might be saved by developing with the same parts a twelve-cylinder to compete with and go one better than the Condor.'

But there was doubt about the Condor too. Experience with the Condor-powered Aldershot night bombers confirmed that the risk of a single engine was too great for peacetime, and multi-engined aircraft must be the future policy, though for day bombers there was reasonable chance of spotting a suitable forced-landing field should the engine fail. In any case, Trenchard favoured the practical merits and simplicity of the air-cooled radial, particularly for fighters, and was convinced that in time of war sufficient aero-engine production could only be attained if the types were capable of manufacture by automobile firms. In due course an Air Ministry official was sent to Derby to consider this possibility. As far back as the early years of the war Claude Johnson, managing director of Rolls-Royce, always insisted their workmanship was so high that no other firm could construct their engines, though later he agreed that Fedden at Cosmos was so outstanding that he could undertake such work. The same attitude still held. Arthur Wormald, in charge at Derby, convinced his visitor that only Rolls-Royce could build the engine. On reading the report Trenchard wrote across it: 'No more Condors.'

It was that qualified and very experienced engineer, Lieut-Col L. F. R. Fell, DSO, the AD/DTD Engines, who directed Montague Napier's inclination towards a twelve-cylinder engine, for he had been immensely impressed by the Curtiss D.12 which Fairey had introduced with his Fox. Frank Nixon, who later joined Rolls-Royce from the Bristol engine department said: 'With the aid of two draughtsmen, AD/DTD (Engines) actually draughted a twelve-cylinder vee layout using Lion components, but Napier wanted to build a different type which did not command the confidence of the Air Ministry, and financial support was withdrawn. Ironically enough for Napier, Fell then persuaded Royce to enter upon a design programme which led to the Kestrel. Fortunately Royce had been making a review of possible alternative types, and was therefore receptive when Fell suggested he should examine the Curtiss D.12 and design a replacement. Rowledge played a leading part in its design. The engine eventually had a capacity of 1,295 cu in with monobloc cylinder head and open cylinders following Puma practice and making a break with Curtiss and Hispano-Suiza tradition.' But before that engine was ready for bench test there was much to be done, including design and development of a gear-driven supercharger, for which they sought the advice of Jimmy Ellor of the RAE, who had become the leading authority in this field.

Unaware of Trenchard's decision on the Condor, Sydney Camm at Hawker Engineering was finishing a design with this engine for a single-seater fighter. Known as the Hornbill, it seemed somewhat like a long-nosed heavier Siskin. To Camm's surprise Sopwith had just appointed him chief designer under Bennett as chief engineer. When I asked Carter how this came about he replied: 'No mystery here – quite a simple story. There was a difference of opinion between myself and Sigrist over organization of the technical division and design procedure. A hitherto pleasant relationship became impaired. Quite likely I was at fault, for I was inclined to be impatient. We had just been told that the Horsley day bomber would be accepted for production by the RAF. I had fulfilled my aspiration and justified my appointment as the firm's designer. I decided it would be good to seek fresh pastures, and never had regret in so doing. On the contrary it has always given me a curious sense of satisfaction to find I had sufficient guts to take a chance and seek adventure.'

Shortly afterwards young Hollis Williams also left Hawkers. 'I eventually got a bit too big for my boots,' he explained, 'and found it necessary to put my hat on, so went to Faireys and asked for a job. I expected to be interviewed by some fairly junior person, but was rather surprised when T. M. Barlow took me over. He then said "Just a minute", and I next found myself ushered in to C. R. Fairey who said "I gather you have done some Service flying?" I replied "Yes Sir". He then said "We shall shortly need a second pilot. Are you prepared to take a job in which you will mainly work in the design department, but be available on call?" I said I would be delighted, and managed to establish myself straight away on £5 a week, whereas about £3 10s was the standard.'

On 2 May Fairey Aviation Co Ltd had been re-registered as a public company with nominal capital of £10,000 in £1 shares 'with the object of making an agreement with the Fairey Aviation Co Ltd (incorporated in 1921) and its liquidator, to manufacture and deal in aircraft of all kinds, and component parts thereof including engines'. First directors were C. R. Fairey of Grove Cottage, Iver, F. G. T. Dawson, C. O. Crisp, Lieut-Col V. Nicholl, with remuneration free of tax of £300 per annum for Fairey, and £250 for the others.

'About the time I joined,' said Hollis Williams, 'Duncanson had a colossal row with Fairey right in the middle of the D.O. Duncanson left and went to Blackburns. Marcel Lobelle was thereupon made chief designer, though he wasn't called that; Fairey reserved this title for himself. I was put with P. A. Ralli who had been a mathematical professor at the Sorbonne in Paris. When any abstruse problem arose he would go home, and on return had the whole solution, written on cheap paper. It was he who brought mathematics to life for me, whereas previously it had been a bit of a chore.

'Some weeks later, Dick Fairey had a breakdown, and was ordered to take a long rest in the South of France. At that time the production line-up was the IIID seaplane. They were used in the Far East and Middle East, and the termites simply loved the wood and fabric construction. The

In early form the fin of the Fairey IIIF was rectangular with horn-balanced rudder, and there was a Scarff ring, but later machines had rounded fin with inset rudder balance and a special pillar mounting for the gun. The lower illustration is a dual-control Mk. IVC. (*Upper view courtesy H. F. King*)

Ministry wanted the job metalized. Old man Fairey wouldn't have anything to do with it. He said wood was the right material for an aeroplane – and went off on his convalescence. The day he left, Colonel Nicholl signed a Works Order for a metal version, intended as an exact replica by converting the wooden components to metal – but it became a free for all as soon as design started, and everything was changed including the engine, though it was all very simple and straightforward. When Fairey returned three months later the fuselage was being erected. He took one look, and then a great roaring started, heard all over the factory. This went on for at least a week whenever Fairey and Nicholl met, for they started rowing wherever they were, shouting at each other – and everybody stood aghast waiting for the storm to abate. Eventually Nicholl gave orders for the job to be packed up "as the old man wouldn't wear it". Then came the lull. Fairey was obviously making a mental assessment of the position. A week later it was back on priority and, as the Fairey Ferret, was the next type to go into production, and led to the very successful Fairey IIIF.'

8

On 8 June the *Daily Mail* announced £5,000 in prizes for a 1926 Light Aeroplane Contest. Again the aircraft were to be dual-control two-seaters with engines not exceeding 170 lb. The main requirement would be not less than 20 flights aggregating 2,000 miles at an average of not less than 50 mph carrying a load of 340 lb. But that was music of the future. Current interest centred on the summer's great pageantry of the RAF Display on 27 June – rivalling Henley Regatta and Royal Ascot. Again it was a royal occasion for the King was present, trimly bearded and bowler hatted, accompanied by his tall Queen in palest blue. Trenchard had ordered the Display this time to be a demonstration of the average work of the RAF rather than an exhibition of picked stars performing individual evolutions – but the formation flying by D.H.9As, Flycatchers, Grebes and Virginias was no less spectacular than before, and there was Sqn Ldr Longton, newly promoted CO of No. 58 Bombing Squadron, throwing great loops and spins and even flying inverted with the twin Jupiter-powered Boulton & Paul Bugle as he 'fought' two Grebes flown by Flt Lieut H. A. Hamersley and Flg Off J. N. Boothman. Dramatically enthralling were live drops by three parachutists, led by Flt Lieut F. O. Soden, DFC, using the newly standardized Irvin parachutes to show that the RAF at last had got them – though before their use in squadrons became universal all machines must be modified by replacing the wicker seats with metal ones having deeper bottoms to accommodate the packs.

As light relief there was a race between six little silver D.H.53 light monoplanes piloted, as one humorist said, by sitting officers of the Air Ministry instead of flying RAF officers. It was won by Wg Cdr W. Sholto Douglas of the Directorate of Equipment; second was Air Commodore C. A. H. Longcroft, Director of Personal Services. 'The way they flew their 53s round the marks showed what good pilots can do with these machines, the Air Commodore merely disturbing the flag as he went by, whereas the Wing Commander must have taken the ear off the man who was holding it.'

As always the new types were of major interest to the industry and RAF, though the latter facetiously named the prototype enclosure the Amusement Park. 'Nevertheless much of the interest in the fly-past was lost owing to the absence of the Hawker Heron,' reported *The Aeroplane*. 'Mr Bulman, who is now the Hawker test pilot, earlier had given an extraordinarily good exhibition of its capabilities. She can stand on her tail and climb almost vertically, as helicopters would like to do. Her controllability, at any rate in Bulman's skilled hands, is extraordinary, and she is quite one of the prettiest radial-engine machines one has seen. Unfortunately she had split the tip of one blade and could not fly again, so the parade was led by the Gamecock, the latest development of numerous rotary and radial-engine Gloucester machines which started with the Mars and are derived from the Nighthawk, Nightjar, and Grebe.

Flt Lieut Paul Ward Spencer Bulman, MC, AFC – known as 'George' – the imperturbable chief test pilot of Hawkers, bald headed but only 29 in 1925. (*Flight*)

'Next was the all-metal Short Springbok II which resembles the original Springbok except that the lower wing is attached to the body instead of below, and the chord of the top wing is increased, but the most striking feature is the excellent body shape. Then came the Hawker Hedgehog, closely resembling the now well known, but still officially undescribable Woodcock. Special features are the neat exhaust collector ring with long exhaust pipes and installation of the two synchronized Vickers guns inside the fairly large diameter body, with muzzles recessed in the sides and line of fire between pairs of cylinders, instead of carried outside alongside the

In stately fashion the Blackburn Cubaroo comes floating in. Bulman said it was as easy to fly as a D.H.9A – but the Air Ministry decided single-engined bombers were too unreliable. (*Flight*)

pilot as specified by Air Ministry gunnery experts. The slotted and flapped Handley Page Hendon torpedo carrier should have followed, but was withdrawn at the Admiralty's insistence as too secret, probably because of the torpedoes, so next was the hugely impressive Blackburn Cubaroo with 1,000 hp Napier Cub, and she floated off the ground with her usual amiability and ease of handling.

'Next came the Bristol Brandon ambulance biplane, which is one of the nicest jobs, with beautifully light and airy cabin more like an up to date operating theatre than part of a flying machine. Following it was the D.H.54, a big 14 passenger biplane with Condor, said to be particularly

Derived from the Commercial Ten-seater, the Bristol Brandon had slight sweepback and deeper gap. Although first flown in March 1924 it took a year of maker's trials before handing to the RAF. (*Courtesy Air Marshal Sir Ralph Sorley*)

It was hoped that performance and carrying capacity of the Vickers Vanguard would lead to adoption by Imperial Airways when it flew in mid-1923, but so many modifications were made, including installation of Condors, that five years passed before delivery for air route trials.

easy to handle, and landing remarkably slowly with full load, so in every way is an improvement on previous passenger machines. Every line is unmistakably de Havilland. Last was the Vickers Vanguard with twin Rolls-Royce Condors, and fitted for 22 passengers, cabin boy and buffet. A novelty is location of petrol tanks in the bottom of the fuselage where they may be dropped clear in the event of fire or forced landing.'

One reason for poor top speed of the Vanguard was the great resistance of the uncowled Condors, but ease of access and better cooling were considered overriding necessities. (*Flight*)

Unfortunately the Atlas Army co-operation two-seater, the Bristol, Handley Page, Hawker and Westland bombers, and the Avro Ava, were too secret to be revealed.

A week later, the King's Cup race was anti-climax. Starting from Croydon, the course via Harrogate and Newcastle to Renfrew, and back via Sealand and Bristol, had to be successively flown on Friday and Saturday. Of the 14 starters the earlier ones were forced down by fog north of the Thames – the Moths piloted by de Havilland and Cobham, and the D.H.51 piloted by Sempill landing near St Albans; T. W. 'Jock' Campbell with the Bloodhound, H. H. Perry flying an Airdisco Avro 548A, and Bulman with the Woodcock descending near Luton – the latter running into a hedge and hitting a tree, breaking his machine badly.

Although three Bristol Bloodhounds, similar to the modified prototype G-EBGG were delivered to the RAF, the constructional system of fabricated steel was considered impracticable for a production order. (*Courtesy Air Marshal Sir Ralph Sorley*)

Meanwhile Charles Barnard, after various landings with his D.H.51A from Bedford onward, was struggling gamely to Newcastle which he reached too late to continue, so returned to Croydon next day. Sqn Ldr Longton and his Airdisco Jaguar Martinsyde disappeared into thick fog and only late at night was it discovered that he had landed on Nottingham Racecourse and after a long wait for petrol managed to reach Harrogate, but realizing his tanks had insufficient capacity for longer stretches, he returned next day. Earlier it transpired that the D.H.34 piloted by Colonel Minchin of Imperial's, flying *hors concours* for the Press, had landed on a sloping field near Durham and smashed into a wood. The universal absence of brakes accounted for two more. Frank Courtney, after making good time with the Armstrong Whitworth Ajax export version of the Atlas, arrived at Newcastle where white crosses had been placed on Town Moor to indicate the landing area. Having passed over one, he landed near the second and gently trickled into a ditch 20 yards ahead, knocking off his undercarriage. Soon Capt J. L. Bennett-Baggs, who would shortly join Armstrongs, appeared overhead with his Siskin V, saw Courtney's machine apparently standing safely on the aerodrome, landed, and seeing

The Armstrong Whitworth Siskin Vs flown by Frank Barnard and Bennett-Baggs for the King's Cup differed from RAF Siskins in having equal-chord wings, like the 1918 prototype. (*Flight*)

the ditch at the last moment, opened up to swing round, but put a wheel over the edge – and he too was out.

That evening, a few seconds before 7 p.m., Capt F. L. Barnard, chief pilot of Imperial's, landed back at Croydon with the other Siskin V. On remarking that the crowd was not large, he was told: 'Never mind! Both of it are frightfully enthusiastic!' An hour later the scratch man, Flt Lieut H. W. G. Jones, swept over the finishing line with the Siskin IV, and found he had been promoted to Sqn Ldr! An hour and a quarter later came Major Hemming with the D.H.37 *Sylvia*, and finally Bert Hinkler with his Avro 504N trainer half an hour afterwards.

For the second circuit next day, Hinkler scratched as he had no chance of winning. The other three flew round without incident, this time taking it right handed. Barnard's silver Siskin, G-EBLQ, flashed past the winning post at Croydon at 4 p.m. having averaged 151 mph, followed by Jones at 5.20 after averaging 141.5 mph, and at 6.10 came Hemming, navigated by Capt Freddie Tymms of the Air Ministry, with the excellent speed of 120 mph. Said Geoffrey Dorman with customary crack at Cobham: 'Barnard's win is all the more creditable as he was suffering from a swollen throat, unlike many star performers in all professions who suffer from swollen heads.' But none could deny that Alan Cobham was the best publicist of flying that civil aviation had, for it was not mere self-advertisement that coloured his pronouncements. Much more lurid was the suit of orange 'plus-fours' worn by burly Bennett-Baggs, rivalling the similar maroon suit favoured by the ever hearty Harold Perrin who was volubly organizing everyone on behalf of the Royal Aero Club.

Clever publicity by de Havillands, aided by the many demonstration flights of Cobham, Broad, and D.H. himself, had resulted in establishment of the Moth as the approved machine for the five clubs selected by the Air Ministry for financial support:– London Aeroplane Club, Newcastle Aero Club, Midland Aero Club, Lancashire Aero Club, and Yorkshire Light Aeroplane Club.

First to open was Lancashire on 21 July, after allocation of Moth G-EBLR, but at the last moment it had engine trouble preventing despatch – so, rather than upset proceedings, Cobham flew there with the well tried prototype G-EBKT to give joy-rides, returning to Stag Lane next morning. The Cirrus in the Lancastrians' blue Moth continued to be reported intractable, though club members muttered that its ill was 'Perrin-itis' delay so that the London Aeroplane Club could be the first to fly when their silver-painted Moth was ready in August. Nor did the Lancashire club get their aeroplane until after the opening of the London rival at Stag Lane by Sir Philip Sassoon with much attendant publicity on 19 August. There was gnashing of teeth in Lancashire when he was reported to have congratulated the Royal Aero Club committee 'on having won for London and rightly for London the first Light Aeroplane Club in the country.' In the first week, Capt F. G. M. Sparks, chief instructor at London, and his assistant G. T. Witcombe, completed 45 hours' flying after a ballot for the first twenty pupils – David Kittel, the winner who drew first place, gallantly giving place to Mrs Sophie Elliott-Lynn, a well known and forceful sportswoman. Less than six months had passed since the prototype Moth first flew. Soon the other Moths were delivered – painted red for Newcastle, green for Midland, red and orange for Yorkshire.

The operating cost was £2 15s per flying hour, and by offsetting the subsidy, clubs need only charge 30s per hour dual and £1 for solo flying. For £20 or less a member could pay club fee, tuition cost, a mandatory three hours solo, and the small fee for his 'A' private pilot's licence. On all this flying Capt Lamplugh kept a hawk-like eye, virtually controlling civil aviation through his monopoly of insurance and assessment of the safety or danger of every pilot.

9

Over the August Bank Holiday week-end, Lympne was the scene of air races, exhibition flying, parachute descents, and fascinating demonstrations of the smoky art of sky writing by Leslie Tait-Cox, devised by Jack Savage of Handley Page Ltd. C. G. Grey was there, as with every aeronautical occasion, greeted by all. 'That so many private owners, all there are in England in fact, and so many Trade firms should take trouble to send or fly their machines to the extreme tip of English soil for such a meeting and such miniature prizes shows the fine sporting spirit which exists among those who aviate,' he wrote, adding the barbed thought: 'But that the Royal Aero Club should ask a number of hard working people to go to so much trouble and expense shows how little the Club understands the problem of promoting the sport of flying, for which purpose it ostensibly exists.'

There was a collection of aircraft and pilots to gladden every enthusiast. Most of the ultra lights came by road, but Bert Hinkler flew up from Southampton with the Avro Avis, the wings visibly flexing in the bumps. Flt Lieut Chick of the RAE was there with the tiny Hurricane racer. A. N. 'Archie' Kingwill, the Beardmore test pilot, had the Wee Bee; James the clipped wing A.N.E.C.; Bulman was there with the Cygnet; Uwins with the single-seat Brownie was another, and explained that the bad visibility *en route* was 'when you try to see as far as you can see, you see that you can't see very far.' Colonel Sempill flew the stoutly streamlined Dutch-built Pander monoplane. Flt Lieut Soden of parachute fame was there with the second Austin Whippet, G-EAPF: the first was my special joy at the Northampton. Sholto Douglas had a D.H.53; de Haga Haig flew the Parnall Pixie III with a lady passenger, and Frank Courtney the single-seat Pixie II with clipped wings. Lankester Parker had the Short Satellite, known as 'The Iron Balloon', and Comper brought a new miniature parasol monoplane, the C.L.A.3, built by the Cranwell Light Aeroplane Club boys. Hurricane and C.L.A.3 easily proved victors in the Saturday races, drawing the inevitable comment from C. G. Grey: 'There is a certain amount of humour in the fact that two out of the three races were won by machines built by private syndicates of officers and men in HM Service. Of course the Trade, busy as it is on manufacturing aeroplanes commercially

Compensation for the Cranwell Club's C.L.A.2 biplane wrecked at Martlesham, plus prize money from the 1924 Lympne competitions, paid for construction of the sporting little C.L.A.3 parasol designed by Comper. (*Flight*)

for the RAF cannot afford the time to play with mechanical toys, so there are no new Trade machines among the light aeroplanes this year.'

Sunday morning was cold and windy, but the afternoon turned sunny with a stiff breeze. 'Being the British Sabbath, racing would have been thoroughly immoral,' wrote one reporter. 'So private matches were arranged, and various machines went for performance tests duly certified by the RAeC.' Broad and Cobham on Moths started simultaneously on two circuits, which Cobham won by 100 yards. Sholto Douglas on the D.H.53 matched Haig on the Pixie III on a single circuit, and Haig won by a foot. Chick on the Hurricane then gave Haig 45 seconds, and won by a couple of yards when he dived under the Pixie's tail. Broad on the Moth next flew against Chick, giving the Hurricane a similar start and won by 50 yards, though the Hurricane seemed to hurtle past with the clatter of a machine-

A great little Australian pilot – the indomitable Bert Hinkler about to take-off with the Avro monoplane. (*Flight*)

gun compared with the smoothly travelling Moth. Finally Broad raced the Condor-powered D.H.54, a giant among minnows, against Cobham on a Moth, to whom he gave 1 min 49 sec start for two circuits, and won by a few yards, handicapped by the big machine sweeping 200 yards wide at every corner compared with the Moth's tiny turning circle. Measured speed runs and climbs followed. It was slyly reported that 'Mr Hinkler was busy all afternoon picking up passengers, nearly all feminine, and drew a certain amount of chaff by disappearing behind clouds with them.'

Dr Whitehead Reid was a tall and distinguished ex-RAF surgeon who during the war was taught to fly by pilot friends. At Canterbury, after the war, he adapted this £5 S.E.5 airframe to take a 90 hp R.A.F. engine. (*Flight*)

Monday was Gala day, but it proved a day of forced-landings amid the excitement of racing. To the wicked delight of some, Cobham, with St Barbe as passenger, landed in a field of standing corn after his engine had suddenly stopped at 50 ft, and the stems wrapped round the axle, dragging the machine onto its back. Sempill landed another Moth due to a choked fuel line. Haga Haig with the Pixie, Boyes on Lord Edward Grosvenor's D.H.53, Bulman on the Cygnet, James on the A.N.E.C., and Sholto Douglas on the other D.H.53, all landed with engine trouble of one kind or another. At least there were no engine failures in the private owners' Handicap, won by Chick on the Hurricane, with Soden's Whippet second, and Whitehead Reid third on his ancient 90 hp S.E.5. Except for the C.L.A.3, the others were equally obsolete – a Martinsyde F.6, Sopwith Gnu and Scooter, and two B.E.2es which scratched. After the race Courtney gave another of his startling displays with the Siskin, ending in a vertical dive which took him beneath the ridge on which Lympne aerodrome stood, giving the delighted crowd all the drama of an incipient crash. C. G. Grey struck the final nostalgic note: 'Pretty well everything about Mr Courtney's flying has already been praised, but these big high-speed fighters, although eminently fitted for their particular jobs, are really not as pretty to watch at this particular game as were some of the older machines like the Koolhoven B.A.T. and the Sopwith Camel.'

But now, once again airships were in the news with tragic import. On 3 September the US Naval airship *Shenandoah*, which had left her base at

America, like Britain, was influenced by the relative success of German Zeppelins, and attempted to copy them with the *Shenandoah*.

Lakehurst, New Jersey, the previous afternoon for a cruise to Minneapolis, encountered a violent line-squall over Cambridge, Ohio, which lifted her some 2,500 ft, then suddenly dropped her, followed by a second squall during which the hull broke in three, the nose section drifting ten miles before touching down, the others dropping vertically. Fourteen were killed, including the commanding officer, but twenty-eight miraculously escaped although two were injured.

The helium-filled *Shenandoah*, 680-ft long and 2,115,000 cu ft capacity, the first American rigid to be constructed, was of somewhat antiquated design based on a captured German airship, L.33, which was built in 1916. Capt Heinen, formerly of the Zeppelin company, but now consultant engineer to the Americans, stated that eight of the eighteen safety valves had been removed against his advice. 'It is probable that the ship was driven up by currents of air to an unintended height, and some of the gasbags from which the valves had been removed then split. They would not burst with an explosion; the fabric would tear open, and the gas pour out. This would mean withdrawal of the interior pressure which helps support the framework; and would not even need the additional embarrassment caused by storm eddies and manoeuvring to bring about collapse of the framework locally. This would occur with sudden loss of buoyancy and rapid descent, the ship becoming unmanageable and out of trim.'

His criticism was backed by an outburst from 36-year-old Colonel William Mitchell, the former Assistant Chief of the Army Air Service and advocate of a separate Air Force, who had been deposed because of his outspoken comments on the vulnerability of warships after he had demonstrated sinking them from the air by bombing. Reported *The Times*: 'Infuriated by the declaration of Mr Wilbur, Secretary of the Navy, that the disaster of the *Shenandoah* and accident to the US PN-9 flying-boat proves that the Atlantic and Pacific are the country's best bulwarks against air invasion, Colonel Mitchell replied that they were the direct result of incompetence and criminal negligence and almost treasonable administration by the War and Navy Departments who have conducted aviation so disgustingly in the past few years as to make any self-respecting person ashamed of his uniform, and he added condemnation of non-flying bureaucratic superiors who dictate policy while preventing flying personnel from telling the truth.' It was a viewpoint shared by many a flying man in Britain.

Of affairs in Britain where the official Airship Works at Cardington was competing with the Airship Guarantee Co, Nevil Shute in *Slide Rule* wrote: 'However satisfactory the competitive experiment may have been to the Cabinet committee, it cannot be said that it brought peace to the competing staffs. Each had its peculiar viewpoint, and the two were irreconcilable. The Air Ministry staff at Cardington considered they were engaged upon a great experiment of national importance, too great to be entrusted to commercial interests, and it was impossible to suppose that any private company could compete with them, backed as they were by the finance and research resources of the Government. The staff of the private company took a different view. In 1916 the principle had been laid down for aeroplanes that all construction was to be in the hands of private enterprise – a decision imposed by bitter experience. In the realm of airships this principle had never been observed, and the bitter experience was not yet at an end. The disaster to the Government-designed R.38 was still fresh in memory, yet these were the very same men, said the private staff bitterly, who were to be entrusted with construction of another airship when by rights they ought to be in gaol for manslaughter.'

All that the Vickers-financed Airship Guarantee Co were given was a great derelict shed in the centre of a desolate heath three miles from the Yorkshire town of Howden. Rain leaked through the rusty roof, the floor was littered with rubbish, and the offices and workshops along each side were uninhabitable; so the first task was to make order from the chaos, and build houses for the staff. Meanwhile Barnes Wallis and his team were designing the ship in the London office of Vickers.

By contrast a small business, the Southern Light Aeroplane Club, was trying to get going in the hands of Eddie Wallace, an ex-seaplane test pilot; Cecil Pashley, well known as a pre-war pilot and amateur constructor; and big Fred Miles, a keen youngster. They had taken over the old RAF aerodrome at Shoreham, and in one of the sheds, alongside their joy-riding Le Rhône Avro, were building a single-seat 6 hp light aeroplane called the Gnat which they hoped to market at £220. At least they had the good wishes of the Air Ministry though their club was refused financial assistance on the score that only six were authorized.

During August the neat, small-span Gloster III biplane (the company name Gloucestershire having been changed, as too unwieldy) and the startlingly novel and beautiful Supermarine S.4 mid-wing monoplane had been tested respectively by Hubert Broad at Felixstowe, and Henri Biard at

Devised at minimum cost from the 'Bamel' design, the Gloster III resulted from a contract for 'an aircraft for research and development at high speed', and was loaned to the builders for the Schneider contest.

Southampton, in preparation for the Schneider contest at Baltimore. The high compression (*sic*), direct-drive Napier Lion of each performed well, aided by the Fairey-Reed propeller of twisted duralumin, and special lead-ethyl fuel in lieu of the standard 70 octane benzol mixture. On 13 September Biard established a seaplane world record by flying at 226 mph. With enormous emphasis on secrecy, the Press were invited to inspect the machine at Southampton four days later, though as C. G. Grey said: 'All the longshoremen at Woolston had been able to watch every time it came out, and everybody down Southampton Water has seen it whenever flown. All the enterprising air attachés must have had their representatives with cameras in the vicinity of Woolston and photographs of the S.4 must already be in the hands of our esteemed competitors.' The cantilever undercarriage particularly interested rival designers, and the duralumin floats, like those of the Gloster, proved clean running and of low drag, and Mitchell's device of offsetting the starboard float to counteract engine torque on take-off, as well as using both for fuel tanks, was vindicated.

An important Air Ministry official dismissed the design leadership of the clean Supermarine S.4 with the pronouncement: 'The only point on which British designers can fairly claim superiority is in float design.'

Certainly the S.4 seemed faster than the pale blue Gloster, but when Broad's machine was compared with its stable-mate the 'Bamel', the advance in Folland's design was clear, and it handled well. Both a second Gloster III and the 'Bamel' on floats were going to the USA as stand-by, with Hinkler as reserve pilot, but not until a fortnight before despatch had he the chance of taking the 'Bamel' for his first experience of a really fast machine, yet lapped the course with vertical turns and characteristic skill. So marginal was the available time that he had no opportunity of even flying the Gloster III, and neither Biard nor Broad had more than two or three flights on the racers, though the Americans had been constantly flying fast machines for the last twelve months.

While at Felixstowe inspecting the pretty Gloster, the Press were diverted by the new Rohrbach cantilever monoplane flying-boat, powered

The Air Ministry showed initiative in breaking away from the RAF theme of biplane flying-boats by ordering a twin-engined Rohrbach, but it was soon evident that British hulls were more sea-kindly.

by twin Lions in nacelles strut-mounted above the wing, which circuited and alighted. 'The machine is very practical, but certainly not beautiful,' recorded one reporter. 'Still it seems to handle well both in the air and on the water. One gathered that the pilot was Herr Landmann, the famous German who in 1914 flew for 21 hours on an Albatros.' Thus was dismissed an all-metal machine, which was far in advance of anything produced in Britain but at least could be regarded as Scottish by adoption as William Beardmores had the British constructional licence.

On 26 September the British team of company directors, officials, pilots and mechanics, left Southampton on the *Minnewaska* with Capt Charles B. Wilson as non-flying leader. But the contest at Baltimore was again disaster for the British. As a preliminary on 22 October, Biard took-off with the Supermarine S.4 for navigability and seaworthiness trials under ideal conditions with little wind and calm sea. As described by one of the Schneider Committee: 'He decided to practise some sharp turns and completed one satisfactorily and then attempted another. It was noticed his aileron was hard down on the lower side. The machine seemed to get out of control and did a falling leaf descent, making a huge hole in the sea. Eye-witnesses seemed certain there was no structural failure of the wing, which in fact examination seemed to verify.' Though *The Times* correspondent at the contest said that the wings of the S.4 fluttered and put the machine out of control, and there was suspicion that the cantilever wing was not torsionally stiff enough, onlookers believed that Biard had stalled through unfamiliarity with the 'g' effect on a high-speed turn. It was a fantastic escape, for he was picked up by launch merely slightly dazed.

Although Broad's Gloster III was ready for the contest, the second had not even been assembled. There was a rush to prepare it in place of the Supermarine so that Hinkler could compete. He took-off next day at 5 p.m., but a wing bracing wire broke, forcing abandonment. Gales followed. Not until the morning of the contest on the 26th was Hinkler able to take out the Gloster in an attempt to qualify. He took-off and circled a couple of times, then alighted for the half-mile taxi-ing test, but the waves were two or three feet high; one of the float struts cracked, the machine tilted, and, before he could switch off, the slowly revolving propeller punctured the floats, but the machine was taken in tow just in time to save her.

Broad, as the only remaining British competitor was flagged by the starter that afternoon three minutes after Lieut Doolittle of the US Army Air Service had taken off with his Curtiss R3C-2, and was followed by the other two Curtiss racers flown by two US Naval officers, and finally Signor G. de Briganti lapped with the Italian Macchi 33. Doolittle was an easy winner, flashing round the course with full-out vertical turns, achieving an average for the seven laps of 232 mph.

Winner of the 1925 Schneider, at Baltimore, was Lieut 'Jimmy' Doolittle with his compact, clean Curtiss R3C-2 Army racing floatplane. (*Flight*)

'By hook or by crook, we've got to win next year and bring the Trophy back to Europe, and win again in 1927 to prevent it becoming permanently domiciled in the USA,' urged *The Aeroplane*'s representative. 'If the Air Ministry, engine makers, and aircraft constructors get down to it and start work we may do something next year. The Gloster people began their 1925 challenge as soon as their 1924 machine crashed, and it is known they are already at work on a monoplane with wing radiators for 1926 provided they get reasonable support from the Air Ministry.'

But there did not seem much hope, for Sir Samuel Hoare said: 'We are all sorry we had such bad luck in the race, but we shall not give up trying, and I intend at once to take up the question of next year's entry with the British constructors whom I hope will see the way to enter on their own resources. From all points of view it is better that private enterprise should take the field in these events. If, however, there is no other way of securing a British entry for next year's race, I shall be prepared to consider again the loan of Air Ministry machines under the same conditions as this year.' It was the old game of cat and mouse.

'One fact stands out from all the talking and writing about the contest,' wrote C. G. Grey. 'It is that now is the time for the British aircraft industry to make its great effort to capture the world's aviation as in the past we captured the world's sea traffic in the shipbuilding trade. At no time in the history of aviation has there been so much public interest in flying as today. In every country the newspapers continually publish long and more or less informed articles on aircraft, either as private vehicles or commercial transport or engines of war. The mass mind of humanity is moving towards aviation and now is our opportunity to make use of that psychological momentum.'

10

Despite continuing and costly work on the Brennan helicopter it was shaping towards another British fiasco. *The Aeroplane* reported: 'On Friday October 3, the secretest of our secrets was wheeled, pushed, or otherwise transported from the old airship shed at Farnborough, wherein it has incubated from time immemorial, out into a very secret enclosure within the precincts of the RAE. After a preliminary ground test, the machine, which was piloted by Mr R. Graham, succeeded in rising a foot or two off the ground. It then lost stability, toppled over, and crashed sideways on the ground. The time seems ripe for forcing an enquiry into this waste of public money of something like £100,000 expended on this precious helicopter. Apparently Mr Lloyd George was personally responsible for the contract under which Mr Brennan has been enabled to draw such immense sums for an experimental machine in which he, Mr Brennan, is probably the only believer.'

However Bob Graham tells that after inevitable teething trouble and development work, outdoor trials had begun as far back as 16 May, 1924. 'This was the first opportunity of trying out the translational controls and, to ensure there was no accidental take-off, the car was loaded with 750 lb of lead shot. It was found there was insufficient moment of inertia in the plane at right angle to the rotor blades, the use of either feathering or

Definitely airborne: Bob Graham becomes No.1 helicopter pilot with the Brennan at Jersey Brow, Farnborough, lifting from the point where Cody started several early flights.

ailerons causing the blades to lose incidence, so it was decided to fit two additional smaller blades at right angles to the existing ones, and their incidence could be altered collectively with the others. This modification took some time, and not until 2 May, 1925, was the aircraft on the airfield again and several ascents made, but with poor measure of lateral control. This was steadily improved, height being limited to 10 ft for safety, though most flights were at 3 to 6 ft. All were attempts at steady hovering, there being less than 100 yards available, but when hovering the aircraft was unstable, and if it drifted any distance had to land. A gyroscopic control was then designed which it was considered would solve the instability, but the Air Ministry insisted that the aircraft must first be proved on manual control.

'According to RAE records, some 200 flights at an average of three minutes each were achieved, and a payload of 600 lb with one hour's fuel was demonstrated. On 2 October, 1925, when giving a demonstration to Government officials, a fault occurred in the control system, and when landing the rotary structure tilted to such an angle that the propellers hit the ground. The aircraft was badly damaged, but could have been repaired in relatively short time and at reasonable cost. However the Air Ministry decided that the whole question of helicopter development should be referred to the Aeronautical Research Committee – who advised that no further work be carried out on helicopters of the Brennan type, but that if the Air Ministry considered rotary development should proceed, then it should be based upon the autogiro. Brennan had neither funds nor facilities to continue by himself so the project had to be abandoned.'

Following cancellation of his contract, Brennan secured one for a gyroscopically-controlled large tandem two-wheel military truck which he developed in similar secrecy in a shed at Jersey Brow on Farnborough aerodrome. Meanwhile, encouraged by the simpler mechanical arrangements of the autogiro, the Air Ministry announced that the long-established helicopter competition was now closed, for although there had been 34 competitors, only one of them sent a machine to Farnborough but did not carry out any of the tests. The future of the helicopter seemed doomed.

The Cierva Autogiro was more immediately promising. Early in September Cierva's latest, consisting of an Avro 504K fuselage, with four-bladed freely revolving rotor on a pylon above the forward seat and ailerons on a spar below, had been received at Farnborough through the friendly co-operation of the Spanish Government who had lent it to the inventor so that he could demonstrate to the British Air Ministry.

In a letter to Reggie Brie, later well known for his Autogiro piloting, Frank Courtney recorded: 'I first saw the Autogiro in Madrid, where I had delivered an aeroplane. A Spanish officer showed it to me in the hangar as a great joke perpetrated by a nut named Cierva. It seemed, however, that Cierva was already in contact with Major H. E. Wimperis, Director of Research at the Air Ministry, and had been offered terms for a series of demonstration flights. Wimperis, long a friend of mine, suggested I was the chap for that job, so Cierva got in touch. I was much impressed with his line of thought, so it was arranged I would do the flying.'

The machine was the latest of six Cierva had built since 1921. The first three had rigidly attached blades, but with the fourth he hinged them to the hub with freedom of movement up and down, centrifugal force keeping them extended when rotating. The flexible attachment prevented transmission of gyroscopic forces to the axis, and the machine no longer tended to fall over sideways. Cierva stated: 'Lateral control was provided by tilting the axis of the windmill to right or left, but it became immediately evident that the pilot's strength was insufficient to work the control. After many months of exploration, crashes, and rebuilding 15 times, I fixed the windmill axis rigidly and built into the machine two small non-lifting ailerons at the end of streamlined spars. On 17 January, 1923, piloted by Lieut Gomez Spencer, of the Flying Corps of the Spanish Army, the machine flew right across the aerodrome at Getafe at a height of several hundred metres, and after transporting to Quatro Vientos near Madrid it flew four minutes on 31 January, 1923, at more than 25 metres in a closed circuit, officially observed and controlled. That was the first real flight ever carried out by a machine, heavier than air, differing from the conventional aeroplane.'

Frank Courtney taxies out at Farnborough with the Spanish-built Cierva C.6A Autogiro on 15 October to become the second Briton to fly a rotary-winged aircraft. (*Flight*)

The next machine was a three-blader of improved design, but crashed while taxi-ing. The Spanish Army Aeronautical authorities then undertook responsibility for these experiments, and sanctioned construction of Autogiro No. 6 in their workshops. On 9 December, 1924, their Capt Loriga made his first flight, rising to 200 metres and landing almost vertically; three days later he flew 12 kilometres between the aerodromes of Quatro Vientos and Getafe, but serious illness prevented him demonstrating in England.

Pilot and inventor – Capt F. T. Courtney, despite spectacles, made more prototype initial flights than any of his contemporaries. With him is the Autogiro inventor, Don Juan de la Cierva. (*Flight*)

Recollecting his impressions of piloting the Autogiro, Courtney wrote: 'In my first flights the chief problem was starting the rotor. There was a comic process of spinning it with a bunch of men tugging a rope wound round its spindle, and the machine would then take-off in about five yards on opening the throttle. For the first few landings I kept some forward way, landing with a brief run. After a few landings I learned to store up energy in the rotor by a sharp pull up a few feet from the ground, which enabled a landing with no run at all. My first impression was the strangeness of going through the air with no wings; it was a weird thrill with sensations that I have never forgotten. As to the Air Ministry tests, they were in four parts. The first three were various manoeuvres and landings which gave no trouble. The last was a vertical drop of which everyone, including me but not Cierva, was very doubtful. I managed this but broke the machine simply because the rate of descent was much higher than Cierva predicted, but everyone was satisfied, and he received his money and an order for two machines from the Ministry.'

Wimperis was convinced the Autogiro was one of the most important aeronautical inventions of recent years. It was solely due to him that the machine was brought to Farnborough, for almost everyone had considered the idea of an unpowered rotor just another crazy notion. 'When I first saw the Autogiro in the air I realized what courage must have gone to making the first flight with machines of this character,' said Wimperis. 'I certainly felt that Capt Courtney showed uncommon courage. We had to have someone to take the place of the Spanish pilot who fell ill, and Courtney did it with complete success. When I saw this machine in flight with no wings, looking rather like a rotating St George's Cross, and felt that I did not know what was holding it up, it reminded me of the saying of that other famous Spaniard, Sancho Panza, that behind the Cross stands the Devil.'

There was a word of warning from Professor Bairstow: 'It seems to me one of the rare occasions when nature has presented gifts to the inventor. A screw is produced which autorotates and is free from high stresses. It proves very stable and the flapping essential, nor does the stress problem adversely affect aerodynamic behaviour, but as usual some gifts are withheld, for the possibility of hovering in still air or of rising vertically from the ground does not come within the device's ability.' One could almost hear Mr Handley Page say 'Hear, hear,' lest these rotating wings harmed financial returns from slots.

As usual C. G. Grey had the last word: 'Meantime one does not recommend members of the SBAC to shut down all work in progress on ordinary aeroplanes. Considering that the Air Ministry has just placed an order for 100 of the dear old D.H.9As of 1917 design with Liberty engines from the inexhaustible store of the Aircraft Disposal Co (now A.D.C. Ltd), one may safely assume that even if the Autogiro fulfils and surpasses its inventor's expectations, we shall still be ordering Vernons and Southamptons and Fawns and Grebes in 1933, and machines on the present Secret List will still be going strong in 1940. Therefore, let us wish all success to Senor de la Cierva, if only because his Autogiro has completely put the lid on the sempiternal Brennan helicopter. And let us proceed to build and advertise British aeroplanes rather more than usual. They will certainly be useful in the next war, even if we use something on the Autogiro principle in the war after that.'

In contrast to rotary flight, Farnborough aerodrome on 28 October was the scene of the first power trials by Geoffrey Hill of his tailless light plane, the Pterodactyl. To the wing, tested as a glider the previous winter, he had fitted a polygon-section nacelle encased with true Hill ingenuity in special

Hill's Pterodactyl is weighed in the RAE Airship Shed to check centre of gravity before flying. The simple undercarriage was braced with wires fore and aft. Vickers Venture J7277 is behind the wing.

plywood made of two veneers of birch and a core of the almost unknown balsa wood, resulting in a weight half that of normal ply. At the aft bulkhead a steel-tube engine mounting weighing only $3\frac{1}{2}$ lb carried a Cherub engine cowled with light aluminium. Within the nose a steel-tube hoop was fitted to save the pilot's legs in a nose-over.

More shock absorber was added to the simple cross-axle undercarriage after initial taxi-ing, for the 5-ft wheel-base gave an unpleasant ride on rough ground. Five days later conditions were good enough for a brief flight in a 15-mph wind. Sensitive though the machine was fore and aft, it proved stable on all three axes, though laterally out of trim. Later flights confirmed fundamentals he had sought, for there was no definite stall and good control was retained even at 45 degree incidence, the attitude to the horizon remaining constant, but there was an extraordinary inertia effect as though a heavy weight was on top of the stick.

Airborne uncowled on its first flight at Farnborough, the Pterodactyl flies low down to show all is well.

Said Hill: 'What of the future? The first step is to prove the ideas demonstrated on the light aeroplane will work equally well on larger scale and in biplane form. What I think encouraging is that though normal design is moving but slowly through the bog of convention, the ideas on which I have worked form an avenue of escape as there will be additional advantages in tailless design, for it can be built for a smaller percentage structure weight than normal, and reopens possibilities of the pusher without loss of performance.'

Non-flying technicians of the RAE, brilliant men all, were preoccupied with this idea that stalling must either be prevented or controls remain fully effective when stalled. One of the greatest exponents was McKinnon Wood who was responsible for Farnborough's development of interlinked aileron and Handley Page slot. In the past year 339 aircraft had been written off by accidents, most of which were due to stalling and spinning in, resulting in a fatality almost every week. Nobody queried whether the considerable sum spent on stall research would have been better used for more rigorous flying training and development of blind-flying instruments instead of dependence on the visual horizon which changed with mist and storm and darkness. To make the stall safe was all that seemed obvious.

Said McKinnon Wood: 'In the slot and aileron control we have got rid of the objectionable and weak features of ordinary ailerons, and provided ailerons which retain full effectiveness when the aeroplane is stalled – but one has done nothing to stop the aeroplane passing into a spin when stalled, whereas in Capt Hill's aeroplane one has wing-tip surfaces at a fine angle to the wind, and his aeroplane is so proportioned, according to the wind-tunnel, that it has no tendency to autorotate. I believe that provision of really effective control constitutes nine-tenths of the battle. Certainly Hill's aeroplane opens new vistas in design which might enable us to secure other advantages, but it is his feature of safety in slow speed flight which particularly appeals to me.'

The RAE device, whereby the downward moving aileron opened its adjacent leading-edge slot but kept it closed for upward movement, had been publicly demonstrated at Croydon with an RAE Avro 504K by George Bulman before joining Hawker Engineering, and was designed to give positive lift to the dropped wing, compared with Handley Page's method of Patent 223,292 in which a spoiler was lifted behind a fixed slot on the upward wing to disturb the flow and reduce lift to help pull the wing down. Of the two, the RAE method felt far safer.

When Bulman flew at Croydon, Anthony Fokker brought his latest F.VII to demonstrate the effectiveness of its much publicized blunt-nosed 'stable cantilever wing.' Following the official show by Bulman of steep turns at such slow speed that an ordinary Avro would have spun in,

Grouped at the Avro 504 display of interconnected aileron and slot are (*left to right*) Alliott Verdon Roe, J. Brennard (*News*), Frederick Handley Page, Air Commodore Samson, Harry Harper (*Daily Mail*), Sir Sefton Brancker, C. C. Turner (*Observer*), C. G. Grey.

Fokker took-off with nine passengers, lifting his machine at an angle matching Bulman's, then at 1,500 ft throttled back, pulled the nose to stalling point and let the machine sink steadily, apparently under full control, though a wind-tunnel model of it made by the RAE showed that an aileron deflected at 20 degrees had only half the power of a Bristol Fighter's at the same angle. But for both Dutchman and Englishman the demonstration became almost laughable when a French pilot took-off with his lightly-loaded Farman Goliath in as short a run as the Avro, and far shorter than the Fokker, then hovered much as Bulman had, and at remarkably slow speed brought her round on a steep turn with full rudder and reversed aileron movement to drag it round. Though truly Gallic, it was strictly unconventional and showed what a good demonstration pilot could always do in foxing spectators into believing an aeroplane could safely perform the tricks he made it do.

That pet of RAF fighter pilots, the Gloster Grebe, began to reveal dangerous wing flutter, so was modified with matched ailerons and V-struts bracing the overhang to obviate it.

There could be no trickery with a test dive made by Flt Lieut D'Arcy Greig at Martlesham with a Gloster Grebe experimentally fitted with modified ailerons and top wing overhang lift struts to overcome flutter experienced with production machines. Now that the first Irvin parachutes were available, they had been issued to Martlesham and Farnborough test pilots and, taking advantage of this new safety precaution, it was decided to test the Grebe in a terminal dive at which the drag equalled thrust plus gravity. Nominally its diving speed was regarded as 50 per cent above the top speed of 152 mph, but D'Arcy Greig attained 240 mph and levelled out with no sign of flutter at the maximum rate he could apply. No machine had been deliberately dived like this before, but henceforth it would be standard practice.

In different way, demonstration at Northolt of a Fairey Fox in October to Sir Hugh Trenchard and Sir Geoffrey Salmond was equally important, for it turned the conventional outlook of the RAF on two-seat bombers to new conception of speed and aerodynamic cleanness, for the Fox was faster than the latest fighters. Secretly Fairey was building a single-seater, the Firefly, to beat his Fox. It hardly needed the businesslike charm of Fairey to impress the rock-like figure of Trenchard with the merit of what he was witnessing.

'I remember the occasion as if it were yesterday,' recorded Norman Macmillan. 'After the flight the CAS asked me to accompany him apart from the others, and we walked onto the grassy airfield well out of earshot. There he asked what I really thought of the Fox; did I think it was an aeroplane that could be handled safely by young and less experienced pilots of the RAF? I told him frankly it was one of the easiest and most viceless aeroplanes I had ever flown. We walked back to the hard stand, and Sir Hugh looked at Dick Fairey and said in his booming bass: "Mr Fairey, I have decided to order a squadron of Foxes." '

Commented *The Aeroplane*: 'Certainly it is a pity that in order to acquire a squadron of Foxes the Air Ministry has to order some £60,000 worth of American engines for it. When the Fox was designed no suitable British engines were available. Either this squadron of Foxes has to be produced and put into the air within a few months with Curtiss engines or it must be postponed another year or so until redesigned to take a direct-drive Napier or Rolls-Royce after the engine firm concerned has got them through their type test.' Here was echo of Trenchard's decision to abandon the Condor, and Fell's initiative in getting Rolls-Royce to design a powerful twelve-cylinder inspired by the Curtiss D.12. But no further Foxes were ordered. That was not due only to the American engine. In his log book Wg Cdr Harry Busteed dismissed the Fox with the comment 'Controls heavy.'

11

Autumn, and the trees becoming bare. So too were some of the aircraft factories. I had seen the relative emptiness of Westland with its small batches of D.H.9As for reconditioning, but now there was real concern about the future because the Yeovil bomber had been turned down in favour of the modified Hawker Horsley. Even so it would take many months to get the Horsley into production; meanwhile the company was dependent on building lower wings of the Grebe under sub-contract. Avros similarly were short of work at Manchester and building Grebe top wings. 'The main shop where wings, fuselages, and so forth are built looks as if it could hold about a thousand hands but was almost uninhabited,' wrote a visitor. 'The reason is that the firm has struck one of those awkward periods when contracts are almost finished and new contracts not yet placed. Aldershots and Bisons are approaching completion; all wood and metal parts have been finished and passed to the erecting shop, leaving the metal and woodworking shops empty. That is the worst of the aircraft trade: unless contracts are placed so that a new one overlaps the end of a previous contract, there must be periods when many departments are idle.'

The Blackburn shops were more fortunate with Darts produced at two a month, including three dual-control civil seaplane versions for their subsidiary RAF Reserve Flying School, and were beginning twelve Blackburn Mk II reconnaissance biplanes with top wing raised above the fuselage instead of flush with the top decking like the eighteen Mk Is built the previous year. Meanwhile Major J. D. Rennie was supervising design of a great three-engined biplane flying-boat of 95-ft span and 20,000-lb tare weight, compared with 75-ft span and 14,300 lb of the already popular Supermarine Southampton, and it had a wood hull which was a compromise between the latter's Linton Hope construction and the Felixstowe-designed boats with which Rennie was particularly familiar.

At Brough, three Blackburn Darts were switched from the RAF production line to civil guise and fitted with elegant floats of mahogany Consuta ply carrying retractable beaching wheels.

Oswald Short was furious. 'Major Rennie was a naval architect, employed by the Technical Department of the Air Ministry in consultant capacity,' he told me. 'Holding that position he would be fully aware of how earlier Short flying-boats were built – in fact all the "know-how". This information had to be supplied by designers when tenders were submitted. In his official capacity, he visited our works and I showed him the drawings of the Short Singapore I. Shortly afterwards he left the Air Ministry and joined the Blackburn Aircraft Co as hull designer, and this led to the three-engined Iris.' But Rennie, a delightful character, could afford to smile.

Disaster to Short Bros had only been averted by building light-weight aluminium omnibus bodies for the new fully enclosed double-deckers adopted by the London General Omnibus Company and Thomas Tilling

Ltd. 'To keep the Works at Rochester a going concern we required a turnover of at least £400,000 per annum,' said Oswald. 'During this period I had to deposit with the Westminster Bank Ltd all my personal stocks and shares as guarantee. Meanwhile the flying-boat built by the English Electric Co with all-metal hull and parallel fluted bottom was completed late in 1925, and had closely followed my first all-metal hull of 1924. It seemed a strange coincidence that an unknown (!) designer reproduced an exactly similar hull to mine, embodying both my patented systems of construction. Had the E.E.C. originated it they would certainly have claimed so, but there is no evidence to show they did.' This duralumin hull with stainless steel bottom had been designed by Manning conforming with shapes developed at the NPL by Baker, and had a high, stem-like prow and emphatic rear step, with deeper body than the pre-production wooden hulled boats, but utilized the same wood wings and tail surfaces.

In a very different specialized field, A.D.C. Aircraft Ltd had been outstandingly successful in disposing of millions of pounds of stock, repaying the Government, exclusive of taxation, some £$1\frac{1}{4}$ million as its share. The company's premises at Croydon were larger than that of any aircraft manufacturer's. Not only reconditioned aircraft and engines but machine-guns, Scarff rings, and bomb-sights could be bought – a telling point to countries unable to afford expensive armaments. Allied with their widespread sales was a tremendous spares service. With the advent of the Moth and its need for Cirrus engines, the manufacturing future of the company also seemed assured. But even de Havillands, who bought these engines, were glad to boost private venture Moth production with a Government contract for Grebe ailerons and miscellaneous parts.

Nor was Handley Page, the originator of A.D.C., happy – for the long shed at Cricklewood had a busy air only because the three-engined W.9 Hampstead, and the twin-engined W.10s being erected, occupied so much space. The Hampstead had just completed official trials. At full load, with

Painted in Imperial Airways original livery of deep blue fuselage and silver-doped wings and tail, the Jaguar-powered three-engined Handley Page W.9 Hampstead was not quite the success it was hoped.

one Jaguar stopped the climb at 2,000 ft was 200 ft/min, and with two out it descended at an angle of 1 in 20. Surprisingly, this was regarded as rather good.

'The Handley Page Hampstead is suffering from trouble with the rudder,' reported Geoffrey Dorman from Croydon. 'This was originally observed by Imperial Airways pilots on first testing it, and confirmed by Farnborough, so it is not in service yet. Ever since the first W.8 of 1919 there has been this trouble. The tail appears to bounce constantly between the slipstream of the wing engines, and unless the pilot works at the rudder the whole time it is uncomfortable for the passengers. The W.8bs behave much as the original W.8 did, and it looks as though the W.10s will have the same trick.' But would they? These were virtually civil versions of the Hyderabad powered with Napier Lions, and would have the triangular fin and rectangular rudder of the model I had wrongly drawn in the Handley Page wind-tunnel – and Reynolds had said it cured the faults.

A month after the Hampstead appeared, H.P. sought out MacRostie, and told him Imperial Airways had accepted the tender for their new airliner – the W.10. 'But they stipulate we deliver by the end of March, and if we can they will order four. Can you do it?' MacRostie agreed, provided the D.O. merely kept track of what he did instead of issuing design drawings. That was November; by the turn of the year the machines were taking shape in the shops, relentlessly urged by H.P.

With equal zeal he attended the first House Dinner of the Royal Aero Club on 26 November, where Fairey as speechmaker scathingly compared what he described as 'the ostrich-like British aeronautical outlook' with America where 'they concentrate research like a searchlight on one problem, and solve it'.

Said Handley Page, who still regarded Fairey as a bit of a newcomer: 'When our Associates – I understand they were not legally Allies – from America came over and won the war, we gave them all our aeronautical experience, for which the post-war American Commission allocated a mere £100,000 in all to British aeronautical designers. Since then America has made wonderful progress, but we could similarly have won trophies and records had we set our minds on it. What we needed was not to reduce the frontal area of our machines, as Fairey has said, but the obtuse resistance of officials in this country. We need to move along new lines of approach to get recognition. And may I add – many a flower is born to blush unseen at Cricklewood!'

Somehow H.P. failed to comprehend that he had the real key to reducing the danger of inadvertent stalls, though he casually referred to the possibility when Professor B. Melville Jones lectured on that subject a few weeks later, for he opined: 'There are two aspects which we have to study: one is the question of lateral control, and the other is preventing the stall taking place. I suggest one could have a completely slotted machine with an emergency handle to unlock the slots when in a stall. The resultant forces on the forward aerofoil are such that it should open automatically when you approach a very large angle of incidence.' The importance of this comment was overlooked, so intent were the ARC and RAE on developing interconnected slots and ailerons in combination with a slotted and flapped central portion of the wing, though Melville Jones made it clear from his own experience as a pilot that a powerful rudder was also vital, and warned that: 'In general it may be said that a low fin and rudder, which does not project behind the body, or is directly over a continuous elevator, will be defective when the aeroplane is stalled, and that defect will be accentuated if the body is short and thick, broad in plan near the tail, and has its upper line much curved downwards.'

Men in scores, hundreds, thousands, would continue to die from stalling, whatever the palliative, however great the tragedy, but with the death from more natural cause of the Queen-Mother Alexandra on 20 November an affectionately regarded link with Edwardian and Victorian days was broken. Yet with the signing of the Treaty of Locarno on 1 December a fresh link was forged with the future, whereby the Franco-German frontier was guaranteed by Britain and Italy. It was a big step towards re-establishing balance of power in Europe. Russia remained the great outcast, against whom it was deemed desirable to bring Germany, now experiencing economic recovery as the result of the Dawes Plan, into closer relation with her Western neighbours. Hunger, poverty, despair had been the lot of the Germans: theirs had been a terrible penance for war. Yet there was no real confidence that they would regard the Treaty of Locarno as more than another scrap of paper if pressures mounted – but for the time being it seemed to promise a bastion against the Communism of Russia under the rule of the uncomprising, murderous Stalin who was exterminating the richer peasants known as kulaks.

The Treaty afforded no reason for slowing down expansion of the RAF, but initially extensive ground reorganization with more bases must be established. Thus the Air Ministry bought the great house of Bentley Priory and its park at Stanmore with intention of making it HQ of the Inland Area when Uxbridge became HQ of the Fighting Area, Air Defence of Great Britain. Almost immediately there followed an announcement in the Press: 'After protracted negotiations, an agreement has been entered into between Mr Claude Grahame-White and the Government for acquisition of Hendon aerodrome, which will now be an important centre of defence.' Undisclosed was the long unhappy story in which the Treasury had managed to postpone for four and a half years the final hearing of a writ the great, but discredited, pioneer airman had served. It was only concluded when Lord Northcliffe offered the front page of the *Daily Mail* in which to shout the scandal of his treatment. But Grahame-White's solicitor, Merkel, felt this was undignified and managed to inform the Treasury of this intention. There was immediate capitulation. Within seven days a deposit was paid, with balance to follow in three months, and all claims on the Grahame-White Co were annulled.

The end of the year found Cobham in the Sudan, halfway on his flight to the Cape, accompanied by his faithful, cheerful engineer, Arthur Elliott, with B. W. G. Emmott of Gaumont as ciné photographer. Theirs was a survey flight to report on the possibility of airline operations to South Africa. They had left Croydon on 16 November, flying the second prototype D.H.50, G-EBFO, re-engined with a 385 hp Armstrong Siddeley Jaguar, and now owned by Imperial Airways. Progress was followed with intense interest by Fleet Street, for Cobham was ever regarded as 'a good story', one reporter describing his departure as 'Pen in one hand and control stick in the other, bound for Cape Town.'

While he flew into the sunlight of yet greater publicity, the year finished with vast floods all over Europe.

CHAPTER VII
TOWARDS UNDISCOVERED ENDS
1926

'According to your Ma^{ts} command we have examined the particulars of the draught and the discussion presented to your Ma^{ts} and by comparing the rules of Art and Experience together we have agreed to the proportions under written, which we most humbly submit to your Ma^{ts} further pleasure.'

Phineas Pett (1635)

1

On the evening of the Royal Aeronautical Society's Diamond Jubilee on 12 January, a Conversazione was held in the aero section of the Science Museum. Into that cavern of antiquities, welcomed by Sir Sefton Brancker as chairman of the RAeS Council, 'came almost all the living pioneers of aviation, such as the aged gentlemen who had to do with balloons and queer flapping wing machines long before the Wrights persuaded their first machine to keep off the ground for a few yards, and still longer before A. V. Roe and the late S. F. Cody hopped from the earth in England. There were also people who have made modern flying what it is: the constructors and designers of our modern highly efficient – so far as the Air Ministry experts permit – aeroplanes and engines. And there were also those whose interest in aviation is rather in the direction of limelight oratory than practical science.' So wrote C. G. Grey after viewing the white ties and starched shirts, the sweeping dresses, jewels and cigars, circulating among the glass-covered exhibits. Though in financial straits, the oldest aeronautical society in the world was proud of the occasion. But there was hope. Laurence Pritchard, the newly appointed secretary, had persuaded the Air Ministry to change its mind and grant £250 a year for five years, and he now was extracting donations from the SBAC, individual manufacturers, oil companies, and people of note.

Voices at RAeS lectures were mainly those of the ARC, RAE, NPL, Airship Establishment, and Air Ministry researchers and officials. Only the indefatigable Fedden, pushing the advantage of his particular design of radial engine, represented any attempt by manufacturers to participate in

The Beardmore XXVI for Latvia was a unique break-away from the conventional biplane structure, but the Air Ministry turned a blind eye to a development order.

the Society's programme. It was significant also that the aircraft industry was no longer represented on the ARC nor the twenty sub-committees dealing with accidents, airships, aerodynamics, engines, light alloys, and not even on the design panel which was composed entirely of scientists headed by Professor R. V. Southwell, FRS.

Aircraft designers preferred a closed shop. Armstrong Whitworth, Boulton & Paul, Bristol, Handley Page, Vickers, and Westland had their own wind-tunnels, and others used NPL or RAE facilities for aerodynamic or hydrodynamic problems, except Shorts with their pendulum and water-tank. All had been mildly venturesome, and during the last few years each had tendered at least one monoplane instead of a biplane as the answer to one or other of the many Air Ministry specifications, but unfortunate post-war experience only endorsed the pre-war attitude which led Government officials to select biplanes as a matter of caution. True, Beardmores had a

The rugged Hawker Hornbill fighter prototype awaits test alongside the ultra-light Cygnet at Brooklands.

contract for an all-metal giant monoplane, the Inflexible; but if it proved another disaster it could be safely excused as the work of the German designer Adolph Rohrbach, who at least had demonstrated that his cantilever flying-boats flew. Bill Shackleton at Dalmuir had over-all responsibility for the details, and was certainly displaying commendable initiative. Early in February his unique Beardmore W.B.XXVI biplane, powered with 360 hp Rolls-Royce Eagle IX, was flown by Archie Kingwill, who demonstrated not only its virtuosity in aerobatics but also that at the stall the controls remained operative and there was no tendency to spin.

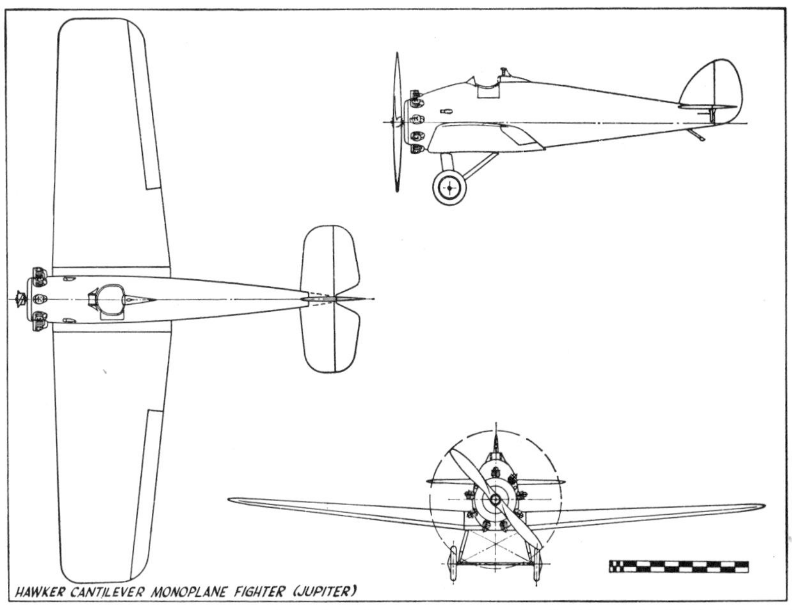

If the all too cautious Air Ministry had found courage to back this early 1925 design of Camm's, Britain might have gained a vital armament lead.

At Hawker Engineering, Sigrist had almost completed the Carter-initiated V-strutted Hornbill which weighed 3,000 lb empty; but even before it flew, DOR regarded it as too heavy for current ideas of a light and manoeuvrable fighter, and in any case Trenchard's edict that Condor production should be matched only to the projected life of the Horsley gave it little chance. So Camm was busy with a more practical and advanced fighter conception than even the H.P.21, for it took the form of a low-wing cantilever monoplane powered with a Jupiter, and mounting twin guns enclosed in the rounded fuselage with breeches accessible alongside the pilot. It could have put Great Britain years ahead of any other fighting nation. Instead, it was known that the Air Ministry proposed to issue a new fighter specification for a biplane with Jupiter engine.

Nevertheless, Bristol had been successful with a tender to specification 4/24 for a twin Jupiter VI powered, all-metal monoplane of 70-ft span, with shoulder-wing cantilevered from the wide centre-section on which the engine nacelles were mounted. Structurally it was the result of Pollard's development of high-tensile steel techniques, and featured a triangular-section fuselage faired with stringers and formers into rectangular shape forward, with wings having two steel spars, as in the Brownie, and duralumin ribs. The wide-track undercarriage set the machine abnormally close to the ground at relatively shallow angle, and the axle was faired with a wide-chord tapering aerofoil of sufficient lift to sustain its own airborne weight. Although details of armament were not initially disclosed, the intention was to carry a 37-mm shell-firing Coventry Ordnance Works gun, known from the intials as the 'C.O.W. gun', in the nose and another in the aft cockpit. They were 6 ft long and fired 1½-lb shells at the relatively slow rate of 100 per minute, making it necessary to allow for a recoil of some 2,000 lb.

First conception of an aeroplane is always subject to an initial evolutionary process of aerodynamic changes in the wind-tunnel. Thus the Westland 'Twin-engined fighter' with diverging gap eventually became the Westbury.

Since the Bristol Bagshot, as it was presently named, was aerodynamically a gamble, a contract was placed with Westland for a competitive twin-engined biplane to the same specification, eventuating in a high aspect ratio 68-ft span, three-bay design which was the result of wind-tunnel development tests beginning with a cleaner, but lower aspect ratio, two-bay design. It was remarkably similar to North's machines, though of more massive appearance and with greater gap. The undercarriage legs were in fact subject to B & P royalties, though designed by Bruce based on 'buffers' he had devised pre-war for Brennan's monorail. Two prototypes were ordered in June 1925 – the first with central and fore fuselage structured in steel tube, and wood for rear fuselage, tail, and wings; but the second had duralumin box spars instead of wood. But although the pundits forecast that a monoplane equivalent of a biplane would weigh very much more, the Bagshot eventually was only 350 lb heavier than the 7,875 lb of the Westland Westbury.

Another attempt at an up-dated replacement for the much loved Bristol Fighter was the Boarhound, but, after briefing by Uwins, Barnwell briefly flew it and decided the Atlas was better, though felt his Boarhound had possibilities as a D.H.9A successor. (*Courtesy Air Marshal Sir Ralph Sorley*)

Barnwell was busy with other projects. As a Bristol private venture, he had produced several designs early in 1926 for a two-seat fighter based on specification 8/24, to compete with the D.H.42B Dingo II, Vickers Vixen, and Armstrong Whitworth Atlas. Ultimately, as the Boarhound, it had a deep fuselage framed with longerons and struts made from pairs of semicircular section flanged high-tensile steel strips riveted lengthwise and incorporating gusset plate attachments. In the current mode, the top wing was of greater span and chord than the lower, using a similar but more complex fabricated steel spar to those Pollard had developed for John North at Boulton & Paul, but with lipped and wider flanges so that external fittings could be added after the long box had been riveted up. All such steel spars, with varying complication or simplicity, were the result of similar basic rolling and riveting methods, and usually infringed the many patents held by Major H. N. Wylie or D. J. Mooney, or both. Thus the steel spar developed by Hawkers, comprising seven strips forming a dumbbell section, necessitated royalty payments to Wylie, Boulton & Paul, and Armstrong Whitworth.

With all such spars, success depended on the rolling-mill tool designer and his skill in allowing for spring-back and stretch of the tough material. Where there were bends of small radius it was important to form them in the first pair of rolls; the groove thus made acted as a guide through subsequent rolls, during which progressive forming on every part of the strip width was necessary. Even so it required trial and error adjustment of dies and rolls to overcome the formed strip buckling into sinuous shape as it emerged. Sometimes the sample spar might prove $\frac{1}{16}$-in under or over the requisite dimension, and rather than make fresh tools to produce exactly the originally calculated size, it was often preferable to amend the drawings to conform with the specimen! In every aircraft firm the art of rolling metal corrugations and bends became a private empire of absorbing interest to its operator, but firm management was required to prevent costly experimentation, for contract prices must cover discovery of techniques as

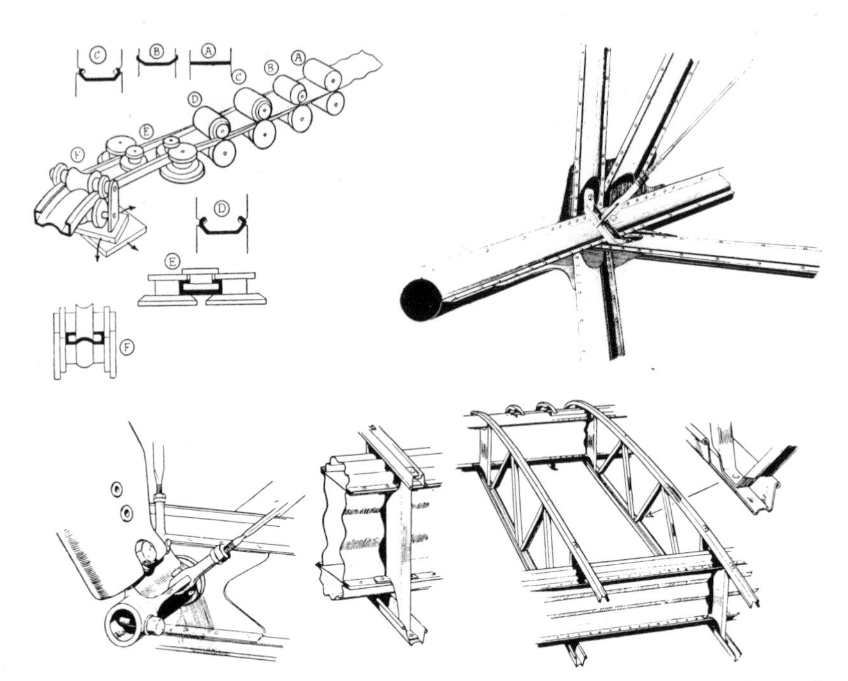

First requisite of steel strip construction was skilled rolling-mill technique as illustrated diagramatically with lettered sequence of shaping A–F (*left*). Although a steel assembly was at first believed by the RAF to be too complex for Service maintenance skills, the Boarhound was repaired in a week after crashing at Odiham, despite fuselage joints (such as *top right*) and light gauge wing spars with eccentric strut attachment (*bottom*).

well as building the machine. Even in the case of so novel a design as the Bagshot, only £14,750 was payable by the Air Ministry as the all-in price.

This monoplane was not the only one of Barnwell's problems. In *Bristol Aircraft since 1910*, C. H. Barnes stresses that: 'In his struggle to establish the Bristol Jupiter as a more worthy successor to the Napier Lion than the Armstrong Siddeley Jaguar, Roy Fedden placed high value on the prestige gained by record breaking and racing. He had pressed Barnwell to design a Jupiter-engined racer in March 1924 but the project was rejected by the directors. Fedden returned to the attack in November 1924 with a small-diameter engine proposal, comparable with the Jaguar, and asked for a racing single-seater to be produced for the 1925 and 1926 seasons at an estimated cost of £10,000. Again the directors rejected the project, but at last in September 1925 Barnwell was allowed to design and build a fast biplane, the Badminton, with a landing speed not greater than 60 mph, and an estimated speed of 180 mph with a 510 hp supercharged Orion engine.'

Unmistakably it had the Barnwell trademark, with elliptical wing-tips and almost finless balanced rudder. Of conventional spruce framework, it was only 24 ft in span, with upper wings attached to a streamlined pylon structure above the fuselage centre-line somewhat similarly to the original 'Bamel'. By using a wide-track undercarriage and the wing loading kept down to 11 lb/sq ft, it was intended to be safe enough for emergency landing in any reasonable field when used for a race such as the King's Cup in July.

Although Barnwell was often antagonistic to using a radial for military designs – preferring the neater low-drag shape of water-cooled engines such as the Lion, or even the Fairey Felix Curtiss D.12, unpolitic though it was to buy from a rival airframe manufacturer – Fedden had a radial protagonist in Colonel W. A. Bristow who offered to design a racing seaplane using a more powerful, short-life version of the Bristol engine. Of this project Carter wrote: 'After leaving Hawkers I would have preferred a short holiday, but Roy Fedden proposed I should prepare this design for the Schneider contest and he would build a special engine. I was nominally working for Colonel Bristow, who had independent means, and the project was backed by Oswald Short. I schemed a low-wing semi-cantilever monoplane with wings braced by V-struts from mid-span to the float undercarriage, and a model was tested in the 14-ft NPL wind-tunnel. We submitted the design to the Air Ministry who eventually issued a contract to build it.' But before that could be undertaken, much research and development was necessary both on the racer's design and its engine, for NPL results were disappointing and drastic revision was required to reduce resistance. Streamline wires replaced the struts, and the projecting cylinder heads were enclosed in helmets, entailing further experiments in cooling.

2

Napiers, in the competing camp of water-cooled engines, were enjoying further triumphs. While Bristols were proceeding with endurance flights with a sealed Jupiter in their unwanted Bloodhound, a big Dornier Wal flying-boat, *Plus Ultra*, with two Napier Lions in tandem above the centre of its parasol wing, completed a flight across the Atlantic on 3 February, piloted by little Commandante Ramon Franco and Capt Ruiz de Alda, flying from Palos de Moguer, near Gibraltar, to the Canaries, Cape Verde Islands, Pernambuco, and Buenos Aires, covering 6,259 miles in 60 flying hours. Ramon was the madcap younger Air Force brother of 33-year-old Francisco Franco who – with thought of ultimate dictatorship quite foreign to his mind – had been promoted Brigadier-General on the same day newspapers acclaimed the great flight which made Ramon a hero of Spain.

By using twin Lions in his latest 14-passenger airliner, Handley Page was sharing the lustre surrounding Napier engines. MacRostie had kept his production date: in the first week of February, only two and a half months after starting the W.10 conversion, the machine was tested by Hubert Broad – the greatly increased power and Hyderabad fin with inset hinged rectangular rudder making it a very different proposition from the early W.8. Nevertheless, there was a growl from C. G. Grey: 'The engines still stand unhoused and uncowled; control cables and pulleys and a species of

Handley Page's W.10 twin Napier airline version of his Hyderabad bomber was typically British with wires and struts, but it was sound business to develop a proved design rather than risk revolutionary failure.

sheet-metal lever, commonly called a bloater, stick out all over the machine in the best war-time fashion. Even the cabin chairs are of the old straight-up, short-back type of wicker which provides about the least comfortable seating imaginable.' Of its engines the *Daily Sketch* fulsomely stated: 'To-day the Napier aero-engine is the best, and the leading aviation motor of the world. According to the latest available statistics, Napier engines come out on top for mileage flown with minimum troubles. The Napier Lion is indeed well named, as it truly impersonates this heraldic symbol of Great Britain.' Meanwhile its major rival continued to advertize in dignified manner 'Rolls-Royce aero engines – the first and best in the world', for had they not been the first to make every great flight – the Atlantic nonstop, and from England to India, Australia, South America; to cross the South Atlantic, fly round Australia, and lead the way from Holland to the East Indies, and Brussels to the Belgian Congo?

With the arrival at Cape Town on 17 February of Cobham with his D.H.50, Armstrong Siddeley also gained a great share of world advertisement. Indefatigably, Cobham had sent reports to newspapers from most of his stopping points, so all Britain had been aware of his brilliantly organized progress. Though the journey of over 8,000 miles in ninety-four flying hours had occupied four months there had been no intention of setting a record, but instead this was a careful survey of twenty-five landing places *en route* which might be used by Imperial Airways, and films had been taken from the cabin by Bill Emmott to entrance the public in Cobham's unremitting preaching of the Gospel of Aviation.

What might be done for aviation, whether civil or military, was entirely dictated by the Air Estimates for 1926–27. The *Daily Telegraph* commented: 'The estimates of £16 million for the current financial year are now under consideration, and the outcome is of exceptional concern to the public because of the need, yet admitted difficulty, of balancing reduction of expenditure with maintenance of national security. Our impression is that we are to mark time for two if not three years in the air programme that, in 1923, was adopted as the minimum for defence of the country. In that year Mr Baldwin set forth the principle on which air policy was to be based – "A Home Defence of adequate strength to protect against air attack by the strongest air force within striking distance of this country." Subsequently Sir Samuel Hoare announced that a force of 52 squadrons would be created with as little delay as possible and organized to make further expansion possible. These 52 squadrons were the accepted minimum, and it was generally understood that they would be completed in about five years. In 1924 the Labour Government declared adhesion to the policy, and in March 1925, with the return of the Conservatives, the principle was reiterated. Successive governments of opposing politics have thus endorsed and confirmed both policy and principle. Are there now adequate reasons for throwing it overboard?'

The consequences of cuts were indicated at the annual meeting of Rolls-Royce. 'The result of any important curtailment of the Air Ministry programme for new aeroplanes and engines must necessarily involve the discharge of a considerable number of workmen who are experienced experts and could not be replaced. This would place the British Empire in a position of danger. Obviously engine designers could not spend time and money in bringing out new aero engines unless there were sufficient orders to justify expenditure. The aircraft industry, after almost collapsing after the war, has been built up again – without the lavish subsidies given by foreign governments – because it was assured of a continuous and clearly defined air policy. If this sure support is removed, the danger is not only prolongation of our defencelessness in the air, but the hopelessness of attempting expansion in emergency if the foundation is destroyed.'

In fact the reduction proved only £454,810, bringing the total Estimate back to the approximate level of 1924–25. This ensured continuation of production contracts and experimental work at almost the current rate, but certainly confirmed that the promised major expansion of the Air Force, and therefore of the industry, was being brought at least to a temporary slow-down.

The Air Ministry policy of placing competitive orders with two or more firms for aircraft of identical specification had become firmly established. Thus the D.H.56 Hyena, flown by Capt Broad in February, was a re-design of the all-metal D.H.42B Dingo III to overcome deficiencies of the pilot's view, so his cockpit was now in line with the trailing edge instead of under it, and, to rebalance the weight distribution, the top wing was moved rearward by raking the rear interplane struts back and swinging the front ones vertical, and the nose was considerably extended to bring the weight of the 385 hp Jaguar III forward. The result was untidy, and, as performance was worse, a 422 hp Jaguar IV was fitted for assessment trials at Martlesham against the Armstrong Whitworth Atlas and Bristol Boarhound, which had been there since the previous summer, and the private-venture Vickers Vespa which had first flown in September 1925 and was delivered to Martlesham during February in civil guise. To obtain the requisite low landing speed, all except the Atlas approached 45-ft span, but

Developed from the Dingo with intention of giving better pilot view, the untidy looking D.H.56 Hyena seemed an anachronism compared with its successful competitor the Armstrong Whitworth Atlas. (*Courtesy Air Marshal Sir Ralph Sorley*)

There was a similarity of thought between the war-time Parnall Zeppelin fighter and the two-bay, lightly loaded Vickers Vespa (*illustrated*), for both had been draughted by that pioneer-designer Camden Pratt. (*Courtesy Air Marshal Sir Ralph Sorley*)

Of typical Blackburn geometry, the single-bay 34 ft 9 in span Sprat was the more conventional of the three trainers competing to specification 5/24, but the requirement was ultimately abandoned. (*Courtesy Air Marshal Sir Ralph Sorley*)

Second of the contenders was the distinctive Parnall Perch, but the side-by-side arrangement was criticized for poor view diagonally across the cockpit although otherwise superb, and the symmetrical-section interchangeable wings had too sharp a stall.

the Armstrong machine relied on smaller, wide chord, stable C.P., high-lift, low-drag wings to give the same slow landing but better manoeuvrability. For this, Lloyd had selected an R.A.F.28 aerofoil derived by the new centre-line curvature technique, and developed from the symmetrical, 10 per cent thick, R.A.F.27 with stationary C.P. which had been recommended by the Farnborough investigators A. S. Hartshorn, BSC, and H. Davies, BA, as aerodynamically better than R.A.F.31, and suitable for biplane construction – but it required care in landing to prevent a wing dropping.

Similar competition with specification 5/24 for a small two-seat trainer having interchangeable wheel and float undercarriage resulted in the Blackburn Sprat, Parnall Perch, and Vickers Type 120 Vendace. The

Following the Vespa to Martlesham came the single-bay Vickers Vendace after naval trials the previous year when it had been recommended as a deck-landing trainer. (*Courtesy Air Marshal Sir Ralph Sorley*)

Sprat looked like a scaled-down version of the as yet uncompleted Blackburn Ripon torpedo biplane, but the Perch was neater and lighter, and though similar in having a swept-down top cowling, seated the pilot higher, his eyes level with the centre-section. Altogether different was the Vendace, with slim fuselage and straight top cowling, conventionally high centre-section, and wings of 10-ft greater span than the bare 35 ft of the other two, but it was some 200 lb heavier.

In the same manner the Blackburn Ripon single-bay torpedo biplane to specification 21/23 (as a Dart replacement with Lion VA) was matched by the Type 571 Buffalo designed by A. V. Roe, and Handley Page's Harrow which had full-span mechanically operated slots. They were almost ready for testing.

Wings of the Blackburn Airedale tilted and folded back by winch and cable, on a diagonal from the rear spar centre-line to forward strut trunnion, until leading edge down and parallel with the fuselage. (*Courtesy Air Marshal Sir Ralph Sorley*)

Bumpus's previous design, the Airedale three-seat high-wing monoplane to specification 37/22, was aerodynamically wrong, because of high induced drag in its ingenious attempt with low aspect ratio wings to break from convention, and the prototype had crashed beyond rebuilding on taking-off for delivery to Martlesham Heath at the end of 1925. A second version with stronger undercarriage and auxiliary fin under the tailplane to improve directional stability was now far too late, and did not reach Martlesham until June, though intended as competitor to the private-venture Hawker Hedgehog which flew in February 1924 and was delivered for official testing in September of that year. Meanwhile orders had been placed for the Blackburn Mk II biplane with wings raised above the fuselage, and for the similar rearrangement of the Avro Bison Mk II.

In contrast, construction of the Blackburn Iris was proceeding well. Rivalling her, to specification R.14/24, was the Saunders A.3 Valkyrie, similarly powered with three 650 hp Rolls-Royce Condor IIIAs. Its beautifully proportioned monocoque hull was skinned with Consuta, and though the 97-ft wing span was only 1½ ft more than the Iris, it appeared a bigger machine because of the smaller gap; but it was still a long way from completion because of priority for the twin-Jupiter powered Medina marine airliner built as rival to the Supermarine Swan. Thompson chal-

When it came to assessing conventional against unconventional design the latter usually had snags, so the Saunders A.4 Medina became the last of Thompson's aeronautical work.

lengingly strut-braced its biplane wings, the lower of which had greater span and chord, and only the top had dihedral. Unlike the Swan the engines were cowled and directly attached under the top wing. Although smaller and weighing 340 lb less than the 11,900-lb Supermarine, the wing loading was 1 lb/sq ft greater, so take-off would be worse, but the estimated speed was 115 mph compared with 92 mph achieved by its rival when flown a year earlier.

S. E. Saunders had also been building the floats for the Parnall Peto, which in early spring made its first flight. Built to one of the most difficult requirements issued by the Air Ministry, specification 16/24, it was a compact two-seat twin-float biplane which folded into the restricted space of an 8-ft wide hangar on the deck of submarine M.2. Not surprisingly there were no competitors for a gamble likely to give only small returns. The staggered unequal R.A.F.31 wood-structured wings had Warren interplane bracing of steel tubes, and the fuselage was a spruce-framed Warren girder. The pilot's cockpit was the aft of two located behind the wings, but the Armstrong Siddeley Mongoose engine was started with hand-turning gear in the front cockpit. A novel introduction was stainless steel for metal

Initially with a 128 hp Lucifer, the Parnall Peto, re-engined with a 169 hp Mongoose, fulfilled its difficult role provided the sea was calm, but was eventually lost when the M.2 carrier submarine sank with all hands in Lyme Bay, Dorset. (*Flight*)

fittings to eliminate corrosion. Floats were a splendid example of boat-building art, but Felixstowe found the machine so light that in a choppy sea it tended to bounce into the air prematurely.

Bolas was now wrestling with specification 1/24 for a reconnaissance floatplane/landplane named the Pike, intended to compete with the Short Sturgeon as a three-seat fleet reconnaissance replacement for the Fairey IIID. Again he used a greater-chord top wing and Warren bracing for a 45-ft span biplane, locating the pilot in line with the deeply recessed leading edge of the centre-section, placing the gunner equally high in a rear cockpit conversely recessed into the trailing edge. Though these positions seemed to give the crew an outstanding view, the pilot's was blanketed rearward, and the gunner could not use his gun forward and down – yet in the conventional Perch he had given the pilot an excellent vista.

Ready for its first flight at Yate, Gloucester, the Lion-powered Parnall Pike proved slow on the controls and required very large horn balances, and the cockpit was appallingly draughty.

The snags of pilot location had been experienced by Arthur Gouge at Shorts with his single-bay Chamois derivative, for he had placed him well forward, with the centre-section crammed on his head necessitating a hole to see upward and destroying the wing's efficiency. The Sturgeon was therefore much more conventional and given a $7\frac{1}{2}$-ft gap instead of the $5\frac{1}{4}$ ft of the Springbok, the pilot gaining a steeply oblique upward view beyond the leading edge. The single-bay fabric-covered wings had duralumin box spars and were of equal span but unequal chord, using the currently popular R.A.F.32 aerofoil section. Fuel was in a centre-section tank to give the simplicity of gravity feed. Like its predecessors, the Sturgeon had a duralumin monocoque fuselage – but as yet it existed only in sub-assemblies and components.

The big Singapore hull was being riveted up in No. 1 shop, and a little monocoque two-seat monoplane twin-float seaplane was being finished alongside. The Mussel, as it was named, was a semi-cantilever low-winger, with box spars of corrugated duralumin strip, and wooden ribs of R.A.F.33 section which had slightly reflexed trailing edge, aiding stability by limiting C.P. travel. Intended for full-scale correlation with the water-

Although the Armstrong Whitworth Wolf failed to gain production orders it led to a fascinating research variant, the Ape, shown with tail steeply angled for stalled flight investigation.

tank experiments, its floats were scaled-down reproductions of those developed for the Gloster III racing seaplane. Although designed for a Lucifer, this engine had been criticized for inducing heavy vibration in other machines, and as the RAF already considered that resonance and noise were excessive in Short metal fuselages, Oswald decided to use the Cirrus as it was so successful in the Moth.

Others were also using full-scale aircraft to obtain aerodynamic information. Armstrong Whitworth were building an experimental version of the Wolf, named the Ape, which had alternative wing positions, a variety of tails, and hinges enabling the rear fuselage to be set at varying angles.

Endeavour to obtain full-scale information was exemplified not only by the Ape but the *ad hoc* Bristol Type 92 which flew initially with 3-ft fuselage depth and two years later with 5 ft.

Bristol were flying an even uglier biplane, Type 92, which had unstaggered square-tipped wings at the huge gap of 9 ft to reduce interference, and poised a narrow circular fuselage midway on a centre-line pair of struts. Basically a square-section plywood box 2 ft by 2 ft at the cockpit, the fuselage tapered conventionally to a horizontal knife-edge, and was initially faired to a circle of 3-ft diameter. This would be progressively increased for later tests until its Jupiter was fully enclosed and the fuselage followed a deeply streamlined form – but initial tests were disappointing: control was sluggish, and single ailerons on the lower wing would require augmenting before Fedden could get data on a wide range of cowlings for his radials.

Nor was rotary-wing flight overlooked. Following the successful demonstration of the Spanish-built C.6A Autogiro, two identical machines, using 504 fuselages and undercarriage, were ordered from A. V. Roe Ltd, powered with 130 hp Clerget instead of the 110 hp Le Rhône, and known as the C.6C in single-seat form, and the C.6D as the very first of all two-seat Autogiros.

This development was supplemented with contracts to A. V. Roe and Parnall for light single-seat Autogiros, respectively the C.9 and C.10, with 30-ft rotor and powered by the new 70 hp Genet which Armstrong Siddeley were developing as a rival to the Cirrus. Chadwick sketched for the Avro a slim plywood-covered rectangular-section fuselage, terminating in a typical Avro tail, and fitted with a wide-track oleo undercarriage for greater shock absorption; ailerons would be mounted on a braced lower winglet instead of the outriggers used for the larger C.6C. Within a few months Avros received a further order to fit a rotor to the wingless fuselage of their five-year-old biplane, G-EAPR, which was powered with a 180 hp Wolseley Viper, and though known as Type 552, with many rearrangements of undercarriage and float, was essentially a 504. In Autogiro form its fin would be retained, and a winglet fitted to carry ailerons. Matching it was a second Parnall Autogiro, known as the Gyroplane, ordered by the recently formed Cierva Autogiro Co Ltd as the C.11. The dynamo behind the new company was its chairman, Air Commodore J. G. Weir, brother of Lord Weir of war-time aircraft production fame.

3

There was increasing tempo of hope. Symptomatic were the full-page advertisements of aircraft and aero-engine manufacturers in the aeronautical Press reflecting a sense of financial security. Every achievement was turned into publicity. There were photographs in the newspapers of three Danecock versions of the Woodcock leaving Croydon on delivery flight to Denmark, and even more had been made of the Beardmore W.B.XXVIs built for Latvia. The first of a batch of six Supermarine Lion-powered Seagulls for the Australian Air Force had still greater publicity on acceptance by Sir Joseph Cook, the High Commissioner, whose wife 'christened' it with champagne in true ship fashion, after which Henri Biard took him for a flight around Southampton Water.

Glossy magazines featured celebrities of the flying clubs grouped by their D.H. Moths, or nonchalantly garbed in flying kit. Every club had been equipped with these attractive little machines of which twenty had been built in 1925, and already one hundred had been ordered for 1926. Encouragingly the first of the New Year's batch had been purchased by the Air Ministry, given RAF roundels and numbered J8030.

Alan Cobham and Sir Sefton Brancker after the African flights with their famous D.H.50, G-EBFO, painted in the colours of Imperial Airways. (*Flight*)

De Havillands sailed onto another wave of publicity when Alan Cobham arrived back at Croydon on 13 March, having beaten by two days the ss *Windsor Castle* which he had seen leaving Cape Town when he took off on 26 February. His return was a gala occasion. Roads to the aerodrome were lined with cars; the public enclosure was crammed. Shortly before 3 p.m. a small formation of six Moths and a D.H.9 left Croydon to meet him over Kent; more D.H.9s followed with Press photographers: all failed to find him. Meanwhile Cobham's blue and silver D.H.50J came gliding in and landed. Instantly a seething crowd broke through the barriers to get a close view. Only with difficulty were the police able to rescue pilot and crew. Cobham was carried shoulder-high to Customs, from which presently he emerged, to the delight of the multitude, and Sir Philip Sassoon, Under-Secretary of State for Air, made a brief speech of welcome on behalf of the Air Ministry. 'Subsequently, as recorded on all the news sheets of the world,' wrote C. G. Grey, 'Mr Cobham was commanded to Buckingham Palace and granted an interview with HM the King, who, on learning that Mrs Cobham was waiting in the car outside the Palace, promptly sent for her and congratulated her on her husband's fine performance.

Thereafter Mr Cobham in spite of the fatigues of his journey delivered a little broadcast talk, which was among the few interesting things which the BBC has done within living memory.' Promptly Gordon Selfridge, with his inimitable flare for publicity and the money to pay for it, got hold of the D.H.50 for exhibition at his great Oxford Street emporium.

Within days it was known that Cobham would next attempt a flight to Australia and back using the same well-tested D.H.50. The original proposal was that on reaching Calcutta it would be fitted with Short-built floats similar to those of the Gloster III for the journey to Sydney and back to India, but, in the event, it was used throughout as a seaplane, except from Darwin to Melbourne and back.

The RAF were not to be outdone. Four Fairey IIID landplanes, led by Wg Cdr C. W. H. Pulford, left Cairo on 1 March for Cape Town, and made steady but slow progress, reaching Kisumu in Kenya on the 13th,

The Fairey IIID with various Marks of Lion engine was the solid, dependable mainstay of the RAF's maritime duties and easy to fly, although lateral control was poor when the ailerons-cum-flaps were lowered. (*Flight*)

and Pretoria on 2 April. Three days later they continued to Johannesburg accompanied by fourteen South African Air Force machines and a fifth RAF Fairey IIID which had arrived by sea. On the 12th they reached Cape Town. Felicitous messages were exchanged between Sir Samuel Hoare and the South African Government. An equally cautious return trip to Cairo was now being planned, and from there it was proposed to fly them as seaplanes to England.

That was another good boost for Napiers, but Bristols, with a series of shuttle flights between Croydon and Filton, had secured even greater publicity for the reliability of their sealed Jupiter in the Bristol Bloodhound – piloted in turn by F. L. Barnard, on leave from Imperial Airways, and Lieut-Col F. F. Minchin, a free-lance pilot established at Brooklands – achieving 226 hours, with 25,074 miles covered without any replacement at an average fuel consumption of 21.9 gallons per hour and oil at 3.95 pints per hour. Afterwards the engine was transferred without adjustment to a Heenan & Froude test brake and found to deliver 440 bhp at full throttle compared with 452 bhp prior to the flight. On dismantling, the only replacements were one exhaust valve and one inlet valve spring.

While the RAF flight was jogging along as a well-organized operational squadron exercise under extremely difficult conditions, Schneider hopes for a British entrant were dashed by the announcement: 'At a meeting on 19 March, at which representatives of the Air Ministry, Royal Aero Club, Society of British Aircraft Constructors and others interested in the Schneider Cup were present, it was unanimously decided that it was inexpedient for the Royal Aero Club to make a challenge this year.' Perhaps the fact that the RAeC had not set eyes on the Trophy for some years explains their reference to it as a cup, but the danger that the way was now open for America to score a third successive win and gain permanent hold of the Trophy was due to procrastination on the part of the Air Ministry who in the course of four months had not been able to decide whether to financially support construction of two new racers and engines of boosted power. That there was some activity in connection with Schneider machines was disclosed when a keen journalist discovered that the Gloster 'Bamel', with Courtney as pilot, had turned over at Felixstowe while being towed. Although still a free-lance, Courtney had just been appointed technical manager of the new Cierva Autogiro Co Ltd.

Surprisingly it was not he, as the customary Armstrong Whitworth pilot, but F. L. Barnard, on leave from Imperial Airways, who made the initial flight at Coventry of the first Argosy three-engined airliner in mid-March. After being airborne a few minutes, one of the Jaguars failed. Nevertheless Barnard seemed enthusiastic – but he was not a test pilot. Yet despite the angularity of parallel-sided cabin, abruptly tapered rear fuselage, and conventional multiplicity of interplane struts, the Argosy certainly looked efficient though could not compare for cleanness with the latest Fokker tri-motor monoplane, nor was there much intrinsic aerodynamic improvement in relation to the three-engined Handley Page

Comparable with the Handley Page Hampstead in size and power, the Armstrong Whitworth Argosy carried twenty passengers at 95 mph and had a range of 520 miles. (*Courtesy Air Marshal Sir Ralph Sorley*)

Hamilton and Hampstead except that the undercarriage had been simplified to a single wheel each side instead of a double undercarriage, but retrogressively a biplane tail was used instead of the monoplane surfaces of the H.P. The engines were uncowled and the central engine looked like a small rosette on the bulky nose, though those between the wings had light streamlined tail fairings of fabric on formers and stringers.

One of the younger aircraft technicians, Frank Radcliffe, BSc, of Glosters, at that time said: 'I recently flew in both the Hampstead and the Argosy and can say neither offers the comfort of an old Ford car. As for noise, even the cotton wool did little good. The side bracing wires come through the woodwork in the Argosy with a good big clearance for the

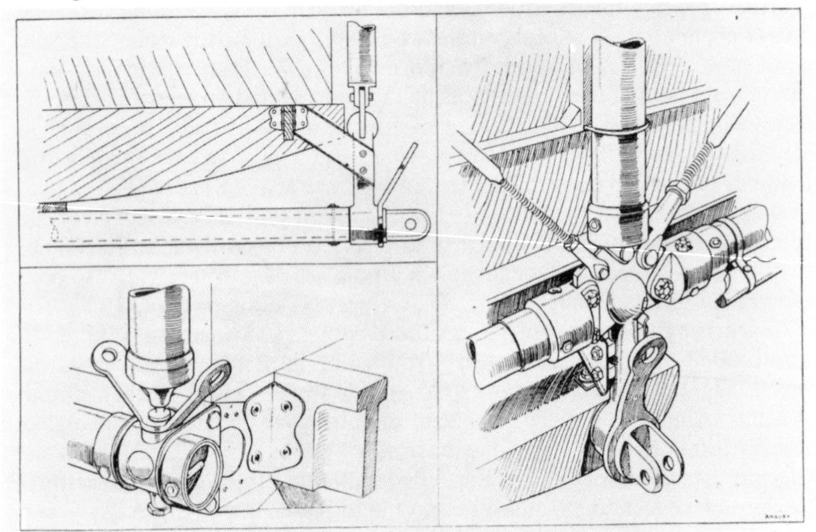
The Argosy led the way with British airliners in the change from wood to metal, for the fuselage was built with steel tube, but the cabin was a ply box on wood beams mounted within this structure. (*Flight*)

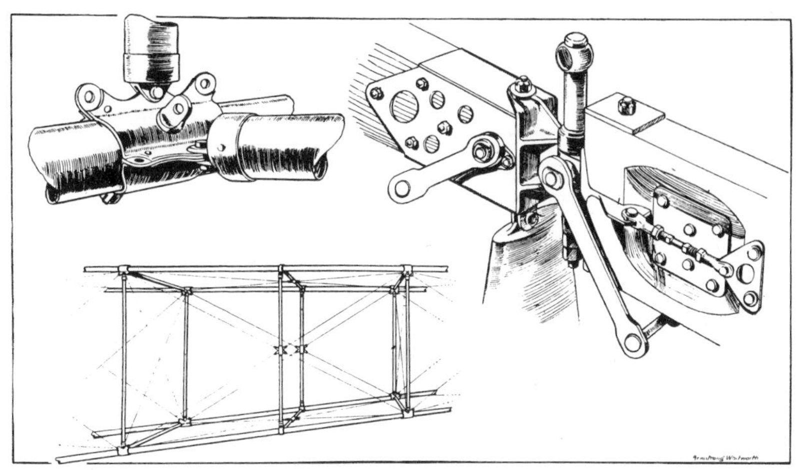

Construction of the Argosy was similar to the much bigger Armstrong Whitworth Awana troop carrier, of which wooden wing and tube-structured fuselage details are illustrated. (*Flight*)

draughts to come through. In my opinion the engines should not be attached to the body, but placed on the wings: the noise is far too great with them as they are, and it would appear that a better position would be nearer the trailing edges because the noise would be behind the passengers. Further, the nose engine makes you aware of its presence by the vibration it transmits to the passenger cabin, especially if there is irregularity in cylinder firing. "Ripping panels" on top of the cabin ought to be waterproof, so that when there is rain they do not become "dripping panels" to the discomfort of the passengers. Then there is the design of chairs: the Hampstead's cause cramped knees, and the Argosy a stiff neck as one's head cannot rest against the chair back. With the present wide wings of a biplane all that some of the passengers see is an expanse of fabric, which is a dull sight. A cantilever monoplane wing seems an absolute necessity; if

Built to the same requirement as the Vickers Victoria, but with a span of 105 ft 7 in compared with 86 ft, though the area of 2,300 sq ft was little greater, the Awana first flew on 28 June, 1923, but latterly was relegated to experimental flying as production cost would be greater than its rival which had Virginia wings. (*Courtesy Air Marshal Sir Ralph Sorley*)

we place it above the cabin the passengers have a clear view and flying becomes a little more interesting, but when the novelty goes, flying in a commercial airliner is dull.'

With various difficulties to iron out, Armstrong Whitworth were slower than usual with publicity. As a result Handley Page stole the thunder. By the end of the month he had managed to deliver the four Napier-powered W.10s, and at Croydon on 31 March stage-managed a picturesque ceremony for their christening by Lady Maud Hoare. At the splendid lunch which followed at the aerodrome hotel, H.P. was able to prompt Sir Eric Geddes into saying that Imperial Airways, starting with a nondescript fleet, had now 75 per cent of one type – the good old Handley Page! It was no small triumph that a machine basically designed in 1919 was regarded as the latest airliner.

An interested guest was Lester D. Gardner, proprietor of America's *Aviation.* The previous evening he had given a resumé of American aviation progress. There were over 1,700 civilian owners of aeroplanes compared with a score in England, though there were no US passenger airlines nor government control over aviation: it was a free for all. However, the US Post Office operated an Air Mail between San Francisco and New York, regularly covering 3,000 miles in a period of 30 hours of day and night flying. Was it significant, people wondered, that Sir Sefton Brancker on returning from a visit to the States publicly said he was in favour of removing restriction wherever possible from the design, construction and use of civil aircraft?

No intricate fittings for the Fokker F.VII: straight-forward but skilfully welded joints for the fuselage metal-tube structure and engine mounts; clean cantilever wing of wood box spars, ply ribs and skin – yet here was the best aerodynamically clean airliner, cheaper than others, and pleasant to fly. (*Courtesy Air Marshal Sir Ralph Sorley*)

There was certainly evidence of broad outlook on the part of Major Wimperis, who induced the Air Ministry to purchase one of the new three-engined Fokkers, and on 30 April it was flown to the A & AEE at Martlesham Heath direct from Amsterdam in $3\frac{1}{2}$ hours by V. Grasé, chief pilot of the Fokker works, accompanied by his managing director B. Stephan. To everyone's astonishment, 'Mr Grasé, after sundry side-slips and Immelmann turns, ended with a complete loop, and on landing subtly emphasized the demonstration by stepping from the machine immaculately clad in light summer clothes instead of helmets, goggles and leather coats or Sidcots used by all British pilots.'

Already dated compared with the Supermarine Southampton, the English Electric Co's Kingston prototype first flew in 1924 and featured a gunner's cockpit at the trailing edge of each nacelle.

Neither civil nor military aviation could save the English Electric Co Ltd of Preston from abandoning its aircraft department in March. The last of the batch of Kingstons, N9713, had just been delivered with wooden hull featuring a high vertical bow and deep forebody, like the Supermarine Swan, and had proved far more efficient than the others, reducing take-off from 25 seconds to 12. In desperate hope, the Mayor of Lytham St Annes telegraphed Sir Samuel Hoare that closure would entail serious local unemployment, but was told there was no alternative and the decision was final. After more than a decade and a half it was the end of W. O. Manning as a designer though he cloaked his professional dismay by adopting the guise of 'consultant'. Gibson Knight, the manager, transferred to another section of the E.E.C., and his assistant V. S. Gaunt joined Westland; Marcus Manton, their pioneer pilot, became an area manager for Shell.

April saw a sequence of prototypes on their first flights. On the 6th the Short Mussel, piloted by Parker, was tried from the Medway, but proved a disappointment, with struggling take-off and sluggish climb. The engine was checked for power, the airscrew changed, but no improvement. Buffeting at the stall led to fitting a wing-root fillet of fabric and ply, and on the next flight it was a very different machine. Gouge heeded the warning and added a hand-beaten aluminium fillet to the wing root of the Chamois, and

On its first flight the twin-float Short Mussel seemed a dud, but when demonstrated to the Press, after fitting a wing-root fillet, it seemed second to none. (*Short Bros.*)

modified the design of the Sturgeon to incorporate a still bigger one. But there was high optimism at Shorts, for Oswald had just received a contract for two of his three-engined version of the Singapore – the Calcutta.

The Avro Buffalo, registered as G-EBNW, flew next, but, as with all prototypes, there was trouble getting fin and rudder areas right and adjusting the control balances. Its rival, the Blackburn T.5 Ripon, was flown on 17 April at Brough by George Bulman, who had been loaned as a friendly gesture to Bob Blackburn by Tom Sopwith. Longitudinal stability proved

Designed to specification 1/24 for a Fairey IIID replacement, the Short Sturgeon used Springbok experience in design, and to meet criticism of reverberating noise had an exhaust manifold. (*Short Bros.*)

Built as a private venture deck-landing torpedo-carrier, the Avro Buffalo was designed by making an overlay drawing of the Bison with pointed nose but shallower fuselage, initially using the same tail, and had slightly modified wings of 6-in greater chord. (*Courtesy Air Marshal Sir Ralph Sorley*)

deficient, so increased sweep-back was necessary to bring the C.G. forward, and in hope of increasing top speed the cowling was reshaped to lessen the drag of the exposed three blocks of the Lion engine.

On the 24th Bulman flew the scaled-down trainer version, the Sprat, designed for deck-landing practice and seaplane conversion courses. Almost immediately it went to Martlesham for comparative trials with the Vickers Vendace I and Parnall Perch, and similar trials as seaplanes followed at Felixstowe. On the same day that Bulman flew the Sprat, Broad tested the Handley Page Harrow Mk I, a cleaned-up two-seat redesign of the Hanley/Hendon slotted-wing torpedo-carriers. In size it was almost

Official rival of the Avro Buffalo was the single-bay Blackburn Ripon, to specification 21/23, which initially had a blunt nose characteristically of Bumpus design, but in the Ripon II it became remarkably like the Avro. (*Courtesy Air Marshal Sir Ralph Sorley*)

Third of the torpedo contenders was the conventional, still unslotted, two-bay Handley Page H.P.31 Harrow. But it was the Ripon which won the contract.

identical with the Ripon, but the 44-ft span wings had two bays instead of one, and only the top wing had dihedral. The original blunt noses of both Harrow and Ripon were presently redesigned as Mk II versions having more pointed form like the Avro Buffalo.

4

A new dislocation to economic growth arose. French Occupation of the Ruhr and a retaliatory strike of German miners in 1923 resulted in a flush of prosperity for British coal pits, but as soon as the Ruhr industries restarted, demand for British coal exports began to lessen, causing mine owners in 1925 to give notice that wages must be drastically reduced in July 1926. Miners refused any cuts, and were supported by the TUC who threatened to call out railwaymen and transport workers; but Baldwin gained time by appointing a Royal Commission to examine the problems, during which pit owners paid the cut rate and the State made up the difference of some £24 million. The Commission's Report of March 1926, though aiming at better eventual conditions, recommended reduced pay and no subsidy under the reigning circumstances of depressed trade. The Miners' Federation refused to agree; there was deadlock. 'It would be possible to say without exaggeration that the miners' leaders were the stupidest men in England,' wrote Lord Birkenhead, slyly adding, 'if we had not had frequent occasion to meet the owners' – for the latter were unwilling even to accept proposals for reorganization. 'Not a cent off the pay, not a minute on the day,' was the miners' slogan. Backed by the TUC, they began a strike at midnight on Monday 3 May: railwaymen, transport workers, and printers followed next day. The Government refused to negotiate.

The British public regarded the entire absence of transport with good humour but as an unfair involvement in someone else's quarrel. Long premeditated emergency plans were put into operation by the Government. To we who were students, safeguarded from the harshness of world problems, it was great fun, for we volunteered to drive buses and tube trains, or guard dumps of supplies – in fact anything to dislocate the strikers. Few people were sympathetic to the fact that miners' wages offered the meanest standard of living, or that unemployment benefits had been extended in 1924 only from 15s to 18s a week for men, and the children's allowance was 2s.

Churchill maintained printed news by instituting the *British Gazette*, but the new-fangled wireless proved a great boon, for by now many a home had a 'crystal set'. Newspapers struggled back after a few days, and *The Aeroplane* issued a special typewritten edition in which I read with interest: 'Capt G. T. R. Hill, who has become world famous as the designer, constructor, and pilot of the first really successful tail-less aeroplane, has joined the staff of the Westland Aircraft Works at Yeovil.'

Coming in to land at the small aerodrome alongside the combined Westland Aircraft and Petter engine factories at Yeovil.

Westland aerodrome, like others strategically placed across the country, was co-opted by Harold Perrin of the RAeC as a distribution centre for the *Daily Mail*, of which over 300,000 copies printed in Plymouth were brought by road at night, and flown at dawn to such places as Cardiff, Norwich, Birmingham and Bournemouth by a motley collection of Avros of Berkshire Aviation Tours, D.H. Moths from the London Aero Club, D.H.9s owned by A.D.C. and de Havilland, a Handley Page W.8 and a D.H.50 owned by Imperial Airways, some Bristol School machines, and the Westland Limousine.

The Aeroplane reported: 'Judging by the aircraft factories in the London district, about 50 per cent of the men were out at the end of last week. The Fairey Co reported 60 per cent still at work, and the Hawker Co about 55 per cent. Allowing for a good many out because of the difficulty of getting to work, this is satisfactory from the viewpoint of law, order and decency, but eminently unsatisfactory from the view of the TUC. Naturally the majority at work were non-union men, and even the pickets were quite polite to the staff and to the men who remained at work.

'Woodworkers incline only to strike in a quiet kind of way; there is something soothing about wood which allows them to think quietly, even if they breathe the mental attitude of philosophic Bolshevism, and they are seldom inclined to actual violence. Sheet metal workers on the other hand, are inclined to be violent anarchists; anyone who has ever had anything to do with a tin-bashing shop, especially where machinery is used, must understand, even if they do not sympathize with, the attitude of the workers, for anything more nerve destroying than the din one cannot imagine. With the engineers one has every possible sympathy. When a man has served years of apprenticeship to become a skilled workman it is ridiculous that his wage should be about £3 a week, when a lump of a railway porter can get £4 plus tips. Whatever may be their personal grievances, there is no doubt that the general feeling among the workers both outside and in the aircraft industry is that the coal trade might very well be left to settle its own affairs without interfering with others managing to make a decent living. We have been brought up to believe coal is essential, yet we can do perfectly well without so long as we can get oil. If coal is to continue as a commercial proposition it must compete with oil on a cost basis, but no amount of bolstering with subsidies or boosting by strikers will do that: it has got to commercialize itself.'

As a 'National stoppage' the strike lasted only nine days, though it seemed a lifetime. On 12 May the TUC called it off as a complete failure. Obstinately the miners remained out, adding a million to the workless. By the time they returned, six months later, £60 million had been lost in wages. The Government cashed in with proposals for a *Trades Disputes and Trades Union Act*, attacking political use of union funds and making strikes in sympathy with other unions illegal. Yet the fact that the public had carried on, regardless of what seemed revolution, managed to raise the country's prestige in the eyes of the world.

Meanwhile the strike delayed Alan Cobham's departure to Australia. He had been due to leave on 3 May, but overhaul of his engine at Armstrong Siddeley ceased when Coventry workers struck in sympathy with the miners. This gave scope for a malicious commentator to say: 'It is understood Mr Cobham will not start his flight until the daily Press is in full and complete operation again. The delay will cause a collision with the monsoon as there is no room both for him and the monsoon at the same time in India.'

Blazing the trail for future airline routes were many other great flights by all nations. Piloting a Savoia S.16bis flying-boat re-engined with a 450 hp Lorraine-Dietrich, the Marquis de Pinedo flew 33,000 miles from Italy to Australia, Japan, and back. More spectacular, but almost unreported to the British, because of the strike clamp-down on newspapers, was the first ever attainment, on 9 May, of crossing the North Pole by air in the course of a $15\frac{1}{2}$ hour out and return flight from Spitzbergen by Lieut-Cdr Richard E. Byrd and his pilot Floyd Bennett. Byrd had persuaded Fokker to sell him the prototype Fokker tri-motor F.VII which had been converted from a single Wright-engined machine, and Byrd named it *Josephine Ford* as Edsel Ford had financed him. It had needed three attempts to take-off from a 600-ft runway of deep snow, and they had relied on an earth inductor compass for an accurate course. Laconically Byrd reported: 'On reaching the estimated position of the Pole, the machine was flown around the location in smaller and smaller circles for 15 to 20 minutes and then the return was begun. No flags were dropped, as Peary had already planted the Stars and Stripes. The Polar Sea was a waste of snow, and no land or life was discovered.'

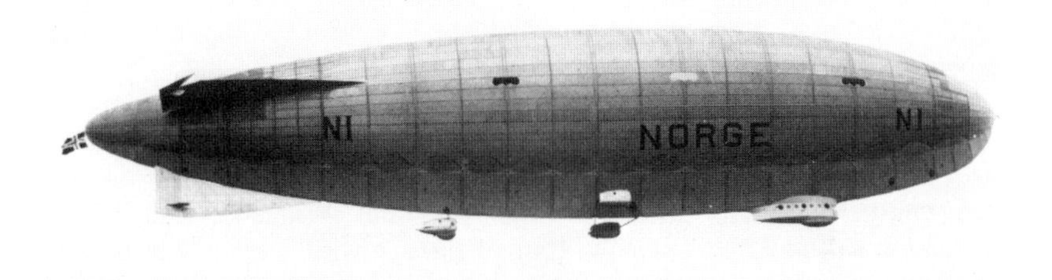

On her flight from Italy to Spitzbergen, the *Norge* stopped overnight at Pulham on Sunday, 21 April, 1926, after flying 32 hours nonstop. (*George Swain*)

Two days later, Capt Roald Amundsen, accompanied by the American explorer Lincoln Ellsworth, and the Italian General Nobile, departed from King's Bay, Spitzbergen, by airship and guided by radio, reached the Pole at 2 a.m. on 12 May. Descending to a few hundred feet they dropped the Norwegian, American and Italian flags on steel-pointed rods into the ice. Though they confirmed that there was no land at the Pole itself, much open water was observed, leading to speculation as to whether aeroplane and airship had even been to the same area. A difficult return, impeded by cloud and snow and with wireless out of action, ended with the ballonets punctured by ice from the propellers, necessitating landing at 8 a.m. on 14 May at Teller, 90 miles short of their destination, and the airship was deflated and packed for transport to the USA. In the lighter-than-air coteries of America, Germany, France and Britain there was jubilation at this special vindication of airship possibilities.

With a 60 hp Genet engine, the Blackburn Bluebird prototype at once became a valid though costly rival of the D.H. Moth, but it would be a year before the first production machine with 80 hp Genet could be available.

5

Pilots, aeroplanes, and engines had to be British for the *Daily Mail* £5,000 prize two-seat Light Aeroplane Competition of 1926. Similar eliminating tests to those of the previous year were specified, but this time competition flights were to follow routes having turning points at popular seaside resorts to interest the public.

The engine weight limit of 170 lb debarred the Cirrus, so the lighter, but still unproved, 60 hp five-cylinder Armstrong Siddeley Genet radial was of particular interest. Reported *The Aeroplane*: 'It is understood that one such engine is being fitted into the Westland Widgeon, and Major Openshaw will make an extended tour of Europe with it by way of finding out teething troubles which are always present in any new type of engine.' Instead, the first was fitted to last year's Blackburn Bluebird prototype

The prototype Avian was no beauty but undeniably Avro in conception. Fuselage three-ply was only 1 mm thick, and the nose cowl was faired by a similar conical blister to that of the original 504.

after its failure with the unsatisfactory three-cylinder Thrush, and the seats were slightly staggered to comply with the cockpit dimensional requirements. When Flt Lieut L. H. Woodhead flew it on 4 June, there seemed every indication that it could prove a winner – but there were close rivals. Geoffrey de Havilland was making a simple redesign of the Moth, lightening the structure and lengthening the nose so that it also could take a Genet when one was available. Even more favoured was an I-strutted light biplane designed for the A.N.E.C. firm by John Bewsher, once a prominent member of the Sopwith team. And at Hamble Roy Chadwick had devised a light-weight biplane, Type 581, by using an identical fuselage to that designed for his C.9 Autogiro and fitting square-cut 32-ft span wings with exaggerated gap of 5 ft 3 in – but despite its Avro hallmark and typical finless rudder, its conception was clearly influenced by the Moth.

Although the uprated Cherub III might give the 1925 ultra lights fair prospects, it was a seven-cylinder geared radial designed by Capt Douglas Pobjoy, Education Officer at the Apprentices Wing, Cranwell, which could have made them winners had it been started earlier, for its extremely light weight of only 100 lb for 50 bhp matched anything yet built except the Schneider engines. As a complementary team effort, his brother officer, Nick Comper, had designed a new two-seater for the Cranwell Club, possibly inspired by Bolas's unconventional biplane conversion of his Pixie, for it had an upper wing of smaller span and chord than the lower, and was intended for the Pobjoy 'P' prototype currently under test at Cranwell before being sent to the RAE for type tests.

Encouraged by the success of the Cranwell apprentices, Halton had formed a similar club in December 1925, and their Education Officer, Capt C. H. Latimer-Needham, ex-RFC, had designed an equally distinctive Cherub-powered two-seat biplane which had X-shaped interplane struts. Like their Lincolnshire buddies, the Bedford boys were building it partly as official training and partly in spare time, but were a long way behind programme.

Amateur or professional, these were the fun aeroplanes, regarded as of little importance compared with military machines. Of the latter, Bert Hinkler flew the Avro 566 Avenger private-venture rival of the new Fairey Firefly single-seat fighter at the beginning of June. With its smooth oval section, silver-painted semi-monocoque fuselage of fabric-covered double-planked mahogany on a light system of frames and stringers, it was somewhat reminiscent of German war-time aeroplanes, but its direct-drive 525 hp Napier Lion VIII, sired by the Schneider racers and closely cowled in a shining pointed nose, at least ensured 50 per cent greater speed, clocking nearly 180 mph, and climbing some 2,000 ft per minute, marginally beating the Firefly, but was surpassed a few days later when George Bulman flew Camm's rugged looking V-strutted Hawker Hornbill of similar size but with 170 hp more in hand from its big Condor IV. Both Avenger and Hornbill had their snags. Though the Avenger had ailerons only on the top wing like the Hornbill, it was heavier and slow in roll. By

Radically different from equivalent fabric-covered fighters was the neat Lion-powered Avro Avenger with inset wing radiators, but its shapely semi-monocoque wood fuselage caused it to be turned down by the RAF on the score that equipment was not readily accessible. (*Flight*)

contrast the Hornbill exhibited undamped directional oscillations if power was suddenly changed, and the rudder was too small to hold up the nose in steep turns above 150 mph. Compared with the roomy Avenger, its cockpit drastically restricted the pilot's movements. Adding to Camm's worries were cooling problems with the big engine.

He began to feel that his first solo effort at military design was disappointing, but already he was working on specification F.9/26 for a fighter replacing the A.W. Siskin and Gloster Gamecock, as these were now at the mercy of two-seat fighters of the calibre of the Fairey Fox. In March an initial order had been placed for twenty Horsley bombers of wood construction as approved with slightly swept-back wings, and for later contracts Camm was re-designing for partly metal construction based on his

Described by Norman Macmillan as 'a delightful single-seat stable companion to the Fox', the private venture Felix-powered Fairey Firefly I of composite construction, though pointing the way for Camm at Hawkers, failed to get the RAF's blessing and orders.

Test flown later in the year by Macmillan, the Flycatcher II was not a derivative of the Service naval fighter of that name, but a redesigned, metalized Firefly handicapped by its undeveloped Bristol Mercury later changed to a Jaguar.

square-tubed Heron, as well as modifying the bomber form to meet additional load requirements as a torpedo-bomber capable of launching a 2,800 lb torpedo. He was driving his staff at maximum pressure. Years later Sir Thomas Sopwith remarked to me: 'I can't imagine why his men put up with him. He was a genius – but often quite impossible.' Why they stayed, as with every underpaid technician in those days of scarce employment, was because they were usually married and daren't risk dismissal. 'In Camm's favour,' said one of those draughtsmen, 'I would say he was a perfectionist where detail design was concerned, personally vetting every drawing. He hated to see a bent plate if a flat plate could do the job, and I once heard him say that welding was the last resort of failures. When I left the firm, voluntarily but just in time, it took considerable readjustment to work for people like Henry Knowler of S. E. Saunders and H. P. Folland, both of whom treated their staffs as humans and colleagues.'

June saw much flying by prototypes. On the calm and sunny morning of the 19th the single-seat Avro C.6C Autogiro made from a 504K fuselage was wheeled onto the grassy little aerodrome at Hamble on the edge of Southampton Water. Instead of the 110 hp Le Rhône of the Spanish C.6A it had a 130 hp Clerget. Engine tests and rotor spinning had been under way for some days. Now Frank Courtney flew it. Rotor starting was still by rope followed by fast taxi-ing round the aerodrome to speed it. When Courtney applied full power, the take-off was visibly even shorter than that of the prototype, and climb steeper. Elevators, and outrigger ailerons mounted behind small fixed surfaces, were bigger than before, and Courtney found they gave more effective control. By now he was used to the heavy thumping vibration as the blades rotated and could ignore the apparent tilt of the rotors as the forward moving blades lifted and the top wires, normally holding them level when static, hung slackly. All seemed well enough for the machine to be presented at the forthcoming RAF Display on 3 July.

Major Rennie's impressive, sonorous, triple Condor Blackburn Iris flying-boat, to specification R.14/24, took over two years to build – a long time for those days – but already a metal hull was under construction, for Shorts had shown that water-soaking wooden hulls were obsolete. (*Flight*)

On the day that the Autogiro first flew, the Blackburn Iris, after a launching ceremony when it was 'christened' by Mrs Blackburn, made its initial flight piloted by Flt Lieut H. G. Sawyer, RAF. Although this flying-boat was on the Secret List, it could hardly be hidden from the public with the Humber as testing site, so the Press made much of the trials of 'the Giant Flying Ship', stating it had taken two years to construct at a cost of what seemed the enormous sum of £60,000. She was moored off the slipway over night, and next morning a great crowd watched her take-off for the MAEE at Felixstowe where it was intended that the RAF should take responsibility for performance and handling, as Major J. D. Rennie, after his long experience at Felixstowe, decided they were more competent to do this than any civil pilot Blackburns might engage.

Meanwhile the ubiquitous Mr Courtney went to Cowes where he tested the three-engined Condor-powered Saunders Valkyrie flying-boat, rival of

Why the Blackburn Iris was preferred to the somewhat cleaner, equally seaworthy Saunders Valkyrie (*illustrated*) is not on record, but practical considerations of maintenance, such as easy removal of engines and mountings without disturbing the wing structure, could outweigh performance.

the Iris. Though its white-enamelled wooden monocoque hull was somewhat similar to that of the Iris, the small mainplane gap was very obvious yet afforded the same vital propeller tip clearance, and, instead of the biplane tail and triple rudders, it had a single large fin and big balanced rudder and monoplane horizontal surfaces.

Later in the month the first Vickers experiment in completely all-metal construction, the British-built Type 121 Scout version of the Wibault 7.C1, was flown by Flt Lieut E. R. C. 'Tiny' Scholefield, a giant of a man recently appointed Vickers chief pilot. Habitually he favoured a tiny Austin Seven two-seat open runabout, his head higher than the windscreen, and would escape a traffic jam by picking up its front end and trundling the car away. Luckily for his test flight he had obtained the first Irvin parachute released for civilians. Accustomed to the French prototype, F-AHFH, he did not hesitate to aerobat the Vickers replica, but got into an

The finely corrugated metal-skinned Jupiter-powered Wibault parasol attracted Vickers by its relative simplicity; but French strength requirements were less than British, so the twenty-six Vickers-built production aircraft for Chile required stiffening.

uncontrollable inverted spin and baled out just in time at less than 2,000 ft. The machine smashed into the Vickers Sports Club grounds at Byfleet, but 'Tiny' landed in a garden, and strolled back to Brooklands with his parachute bundled under his arm. Investigation by Vickers and the RAE presently revealed that the considerable positive tailplane setting and rear C.G. position were the culprits which held the machine inverted – though nobody remembered similar war-time accidents when the dangerous D.H.6 was inverted. Nevertheless twenty-six Vickers-Wibaults were built for Chile, and the virtues of the all-metal constructional system sufficiently warranted Vickers to go ahead with a single-seat low-wing all-metal shipplane to specification 17/25.

At the end of June, multi-millionaire Harry F. Guggenheim, president of the Daniel Guggenheim Fund for the Promotion of Aeronautics, announced the offer of substantial prizes for an International Safety First Aircraft Competition to be held in the USA. Boldly he asserted: 'Flying is now entering upon a new era, and the world is on the verge of developments which will make it clear that not only is flying over great distances mechanically feasible, but that the whole art and science could be made so nearly accident proof as to place flying on a basis of safety comparable with travelling by rail, steamship or automobile.

'Research work in progress, and construction of certain new types of airplanes involving radical aerodynamic departures from conventional, justifies faith that complete solution of this problem of safety is possible. Due to the weak financial status of the infant airplane industry all over the world, manufacturers are unable to finance development of airplanes that have no definite commercial value; this restricts them for the most part to war types. To give the greatest possible encouragement to every endeavour to make the airplane safe, the Board of Trustees have determined to organize an International Competition, and will appropriate $150,000 to $200,000 for this purpose to make it interesting and attractive to the best designers and manufacturers of aircraft throughout the world.' As an indication of what might be done he mentioned Professor Melville Jones's

Formation flying at the RAF Display was always a great attraction. Here D.H.9As and Fairey Fawns make patterns during the morning flying. (*Flight*)

Not since the S.E.5a had Folland promoted a water-cooled engine for his fighters, but 'Bamel' experience with the Lion led to its installation in the Grebe-like Gloster Gorcock. (*Flight*)

work at Cambridge on stability, Handley Page's slot and aileron control, Geoffrey Hill's tailless aeroplane, and Cierva's Autogiro.

But safety was the last thing the British public had in mind at that great spectacular event, the annual RAF Display on Saturday 3 July. The sun was hot and brilliant, and cumulus sailed the summer sky, but although this was the last day of Henley Regatta and of Wimbledon week, Hendon rivalled them with womenfolk in prettiest clothes and wide-brimmed straw hats, amid the great crowd of 120,000. Yet the Display did not go with quite the swing of its predecessors, for there were longer but fewer events, and protracted intervals between. There were the usual beautiful aerobatics and tight formation flying. There were Sqn Ldr Harris with Virginias from Worthy Down, Sqn Ldr Slessor and Sqn Ldr Coryton flying Bristol Fighters, Flg Off Atcherley performing solo with a Gamecock, Sqn Ldr Collishaw, the Canadian war ace, with a formation of them, and leading the sole squadron of D.H.9As was Sqn Ldr H. V. Champion de Crespigny. Among those in the Handicap Race for members of Air Ministry Directorates were Air Commodore Longmore on a Fairey Flycatcher, Wg Cdr Sholto Douglas on a Hawker Woodcock, and Sqn Ldr Sir Quintin Brand on a Gloster Gamecock. Sqn Ldr A. H. Peck led Gloster Grebes manoeuvring to his radio-telephone instructions. All this was routine, but where the Display differed from its forebears was in the preponderance of aircraft designed and built since the Armistice.

Of the galaxy in the Experimental Aircraft Park, Capt Sayers commented: 'For the first time they really give the impression that at last British aircraft designers have been permitted to produce aircraft such as the RAF deserve. It will be necessary to go far to find cleaner design and more delightful lines. The Avro Avenger, Fairey Firefly and Fox, the Gloster Gorcock and Hawker Hornbill are, to eyes accustomed to the

Flt Lieut R. H. Horniman displayed the Hawker Hornbill with thunderous aerobatics and an impeccable landing. (*Flight*)

officially supervised designs of the past few years, almost unbelievably refreshing in the obvious fitness of their form for the business for which they are intended. Having expressed one's pleasure at the beautiful aircraft, one may be permitted to welcome what must have struck most spectators as the two ugliest – the Pterodactyl and the Autogiro.'

Sixteen silver-doped aircraft glittered in the great fly-past, led by the Cherub-powered Pterodactyl flown by Flt Lieut Chick, whose brilliant display of stalled flying made the machine seem easy and fun to fly. The Blackburn Sprat trainer and its rival the Vickers Vendace followed. Then came the smooth-nosed Gloster Gorcock powered with 525 hp Lion VIII, demonstrating a performance that seemed another step forward in design,

The Heron was flown from Martlesham, where it was on test, to Hendon for the Display. (*Flight*)

but was matched by the Hawker Hornbill which 'showed that it could do most of the things to which Scout pilots were accustomed'. Close on its tail came the Avro Avenger and Fairey Firefly, and those appraising them considered the Firefly was the fastest of them all, though this may have been coloured by Dick Fairey's offer to let it race a 200-mile course, against any other carrying full military equipment, for a stake of £1,000 a side – but there seemed no takers.

After the fighters came the squat Armstrong Whitworth Atlas in no way different, except for roundels and tricolour, from the civil-registered Ajax of the King's Cup race the previous year. Closely following were its rivals, the Bristol Boarhound and Vickers Vespa, but the de Havilland Hyena, after a struggling take-off, disappeared over the far side of the railway bank, but landed without damage. Clearly they all were faster than their Bristol Fighter predecessor, yet the advance made in a decade was hardly spectacular.

The lithe Fairey Fox now took the stage and, although revealed at Andover the previous summer, remained the focus of attention, for her lines seemed perfection for a biplane, and the Curtiss engine screamed like a banshee when she dived. Clearly she should have been a two-seat fighter, but the Air Ministry, with ingenuous malice at Fairey's initiative in making so clean an aeroplane, decreed the Fox must be a day bomber, and at that fitted the bombs externally.

The big Hawker Horsley bomber clambered after her, to the immense respect of every pilot present, for the RAF remained conventional on matters of design after being nurtured so long on war-time aeroplanes. Yet as C. G. Grey said: 'The Horsley is so beautifully proportioned that at a distance she looks like quite a small machine. Mr Camm, their new designer, has proved he is quite in the front rank, and with Mr Fred Sigrist to look after construction and production this machine is bound to be a success.' But it was Carter who had been responsible, and it is in this manner that credits go to the wrong man in later recorded history.

After the Horsley the big Avro Ava night bomber/coastal torpedo landplane piloted by Flt Lieut Webster from Martlesham thundered into the air, but a cloud of steam streamed from the port engine, and 'Webby' turned down-wind using starboard engine only and, over Colindale Avenue, threw it into a steep side-slip and spectacular swish-tail landing that made it seem the controls were perfection. While this was happening the Armstrong Whitworth Argosy, G-EBLF, painted in Imperial Airways colours, rumbled serenely around the aerodrome, attracting respect as the newest example of three-engined airliner, though its long nose was reminiscent of the war-time Blackburn Kangaroo.

Ending the procession with a touch of comedy came the Avro-built Autogiro flown by Flg Off Frank Courtney of the RAF Reserve, who joined with Chick on the Pterodactyl for an entertaining *pas-de-deux* – but as an onlooker said: 'If one of them is right the other must be wrong!' When His Majesty visited the Aircraft Park, it was Señor de la Cierva, not

Because it was a strange shape in 1926, the tailless Pterodactyl seemed merely an interesting experiment rather than a pointer to the future.

Geoffrey Hill, who was presented to him, yet Hill's machine equally foreshadowed the shape of things to come.

Two days after the RAF Display came headline news that an explosion in Alan Cobham's D.H.50J seaplane had occurred on his flight to Australia, injuring his famous little mechanic A. B. Elliott. They had taken-off at dawn from the Medway at Rochester on 30 June, reaching Naples the same evening, and next morning flew to Athens where Cobham felt too ill to continue until next day. On 3 July he reached Alexandretta where the metal propeller was replaced with a wooden one because it had better static thrust. On 4 July he took-off for Baghdad where 'on alighting it was like going from the cold air of the street into the hottest room of a Turkish bath. The problem to be faced is whether passengers will stand it, so the medical profession will have to discover how to fortify us against these sudden changes. At any rate, it nearly bowled me over.' Thence he flew on, encountering severe dust storms, but hugged the banks of the Euphrates for safety, flying at 30 ft. Later, 'while passing over a dryish stretch there was a sudden bang in the cabin as though a Very pistol had exploded. Realizing that would be dangerous, I asked Elliott if we were on fire. He said "No, the petrol pipe burst, and I'm hit." I wondered what I had better do. It was a lonely spot and frightfully hot. I felt the best thing was to race off for Basra.' There he alighted 45 minutes later, and discovered that Elliott, who typically said 'Don't forget to turn the oil off', was very weak. He was rushed to hospital where it was found a bullet had pierced his arm, both lobes of his left lung, and into his back. Clearly a Bedouin had fired at them. For a time it seemed Elliott was recovering, but a relapse followed and he died. 'For a while I wanted to give up the flight, for I had been with Elliott so long and knew him so well, but I received telegrams urging me to go on. The RAF lent me Sgt Ward, who had never seen a Jaguar engine before but understood it in theory, and we set off down river towards Bushire.'

Cobham and Elliott refuelling the D.H.50 from a pontoon on the Tigris near Baghdad.

While Cobham's hard tried D.H.50 was heading into the sultry East, its little stable mate the Moth, flown by Hubert Broad, was adding further lustre to the de Havilland name by winning the King's Cup race. This time, instead of circuiting Britain, there was a series of local courses – four triangular circuits on 9 July from Hendon respectively to Coventry and Martlesham, twice to Martlesham and Cambridge, finally to Coventry and Cheltenham; next day four from Hendon to these places in different order. Because competitors passed through Hendon four times each day it attracted many more spectators.

At Fedden's persuasion, the Bristol directors reluctantly agreed that another racer could be built to demonstrate the uprated 510 hp Jupiter VI, resulting in the stubby, 24-ft span Badminton shown ready for its first flight.

Significantly only five of 13 starters finished, three of which were Moths flown respectively by Hubert Broad, F. G. M. Sparks of the London Aeroplane Club, and W. J. McDonough of the Midland Aero Club. Sqn Ldr H. W. G. Jones had fastest time with a Jaguar A.D.C. Martinsyde at 151.9 mph, pursued by 'Tiny' Scholefield with the Lion-powered Vickers Vixen at 142.2 mph. There were the usual forced landings. Geoffrey de Havilland, flying his own Moth G-EBNO, which had the latest hotted-up Cirrus, broke an oil pipe; W. L. Hope with another Moth had an obstructed fuel supply; the Master of Sempill with his 120 hp Airdisco-powered D.H.51 had a broken rocker arm; Nick Comper's Cherub-powered Cranwell C.L.A.3 monoplane had engine failure and turned over in a corn field; Alan Butler with his resplendent D.H.37, renamed *Lois* instead of *Sylvia* and re-engined with a 300 hp A.D.C. Nimbus, had inlet valves jammed with broken pieces of aluminium intake casting; Frank Courtney with a Nimbus-powered A.D.C. Martinsyde had piston seizure; F. L. Barnard with the Jupiter-powered racing Bristol Badminton had failure of fuel pressure but, despite the high landing speed, managed to get down in the famous Portmeadow at Oxford, once the scene of several pioneer flights; H. H. Perry with another Nimbus-powered Martinsyde had fuel failure, landed safely, but found the field too small for take-off after rectification; and finally the Parnall Plover piloted by Sir Quintin Brand, co-pilot of Sir Pierre Van Ryneveld of the first England—South Africa flight, ran out of petrol because someone had omitted to replace the filler cap after refuelling. So many engine failures were hardly a good advertisement for the sport of aviation; nevertheless the Moth-equipped flying clubs were thriving.

7

Behind the activities of the flying man was the growing army of scientists tackling the many problems of aviation to make it safer: not least was structural strength. Thus stressmen of the industry were benefiting greatly from the co-operation of H. B. Howard, BA, BSc, head of the RAE Airworthiness Department. Paradoxically, in a lecture to the RAeS he said: 'The maximum loads which the structure will be called to carry cannot be stated with certainty, as they depend primarily on how the pilot manipulates his controls. The maximum he can impose is calculable, but is greatly in excess of those he uses. It is neither practicable nor necessary to make the aeroplane absolutely unbreakable at all speeds.' On that hopeful basis some complex calculations were erected, particularly in the case of tapered wing monoplanes which so long had been shunned in favour of the simplicities of biplanes.

The Bristol Aeroplane Co were experiencing these problems in stressing the twin-spar Bagshot monoplane, which had a wing based on the aerodynamically similar half-scale Brownie. Barnwell had not been enthusiastic, predicting that the big machine would be over weight, and recommended abandoning the project as a waste of time. The Air Ministry declined to cancel the contract, and since February work had slowly proceeded. A junior stressman was a youngster named A. E. Russell who had just graduated from Bristol University. In later years he recollected: 'The fabric-covered wing structure was based on two independent spars with short struts from the fuselage to the oleo leg attachment on the wing to reduce the bending moments which were considerably higher than we had met before. The spars were based on corrugated steel-strip flanges with an increasing number of laminations towards the root. For testing, we devised a rig with mechanical linkage and hydraulic jack for loading the spar which was mounted some way above the ground. Instrumentation was by eyeball scanning by me as the most junior, looking for impending collapse while walking a narrow plank – the idea being that I could quickly jump off when the specimen broke. There were many premature failures, often accompanied by a loud report that would have rivalled the simultaneous firing of both C.O.W. guns, and great was the relief when the design load was eventually carried.'

Safety in another form was demonstrated to the Press by that great salesman-engineer Frederick Handley Page, at Cricklewood on 20 and 21 July with his full-slotted Hendon torpedo-carrier. Flown by Capt

Demonstrating leading-edge slots and slotted flaps to the Industry – the modified Handley Page Hanley, renamed Hendon, staggers past the cameras at slow speed. (*Flight*)

Wilcockson of Imperial Airways – who did spare-time test work for H.P. thus saving the cost of a full-time test pilot – a number of slow passes were made, hanging on the slots, while the trilby-hatted, genial Frederick explained the benefits in vigorous terms. As the machine was obsolete, the Air Ministry magnanimously agreed that on the second day it could be tried by the industry's pilots, and it was flown by Piercey for Gloster, Openshaw of Westland, Bulman of Hawkers, Payn of Vickers, Lankester Parker of Shorts, Barnard for Imperial Airways, and even the Bristol designer Frank Barnwell flew it despite a hair-raising reputation as pilot. *The Aeroplane* reported: 'Mr Barnard, who had never flown a slotted-wing machine before, gave a most amazing exhibition of skilful piloting. It was even more spectacular than Mr Bulman's of slotted Avro fame, because he half looped the machine and generally pitched her about in the way which recalled Mr Kipling's phrase about "incredible improprieties". One remarked to Mr Savage, manager of the Handley Page Co, that Mr Barnard seemed to have perfect confidence in Handley Page construction, to which Mr Savage replied that he ought to, considering Imperial Airways had lately been using nothing else.

'The attitude of Air Ministry technical experts towards the slotted wing has been criminal. It is true they have not condemned it, but they have not shown anything like the enthusiasm for its development they should have. The death of every aviator killed through his machine stalling in the last three years can be laid at their door. One can only hope that under the new direction we are promised for Supply and Research, and Technical Development, sufficient progress will be made with the slotted wing before the end of next year to make it as commonly used as pneumatic tyres on motor-cars.'

Though the Air Ministry was criticized, its view was dictated by reports from experienced Farnborough pilots who had found deficiencies. Although the industry's pilots had now flown the Hendon, there was no rush to adopt the system, for they found the interlinked inner slots and lift-flaps cumbersome to operate and caused a big change in trim, and ailerons interconnected with wing-tip slots, though effective in the stall, had a stiffly mechanical feel. In any case the slots could do nothing to save a crash if the machine was sinking fast and level in stalled flight, though this was better than spinning in.

Important changes had taken place in financial control of Handley Page Ltd. Lieut-Col J. Barrett-Lennard duly retired 'in accordance with the terms of an agreement between Handley Page Ltd and the Aircraft Disposal Co Ltd', the obligations to that company having been fulfilled and mortgage debentures from the Royal Bank of Scotland, of which Barrett-Lennard was a director, having been repaid. Accordingly the Board was reconstituted, with S. R. Worley as chairman, 'a man with long dragging moustache and the mien of an old crusader knight, who was an accountant and gentleman by choice and an industrialist by accident'. With him were Frederick Handley Page, the Rev W. E. B. Barter, D. F. Sutherland, Capt the Hon M. Knatchbull, and Wg Cdr Louis Greig. The immediate task was to secure a debenture for £120,000 charged on the company's undertaking and property, excluding the aerodrome.

The Fairey Aviation Co Ltd also underwent transformation, with registration on 3 August as a public company with nominal capital of £10,100, with Richard Fairey as managing director and chief engineer, and F. G. T. Dawson, their solicitor C. Crisp, Lieut-Col V. Nicholl, and Sqn Ldr Maurice Wright as his co-directors.

Such reconstructions might be milestones, but so were the stages of Alan Cobham's journey, for he had Australia almost within grasp. On 29 July he reached Singapore, which he considered would make an ideal seaplane base, stopped a day for engine maintenance, and then 'pushed on along the Sumatra coastline, where the country consists of a lonely swamp jungle, and then followed the Java coastline where the weather was delightful'. On 3 August he reached Bima; on the 4th attained Kupang on the edge of the Timor Sea where 'the water is so clear that you cannot see the surface, and the crest of a wave running across it gives the impression of a bucket of water thrown onto glass. Alighting on the water is therefore very dangerous.' On the 5th he set off across the 500 miles of the open Timor Sea. 'We went on and on, and presently felt we ought to see the coastline in about half an hour; then thought we did see it, but were mistaken. With relief at last it materialized and I found that, after this long journey by compass, we had hit our objective within five miles.' He touched down in Australian waters at Port Darwin to a great welcome 37 days after leaving London. The machine was hauled ashore, floats removed, and a land undercarriage fitted for the journey to Sydney and Melbourne. 'Before leaving Port Darwin we visited the monument marking the spot at which the late Sir Ross Smith crossed the coast on his first flight from England to Australia,' wrote Cobham with the acknowledgment and understanding of one pioneer for another. 'His was a magnificent effort. Another six years' experience intervened before we accomplished our flight.'

Two days before Cobham reached Australia, a forgotten Welsh airship pioneer, Ernest T. Willows, met his end at a Flower Show near Bedford when the netting of his captive balloon broke during hauling down and the basket fell to the ground. Twenty-one years earlier he had launched the first of his airships, yet never managed to become established. As a friend said: 'The real cause was that he had no true engineering training. If as a youngster, instead of building airships by rule of thumb, he had spent those years with a first-class engineering shop, he might to-day have been among our leading aeronautical engineers. As it is he is the victim of a useless type of aircraft, and the type of airship to which he devoted his life has proved of little worth, though it fulfilled some useful purpose during the War 1914–18.' It was the aeroplane pioneers who were winning through.

Quickly the muddy little Stag Lane aerodrome, several fields from the Edgware Road, had become animated with gay Moths. Sir Sefton Brancker's appropriate G-EDCA is on the left. The de Havilland office and drawing office are the wooden bungalows beyond the Moths in the foreground.

On the crest of the wave, with the tremendous publicity of Cobham's flight to back them, de Havilland boldly reduced the price of the Moth to £795, helped by a cut of £40 in the Cirrus price. It was indicative of the popularity of this combination that at the Bournemouth Flying Meeting on 20 August, eighteen of the twenty-six aircraft were Moths in great variety of gay and beautiful colours. To support de Havilland world activities, the company instituted *The D.H. Gazette*, with sixteen pages of information on Moths, the flying schools, clubs, air travel, and equipment. An advertizing competition to decide the name of the big three-engined D.H.66 was won by an Eton boy who proposed 'Hercules'. The prototype, which had been designed and built within a year, was expected to fly by the end of the month. Behind it in the shops was an almost completed, very clean, Lion-powered two-seater, the D.H.65, which utilized many D.H.9A components, and though registered G-EBNJ was clearly of war-like purpose and designed to beat the Fox.

The number of private owners at Stag Lane was rapidly increasing. Capt Lamplugh bought, on behalf of his insurance group, a brown and silver Moth which he flew to meetings; the comedian Will Hay had a silver one; Mrs Elliott-Lynn formed a small group owning a pale blue Moth, but was also the somewhat erratic pilot of an S.E.5. There, and at such meetings as Bournemouth, the carrying ability of the Moth was being nobly proved by genial Ronnie Malcolm who weighed all of 16 stone and could barely squeeze into the cockpit.

As a venue for air racing Bournemouth proved a success, for the summer crowds thronged to the racecourse adopted as its aerodrome and had an exciting close-up view of the aircraft speeding down the finishing straight, but so close was the handicapping by Dancy and Rowarth that on some of the laps the machines seemed uncomfortably close. As a result swarthy young Dudley Watt, known as Dangerous Dan, owner of the Sopwith Grasshopper and the late Harry Hawker's Swallow, came into prominence. Said one onlooker: 'When I first saw Mr Watt performing I thought he was the most dangerous and reckless flyer within my experience, yet if he does come to grief it will not be through lack of skill in getting out of a tight corner, but through being let down by his engine in a tight corner he would never have got into if he were a little more cautious or experienced.' Not long after, I saw him fly his green Grasshopper rather than taxi from the Westland hangars to the petrol pump 200 yards distant, and he crashed into it, wrecking the machine.

The green-painted Sopwith Grasshopper always attracted attention because already it looked an antique, though was younger than many an Avro. Dudley Watt rebuilt it, and eventually Constance Leathart became the venturesome owner. (*Flight*)

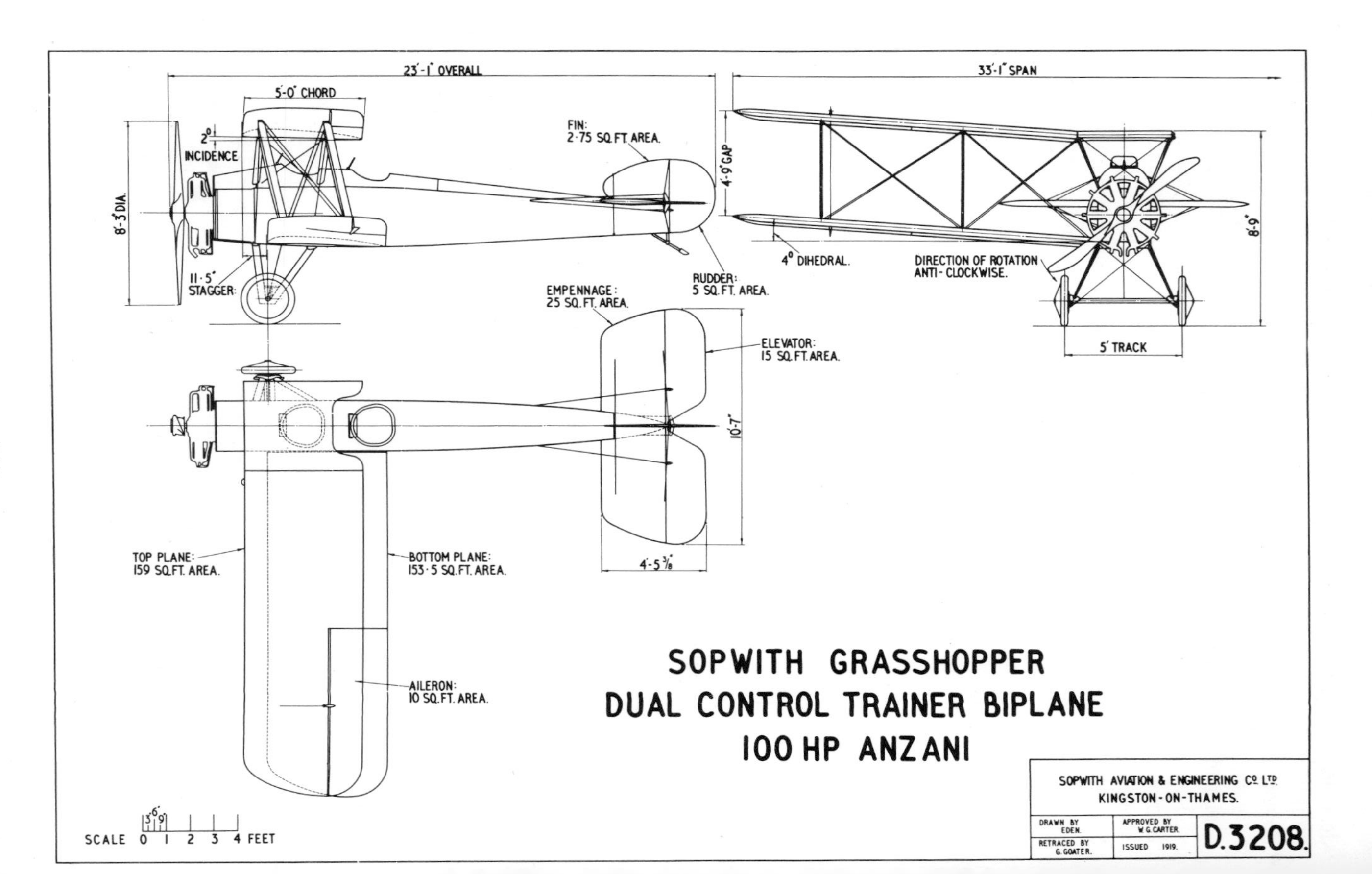
23'-1" OVERALL
33'-1" SPAN
5'-0" CHORD
2° INCIDENCE
FIN: 2·75 SQ.FT. AREA.
4'-9" GAP
8'-3" DIA.
8'-9"
4° DIHEDRAL.
DIRECTION OF ROTATION ANTI-CLOCKWISE.
11·5" STAGGER
RUDDER: 5 SQ.FT. AREA.
EMPENNAGE: 25 SQ.FT. AREA.
5' TRACK
ELEVATOR: 15 SQ.FT. AREA.
10'-7"
TOP PLANE: 159 SQ.FT. AREA.
BOTTOM PLANE: 153·5 SQ.FT. AREA.
4'-5 3/8"
AILERON: 10 SQ.FT. AREA.
SOPWITH GRASSHOPPER
DUAL CONTROL TRAINER BIPLANE
100 HP ANZANI
SOPWITH AVIATION & ENGINEERING Co LTD.
KINGSTON-ON-THAMES.
DRAWN BY EDEN.
APPROVED BY W.G. CARTER.
RETRACED BY G. GOATER.
ISSUED 1919.
D.3208.
SCALE 0 1 2 3 4 FEET

My four years at the Northampton Engineering College of London University had ended, and, armed with diplomas, I had applied to the RAE for a stressing job advertized at £5 a week. After interview and medical there was a long silence. Meanwhile I encountered Arthur Davenport at one of the RAeS lectures, and he suggested I apply to Westland for a job under Capt Hill in the new section being formed to develop the Pterodactyl. I did so, and sent various drawings I had made at college so that they could judge my standard. Still no answer from Farnborough. From Yeovil, R. A. Bruce wrote offering a job as draughtsman at £3 a week, but as the contract for the next Pterodactyl had not yet been received, he said it would be advantageous if I spent a few months on the bench. Still no reply from Farnborough. I bought some overalls and went to Westland. Three weeks later the RAE notified I had been appointed junior stressman Grade II. I stayed at Westland.

At first glance the prototype Westland Westbury could have been taken for one of John North's Boulton & Paul designs. An early modification was extension of the nacelles to get a better streamline and gain a doubtful 2 or 3 mph.

We were less than one hundred all told, half of whom worked at a dozen benches at one end of the Vimy erecting shop behind a row of D.H.9As. Beyond those machines was the prototype Westbury twin-engined C.O.W. gun fighter almost ready for flight, and alongside was the second fuselage. And at last that gravel floor had been concreted. The men were a cheery lot, working steadily at sheet-metal fittings, and in other buildings were carpenters and joiners building wings and fuselages. Whenever Bill Gibson, the ever hurrying ex-pilot works superintendent, appeared the tempo subtly increased, and every morning Stuart Keep, as technical supervisor, stumped around every shop on his tin legs and two sticks, demanding from each foreman the state of progress. Pay was a flat rate known as 'day work' for routine jobs, but for a large batch of identical fittings a 'piece work' price was set so that an industrious man could earn more, but if he exceeded the estimator's time for the job he was debited by that amount. Almost immediately I was in debt – until it was discovered that mine was a fixed weekly pay! For the select few making prototypes there was the 'experimental' rate giving consistently slightly higher pay than the others. Everyone seemed satisfied. They made aeroplanes because they liked them, and it was the same everywhere. London pay was somewhat higher than the rate at rural Yeovil, but a man was lucky to take home more than £5.

Everywhere the same enthusiasm prevailed. When the new Short Singapore was launched at Rochester on 17 August, every employee crowded to the open doors to watch Lankester Parker, with Eustace Short as observer, make the first flight, and Oswald Short tactfully ignored their absence from the benches, knowing the men would make up for the lost time. But it was only a two-minute hop, for some cowling came adrift and Parker alighted straight ahead. Two days later he flew the Singapore again, but water in the bilges revealed considerable leakage, and the machine vanished for a long period into No. 3 Shop for rectification.

Though only 2 ft less than the span of the Iris, the splendid prototype Short Singapore had single-bay outer wings and engine mountings integrated with the wing structure which, for the first time in any British flying-boat, had metal spars and ribs. (*Short Bros.*)

Of more immediate interest to RAF pilots was a 7,000 mile cruise from Felixstowe to Egypt and back by two wooden-hulled Supermarine Southamptons, S1037 and S1038, to prove ability of operation from unprepared bases. In the autumn of the previous year, immediately on delivery to the Coastal Reconnaissance Flight at Calshot, four of these boats had cruised the Irish Sea, alighting at many points to test seaworthiness. Undoubtedly the Southampton was proving the finest flying-boat the RAF had ever had.

Though August and early September had their sporting excitements, such as England regaining the Ashes, Miss Ederle swimming the Channel in record time, and *Coronach* winning the St Leger, aeronautical en-

The two Supermarine Southamptons used for the first long-distance foreign cruise by RAF flying-boats were piloted by Sqn Ldr G. E. Livock and Flt Lieut D. V. Carnegie. (*Grp Capt G. Livock*)

thusiasts had Lympne as Mecca for the *Daily Mail* competition opening on 10 September. Of the 16 entries, 13 arrived on time, but the pretty little A.N.E.C. *Missel-Thrush*, piloted by Colonel G. L. P. Henderson, eliminated itself in mid-week when the axle elastic stretched sufficiently for the undercarriage to dig into the ground, and the machine turned over with damage beyond immediate repair. Of those which failed to arrive, the RAE Aero Club's Sirocco monoplane was finished in time but had an enormous error in C.G. location – a bad mark for the scientists; Halton Aero Club was struggling to complete their novel biplane, and the latest Cranwell C.L.A.4 biplane had a connecting-rod failure which smashed the Pobjoy engine. Unduly strict interpretation of the rules soon reduced the contestants to nine. The stocky 'Sociable' Bluebird was landed heavily by Sqn Ldr Longton, bending the axle and distorting an attachment lug which could not be safely straightened; so Thornton, the young Blackburn designer, proposed fitting an auxiliary wiring plate to take the strain – but the Stewards flatly refused, and rather than risk a pile-up the machine was

The Genet-powered A.N.E.C. IV *Missel-Thrush* takes-off from Hendon on its ill-fated attempt to gain the King's Cup. (*Flight*)

withdrawn. The Stewards also were adamant that the Cherub-powered C.L.A.4 must either scratch or fly with a damaged undercarriage and when it gave way, refused permission to fit a new one. The Scorpion-powered Woodpigeon, now owned by the RAF Seven Aero Club, became the object of contention, for its wooden propeller had been temporarily fitted to their other entry, the Short Satellite, and the Satellite's was on the Woodpigeon when presented for examination. Although it was intended to change them back, the Stewards insisted the machine must be flown as it was, despite the reduced performance.

So strong was feeling against these decisions that a strike by the pilots was proposed but was abandoned as undignified in view of the recent general strike of transport workers and miners. Yet that was not the end of contention. There was much argument about the sticks representing the height barrier for take-off and landing because the aerodrome sloped 20 feet up to them and pilots claimed their machines had to climb 45 feet instead of the requisite 25. Then when one of the Hawker Cygnets broke a shock absorber rubber ring there was dispute over the replacement because it was $\frac{1}{8}$-inch thicker, though it was finally permitted as a minor repair. Even the Avro team became involved over a hand-starting magneto, for the Stewards at first ruled it part of the engine, which would have brought the Genet engine over the 170 lb limit. After witnessing so much bickering the St John's Ambulance men, after a day with nothing to do, quipped they had better go to the Stewards' tent where they would probably find more bloodshed than from crashes.

'As for the pilots, no praise can be too high,' wrote one of the visitors. 'It is no joke to fly a small machine carrying the limited load allowed by its C of A, and have to think whether one is using a few ounces more petrol than necessary – but to do that and surmount or circumnavigate the Sussex Downs in the face of a gale and heavy rain straight off the Channel calls for skill and judgment and endurance in highest degree.' As a result eliminations followed thick and fast. Only four finished, each with a Cherub Mk

With Genet engine, the performance of the Westland Widgeon, now with green fuselage and silver wings, was enormously improved.

III, so it was a triumph for Roy Fedden. The winner was the Hawker-entered Cygnet piloted by George Bulman 'coolest and most imperturbable of aviators'; second was the RAE Cygnet flown in turn by the Farnborough test pilots Flt Lieut Chick and Flg Off Ragg. Third was the Bristol Brownie flown by Cyril Uwins who, 'whether flying the biggest or the smallest machines shows identical steadiness and reliability.' Runner-up was Frank Courtney with the Parnall Pixie III 'whose views on the competition and those who ran it were as concise and informative as his opinions on the aeroplanes he tests in such numbers.' The last day, Saturday, featured four races, and every available machine competed. My special interest was the Widgeon in the Grosvenor Cup race, but though Longton won with the Bluebird, the Widgeon, distinctive with double-tapered silver wings and green fuselage, flown by Openshaw established the fastest time with 105 mph. Hinkler with the rectangular-winged Avian won the Society of Motor Manufacturers and Traders Handicap, though the Genet-engined Moth flown by Broad was $4\frac{1}{2}$ mph faster; and in the Lympne Open Handicap, Ragg was first with the RAE Hawker Cygnet, and Mrs Elliott-Lynn fastest on her S.E.5a, Flt Lieut Waghorn hot on her heels with a similar machine, and Dudley Watt banking vertically with his Swallow. Between races the RAF made a *divertissement* with Grebes in formation, a Woodcock performing aerobatics, and a Gamecock which made a long inverted descent concluding with a spectacular falling-leaf performance.

Whether the *Daily Mail* competition had been of useful purpose led to more argument, for it failed to show the relative advantage of Bluebird, Moth and Avian as potential flying club machines compared with the impracticable, highly efficient Cygnet. If assessed on a figure of merit representing $\frac{\text{useful load} \times \text{miles}}{\text{lb fuel used}}$ the Avian's was 2,092 compared with 2,203 for the winning Cygnet, yet the second Cygnet had only 1,808. The long-winged Avian's load-carrying ability seemed unprecedented, for her pilot, passenger and fuel represented 130 per cent of the empty weight, a figure far better than Moth and Bluebird, yet all three cruised at almost 20 mpg.

8

One of the visitors at Lympne was the charming Larry Carter, but he looked extremely ill and on that Saturday was hurried to a nursing home at Cheltenham, only to die next day from meningitis, aged 28. He had been one of the original AT & T pilots on leaving the RAF, then joined Handley Page Transport Ltd, but in April 1923 became chief test pilot of the Gloucestershire Aircraft Co, winning the Aerial Derby that year on the 'Bamel'. Long illness had followed the crash of the racing Gloster II at Cranwell, and he had not flown since but was acting as liaison between his company and RAF squadrons equipped with their machines.

Meanwhile Hubert Broad, in addition to his work with de Havillands, had been testing versions of the 1925 Gloster III racers, known as the IIIA and IIIB, with modified tails, wings, and radiators, to obtain further data for design of the Gloster IV intended for the 1927 Schneider Trophy. Mock-ups of this machine and the proposed Supermarine S.5 wire-braced low-wing monoplane were ready, but awaiting replicas of the higher powered, reshaped Napier Lion; and Lieut-Col Bristow's Crusader low-winger, though almost complete, awaited Fedden's promised 800 hp Bristol Mercury engine. With no British machine ready, Italy's Macchi M.39 low-winger, which was astonishingly like the secret S.5, became the only rival to the three American contestants. Nevertheless the Air Ministry at least had formed a High Speed Flight at Felixstowe, where Wg Cdr Maycock was C.O., and from its team would be picked the pilots for next year's race.

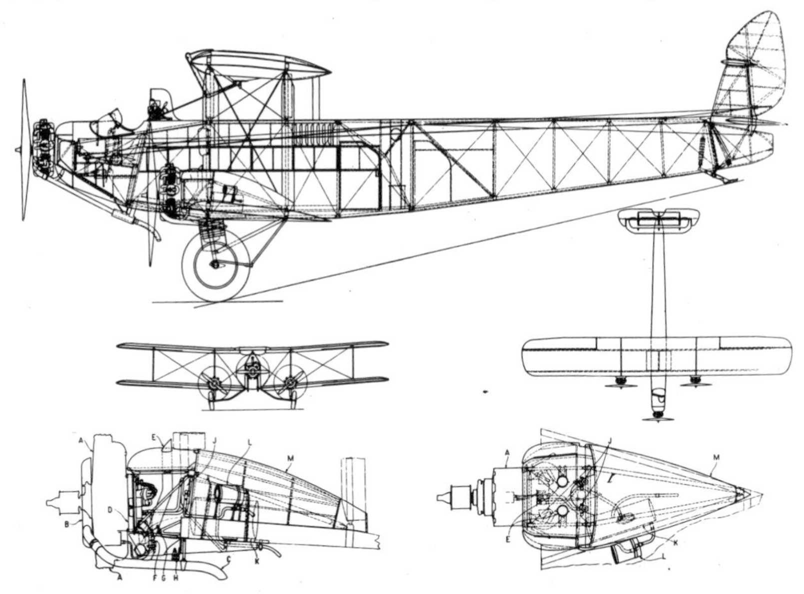

Maker's drawing of the de Havilland Hercules three-engined airliner.

Hubert Broad now became busy with the de Havilland Hercules. On Friday 30 September, at tea-time, everything was ready for the first flight, and as he climbed into the pilot's cockpit the works' hooter was blown so that all hands could watch. She seemed to take-off in her own length. Everyone cheered. Finding the controls responded reasonably, he flew on two engines, and then skimmed the length of the aerodrome on one. Next day it was flown both by Broad and Geoffrey de Havilland, with Walker and Hagg as passengers. But although lateral control was not satisfactory, and the machine returned to the experimental shop to have ailerons fitted to all four wings, it was confidently announced: 'She will be handed over to Imperial Airways in about a month's time, and until 1 January she will probably fly on the cross-Channel services.' By the New Year three would be completed, two of which the Air Ministry proposed positioning in

Cairo, and the third would take Sir Samuel Hoare, Secretary of State for Air, on an extensive tour of the East.

Two days after the first flight of the Hercules, triumphant publicity followed for de Havillands. Cobham arrived from Australia after a flight of 24 days. There had been quick organization behind the scenes; a huge Union Flag was flying from the Houses of Parliament and the riverside terrace was crowded with members and friends when, just before 2.30 p.m., Cobham's D.H.50J, distinctive with long white floats and dark blue fuselage, materialized from the south beneath an overcast sky, circled County Hall and headed upstream, presently returning as the sun broke through, turned above Tower Bridge, side-slipped over the crowds on Westminster Bridge, and alighted smoothly on the Thames. A motor-boat crew from Shorts sped forward to moor the machine, and then transported Cobham and his engineers, Sgt Ward and C. S. Capel, to Speaker's Steps for official welcome by the Secretary of State for Air.

One of the remarkable features of the flight was that Cobham had never flown a seaplane before. Only by trial and error during each stage did he discover the marine techniques, but though he must have severely stressed those splendid duralumin floats which Shorts had built, they magnificently withstood the rigours of so long a journey. Of the enormous publicity contributing to the airmindedness of the British public, Gordon Selfridge had no doubt, for he quickly arranged for the machine to be exhibited in his Oxford Street store, and on the evening of 4 October he held a private view for the Mayors of all London Boroughs, tinctured with a selection of

End of an epic flight: Alan Cobham with his D.H.50 alights on the Thames in the heart of London before thousands lining the Embankment. (*Flight*)

peers and knights having business or political influence. Next day the Department of Government Hospitality gave lunch at the Carlton to Mr and Mrs Cobham – and there Sir Samuel Hoare, having stressed the rareness of a Government extending hospitality to a British subject, announced that His Majesty had been pleased to confer a knighthood on Alan Cobham.

9

The Short Singapore was relaunched on 6 October, with hull leaks cured, and flown for half an hour by Lankester Parker, accompanied by Eustace Short and two others. Next day he began performance trials, attaining 103 kt level – which was faster than the Supermarine Southampton, but the auxiliary trim rudders were insufficient to hold directionally with one engine switched off. Parker recommended having the main rudder operated by a Fletner servo aerofoil of the type he had tried on a D.H.10 at Farnborough. However, the Air Ministry pressed for Service trials, and in typical manner of most manufacturers, the machine was released for critical comment before their own pilot was satisfied. Currently Parker was retesting the Mussel, first as a landplane and then with improved floats, the originals having been fitted to a Moth which was going to the USA for demonstration by Cobham. In the Rochester shops the Cockle miniature flying-boat had returned after six months at Felixstowe and was being overhauled and re-engined with Bristol Cherubs. 'I wanted to scale it up to an airliner,' Oswald Short told me, 'but they insisted on biplanes.'

A monoplane airliner equally appealed to Handley Page, and he persuaded Worley to agree to construction of a 52-ft span experimental high-winger powered with three Lucifers as a half-scale model. He handed design to Harold Boultbee, who was regarded as the monoplane specialist, with instructions to use R.A.F.31 wing section with full-span leading-edge slots and slotted flaps for which the wind-tunnel predicted a sufficiently high lift coefficient for seven passengers to be carried compared with fourteen for the W.8b of almost four times the power. But calculations were optimistic. Climb and take-off of the Hamlet, as it was named, were poor and vibration from the three-cylinder engines was unbearable. H.P. refused to be disconcerted. 'We can always turn it into a twin-engined machine,' he said.

Unchanged as a tri-motor, the Hamlet was one of the attractions at Croydon on 23 October when the annual air demonstration was given to Prime Ministers of the Overseas Dominions. Alongside the Hampstead re-engined with three Jupiters, she seemed a toy, and everyone wanted to try the cosy little cabin. However, it was the RAF with formations of Grebes, Siskins, and Virginias which held the stage, and as examples of progress the stately Horsley and screaming Fox could be compared with the antique war-time Snipe and Vimy still in Service use. The Hamlet was the only representative of aerodynamic progress.

The latest example of that conventional art, the biplane, was the Lion

'If you don't succeed at first, try, try again', thought the great H.P. So the vibration-prone, triple Lucifer-powered Handley Page Hamlet was re-engined with two Lynx and re-evaluated by Tom Harry England. (*Flight*)

VIII powered D.H.65 Hound, tested on 17 November by Hubert Broad, but although able to carry greater military load than the Fox it was initially no faster. De Havillands disappointedly saw their private venturing slipping away in favour of a general purpose two-seater to specification 12/26 using as many components of the D.H.9A as possible but powered with a Jupiter. It was just the type for Westland and Gloster, but D.H. proposed to enter the Hound re-engined with a geared Lion XA, and as second string to rely on their Stag version of the D.H.9A. Vickers similarly took a line of their own with an all-metal version of the Vixen using the Jupiter, known as the Valiant. Bristol abandoned their proposed Type 106 design, and decided on a rebuilt version of the Boarhound renamed Beaver. Fairey, delighted with the IIIF after its initial flight in the spring and promise of orders, proposed to fit it with a Jupiter rather than design a machine with D.H.9A components, arguing that this proposal was another method of standardization.

November also saw the first flight of two other private venture fighters. On the 9th the metal-structured 34-ft span single-seat Vickers Type 123 biplane flew, using a 400 hp twelve-cylinder vee Hispano-Suiza equivalent of Fairey's D.12 Felix. Although the Type 123 had a beautifully pointed nose like the Fox, it seemed of high drag form yet had greater speed and climb than the apparently cleaner Wibault Scout monoplane. The next was a private venture in more than a business sense, for it was a high-wing

Off on its first flight in the hands of Hubert Broad, the private venture D.H. Hound, still of D.H.9A formula, was registered as a civil aircraft. (*Flight*)

Stand-by production for Glosters, as with several manufacturers, was the D.H.9A. The Goral therefore seemed a natural follow up. (*Courtesy Air Marshal Sir Ralph Sorley*)

First visible expression at Westland of a general purpose aircraft using D.H.9A components was this mock-up, eventually to become the Wapiti. Behind is the mock-up of the contemporary Witch monoplane.

Stable companion to the Vickers-Wibault monoplane was the metal-framed Type 123 biplane powered with a 400 hp Hispano-Suiza. Climb was slightly better, speed much the same, and handling typically British.

monoplane designed out of office hours as a 'Racer' by a keen group of Westland draughtsmen including Tony Fletcher of Martinsyde fame. Openshaw was enthusiastic. Mettam did the stressing, and Davenport, the protagonist of monoplanes, gave it his unofficial blessing.

In the hope that R. A. Bruce would decide to build it, the design was kept simple and cheap using a wood structure; the strut-braced wings were untapered, with conventional box spars and Warren girder ribs, and the fuselage comprised a ply monocoque cylindrical central portion containing the pilot's cockpit to which was attached a conventionally-braced rear fuselage of four spruce longerons, spacer struts and wire bracing. Bruce had seen the preliminary design in January. He considered it silently, then started sketching amendments, substituting two cantilever centre-line struts for the usual cabane assembly. It had become his design. 'Tell Widgery to run wind-tunnel tests on a model,' he instructed.

Over 60 aerodynamic tests were made in the next few months, including the effect of slipstream with propeller running and a change from W.4 to R.A.F.34 wing section. Openshaw suggested trying the machine with the Falcon III engine salvaged from the prototype Limousine which had been wrecked when parked at Andover and a Fairey Fawn had charged into it. More wind-tunnel tests with a blunter nose for the Falcon. Construction proceeded. The machine was well advanced in September.

That month Openshaw, on returning from a visit to Martlesham, reported on a promising system of metal construction he had seen in the Hawker Heron, for its fuselage frame consisted of round tubes swaged to rectangular-section at their ends, and joined with fish plates and bolts. Nothing could be simpler. 'Cut off the monoplane's nose forward of the cockpit,' said Bruce. 'We'll try squared dural tubes. It's an easy method of converting carpenters to metal work. The strut ends need not even butt on

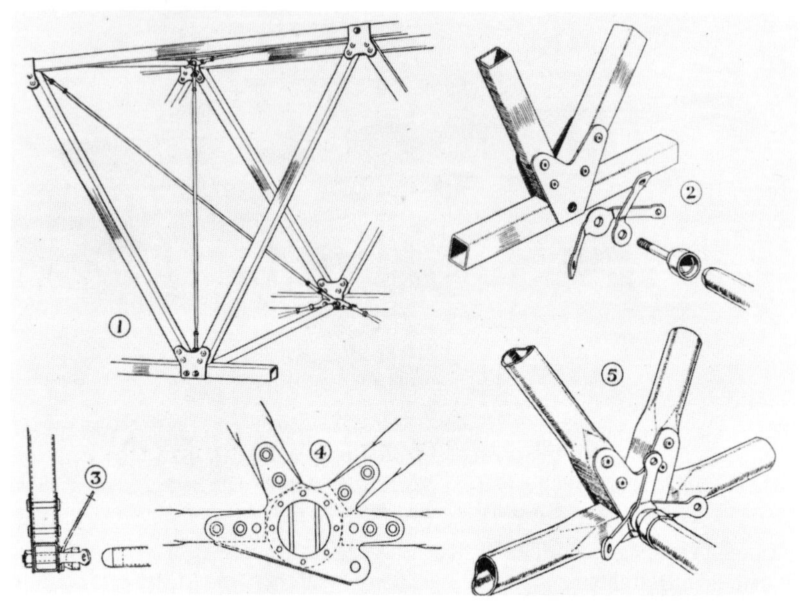

Those self-taught practical engineers, Sigrist and Camm of Hawkers devised a simple Meccano-like method of cheap metal construction compared with the intricacies of drawn steel strip: 1, assembly; 5, typical joint shown in break-down 2 and 3. Even the spar junction, 4, was straight-forward.

Westland were quick to adapt the Hawker square-tube system for the Wizard nose, but instead of butting struts on the longerons left a $\frac{1}{8}$-in gap, and eliminated round tubes and ferrules.

Out for the first engine run, the Westland Wizard, with struts still unfaired and cowling incomplete, had short ailerons, a straight line in the wing-tips, rudder sweeping into the fin and large-chord elevators.

the longerons and the load can be taken through flitch plates like the Hawker system, but use tubular riveting.' Reynolds, the tube manufacturers, readily agreed to draw such a section. In a couple of weeks Fred Rymal, the experimental foreman, had constructed the forward sides in his narrow working bay alongside the Vimy shop.

On the bright Sunday morning of the machine's first flight some fifty draughtsmen and workmen walked or cycled to the factory for the thrill of witnessing their handiwork in the air, nor were they disappointed with the impressive speed, fast climb, and steeply banked turns. There was a cheer when Openshaw landed – but it transpired that the ailerons gave poor response and were too heavy, the rudder ineffective, and the elevators too sensitive. The rear half of the wing-tips was therefore rounded off, the ailerons made longer and their horn balances increased, a crude envelope of plywood was fitted to extend the top of the rudder, and the chord of the elevator was reduced by inserting a false trailing edge leaving the original

The Wizard for its second flight had rounded wing-tips (new dope clear to see on the original photograph), long ailerons with deeper horn, narrow elevators, and primitively extended rudder.

exposed. Flights with the Wizard, as it was presently named, suddenly ended with an airlock just after taking off from the steeply arched 400-yard width of Yeovil aerodrome, but though this clean monoplane was a 'floater' Openshaw managed to get it down on a narrow strip among the spreading houses where it ran into a hedge and turned over. A complete re-build was necessary.

Perhaps the Wizard decided Bruce to redesign the taper-wing Widgeon for greater simplicity with similar 36-ft span untapered wing and Moth type plywood-covered fuselage. With the aid of a carpenter of great skill and experience, mine was the responsibility for its construction. In charge of the design was Tony Fletcher, who since returning from his mission in Japan had been designing propellers for the Falcon Airscrew Co who were now the major producers, though most aircraft builders had a small department making experimental propellers, when it came to trial and error variations to chase missing revs and mph.

Building the Westland Widgeon fuselage upside-down on a jig board was like model making. Longerons were only 1 in × $\frac{1}{2}$ in and the ply cover was 1 mm on the rear portion and $\frac{1}{16}$ in forward.

The Widgeon, as originally drawn, had Martinsyde features of raked wing-tips and tailplane, and a rudder with typical Fletcher outline; he even incorporated duralumin flitch plates for the centre-section strut attachments to the fuselage. But Bruce had his own ideas, which caused Fletcher to assume a look of suffering when the wing-tips were changed to semi-circles and fin and rudder drastically changed, even the flitch plates shrinking to steel. 'My boy,' Fletcher would say mournfully, 'when I was in Japan I had a free hand.' As the weeks passed this lonely, brilliant man turned more and more to himself, presently he vanished, and I missed his rambling accounts of early days interspersed with the beauties of Japan. Thereafter Toby Heal took charge of Widgeon design.

Soon the Westbury C.O.W. gun fighter made its first flight, using Andover as a safer venue than Yeovil. Next day Openshaw flew it back to Westlands. Thereafter my responsibilities for the Widgeon were happily interspersed with duties as observer on Westbury test flights. From its deep cockpit behind the wings I was introduced to the vista of sunlit

The second Westland Westbury had a rounded nose, and metal-covered centre-section guarding blast from the C.O.W. gun when fired forward and up.

England from greater heights than I had ever experienced before, for my task was to record times and altitudes, and after landing I practised the calculations learned at college, recasting the figures to standard atmosphere, helped by Openshaw who was a pilot-scientist like Geoffrey Hill, MA, and Stuart Keep, BSc. It is a misconception to believe that all early industrial test pilots had no scientific training.

At that time single-engine ability for twin-engined aircraft was not mandatory. The need was only to get the machine back in that condition from the other side of our moat, the English Channel. Germany was again the potential danger area, for as a sequel to the Locarno Treaty the German Government had been notified that the Inter-Allied Aeronautical Guarantee Committee had ceased activities on 1 September in accordance with agreement at the Paris Conference the previous May. This removed all restrictions known as the Nine Rules, which previously prevented Germany from building large civil aircraft or importing aircraft suitable for war, though the hopeful law-makers ruled that the aeroplanes must be such that machine-guns, torpedoes, or bombs could not be fitted. However, Germany was allowed to have thirty-six officers of the Army and Navy trained as pilots, and annually for the next six years six more could be trained; nevertheless the door was opened wide by permission for fifty of the Police Force to be trained as pilots – as an agitator later named Hitler noted with interest.

Handley Page growled the warning: 'We are spending £120 million per annum on Defence, yet only £15 million of it on Air Power, though aircraft are the only means by which we could be attacked. By increasing our Air Power we could effect real economy among the other Services. If we could induce people to think of Air Power in terms of bases on the territory of friendly nations, as for example Gibraltar, Malta and Aden, among our sea bases, we could run airlines from those places until civil aviation can fly by itself.'

His advocacy that twin engines were a safeguard against forced landings was badly jolted when a Handley Page W.10, piloted by the experienced Capt F. A. Dismore, with ten passengers, a mechanic and a Pomeranian dog aboard, descended in mid-Channel after one of the Lions stopped. He alighted without turning the machine over, but only by luck were all except the dog rescued, for the machine carried neither rafts nor life-saving equipment. Nevertheless Sir Samuel Hoare at the Imperial Conference was able to say: 'Five million miles have been flown by British services with only four fatal accidents. There is no technical reason why Canada should not be reached in 2½ days, India in 5, Cape Town in 6, Australia in 11 and New Zealand in 13, yet we Nations of the Empire are too hard up for anyone to undertake the cost of an air route to Singapore or Cape Town – but if the RAF undertake Service flights in conjunction with the civil lines, and if the Governments of South Africa and Australia allow their Air Forces to co-operate, we should get valuable military training for Service units and make possible the experimental transport of mails and passengers over a large section of the British Empire.'

As substantiation the President of the Air Council agreed with Imperial Airways Ltd to subsidize, to an annual total of £93,000, a fortnightly service between Cairo and Karachi beginning not later than 1 January, 1927. Three of the D.H. Hercules were ready for this purpose.

To show that the world was at the disposal at least of private owners of aircraft, T. Neville Stack, chief instructor of the Lancashire Aero Club, and his friend Bernard S. Leete of the same club, were each flying a Moth in company across Central Europe to Constantinople and thence to Baghdad, carrying with them their mutually favourite possession, a banjulele. On landing at Hinaidi on 16 December they received a jubilant welcome from No. 70 Squadron of which Stack had been a member when in the RAF. The *Baghdad Times* in a felicitous article expressed the general feeling: 'Perhaps the outstanding impression left in our minds is one of wonder that so much may be achieved with so small a machine and an engine with a nominal horsepower of only 26, for there are bigger engines in many motor-cars.'

10

On 13 November the ninth contest for the Jacques Schneider Trophy took place on the waters of Hampton Roads, Virginia, and was won by Major Mario de Bernardi, representing Italy, at an average speed of 246.5 mph. For the moment the Trophy was saved, and now there was real impetus for Britain to press forward with its Gloster, Supermarine and Short racers. Nevertheless it was clear that the Italians had made a tremendous step forward, for in the previous year's race they had attained only 168 mph, but now their low-wing monoplane, powered with an 800 hp Fiat, made even the American biplanes look somewhat antiquated.

Only the Italians were able to challenge the current American holders of the Schneider Trophy at Baltimore in 1926, and they won with a clean, though wire-braced, Macchi low-winger piloted by the famous Bernardi. (*Flight*)

To Mussolini in Rome, Bernardi cabled: 'Your orders to win at all costs have been obeyed. We have won at a speed never before attained.' To which the Dictator replied: 'To you, to your fellow aviators, to all personnel who have collaborated on the conquest of the dearly sought prize, Italy sends her greatest applause for this superb victory. This Italy, who has followed you and guided you in this perilous and difficult task, is to-day proud of the men who by their genius have created such powerful machines, and is enthusiastic of you magnificent pilots who have brought them to triumph.'

In less ecstatic though equally determined mood, Alan Cobham Aviation Ltd was registered a few days later as a private company with nominal capital of £1,000 and the object of 'manufacturing and dealing in aircraft of all kinds, carrying air passengers or goods, proprietors of aviation, teaching and training schools and aerodromes etc.' It was backed by Lieut-Col Warwick Wright, DSO, brother of the pioneer Howard Wright, and Emile Merkel who had been Grahame-White's adviser. Sir Alan Cobham's world-wide experience had become a valuable asset; he could leave de Havillands secure in his prestige. Few could match his hours as pilot or the safety with which his flying operations were conducted. Concurrently the Gloucestershire Aircraft Co reregistered as the Gloster Aircraft Co, adopting the phonetic name of its racers.

By comparison with civil aviation the record of the RAF was disastrous. Since 1 January there had been fifty-one fatal accidents embracing every type of Service aeroplane. Fifteen occurred with Bristol Fighters and eleven with D.H.9As. One out of 118 wing commanders was killed, two of 236 squadron leaders, three of 664 flight lieutenants – but flying officers, because of limited flying experience, suffered most, with 30 out of 1,100. Most accidents involved spinning. Though Farnborough had shown that a bigger rudder on the Bristol Fighter would give immediate recovery none had been ordered, nor the special wings which Handley Page Ltd had demonstrated with interlinked tip slots and ailerons for they necessitated a redesigned wing structure because the aerofoil section differed from the empirical Bristol shape. It was cheaper to continue with accidents, particularly as presently the Atlas would replace the Fighter.

Having proposed the idea that a slot aerofoil could be kept closed until near the stall and then unlocked, when it would automatically move forward, H.P. pondered whether it really needed mechanical means to close it again. He instructed dapper George Russell, who had been in charge of the wind-tunnel since Reynolds left, to make a detailed study of the relationship between the operating mechanism and the magnitude and direction of forces acting on the slat at different angles of incidence on the main wings. After experimenting with a host of different shapes and arrangements in his wind-tunnel, he at last achieved a slat which was sucked away from the main wing just before the stall. In conjunction with S. C. Ebel of the D.O., a hinged linkage was devised using cardboard models and elastic spring bands. Applying the system to a wind-tunnel model was immediately successful, the slats automatically opening against the pull of the springs which closed them as incidence decreased. 'We'll get the Air Ministry to give us a contract to try it on the Harrow,' H.P. said, confident of overcoming all argument even if it took months.

Perhaps it was his parsimony which explained the absence of any example of slotted wing at the Paris Aero Show which opened at the beginning of December. Nor were there any other British aeroplanes for they were all needed for Government contracts, though there was a splendid display by Armstrong Siddeley with Jaguar, Mongoose, and Genet engines; and Bristol on a somewhat obscure stand showed three versions of the Jupiter and the latest Lucifer; but the water-cooled engine manufacturers, Rolls-Royce and Napiers, had no stand. Nor, by adroit manoeuvre of the *Commissaire Général*, had the Germans, to whom it was explained that all disposable space had been allotted before their machines were qualified for admission to the Grand Palais. Avia of Czechoslovakia, Koolhoven and Fokker of Holland, and Fiat of Italy were the Internationalists, but overwhelmed by twenty impressive French stands each containing several aeroplanes.

French design ideas seemed to have reached saturation compared with the previous show, but most of their machines were aerodynamically cleaner than British standards. Whether bombers or fighters, French priority was speed coupled with multi-guns. They had no lumbering great bombers such as the Virginias on which the RAF was concentrating, but preferred small bomb loads and heavy armament to drive off intercepting fighters. As W. H. Sayers reported: 'Whereas British designers may be suspected of too great a regard for "safety first" and seem loath to abandon the well tried braced biplane, the French try to produce something aerodynamically novel in hope that it will prove efficient. It is difficult in the face of its achievements to believe that the Breguet XIX is anything other than one of the world's very best aeroplanes, and the amazingly large share of records held by French aircraft could never have been secured by poorly designed and badly constructed machines. On the other hand no

Air Ministry stress merchant would pass the Breguet system of carrying the main load wires to the undercarriage, and, worse still, through the centre of the wheels.'

In arrangement of exhibits the French were perfectionists. There was no competition in meretricious decoration; all firms had identical signboards. Even the unsightly glass and ironwork of the Grand Palais roof was hidden by an ingenious cloth giving the impression of a transparent tessellated pavement. Said C. G. Grey: 'The French certainly know how to arrange a show of this kind, and if ever we rise to the dignity of an aero show in England one only hopes the organizers will copy it as closely as possible. We need not make our show as pretty, but we can certainly make it as practical.'

At home the *Daily Express*, the most influential of the lower-priced daily papers, was agitating for an enquiry into the mounting death roll of the RAF. Mr Hore Belisha, Liberal MP for Devonport, asked the Prime Minister whether he would allot a day to discuss the cause of these fatalities. Mr Baldwin, though deeply deploring such accidents, said he was satisfied that every possible precaution was taken and would continue to be taken to safeguard the lives of RAF personnel – and he courteously but definitely refused to give time for discussion.

By contrast, happily strumming their way, Neville Stack and Bernard Leete arrived at Baghdad on 16 December with the banjulele still safe. *The Times* commented: 'The RAF appreciates the difference of this remarkable journey from similar more widely advertized long-distance flights which included Baghdad as a port of call. The ground organization appears to have been comparatively trivial compared with the elaborate arrangements for other memorable flights, and the pilots are carrying out the duties of navigators and mechanics in addition to the routine task of flying the

The de Havilland Hercules prototype, with a delighted Sefton Brancker aboard, sets out for Heliopolis. (*Flight*)

machine.' They were also adding garlands for the Moth. It was a winner everywhere. The clubs were prospering, and recently the Hampshire Aeroplane Club was added to their number, backed by O. E. Simmonds, MA, one of the Supermarine technicians who had played a part in the designing of the Southampton; as secretary there was Bobby Perfect, a very amiable young nephew of A. V. Roe.

Before the year ended, the Fédération Aéronautique Internationale, by unanimous vote of the thirteen countries represented at its Conference in Paris on 16 December, awarded the FAI Gold Medal for 1926 to Sir Alan Cobham for his flight from England to Australia and back.

Winter's cock-crow embarkation in the Hercules. Capt Frank Barnard in leather coat, Lady Maud and Sir Samuel Hoare, with Sir Eric Geddes (*centre*), friends and officials, to see them start on what was regarded as an epic voyage – from England to India. (*Flight*)

Two days later the Hercules, G-EBMY, piloted by Capt C. F. Wolley Dod, left for Heliopolis to inaugurate the desert air mail service. As passengers he had the indefatigable Sir Sefton Brancker, Air Commodore and Mrs J. G. Weir, and Capt T. A. Gladstone who would operate the Cairo—Kisumu Airline. The second Hercules, G-EBMX, piloted by Capt F. L. Barnard and navigated by Sqn Ldr E. L. Johnston, took-off on Boxing Day, in the first grey light at 7.35 a.m., for the first stage of a flight to India carrying Sir Samuel and Lady Hoare.

CHAPTER VIII
AMBITIOUS WINGS
1927

'Now though this industrious Workman should not be able to bring his invention to so great a perfection as some may imagine to themselves; yet those that shall be so happy as to bring it to perfection will be obliged to him for having made known his first attempts.'

Robert Hook (1679)

1

On the morning of 8 January the sun-baked roads leading to the RAF's landing ground at New Delhi were crammed with turbaned Indians walking, running, or riding bicycles amid scattered cars and tongas. All knew that a great mechanical bird was bringing an English Sahib Lord that morning. A Bristol Fighter caused a pitch of excitement but was a false alarm. Fifty minutes later, Sir Samuel Hoare's Hercules came into view, circled, and swept nobly in to land. Waiting to greet him were Lord Irwin the Viceroy, and the C-in-C, Sir William Birdwood. The pilot switched off the nose engine, and taxied towards them, followed by a rush of spectators. Ceremonies and celebrations filled the next few days. Lady Irwin named the machine *City of Delhi*; prominent Indians were taken for flights; then the Hercules was flown to Ambala to be housed by No. 28 Squadron until the return journey planned for February. Meanwhile Sir Samuel and Lady Hoare fulfilled an extensive programme of visits, inspections and discussions. Not until 17 February did they arrive back in London, after completing the last stage from Paris by train and ship because of fog. C. G. Grey expressed the prevalent feeling: 'One tenders the congratulations of all concerned with British aviation on their happy return from a journey which, if not beset with perils, was at any rate infested with rather more than sporting risks. Few people in their circumstances, endowed with wealth, social and political power, and with a family of charming children, would risk their lives merely because they considered their duty was to set a good example to the travelling public.'

Much concentration on Indian air route possibilities followed. To steer matters, Lieut-Col F. C. Shelmerdine, an ex-officer of the Green Howards, recently Controller of aerodromes and licences in the DCA and later Civil Air Traffic Officer (CATO) for the Cairo—Karachi route, was promoted to Controller of Civil Aviation for India. His place as CATO was taken by Frederick Tymms, the Air Ministry navigational specialist who had navigated so many trail-making flights, such as the airline to Berlin, that his knowledge of aerodromes was encyclopaedic.

Meanwhile the ever cheerful Stack of the white Moth and Leete of the red had arrived at Karachi on the same day the Hercules reached Delhi, and were royally received by Wg Cdr Bone, CBE, DSO, and officers of his command at the RAF's Drigh Road aerodrome. Everywhere their flight was widely publicized as 'The world's longest light aeroplane flight'. Five weeks later, they took-off for Delhi, and in March began the journey home.

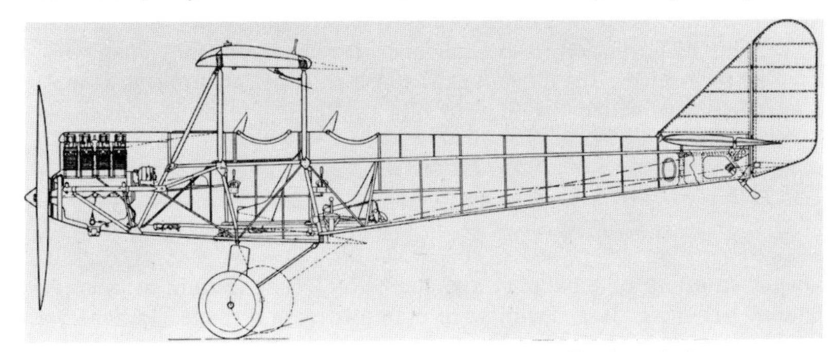

Maker's production drawing of the now Moth-like Avro Avian.

Timed with this publicity, de Havillands announced a further reduction in Moth price to £730 ex-works. It was followed by the first sign of intense competition from A. V. Roe & Co who advertized that the production Avian with Cirrus Mk II would be £675 ex-works, or £750 with Genet. 'Service arrangements have been completed, and spare parts are obtainable from either the Hamble or Manchester Works. Pilots flying an Avian may land at Hamble or Woodford free of charge, and no charge is made for housing their machines overnight.'

It transpired that Bert Hinkler was leaving Avros and planning an ambitious flight to Australia using the prototype Avian. His was the credit for making the first aeroplane flight with a diesel, for he had just completed tests of the ponderous, hard-worked three-year-old Aldershot with 850 hp Beardmore Typhoon inverted six-cylinder inline engine replacing its 1,000 hp Napier Cub. A derivative, the heavy-oil Tornado diesel engine, was intended to power the R.101 because of its reduced fire hazard in so inflammable a device as a hydrogen-filled airship, but it also saved 30 per cent consumption as compensation for weighing three times as much as a petrol engine.

Torsional vibration of the Tornado's crankshaft was a difficulty matched by flutter problems with several aircraft, notably the Gloster Grebe, which Flt Lieut D'Arcy Greig with his 'terminal' dive had proved

As with the Gloster Grebe, it was found essential to brace the overhangs of the Gamecock with V-struts, and to mass-balance and match the new ailerons.

cured with reshaped ailerons, altered control attachment, and additional V-struts to limit deflection of the spar overhang. In quest of solutions, Professor Bairstow mathematically investigated flutter, and corroborated that the C.G. of an aileron must be arranged to lie on its hinge axis by using counter-balance weights as originally proposed by Baumhauer of the Amsterdam Aerodynamic Laboratories in 1923; it next seemed obvious that elevators should be interconnected to avoid differential flutter, as foreshadowed by the Handley Page O/100 ten years earlier.

Flg Off H. J. T. Saint, DFC, recently of the RAE Test Flight, and previously senior pilot of AT & T, had just been appointed Gloster chief test pilot in place of Larry Carter. Modified Grebes were his immediate

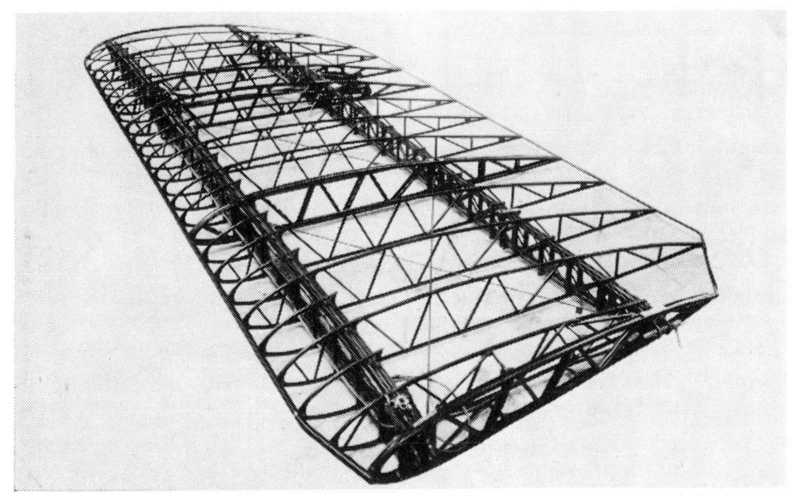
The Steel Wing Co associate of Glosters designed a patented metal wing with lattice spar for the Goldfinch all-metal development of the Gamecock.

task, but soon there would be the Gorrel general-purpose machine and the metal-structured Goldfinch day and night fighter which had a lattice girder spar for the thick-section top wing, and a box spar developed by the Steel Wing Co for the thinner lower wing. On these machines hope of expanding production rested, and even stimulated Henry Folland's intense interest in the three Gloster IV racing seaplanes which the company was building for the 1927 Schneider contest, for intensive wind-tunnel research had resulted in a pretty biplane layout of 40 per cent less drag and 15 per cent more lift than the Gloster III. Design of its major components was by Hubert Martyn, son of the founder director of the company. 'The work was done in Mr Folland's private office to ensure non-leakage of information,' he wrote, 'and was to the instruction and supervision of Mr Folland and his assistant Mr Preston.'

Folland was also co-operating with flight trials of a Bristol turbo-supercharged Jupiter fitted to the Gloster demonstrator Gamecock G-EBOE, but the engine was proving unreliable and the supercharging system unsatisfactory.

Although experimental work on supercharging had been initiated a decade earlier, there had been no great advance despite work on exhaust-driven turbo-compressors by J. E. Ellor at Farnborough, Rateau in France, and later by G.E.C. in the USA; but in Germany and Switzerland there had been research on gear-driven superchargers of multi-stage type. Lecturing on 1 February to a combined meeting of the RAeS and Institution of Automobile Engineers on *The Supercharging of Aircraft and Motor-vehicle Engines*, Roy Fedden explained such things as the difficulty of manufacturing turbine wheels and casing with rotors running at 27,000 to 30,000 rpm in exhaust gases at 650 deg C to 700 deg C; that maintenance of gas-tight exhaust joints had proved more troublesome than any part of the system; and that there were problems of valve timing, the effect of blowers on crankshaft synchronous speed, reduction of thermal efficiency and difficulty of engine and gas cooling, snags with lubrication, and excess noise. He prognosticated eventual better fighter performance with a ground-boosted gear-driven blower engine of smaller capacity than naturally aspirated engines currently used, and that for larger general-purpose machines greatly improved performance would result from supercharged engines throttled on the ground but opened to full power at some pre-determined altitude, while for long-distance bombers, operating at high altitude where fuel consumption was vital, there seemed an important opening for exhaust-driven turbo-compressors.

Nevertheless Farnborough's 'Jimmy' Ellor was convinced that supercharging would become practicable much sooner if designed around the principle instead of attempting to accommodate the principle to a design intended for a naturally aspirated engine. That was what Royce proposed Ellor should do in conjunction with Rowledge, for Lieut-Col Fell became instrumental in securing Ellor's appointment there and he himself joined the company in charge of their London office in Conduit Street.

There had been much flying with the single-seat C.6C Autogiro and the two-seat C.6D, both of which Courtney had tested the previous year, and Cierva had been his first passenger. The two-seater was afterwards sent by rail to Berlin, where Ernst Udet flew it at Tempelhof. On returning, it was fitted as the C.8R with tapered paddle blades, and stub wings were added to relieve the rotor load in forward flight. Earlier the Viper-powered Avro 552A version of the 504K with a similar rotor system had been flown extensively by Bert Hinkler, but now was acquiring a tricycle undercarriage with double front wheels. The stub wings, having proved successful, were applied to the single-seater which successfully flew in this guise at Worthy Down on 1 January, but on 7 February a blade broke away during flight at Hamble and Courtney crashed from 100 ft, 'the machine fluttering down instead of falling direct, the pilot escaping with severe shaking and a broken rib or two. He was taken to Hamble Hospital, and is progressing satisfactorily.'

Frank Courtney flying the C.6C Autogiro with outrigger ailerons before it was fitted with stub wings having conventional ailerons.

In a letter to Reggie Brie forty years later, Courtney wrote: 'I quit my association with Cierva after this, though few know the real story. I had discovered during my Berlin demonstrations that the blade roots bent in the drag direction, so I proposed that a vertical hinge should be added to the horizontal hinge to take care of the drag load. Cierva was very angry at the suggestion, and became even more angry when Parrott of Avros and Bolas of Parnalls supported the idea. He considered that his funny weighted cables between the blades were sufficient to take care of whatever drag differences there might be. When the blade eventually broke, through fatigue induced by drag reversals, Cierva lost his head and tried to blame me and everyone else, but thereafter he included vertical hinges. By then I had had enough of Cierva's indignation at suggestions coming from anyone else, and we parted.'

Courtney was in further trouble on recovery, which perhaps gave him a jaundiced view of his future in England. By the appointment of burly Bennett-Baggs as chief consultant pilot to Armstrong Whitworth and transfer of tall, reserved Alan Campbell-Orde from the flying school to become chief test pilot, Courtney had lost a major source of work and angry words followed. Thereupon Armstrong Siddeley Motors sued him for £347 1s 8d for a car. Admitting liability, Courtney counter-claimed £290 for another car Armstrongs had presented for winning the King's Cup in 1923 and afterwards sold for him, and added £352 as commission owed to him plus damages for alleged wrongful dismissal. At a later hearing Lord Halsbury, his KC, said Courtney did not wish to proceed and withdrew suggestions of dishonesty by John Siddeley and Major Green. 'That comes from a very gallant gentleman, and it does him great credit,' said his Lordship, and made judgment for the plaintiffs with costs. Since all manufacturers now had their own test pilots, Courtney, despite a tremendous reputation for test flying and aerobatics, seemed unemployable, for his glasses precluded a 'B' licence for commercial flying.

Flying certainly remained an art. 'On Friday last Mr Barnard brought off one of the finest landings yet seen at Croydon,' wrote a reporter. 'He came from Paris on an Argosy when the S.W. gale was at its height, and did an autogiro landing, without run, on the tarmac in front of Customs. Twenty men were needed to hold the machine when it was on the ground, and he found taxi-ing to the hangars so difficult that he decided not to take the risk, so took it into the air again from where he landed and put it down in the same fashion outside its hangar door.'

The new Terminal building, hangars and workshops at Croydon were regarded as a great step forward in airline facilities. (*Flight*)

It was not surprising that the number of passengers failed to increase as fast as expected. With a head gale, flights had to be cancelled. In any case it was vital to give passengers greater sense of security. Most of them spent the flight watching vibrating wires and drumming fabric which seemed on the point of bursting, gritting their teeth against the din of unsilenced engines, and tightening stomachs against the bumps. Splendid new terminal buildings under construction at Croydon were certainly reassuring with their air of solidity. The ramshackle group of hangars and temporary offices in the middle of the aerodrome would soon be pulled down to make one large unencumbered flight area. A cone light at Lympne, a neon at Croydon, and beacons at Cranbrook, Tatsfield and at the emergency landing grounds of Penshurst and Littlestone, were proving invaluable on the Paris run except in fog, though the first two were respectively 600 ft and 800 ft above sea level on steel towers clear of surrounding trees, and automatically emitted six beams in two groups of three flashes. There were also floodlights at Croydon and Lympne used at the moment of landing to form an illuminated semicircle of 1,500 ft radius. By contrast there were thirteen great air mail routes flown by night in the USA, and all plentifully illuminated. Thus between Los Angeles and Seattle there were ten revolving beacons of 7½ million candle-power and twenty intermediate beacons of ½ million candle-power. Yet it was radio and not lights that must be the ultimate answer.

What might be done in financing civil aviation additionally to air defence was dictated by the Air Estimates for 1927–28 published on 5 March. Parliament was asked to vote £15,550,000, representing a reduction of £450,000 on the previous year, but the amount for ordinary services in the RAF was little altered, saving having been made on personnel, works, and buildings. Nevertheless there was a heartening increase of £635,000 for technical equipment, including new types of aircraft, for: 'The policy of replacing aeroplanes and engines of war-time design by modern types is making steady progress and it is the intention that in future no more aircraft or engines of war-time designs should be bought.' An allocation of £137,000 was to subsidize Imperial Airways for European services, and £93,600 for the Cairo—Karachi route. Alterations at Croydon Airport, which would be completed in 1928, would absorb £111,000, but a further £10,000 was included for a new wireless telegraphy station, and £8,000 for meteorological services.

Of the Debate on 11 March, *The Times* reported: 'Sir Samuel Hoare has all the scrupulous precision which he proudly ascribed to-day to British aeroplane engines, and, like them on his great Imperial flight last year, he went through the long journey of his Estimates purring like a kitten.' That did not prevent criticism from both Conservative and Labour MPs. Lieut-Col Moore-Brabazon, who recently had resigned his parliamentary secretaryship to the Minister of Transport to become chairman of a firm pioneering distillation of coal, said Sir Samuel's triumphant announcement that we would no longer use machines of war-time design was a cynical reflection on the absence of peace-time progress in the nine years since the war, whereas during the war we changed from one design to another in three or four months, and the Italians with their Schneider Trophy machines did in six months what our technical experts said could only be achieved in two years. Absorbed by his own rhetoric he misleadingly pressed for a considered programme of steady aeroplane construction, similar to that for the Navy's ships – but that was exactly what the Air Ministry was endeavouring to do within limitations of available finance.

The most widely used version of the Armstrong Whitworth Siskin was the 156 mph Mk. IIIA which had a supercharged Jaguar and D.H. profile to its fin and rudder. Entering service in 1927, it eventually equipped eleven squadrons. (*Courtesy Air Marshal Sir Ralph Sorley*)

Though both disarmament protagonists and those who considered the Air Vote disproportionately small compared with the total Defence Bill of £115 million participated vociferously, it was the accident rate which became the major theme. In reply, the Prime Minister said that after studying the accident reports, criticisms, statistics, comparisons of aircraft and place, analyses of causes, and flying history of the pilots, he came to the conclusion that the main cause was the personal equation. Pilots were adventurous, quick brained, with great reserve of nerve; a young man of that temperament became elated, and risks in his training were unavoidable. It drew from C. G. Grey the comment: 'Mr Baldwin's English was as beautiful and his logic as unassailable as is customary in all his public utterances. The only pity is that he should have taken so much trouble personally to investigate this question of RAF accidents and should then have based his arguments and assumptions on entirely false premises owing to lack of essential information. What Mr Baldwin evidently did not learn in the course of his investigations was that a very large proportion of RAF accidents would be prevented if Air Ministry technical experts could be forced to work with greater speed and greater intelligence to equip the Air Force with apparatus which would prevent a large proportion of such accidents.'

With engines in two tandem pairs, the Super Wal was a logical step for Dornier and found an immediate market. (*Deutsche Lufthansa*)

There was no indication in the Estimates of intensive development of commercial aircraft, but the Germans, released from bondage on aeronautical development, were openly displaying the impressively large and remarkable all-metal aircraft they had built in neutral countries during the ban. There was the Dornier Super Wal, with two 650 hp Rolls-Royce Condors in tandem on the centre-section of a braced monoplane wing mounted well above a long, cruiser-like, metal hull which accommodated 21 passengers, pilot and navigator, and had even carried 60 passengers across Lake Constance, flying low in the surface-cushion of air. Equally remarkable was Rohrbach's distinctive Roland three-engined all-metal shoulder-wing landplane which had elliptical wing-tips instead of rectangular like his flying-boat. Flying from Staaken, home of war-time German bombers, it had broken five weight-carrying records for land machines, taking-off in 15 seconds at an all-up weight of 7,100 kg, and could maintain height on two engines. That was followed on 16 March with a world duration record by a Junkers-W33 monoplane which remained airborne almost 15 hours carrying 500 kg useful load. Behind these activities loomed the figure of Major Martin Wronsky, managing director of Luft Hansa, the biggest air traffic combine in the world. On 24 March he lectured to the RAeS on German commercial aviation, drawing a large audience to the Royal Society of Arts.

Luft Hansa had been formed in the autumn of 1925, and in 1926, during which scheduled services had begun on 6 April, flew over 3,800,000 miles, carrying 56,268 passengers, and 950 tons of luggage, freight and mail, compared with 853,042 miles and 11,395 passengers by Imperial Airways in its initiating twelve months of 1924/25. Wronsky stressed that all-metal machines were cheaper to maintain than wooden ones, though Fokker's wooden wing and welded fuselage were in fact cheaper still. Great importance was attached to comfort, and Luft Hansa insisted on relatively spacious cabins, so that passengers could move about freely, and in some cases adjustable seats convertible to sleeping berths were used. Engines were silenced so that passengers could converse without shouting. Pilot training took two years at the German official Air

Traffic School at Staaken, where a 'B' licence was granted after 5,000 kilometres minimum cross-country flying and appropriate examinations; 'C' and 'D' licences followed after further experience and examination. Pay was compounded from a fee per kilometre and marriage and child allowances, amounting to £40 to £50 per month, with a premium of £5 for every 5,000 kilometres without accident. All passengers were insured for £1,250. Said Wronsky: 'International services such as Luft Hansa were only possible by international agreement and co-operation. The first practical example was Anglo-German, and it was Sir Sefton Brancker who was mainly responsible for proposing an International Air Traffic Convention and the ensuing foundation of IATA in 1919. In some countries it was alleged that German effort in commercial aviation was aimed at establishing German air supremacy in Europe. This was a mistaken view: Germany was forced by her position and economic conditions to play a very important part in European air traffic, and to that end Luft Hansa was operating joint services with most of the airline companies of Europe.'

Except for the cross-Channel run, commercial air routes were once more non-existent in Britain, but the flying clubs were steadily gaining ground. At Filton the Wessex Flying Club had been formed as a limited company and two Moths were purchased; the Suffolk Flying Club at Hadleigh, not far from Ipswich, had been trying the prototype Bluebird which recently toured the Continent piloted by the Royal Aircraft Establishment's Flt Lieut Chick, a tough rugger player; Dallas Brett at Lympne was endeavouring to form the East Kent Club, later known as the Lympne Flying Club; Scotland was forming the Glasgow Flying Club. All over the country the number of pilots and private owners was increasing. Among those who had recently learned was Harold Bolas on the Parnall Pixie, de la Cierva at the Hamble Club on a Moth, and three women pilots who had learned on Moths and bought their own machines – Sicele O'Brien, Winifred Spooner and Lady Bailey.

At Westland there was high hope that the Widgeon III, whose construction had been my care, would attract as many owners as the Moth, for its

Laurence Openshaw ready to take the Westland Widgeon III for its first flight before painting. The exhaust stack pipe over the centre-section petrol tank was viewed with suspicion and soon changed.

'Hop in, young Penrose.' It was a novelty to have such easy access as to the front cockpit of the Widgeon III.

36-ft parasol wing gave no obstruction to the great vista of countryside, yet could be more easily folded than any other as there was only a single pin each side to undo, and jury struts were unnecessary though essential on biplanes to hold top and bottom spar roots apart and structurally boxed. The reflexed R.A.F.34 wings were similar to the Wizard's, but instead of the unsatisfactory horn-balanced ailerons there were narrow full-span ailerons built on a torsion tube hinged to a false spar and operated from a strengthened inboard rib by a short cable to the pilot's controls. Because of negligible C.P. travel, both Wizard and Widgeon had remarkably small tailplanes.

There had been a tense moment after the four sets of inverted-V centre-section struts had been bolted between longerons and centre-section, for the fuselage had to be sawn through on the starboard side to make the triangular opening for the door and it seemed that the fuselage inevitably would sag – but the cabane had been designed so that the centre-section in effect became the missing longeron. The engine was mounted; the 'tin-bashers' began forming the cowling by eye and the aid of brown paper patterns which they cut to suit; towards the end of March the wings were fitted, and with the fuselage still in its undercoat, we pushed the machine out for its first flight in the hands of Openshaw. There was a smile when he landed: 'Nothing needs changing – hop in, young Penrose.' But this time it was different from all other flights: this time it was an aeroplane which seemed part of me, mine, for I had watched it grow from baulks of wood and sheets of ply, strips of metal, and a roll of fabric.

The Widgeon's public bow was to be the Bournemouth Easter Meeting on 15, 16, and 18 April. The fuselage was painted Keep's favourite unimaginative green. I made a stencil of a flying duck which was painted on the fuselage. The registration G-EBPW pleased Percy W. Petter, co-owner of the Works, as the last two letters were his initials. My work on the machine was finished; I was transferred to Capt Geoffrey Hill's three-man D.O., designing the Pterodactyl side-by-side two-seat pusher monoplane which was already well advanced, and Herbert Mettam was busy with the involved stressing calculations of the swept wing and its elbowed leading edge. Mine was the job of draughting the tapered cantilever tubular spar of the combined lateral and longitudinal wing-tip 'controller' (elevon), but I was somewhat shaken at having to put into practice the art of stressing I had learned at college.

Openshaw had a narrow escape in the nose cockpit when the Pterodactyl prototype crashed through inadequate 'controllers'.

Following the crash of the Wizard, another near disaster occurred when Openshaw tried Hill's original Pterodactyl I with 'controllers' of reduced span. Because of its short wheelbase, the Pterodactyl on this occasion took-off prematurely after hitting a small ridge, and though its whole conception was to avoid stalling, the left wing dropped despite opposite control to lift it, and the little machine swung abruptly round, dropping from about 10 ft in a manner reminiscent of the mild crash which had ended Ader's pioneer hop in the tailless Avion thirty years earlier. The left wing struck the ground and buckled, and the undercarriage collapsed, but because of the slow impact the pilot was unharmed. Disconcertingly the alleged safety had proved a myth: new reasons for taillessness took precedence – low drag, good view, potentiality as a fighter, even practical superiority as a flying-boat or an all-wing airliner.

3

The Government announced a notable appointment towards the end of March: 'Mr Henry Thomas Tizard, CB, AFC, FRS, has been appointed by His Majesty the King in Council to be Secretary to the Committee of the Privy Council for Scientific and Industrial Research on the retirement of Sir H. Frank Heath, KCB, from that office on June 1st next.' As a scientist-pilot, Tizard had long made his mark, establishing the standard method of aircraft testing and performance reduction when war-time chief technical officer at Martlesham. Since then, as a Fellow of Oriel College, he had been lecturer in Natural Science at Oxford. In manner his was a blend of the military and authoritative academic; but he was also a great patriot and humanitarian, believing not only in the enormous ability of his countrymen but advocating that German and British scientists should resume friendly relations in the interests of human progress.

It was Tizard who was pressing for the amalgamation of the Royal Aeronautical Society and the Institution of Aeronautical Engineers, insisting there was no room for two bodies devoted to the same professional cause. At the RAeS annual general meeting on 29 March he passed a resolution, seconded by John North, 'That this meeting supports the action of the Council in endeavouring to come to an agreement with the Institution of Aeronautical Engineers on the lines suggested in the Beharrell report, and authorizes the Council to carry the amalgamation into effect with such alterations to the draft terms that appear to them advisable.' This, and an even more specific resolution by members of the IAeE, was carried, though there was some dissention in the RAeS because it was proposed that the status of 'Fellow' should be extended to 'Members' of the Institution, for this seemed to imply some lowering of the standards of the RAeS. At least the amalgamation, when presently put into effect, gave hope of participation at RAeS lectures by the practical men of the industry as well as academics and government scientists. Nevertheless those elected to the new Council failed to include Institution members, but comprised Wg Cdr T. R. Cave-Browne-Cave, Sir Mackenzie Chalmers, A. E. L. Chorlton, C. R. Fairey, J. E. Hodgson, Major R. H. Mayo, Lieut-Col Mervyn O'Gorman, T. O. M. Sopwith, Air Vice-Marshal Sir Vyell Vyvyan, and Dr H. C. Watts. Laurence Pritchard remained secretary as well as editing *The Journal.*

On the same day as the RAeS general meeting, Ernest Richard Calthrop, the forceful inventor of the *Guardian Angel* parachute, died, at the age of 70, a disappointed man because the American Irvin had been adopted by the RAF instead of his. Said a friend: 'He was a brilliant thinker and clever writer, using these talents to the full extent, so much so that he probably offended where a less brilliant thinker would be less verbose and consequently more successful from a business viewpoint. Apart from his devotion to parachutes, his particular hobby was breeding Arab horses, and one of the most entertaining sights was to see the mutual understanding between him and his horses as though they spoke the same language. He was also a devotee of painting and photography, but was actually a locomotive engineer, and since 1892 had been consultant until turning to life-saving appliances for aircraft. From 1916 onward he had been the persistent advocate of parachutes; that he had stirred the Air Ministry to action was his memorial.'

A very different pioneer attained distinction that March when the Greek Government conferred the Golden Cross of the Order of the Redeemer upon Robert Blackburn in recognition of his services to Greece. Quietly, methodically, this unpretentious man had been establishing British aircraft abroad. Currently he had sent H. V. Worrall to demonstrate a Velos seaplane in Brazil; Blackburn business in Australia was being established by Sqn Ldr Sandford; Capt Andrews was running the torpedo section of the Spanish RNAS at Barcelona with Blackburn aircraft. To Japan, Blackburn sent Major C. H. Chichester-Smith as the firm's representative. In Greece, Brig-Gen Francis Festing, who had become the South American representative, had played a vital part in negotiating Blackburn's construction and operation of the Greek Naval Aircraft Factory at Phaleron near Athens, and it was now in charge of Major Buck, assisted by C. H. Lowe-Wylde. Recently they had designed a Lynx-powered two-seater remarkably like the Sopwith Dove which Lowe-Wylde and Hollis Williams had resurrected.

Hollis had at length completed the Dove with the aid of H. H. Robinson of Faireys who 'for over a year devoted every Sunday and most Saturday afternoons, provided all the skills I lacked, and his own transport as well. We got the Dove flying with C of A and flew as much as I could afford, but I hadn't the cash to insure, not even third party, and the engine burned almost as much Castrol as petrol, but after almost three years spare-time work it all seemed worth while. I felt I had caught up and could now discuss detail design with experts as I had restressed much of the aircraft due to the Snipe tail, and gained Farnborough's approval. As far as I was concerned the Dove had served its purpose, and I could now devote spare time to other activities. In this state of mind I set off for the Easter Race Meeting at Bournemouth, held on a strip airfield that was really a horse-racing track. I arrived late Good Friday, and pegged down the Dove. Next morning I encountered F. F. Crocombe, who was also at Faireys, and enquired if he had ever flown. On finding he hadn't, I offered him a trip.

'We got the engine going and took-off. As we cleared the end of the strip, the cowling suddenly disappeared together with most of the side panels, all of which were supposed to be held in place by a single cable with strainers which trapped the trailing edge of the cowl into a gully in the supporting structure. This loose cable now wrapped itself round the distributor at the rear of the crankshaft and cut off the engine. There was nowhere to land ahead, so I attempted the dangerous manoeuvre of turning down-wind back to the strip. I finished with plenty of speed but little height; saw an aircraft taking-off towards me with certainty of head-on collision; swung up-wind – and there saw a ploughed field with no obstacles to clear. Unfortunately the furrows ran the wrong way; as we touched down the undercarriage bracing wires broke, and we went base over apex, but apart from facial injuries, we were able to walk away from the wreck having flown two minutes at the most.'

With fuselage painted in Keep's favourite hue of green the newly registered Widgeon III, piloted by Capt Geoffrey Hill, takes Mrs Hill for a circuit before handing over to Sqn Ldr T. H. England.

This air meeting had been preceded with headline news that in a preliminary canter over the course with the Blackburn Bluebird, Sqn Ldr Longton had been shot at by an irate Bournemouth householder in futile protest against the noise. It proved a nationwide advertisement for Bournemouth and its aeroplane races and crashes and hairbreadth escapes. The Hampshire Aeroplane Club instructor, G. I. Thomson, in the first heat of the Bournemouth and District Hotels Handicap Sweepstake had an engine failure as he took-off, and like Hollis Williams found no landing ground ahead so pulled the Moth round to land down-wind but just failed to make the airstrip, and in trying to lift his Moth over the corrugated boundary fence, stalled and charged through it, smashing the wings and breaking the fuselage in two – but neither pilot nor passenger was hurt. Then the Widgeon III flown by Sqn Ldr 'Tom-Harry' England, brother of Gordon England the pioneer pilot and designer, on taking Bruce's 15-year-old daughter, Rachel, as passenger on the 'Pub-crawl' race, had engine failure, landed in a boggy field into which the wheels sank, and the Widgeon turned upside-down. Local yokels watched nonchalantly, making no attempt to help England or the girl, but she took matters coolly, as befitted the daughter of an aircraft designer, and still hanging in the safety-belt discussed the best way to get out. Winner was Bert Hinkler with a Cirrus-powered Avian, and Openshaw second with the Widgeon II. The next event, the 'Kill-joy Trophy', so named after collective protest by churches and residents which led to abandonment of Sunday flying, showed 'Dangerous Dan' Watt with his black S.E.5a attempting to overtake Flg Off Wheeler's S.E.5a on the wrong side as they arrived at the turning point, and as Wheeler banked round, Watt banked steeper until over the vertical, dived under him in what seemed a certain crash, but 10 ft above the ground jerked level and raced on to the next turn. Next the Cygnet flown by Flt Lieut Ragg had a lucky escape when it forced landed with engine failure in the same field as England, but luckily did not turn over. Yet a further engine failure was that of the Cirrus-powered Avian flown by Flt Lieut Gray, but he too pulled off an impeccable forced landing. Even the incessantly joy-riding Hampstead of Imperial Airways managed to create apprehension by coming in to land regardless of all and just missed touching down on Nick Comper's C.L.A.4 which was taking-off.

'Practically all the races provided very close finishes, and some were won by inches,' reported *The Aeroplane*. 'Mr Goodman Crouch and Mr Dancy have evidently got aeroplane handicapping down to a fine art, and have the exact measure of every machine and pilot under every variant of weather. After all, the duty of a handicapper is to provide as close a finish as possible so that if a man is to win he must show the highest skill and judgment.' But many who had seen the bunching at the turns, and the dangerous exploits of skilful Dudley Watt as he weaved through competing machines, felt that Bournemouth, despite the easy gate money, was a bad venue for racing.

Recollecting the excitement of flying in those days, Hollis Williams wrote: 'When I got back to Fairey's after the Easter break, Miss Burns, who was his secretary, came to me in the D.O. and said "I don't know what you've been up to, but the old man wants to see you, and you'd better hurry – he's not in a very good temper."

'Entering C.R.'s office was always a bit nerve-racking. There was this great bull of a man sitting glowering at his desk and one felt the end of the world might come at any moment. He said "I hear you've lost your aeroplane, Williams."

'I said, "Yes Sir."

' "Is it badly damaged?" With a sigh of relief I replied that it was a write off.

'Fairey said, "Send it in, I'll repair it for you. I want to help chaps like you."

'I started to explain that the Dove had been a millstone round my neck for long enough, and that it had served its purpose, but C.R.F. would hear none of it. He signed a works production order which stated "Repair Dove G-EBKY to Hollis Williams's instructions."

'When the mess of ploughed field, castor oil, and wreckage was delivered to the experimental shop it caused a riot. A. C. Barlow, the experimental superintendent, not to be confused with the chief engineer, Major Barlow, was an excitable little man, and he bounced up the office to my desk and said, "They tell me it's your aeroplane. I won't have it – I'll chuck up my job before I touch it," and went on repeating this with increasing emphasis. When I could get a word in, I pointed out that these were the old man's instructions, and that it wasn't my fault if the shop floor looked a little untidy. In the end the Fairey experimental shop made a first-class job of the repairs. Everything had to be referred to me. I prescribed the best of everything with no consideration of cost, and in the engine shop where the Curtiss D.12s were overhauled, the foreman supervised the rebuild of my 80 hp Le Rhône which fortunately hadn't bent its crankshaft in the crack up. When the aeroplane was finished it was far better than it had been originally – and that's why it survives in the Shuttleworth Collection as the Pup.' However it encountered several further vicissitudes before that happened.

The Short-Bristow Crusader was launched on the Medway on 18 April, 1927, for John Lankester Parker to taxi with temporary wood propeller.

Just after the Bournemouth Meeting the Crusader was completed, and despite the projecting helmets of the radial cylinders was remarkably clean and attractive. The Air Ministry would not permit it to be flown at Rochester where it was taxied by John Lankester Parker; it had to be transported to Felixstowe and was tried on 4 May by Bert Hinkler because of an Air Ministry quirk that Parker, despite enormous experience as a civilian test pilot, must be less capable than Hinkler who was a Flight Lieut RAF Reserve, and had had previous experience with the Gloster racers. The flight was near disaster. In relation to the huge helmeted nose, the rudder had looked small to Hinkler, so he had had it increased. At full throttle of the derated engine, flying at the considerable speed for that time of 232 mph, he found that the Crusader hunted directionally but seemed safe. A normal descent was presently made, but on alighting, the port wing dropped, and the machine swung so violently that the float struts buckled, leaving the wing secured merely by its landing wires. Luckily the Crusader could still be towed, and was slipped for repair and reduction of rudder to original outline.

All three 1927 Schneider Glosters had the top wings beautifully faired into the side cylinder blocks, but the prototype IV had 4-ft greater span and conventional fin and rudder.

A few weeks later the pretty Gloster IV biplane was flown by Flt Lieut Kinkead, and on 10 June Flg Off Worsley made the first flight of the Supermarine S.5 – but on both the propellers were proving unsuitable.

By this time entries for the General Purpose Competition for D.H.9A replacements to specification 12/26 were flying. Fairey's IIIF had become the Ferret II with Jupiter, adapted for D.H.9A wings, which were almost the same span but greater chord and were modified to have horn-balanced ailerons. Said a critic: 'Knowing the habits of the Fairey Co, one imagines that the Ferret must have some particular feature to recommend it as

Both Gloster IVA and IVB had fractionally greater gap and a cruciform tail, but the light blue IVB had a geared Lion VIIB instead of direct-drive and was used for the contest.

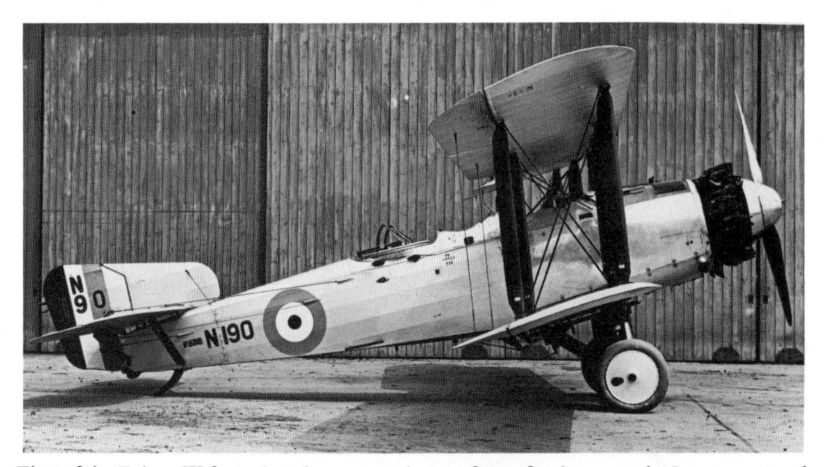

First of the Fairey III formula to have a metal space-frame fuselage was the Jaguar-powered Ferret I intended for Fleet Air Arm requirements.

distinct from the IIIF, for Mr Fairey has a habit of getting it both ways.' The Gloster Goral proved equally conventional, but had a specially designed fuselage of more capacious size than the standard D.H.9A, its fuller lines making a better match for the wide Jupiter than the simple adaptation represented by the de Havilland Stag which offered little more than substitution of a Bristol engine for the hard-worked Liberty of which supplies were becoming exhausted.

The Westland Wapiti was similar to the Goral but with still deeper fuselage in which the pilot sat high and sufficiently aft for excellent view

Flt Lieut Howard Saint displays the Gloster Goral with metal-framed D.H.9A wings, but, like the Westland rival, its 9A rudder proved too small. (*Flight*)

Ready for its first flight, the trim Westland G.P. – not yet named – had obvious D.H.9A lineage, but the fuselage was much more capacious.

above the wings as well as between them. Although wings, tail and rear fuselage were wood, it was metal from engine to abaft the pilot, using the squared tube construction so successfully tried with the Wizard – a prior use which cleared it of Patent 286,482 issued all too late in March to F. Sigrist, S. Camm, and H. G. Hawker Engineering Co Ltd describing 'A joint for tubular framework members comprising connecting plates bolted or riveted to the members which are flattened at the joints or of rectangular section throughout, and are cut at suitable angles at their ends to butt against one another.' That last phrase also freed Westland because the Wapiti spacer struts did not butt on the longerons, the load being taken through the flitch plates, for it is not the principle of an invention but only the application which can be patented; so every designer could validly readapt the original thought of another, leading to almost identical patents for metal construction, controls, hydraulics, and so on.

The Wapiti had made its initial flight piloted by Openshaw on a calm Sunday morning, watched by a keen group of workmen and draughtsmen. It was not expected that there would be any trouble with so commonplace a machine using the standard wings and control surfaces of the long-proved D.H.9A, yet when Openshaw landed we saw that in enclave with Bruce, Davenport and Hill he shook his head. 'Rudder hardly works at all,' he said. They were amazed that the deep fuselage could make such difference. In the next few days the old process of trial and error began, adding plywood area to the rudder, and when that proved insufficient a very large rectangular rudder was made like the Yeovil bomber's but with horn balance faired into the unaltered D.H.9A fin. It was ugly but powerful, and though the fin was now disproportionate it was decided to send the machine to Martlesham as time was short. Meanwhile the draughtsman who had drawn the fuselage happened to measure it to check the tail

When the Bristol Boarhound proved unacceptable, Barnwell revised it as the Beaver with improved controls and neater appearance, but it still did not compare with the Vickers Valiant.

volume with the wind-tunnel model, discovering to his dismay that a complete bay of some 3 ft had been missed out. It was too late to alter the fuselage, for the machine was already receiving favourable comment from Martlesham pilots.

The Bristol Boarhound had already been eliminated, not because its only semblance to a D.H.9A was the shape of its much larger rudder, but because it was found to be a clumsy flying machine, lacking precise control response. It was the Vickers Valiant which the pilots first seemed to prefer, largely at the persuasion of Flt Lieut 'Mutt' Summers with whom young Oliver Vickers was very friendly. The chief engineer, Sq Ldr H. McKenna, a hard-bitten elderly soldier of the old school, liked its all-metal construction and ease of maintenance and arming. Undoubtedly the Valiant had pleasant handling qualities, and was faster than the others, but had the great drawback that it did not comply with the specification

The Vickers Valiant redesign of the Condor-powered Vixen had the now widely accepted Bristol Jupiter VI. Its fuselage was tubular framed, and the two-spar wings had double T-section dural booms with 'wandering' web. (*Courtesy Air Marshal Sir Ralph Sorley*)

An even bigger rudder than the Goral's was necessary on the Westland type, and to lighten the ailerons to cope with greater speed than the D.H.9A, a big inset horn balance was added.

mandate to use D.H.9A parts, and would cost five times as much as the Goral or Wapiti – so it was financial expediency to restrict purchase virtually to sets of fuselages. Neither the Stag nor Hound stood a chance, for their fuselages were too small for bulky general-purpose equipment, and in any case plywood had been proved unsuitable for the tropics. The result was that the Wapiti became the winner, thanks largely to V. S. Gaunt, the new Westland experimental manager from English Electric who organized telling demonstrations of loading stores and armament, and was prompt to ensure that design and operational criticism were countered with immediate action. The prize was an opening contract for twenty-five Wapitis in composite construction, with a modified fin.

The Stag was simply the well-tried D.H.9A with a modern radial engine and the latest rubber-in-compression undercarriage, but was the cheapest G.P. of all. (*Courtesy Air Marshal Sir Ralph Sorley*)

At Hawkers, as earlier at Sopwiths, there was indecision on the value of two-bay wings and good span giving better climb as in the Hawfinch (*illustrated*), or cleaner single-bay wings for speed. (*Courtesy Air Marshal Sir Ralph Sorley*)

The Bristol 101 was also at a discount because of its ply-covered fuselage, yet the compactness of this N-strutted single-bay two-seat fighter gave an aggressive and advanced appearance compared with contemporaries. Though turned down categorically, it proved that a bulky engine was no drawback to speed, for it was as fast as the Fox, as might have been expected by comparison with its similar single-seat predecessor, the Badminton. The latter was being fitted with I-strutted tapered wings of 9-ft greater span than before to gain handicap advantage in the forthcoming King's Cup race. But the Bristol future depended much more on success with a metal-structured biplane fighter which might be powered with either a Rolls-Royce F.XI or Bristol Mercury to meet specification F.9/26 for a day and night fighter. Already in March Hawkers had stolen a lead with the first flight of their two-bay Hawfinch, but on 17 May this Bristol 105, later known as Bulldog, fitted with a Jupiter VI, made its *début*, piloted by Uwins, and a month later was flown to Martlesham for preliminary

With the Bristol Bulldog, Barnwell assessed that a single-bay configuration had the advantage, and it also helped to present an air of solid dependability. (*Courtesy Air Marshal Sir Ralph Sorley*)

handling. Both Bulldog and Hawfinch had a long lead over the Boulton & Paul Partridge, Gloster S.S.18, Armstrong Whitworth A.W.XIV, and Westland Wizard.

By now design staffs were double the size of five years earlier, and were handling at least two projects simultaneously, as well as dealing with the vagaries of prototypes and modifications for production aircraft. Thus Westland had simultaneously been designing the Wapiti, considerably altering the Wizard, preparing the Widgeon for production, evolving the side-by-side two-seat Pterodactyl in a section under Geoffrey Hill, while the bulk of the main D.O. was draughting a 64-ft high-wing outrigger-strutted monoplane bomber, the Witch. Because Wapiti construction required little more than building a fuselage, it had been made in six months, but the more complex Witch was still only a skeleton standing alongside its mock-up which, with that of the Wapiti, was retained long after completion of the machine in order to try-out new equipment or essential structural changes.

Similarly at Hawkers there was not only design work on the Harrier biplane to the same specification as the Witch and Gloster Goring, but even more important was specification 12/26 issued in May for a two-seat bomber to rival the Fairey Fox but powered with the newly developed 450 hp Rolls-Royce F.I. Among several draft exercises, Camm provisionally settled for an I-strutted single-bay biplane with split-axle undercarriage, so a mock-up was being built.

Meanwhile the RAF, impelled by the many long-distance flights of the Americans, Italians and French, decided to attempt a record using a

Flight Lieutenants C. R. Carr and L. E. M. Gillman ready to take-off with the Hawker Horsley specially modified for extreme range. (*Flight*)

Horsley with tankage increased from 230 gallons to 1,100 and take-off weight increased from 9,000 to over 14,000 lb. With an empty weight of 5,000 lb, this was an unheard percentage of disposable load. On 7 March the BBC announced that an attempt would be made to fly from England to India nonstop in an endeavour to beat the record of 3,345 miles made by Capt Costes and Capt Rignot the previous October when they flew from Paris to Jask on the coast of Persia. By May, J8607 was ready, and taken to the South Field at Cranwell as this, though undulating, gave the longest take-off length in the country and, like all others, was of grass. Flt Lieut C. Roderick Carr and Flt Lieut L. E. M. Gillman were selected as pilots. On 20 May they took-off at 10.38 a.m. after an alarming run of 800 yards during which it was touch and go whether they would hit the boundary stone wall, but slowly the machine climbed away south-east. They were last seen over Wiesbaden; then no more for days.

Next morning there was very different headline news. An unknown, lanky and boyish-looking American, Charles Lindbergh, had passed above St John's, Newfoundland, at midnight, having already flown 11 hours in the first ever attempt to fly the Atlantic alone. Evening papers blazoned news that just after midday he was sighted by ss *Hilversum* 500 miles from the Irish coast. Hour by hour bulletins from the BBC worked up excitement. At 5.20 Lindbergh crossed the coast of County Kerry, and two hours later was above St Germans in Cornwall. At 8.30 he was over Cherbourg and at 10.22 landed at Le Bourget, welcomed by a vast cheering crowd.

So overwhelmed was the British Press that it almost overlooked news that the Horsley had ditched in the Persian Gulf, having covered 3,420 miles. But Lindbergh had beaten them with 3,590 miles from New York. He had also beaten them in sheer outrageous confidence in himself and the Wright Whirlwind that powered his 46-ft span Ryan NYP high-wing monoplane *Spirit of St. Louis*. Except for a periscope for limited view ahead and an angled vista from the window, where he sat behind a 425-gallon tank blocking the entire cabin, he had flown blind through hour after long hour of suspense in an epic of epics that won £5,000 offered by Raymond Orteig for the first flight between New York and Paris – and inevitably there would be more from the publicity. Yet those who knew the penniless Lindbergh were aware it was not money but sense of conquest and fulfilment which he valued most. Nevertheless, to commemorate his magnificent flight, Madame Deutsch de la Meurthe, presented 350,000 francs to the Aéro-Club de France, of which her husband had been former president, so that 150,000 francs could be given to Lindbergh and the balance to the families of the late Captains Coli and Nungesser, famous war-time aces, who had been drowned in an attempt on 8 May for the same prize.

On Sunday 29 May a crowd, enormous as that which greeted Lindbergh at Le Bourget, were waiting to see him arrive at Croydon. Long before the Ryan and accompanying aircraft hove in sight they swarmed onto the aerodrome in a tidal wave. Only on his third attempt at landing was a

The arrival at Croydon of Lindbergh with his 'blind' Ryan *Spirit of St. Louis* was an exciting and risky occasion both for pilot and the uncontrollable crowd. (*Flight*)

narrow pathway cleared, and 'the whole mob broke loose from all the enclosures, headed by the privileged persons who were admitted with special passes to the Customs area and had been particularly requested to keep their line and set a good example. A number of the regular aviation people fought their way to the machine and held the crowd off by sheer strength of arm and fist. Lindbergh wisely stopped inside the machine or he might have been torn into souvenirs. Eventually a much battered car with Aero Club officials and police arrived, extricated him, and ploughed to the control tower where he mounted the ladder and showed himself to the crowd, looking like a Royalist on the platform of a guillotine during the French Revolution. Thereafter he was transported to the American Embassy.'

Such was Lindbergh's instant world fame that soon it became universally believed his flight was the first nonstop aerial crossing of the Atlantic – but that was Britain's laurel crown when Alcock and Brown linked North America and Ireland with their Vickers Vimy in 1919.

Air Commodore Samson's Fairey IIIF seaplanes performed well due to careful maintenance and splendid piloting, but twin floats had their dangers compared with the seaworthiness of flying-boats.

By contrast, only the RAF ensign flew a welcome on 22 May when the four Fairey IIIFs led by Air Commodore C. R. Samson had quietly touched down at Heliopolis, Cairo, which they had left on 30 March to fly to the Cape and back, for it was regarded as no more than a practice exercise of difficult conditions. In the Commons there were growls about the cost when the Secretary of State for Air revealed that £55,000 had been spent on landing grounds for Van Ryneveld's 1920 Cairo—Cape flight, and the cost of reconditioning them for the RAF was some £3,000 together with an annual maintenance contribution of £300 to the Sudan Government.

Two days later came more specific news of Carr and Gillman. The predicted favourable winds had failed to materialize. After two nights, strain began to tell for, as they said, 'the instruments and gauges seemed to grin and make faces.' Over the Ormuz Strait of the Persian Gulf, the engine, with very little warning, began to show signs of distress, heating excessively, and suddenly cut out. In the moonlight Carr chose what he imagined to be a long stretch of beach but was moonlit water on which he touched down at 3.15 a.m. local time on 22 May, and the machine nosed over, flinging them out. The big, almost empty petrol tanks kept the Horsley afloat, and they succeeded in swimming back to it. The keeper of the Qoin lighthouse, three miles away, heard the engine cut, and set out in a small boat certain that an accident had occurred but at first could not find them; only on his second attempt at dawn were they rescued. At dusk next day the captain of a steamship a mile distant saw signals from the lighthouse. He sent a boat ashore to investigate and in it they returned to the steamer, which landed them at Abadan whence they were flown back to Cranwell in a Victoria.

Adverse winds delayed a further attempt, but on 18 June, Carr, accompanied by Flt Lieut Mackworth, took-off from Cranwell on the reserve long-distance Horsley loaded with even more petrol than that of the first attempt. It ran what seemed the enormous distance of 1,000 yards before lifting. Slowly it headed south. Eighty minutes later Martlesham Heath crash party were alerted on seeing two Horsleys formating with Carr's machine, which trailed what looked like exhaust smoke though the keen-eyed said it was a stream of liquid. Clearly Carr was going to land. Everyone watched anxiously, for there was grave risk that the undercarriage or a tyre would collapse under such great load. Carefully Carr brought the heavy Horsley in with considerable engine, then eased off the power slowly as he wheeled onto the ground and gradually stopped – the first time this had ever been done with a record breaker at full load; all previous attempts had ended in flames.

Civil flying had not been without its excitements and disasters. In mid-May a typical flying meeting had been held by the Hampshire Aero Club at Hamble, attended by so unexpected a crowd that jams of cars and pedestrians were three miles long. Though clouds were low and wind gusty, there were aerobatics by a Gamecock and an Avro Gosport, and a great fly-past ranging from D.H.53 to Avro Ava, including a Supermarine Sheldrake amphibian looking strangely out of place on land. Because the public footed the bill for the RAF expansion, the High Command considered it expedient to help make the public airminded – so not only were RAF landplanes on view, but a formation of three Supermarine Southamptons from Calshot circled with an air of omnipotent power low around the aerodrome, and five Gamecocks followed with a pretty display. But the essence of the meeting was the enthusiasm and friendliness of the many private owners, instructors, and keen Service pilots typified by young Sholto Douglas, Le Poer Trench and Longton, together with test pilots and such men as Jerry Shaw with Shell's red and gold Moth and Capt Lamplugh with the brown Moth of his insurance group. One result was a miscellany of twenty-four varying aircraft for the handicap race, but the mix-up at turns was disquieting. To all visitors the petrol companies were benevolent in giving a tankful of free petrol. 'That's why I fly an S.E.5a,' said Dudley Watt. 'It has got such a big tank that I go without refuelling until the next meeting!'

At the subsequent Whitsun Meeting at Bournemouth he was less fortunate, breaking his S.E.5a's connecting rod the evening before the races. Several others failed to arrive. Major Henry Petre, the pioneer flyer known as 'Peter the Monk', force-landed the London Aeroplane Club's Brownie near Southampton; Flg Off Allen Wheeler's S.E.5a had engine trouble; Mrs Elliott-Lynn crashed her S.E.5a at Brooklands. Then on the opening morning Major Harold Hemming, managing director of the aerial-surveying Aircraft Operating Co Ltd, was flying Alan Butler's D.H.37 *Lois* when he stalled at 60 ft, the wing-tip striking the top of a steel-framed number board which tore out the interplane struts, collapsing the wings that side and turned the machine over so that it hit the ground inverted, so gravely injuring the passenger, Claude Plevins, undergraduate brother of St John Plevins of the Anglo-American Oil Co, that he died in hospital. Accepting RAF tradition, racing proceeded, and Mrs Elliott-Lynn, flying the Westland Widgeon III from scratch, won the Ladies' race. Hinkler with an Avian won the Private Owners' Handicap, and the Bournemouth Hotel Sweepstake was won by Dudley Watt with another Avian.

Flying continued on Monday with a spectacular opening display by Flt Lieut D'Arcy Greig on a Genet Moth and, for the first time since the prewar flying of Pégoud, performed a 'bunt', diving beyond vertical onto his back, then rolled to normal flight. Yet, compared with the Easter Meeting, attendance was sparse to watch the first Handicap race at 2.30 p.m. Twelve started, but Flt Lieut Gray's Avian dropped out. When the remaining eleven rounded the aerodrome turning point, they seemed bunching dangerously, and as they straightened in the distance the Widgeon flown by Openshaw and another machine were seen to collide and fall. Black smoke shot up: there could be no hope for either pilot – but there was doubt as to identity of the second for neither Sqn Ldr Longton, Capt de Havilland, Capt Broad, nor Lady Bailey re-appeared. Capt Sparks of the London Aeroplane Club turned his Moth and landed on the Racecourse to say it was Longton's prototype Bluebird. He had been instantly killed, but Openshaw died after long agony of burns. Though none felt like taking interest in further proceedings, it was decided to continue, providing remaining races were run in heats of not more than four – a decision that should have been made after the visual proof of danger apparent both in the previous Bournemouth Meeting and the Hampshire Pageant races.

So much for a Roman holiday. Not only young Plevins but two superb pilots had gone, both of whom were engineers. Walter Longton had been an engine tester with the Sunbeam Motor Co before the war when he was well known as a racing motor-cyclist. C. G. Grey wrote: 'Few have done so much for the improvement of aircraft, whether as test pilot of fighters, light aeroplanes, or big bombers, for every aircraft constructor knew that "Longton's opinion" was worth having on anything new, and he gave freely advice worth big money had he been a consulting engineer. He was beloved by his men as by his fellow aviators. One had only to watch the wild enthusiasm of RAF people when he won a race or put up a good show to realize how highly they thought of him. And their esteem was rather increased by the fact that he made his men work like slaves – and worked with them harder than they did.'

Laurence Openshaw's hobby similarly had been motor-cycle and car racing. Like Colonel Lawrence in the guise of Aircraftsman Shaw, he owned a splendid silver-plated Brough Superior, but his Fiat racing car had been exchanged for a four-seat silver and blue Morris tourer on recently marrying Jean Bruce, elder daughter of Westland's managing director. Openshaw, the son of a distinguished Army doctor, had been Sqn Cdr Busteed's deputy of the war-time Experimental Flight at the Isle of

Six special Moths, with extended nose to compensate for the light-weight 75 hp Armstrong Siddeley Genet, were purchased for the Central Flying School aerobatic team. (*Flight*)

Laurence Openshaw takes off on his ill-fated flight with the Westland Widgeon III prototype, at the Bournemouth Whitsun Meeting in 1927. (*Flight*)

Grain and had a Cambridge MA, resulting in appointment after the war as chief engineer at the Carrara Quarries in Italy where he had represented England in the Gordon-Bennett motor race. To Westland he seemed irreplaceable.

4

Like Bulman, the Martlesham pilots found the two-bay Hawker Hawfinch delightful on preliminary handling trials. With Sopwith and Sigrist still playing the master-hand in approving designs, inevitably there was close family likeness to war-time Sopwith machines, yet it was clearly by the same hand which had originated the Cygnet. Though Camm's artistic eye had ensured perfection of shape, the Hawfinch was virtually a redesign of the all-metal Heron with both span and length increased by $1\frac{1}{2}$ ft, but wing areas were almost identical, and with an all-up weight of 2,910 lb it was 210 lb lighter because the two-bay bracing had been selected only after much mathematical investigation to find the lightest possible structure. Next in importance was better view, so the fuselage was 1 ft shallower, with forward coaming sloping more steeply to tighter engine cowling, centre-section lowered in relation to the pilot's eye, and heavy stagger increasing downward visual angle over the lower leading edge. With the temporary engine replaced by a Jupiter VII it seemed they had a winner; but when flown in July to Martlesham for official trials it was found that its nearest rival, the Bristol Bulldog, had the edge on the Hawfinch by 2 mph, with a speed of 173 mph, and surprisingly a ceiling 3,000 ft greater. The Boulton & Paul Partridge, Gloster S.S.18 and Armstrong Whitworth

A.W.XIV remained some months behind. Like the Hawfinch, the Gloster, later to become famous as the basic Gauntlet, had two-bay bracing, but the other two had single bays like the Bulldog.

Although aircraft design seemed making but stereotype progress, knowledge of spinning, stability, and balancing controls had progressed enormously. In early years designers often found it impossible to correct defects in prototypes partly because of ambiguous description by their pilots, resulting in many a machine being scrapped unaltered as being no good. Now, with experience of a further decade, there was more trial and less error in altering shapes evolved in wind-tunnels, for errors were usually of secondary order. Aerodynamic theory at last could be treated mathematically, developed from the work of that great British pioneering engineer, F. W. Lanchester, who in 1879 first proposed the circulation theory of lift, basing it on the research of ballisticians who from 1760 onward had investigated the flight of projectiles and the value of imparting a spin to shells and bullets. His views had been received with polite scepticism by members of the RAeS, and Wilbur Wright wrote to him in 1909: 'I note such differences in matters of information, theory, and even ideals, as to make it quite out of the question to reach common ground with you by mere talk, as I think it will save me much time if I follow my usual plan, and let the truth make itself apparent in actual practice.'

Yet by the end of the war Prandtl in Germany had established Lanchester's theory. Not until 1926 had the RAeS made shamefaced amends by presenting Lanchester with the Society's highest award, the Gold Medal – previously awarded only to the Wrights in May 1909; Octave Chanute in July 1910; Professor Bryan and Edward Busk in May 1915. As one of the results of this theory, Lanchester established the conclusion that: 'The dynamic expenditure of power in sustentation is independent of the wing chord and wholly dependent upon the span.'

His work was endorsed in the Wilbur Wright Memorial Lecture on 16 May, 1927, when Dr Prandtl was awarded the fifth Gold Medal. He described his own work: 'We in Germany were better able to understand Lanchester's book when it appeared in 1904 than you in England. English

The twin-Jupiter Boulton & Paul Sidestrand day bomber had relatively very efficient high aspect ratio wings designed on the Lanchester/Prandtl theory. (*Courtesy Air Marshal Sir Ralph Sorley*)

scientific men, indeed, have been reproached that they paid no attention to the theories expounded by their own countryman, whereas the Germans studied them closely and derived considerable benefit. Lanchester's treatment is difficult to follow, since it makes great demand on intuitive perception, and only because we had been working on similar lines were we able to grasp his meaning at once.' Turning to practical application he said: 'In order to prevent the formation of turbulence and to persuade a real fluid to

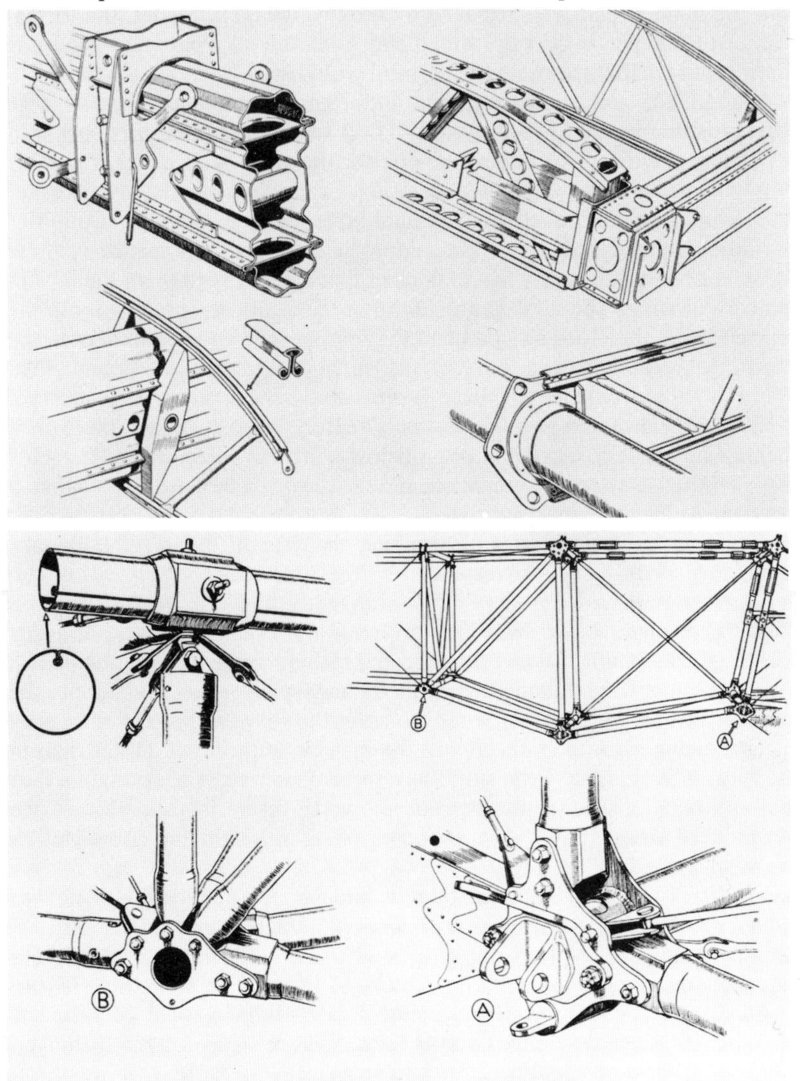

Steel strip construction, as pioneered by North, was used for the Sidestrand's wing frame, but the fuselage was more practically structured with steel tube and tie-rods. (*Flight*)

imitate the smooth flow of non-viscous fluid all that is necessary is to get rid of the thin layer of fluid which is slowed up and brought to rest by viscous friction. This can be done by sucking the boundary layer away into the inside of the body, or by providing some method of preventing it from slowing unduly. In the Handley Page slot-wing a jet of air with high velocity is led to the boundary layer at a point where its velocity normally would be very low. The nearly stagnant boundary layer is thus blown out into the main stream and replaced by moving air from the jet, the formation of turbulence is thus prevented and a smooth flow past the aerofoil is maintained where otherwise the aerofoil would stall.'

Prandtl had also discovered the important phenomenon that at large Reynolds numbers of 100,000 to 300,000 sudden diminution occurred in resistance coefficient to about one quarter the value with lower Reynolds numbers, due to rapid intermixture of frictional boundary layer and outer flow which re-accelerated the frictional layer and prevented separation of the flow from the boundary. He said: 'The technical importance of these turbulent boundary layers lies in the fact that due solely to them the flow in the case of wings and airship hulls follows the boundary practically to the rear end, and therefore on one hand the resistance is very small and on the other the ideal flow can be used as an approximation of actual flow. Without this turbulence all these bodies would possess worse properties, greater resistance and less lift. For this reason, when carrying out experiments with models, we must not go below a critical value of the Reynolds scale otherwise all hydrodynamical similarity with a larger object is lost, a fact well known to experimentors.'

The practical possibilities of applying the Prandtl theory of boundary layer were already being investigated by the Americans at McCook Field. A jet of compressed air was directed tangently to the upper surface of a model aerofoil to blow away or accelerate the boundary layer, and this increased maximum lift and deferred the stalling angle as prognosticated. Similar results were obtained when the boundary layer was sucked into the aerofoil through a series of small holes in the forward part of the upper surface, using a pump connected to the interior of the wing. Other than in the form of a Handley Page slot, there seemed no way of adopting the first method, but the second was considered worth future investigation to see whether the power required for pumping offset in weight and complication the benefit of added lift.

Frederick Handley Page was by now intent on the application of slots to enhance safety, for the demand was clear from the constant stream of RAF accidents, and its solution not only might save lives but should prove a valuable source of income for his company. In any case something had to be done about the Harrow because pilots reported that manual operation of its full-span slots was impossibly heavy. Russell's type of freely hinged wing-tip slats were therefore fitted to it. Scuffham records: 'It was on a Saturday afternoon that it was ready for testing. H.P. with a small crowd of technicians and fitters gathered round the aircraft in front of the running

Although outclassed, like the Avro Buffalo, against the winning Blackburn Ripon, the Handley Page Harrow became the first aircraft to have automatic slots. (*Flight*)

shed. The pilot was Sqn Ldr T. H. England, who had joined the company a few months earlier as chief technical adviser on the flying side, his job in effect combining those of sales manager and test pilot. The Harrow first made a short flight during which he tried the slots by manual control in the old way from the cockpit, checking that flying characteristics in open and closed positions were as expected. All being well, he landed, and in an atmosphere of great expectancy the mechanical gear controlling the slats was removed so that they were free to move according to the dictates of the air stream alone. England took-off again, climbed to a safe height, and pulled the Harrow to the stall. Out they came of their own accord – the Handley Page automatic slot was an accomplished fact! There was no hesitation about their working. Until the critical stalling condition arose, air pressure around the leading edge held them firmly in place. Changes in dynamic pressure as the stall approached pulled them out with equal sureness. The performance was so tremendously encouraging that after incorporation of a few detailed improvements, a demonstration was arranged for Air Ministry technicians. This took place about a month after the first test, and officials were so impressed that they sent a Bristol Fighter to Cricklewood to be equipped in the same way. Several RAF officers holding important appointments either themselves flew the Fighter with its automatic slots or were taken as passengers. Brancker was given a demonstration and so was Sir Samuel Hoare, both invariably keen to try out advances in flying for themselves. Said Sir Samuel: "The machine safely did all the things one should not do, including some which in other circumstances would have meant certain death." Capt Frank Barnwell, the

Bristol designer, tried it, but came in to land too fast, just missed a flag-pole, roared across the aerodrome, and began another circuit. The second approach was low and equally fast, but at the last moment he touched down at the far end of the field and managed to stop. When everyone rushed to him, it transpired he'd been working the altitude lever instead of the throttle!' – but that is improbable.

The Bristol Bagshot was distinguished by its triangular fuselage and 3½-ft deep wing cantilevered from the strut-braced shoulder centre-section, with wide-track wheels mounted on a triangular aerofoil axle.

Barnwell was having other technical difficulties. When the Bagshot was flown by Uwins on 15 July he found trouble. It so happened that I was at Filton learning to fly in the Reserve of Air Force Officers – instructed by that confidence instiller, Cyril Holmes, earlier of AT & T – and saw Cyril Uwins, who was CFI as well as chief pilot, returning on a long low straight approach with the Bagshot on its third flight, and wondered what was wrong. Bristol's later chief designer, Dr A. E. (later Sir Archibald) Russell, reminiscing on his youthful days as stress assistant, said: 'On that flight, when checking speeds, there was complete roll reversal in response to aileron movements at about 110 mph. This was at first puzzling, but quickly the RAE provided the explanation – wing twist. Modifications were tried to increase torsional stiffness by increasing the size of bracing wires and duplicating them in the plane of both spar flanges. It soon became obvious from the inadequate improvement that, if monoplanes were to have any future, a fundamentally new form of structure must be found. A research project was initiated with a multi-spar design employing much improved sheer bracing in plan, and a test wing, incorporating seven spars, was designed to suit an aircraft of 12,000 lb all-up weight. Professor Sutton Pippard, the most eminent authority of the day on aircraft structures, was appointed to recommend a method of stress analysis as calculation of so "redundant" a structure was intractable by available methods, and he decided that for stiffness in torsion all spars must have substantially equal flexural deflections. The share of total bending moment carried by each spar would then be mainly dependent upon its relative moment of inertia and independent of the location of the centre of pressure, and torsion about the flexural axis could be isolated and treated separately. When the wing was tested it carried full load at the first attempt and achieved the necessary stiffness. Concurrently, thin Alclad sheet was being developed in America, and a subsidiary company, Northern Aluminium, was formed in England.

Rigorous weight control was Hill's method when designing the Westland Pterodactyl IA. There were no 'reserve' factors. A Cherub engine was fitted for early flights.

Bristol obtained an experimental batch of sheets, and these were used as covering to replace the Warren bracing across the spar flanges of a multi-spar test specimen, resulting in still greater stiffness and opening the door for adoption of a two-spar box wing.'

A similarly involved problem in monoplane design had been tackled by

Experimental work of this time was ingenious, but crude for cheapness. Here the Pterodactyl's mono-wheel system is tested by towing a rig devised by Stanley Seager of Hill's three-man staff.

Herbert Mettam at Westland in stressing Hill's latest tailless monoplane, the Pterodactyl Mk IA, for the swept-back wing required investigation of a dozen different load distributions. Though initially a short funk strut was fitted to prevent twist at the unsupported elbow, such was the faith in Mettam's mathematics that no full-scale rig loading was deemed necessary. Instead, there was much experimentation to check steering and stability of the single-wheel undercarriage.

Still in hope of meeting the Bristol Fighter replacement specification 30/24, Oswald Short drastically revised the Springbok II to become the Chamois by fitting single-bay wings.

5

It was a day of drizzling rain with gusty north-east wind for the RAF Display on 2 July, but because it had become one of the great events of the year, the British public turned up in scores of thousands. The new grandstand seating 3,000 was packed, and car parks were full. At 3 p.m. precisely, while bombing squadrons were taking-off for group evolutions, the King and Queen arrived, accompanied by the King of Spain, the exiled King and Queen of Greece, and the Duke of York. Attending them were Sir Samuel Hoare, Sir Hugh Trenchard and Sir Philip Sassoon. The ensuing programme ran with the accuracy of a chronometer. One by one the new types took the air in striking contrast: Atlas, Chamois, Hinaidi, Ava, Goring, Horsley, Sidestrand, Fairey IIIF, Hound, Valiant, Wapiti, Bulldog, Hamlet in new guise with two engines, and the rebuilt Pterodactyl prototype.

There was the thunder of night bombers taking-off; scream of Fairey Foxes diving across the crowd; six parachutists dropping from three Vimys; destruction of a kite balloon from which that stuffed dummy, Major Sandbags, descended by parachute; squadrons of Grebes attacking

Production Westland Wapitis soon were fitted with steel-framed wings constructed by Glosters, but covered and doped by Westland. (*Flight*)

Hyderabads amidst intense anti-aircraft fire; finally a tremendous set-piece portraying men isolated in a desert fort rescued by troop carriers after a squadron of Foxes beat off the attacking forces with bombs and machine-gun fire. But for me the outstanding event was the first-ever demonstration of close formation aerobatics by five Genet-powered, scarlet-topped de Havilland Moths from the Central Flying School, Wittering, flown by Flt Lieut D'Arcy Greig, the chief instructor, and his assistant instructors,

The Felix-powered Fairey Fox was always a thrill because of its banshee scream as it flattened from a dive. (*Flight*)

Flying Officers A. E. Beilby, G. H. Stainforth, R. L. R. Atcherley and H. R. D. Waghorn. They looped in formation, rolled in formation, flew on their backs in tight formation, turned inverted, spun and recovered in formation, finishing with a bunt. It was breath-taking, marking a new peak in co-ordinated flying.

There were growls from the French who complained that fourteen German officers, six of whom formerly served with the German Flying Corps, were present as guests of the British Government, and bitter was the argument that the Treaty of Versailles not only banned Germany from having any military aviation, but forbade the German Government to send abroad any military, naval or air mission. Not unnaturally the Germans insisted that these particular officers did not represent a mission, but went as spectators, as could anyone who paid at the gate. Among these affable and dignified German visitors was a certain Capt Hermann Göring, the war-time successor to command of the Richthofen Jagdstaffel after its leader was killed; those who met him at Hendon said he was 'singularly like some of our own fighting pilots, and very entertaining company; altogether our German visitors were a very good lot, and we all hope to meet them again.'

It was surprising that the French Press did not complain as bitterly of the presence at Hendon of General Italo Balbo, the Italian Under-Secretary for Air, who piloted his own Service aeroplane to England for the event. This bearded, dynamic man of 31 was one of the 'big four' in the Fascist movement. The Air Council lunched him at Claridge's; the RAeS, SBAC and Air League dined him next night at the Savoy; he was taken to the Cadet College at Cranwell, the Apprentices School at Halton, the Napier engine works, and the factories of Fairey, Handley Page, Supermarine, and de Havilland where he piloted a Moth. To wish him farewell when he left Croydon at the early hour of 7 a.m., on the 6th, there were attendant representatives from Sir Samuel Hoare, Sir Hugh Trenchard and Sir Sefton Brancker, but he was cautious in heading for Germany via Amsterdam since he was an object of suspicion in France, where politicians regarded Italy under the leadership of Mussolini as potential aggressors nearly as dangerous as Germany.

General Italo Balbo (flying suit unbuttoned), the dominant figure in Italian military aeronautics, in front of his Fiat reconnaissance two-seater. (*Flight*)

England and France were the object of very different attack by Americans that summer in the form of transatlantic flights emulating the successful solo of Lindbergh. Using a Whirlwind-powered Bellanca, Clarence Chamberlain on 7 June flew his financial backer, Charles Levine, following fantastic quarrels with him, from New York to Berlin, thence to Prague, Vienna, Paris, and ultimately to London a few days after the RAF Display. As a reporter said: 'It seems difficult to explain why that enormous crowd should have turned up to see Lindbergh at Croydon and less than 100 came to see Chamberlain.' After yet another quarrel, Chamberlain refused to fly back with Levine, who thereupon engaged M Drouhin, chief pilot of the Farman company, who flew him from Croydon to Le Bourget, but refused to fly him back to Nottingham for the King's Cup race, stating he had been hired to fly the Atlantic and not to become a taxi driver. One-eyed Hinchliffe offered to pilot the aeroplane for the honour and glory, whereupon Drouhin began serving writs binding Levine to keep his contract. In a last fantastic episode Levine got hold of his Bellanca and, although not a pilot, flew it to Croydon where he made ten heart-stopping attempts before landing in a series of gigantic bounces. Then he went home by ship.

Three weeks after the Chamberlain-Levine flight, Bert Acosta and Bernt Balchen, with Commander Byrd as passenger and Lieut Noville as navigator, flew the tri-motor Fokker F.VII *America* from Roosevelt Field, Long Island, across the Atlantic in an attempt to reach Paris, but, after sighting Cape Finisterre as night fell, they became lost in fog and rain. Fuel ran low. At 2.30 a.m. they desperately alighted on the sea at Ver-sur-Mer, near Le Havre, reaching the shore in a pneumatic raft.

First of the transatlantic adventurers to reach England direct were William Brock and Edward Schlee, who arrived at Croydon at 10.33 on Sunday morning 28 August having flown from Harbour Grace, Newfoundland, in their yellow-painted Whirlwind-powered shoulder-wing Stinson Detroiter, which was similar to Lindbergh's Ryan except that they sat in a front cabin with adequate view. Though they had three compasses, their only map of England was a page torn from a school geography book, and on reaching Britain's shores had no idea whether they were over Ireland, England or France, so flew onward an hour or more, skimming the cliffs of South Devon, then followed the coast to Sussex and guessed their way to London. Next day they left Croydon at 09.35 hours for Munich.

Even long-distance flights with light aircraft were becoming the rule. The 60-year-old Duchess of Bedford, piloted by Charles Barnard, had flown with her Moth 4,500 miles from her home of Woburn Abbey across France, Spain, and North Africa, crossing the three great mountain ranges, the Pyrénées, Guadaramas, and Sierra Nevada, in the course of three weeks. Lieut R. R. 'Dick' Bentley of the SAAF was flying to the Cape with a Moth. Ivor McClure of the Automobile Association and Sydney St Barbe had been touring Europe by Moth, intent on visiting fourteen countries in fourteen days, but after achieving six, having crossed the Pyrénées and the Alps twice, they landed in a small field to enquire the way and smashed the undercarriage while taking-off – but this gave opportunity for de Havillands to display the efficiency of their servicing arrangements which their astute salesman-promoter Francis St Barbe had established, with bases having adequate spares whether in England, the European Continent, Africa, Australia, or America. All over the British Isles keen private owners and club members were less ambitiously flying their Moths, visiting friends, landing at small fields and the rapidly increasing club aërodromes. Thus Juan de la Cierva, now a keen member of the Hampshire Aeroplane Club, flew to the Isle of Wight for lunch, landing at Seaview on the same sandy stretch which Gordon England and his friends had used for their lumbering Bristol Boxkites fifteen years earlier. Even at that time there had been almost 400 pilots with RAeC certificates, so it was not really surprising that in the course of two years the London Aeroplane Club had flown 2,900 hours, and 70 members had qualified for the 'A' Licence with only one serious accident when currently the prototype Moth G-EBKT was written off.

De Havilland Aircraft were steadily improving the Moth, and now announced Type X as 'the product of two and a half years and one million miles experience'. Span was a foot greater with less chord for better climb efficiency; the more powerful 85 hp A.D.C. Cirrus II made it faster. These changes gave tactful opportunity to restore the price to £730. Concurrently Blackburn and Avro began pushing the Bluebird and Avian, and presently Westland followed with the Widgeon. The Bluebird cost £795 with a Genet; the Avian with Cirrus II was £695; and the Widgeon £750. There was a game of tit-for-tat: when Lady Bailey, accompanied by Mrs Geoffrey de Havilland, hit the headlines with a world record climb to 17,283 ft with the D.H.60X Moth on 5 July, A. V. Roe & Co replied with Mrs Elliott-Lynn climbing an Avian Mk II to 19,200 ft, and making a journey of 1,300 miles in a single day during which she made 79 landings.

The King's Cup race on August Bank Holiday week-end had a prelude of tragedy, for on the preceding day Frank Barnard, that energetic, sporting chief pilot of Imperial Airways, was killed at Winterbourne, near Filton, while testing the Bristol Badminton. Just after take-off, when at only 200 ft, the engine, specially tuned to give 525 hp, had seized due to insufficient clearance between cylinder skirts and crankshaft balance weights. In attempting to reach a field, Barnard tried to turn into wind, stalled, and the Badminton dived in from 80 ft. 'The contrast between his extraordinary caution as a pilot of passenger machines and his daring and even recklessness as a pilot of single-seaters was remarkable,' wrote C. G.

The Royal Aero Club's 1927 handicap formula favoured large span, so the Bristol Badminton was given tapered 33-ft wings and a special uprated, small diameter Jupiter.

So small that it could stand under the wing of a Cirrus Moth, the de Havilland Tiger Moth was a startling demonstration of what could be done solely for racing. (*Flight*)

Grey. 'He had an unusual gift for imparting information in humorous and striking phrases which stuck in one's memory. There are few pilots who can tell the story of a flight as graphically as he could and add so many pointers, for his ability as an engineer was fully equal to his skill as a pilot.' For Fedden and Barnwell there were lessons – those of high centrifugal loading being immediately applicable to the uprated engine in the Crusader, and that so skilled a pilot would not have been caught out had not tapered wings been more sensitive to stalling than blunt. There would be no more Bristol racers.

For de Havillands it was different. They had sprung a surprise with the innocuously named Tiger Moth, for this new D.H.71 turned out to be a miniature low-wing racer of 22 ft 6 in span. The fuselage had literally been contoured tightly around Hubert Broad's slim form to get the drag as small as possible. Two prototypes were built in greatest secrecy. The first, with an 85 hp Cirrus II, had its initial test by Capt Broad on 24 June. Control, though sensitive, was satisfactory, so it was fitted with the prototype 135 hp four-cylinder Gipsy which Halford had designed as a rival to the Cirrus after resigning from A.D.C. Ltd earlier in the year and setting up as consultant at Victoria, London. Inevitably there were development snags, so for the race the black-painted second prototype had a Cirrus II installed. The tiny monoplane became the focus of fascinated attention by every pilot and enthusiast among those gathered at Hucknall, near Nottingham, for the start of the race, in which every type of light-plane was competing together with a Horsley, Vespa, Vixen III, and Avenger.

Rough weather proved too much for Broad in that cramped cockpit, for in bumps it was impossible to hold the sensitive little machine steady, so he landed at Spittlegate, after achieving 162 mph over the lap of 26 miles, beating the handicap formula by almost 48 mph. Winner was the ever smiling W. L. Hope flying the scarlet-painted two-year-old Moth G-EBME of Air Taxis Ltd – but, to beat the handicap formula based on horsepower and span, it had been fitted with extended tips to the upper wing, narrow wheels, conical windscreen – and a pressure tank in the enclosed front cockpit replaced the bulky centre-section tank. Second was Capt W. J. McDonough, instructor of the Midland Aero Club, flying the demonstrator Widgeon III; third and easily fastest was the Vickers test pilot, Scholefield, with the Vixen.

6

To spend money on such racing was the last thing to interest such a man as H.P. It was far more satisfactory to be able to announce at the annual general meeting of his company that the Air Ministry had directed all Mk III Bristol Fighters in service to be fitted with automatic slots, and that the value of related patents stood at £75,000. 'We have had to turn more and more to building for Service requirements,' he said with delighted sadness. 'The directors also regret the increasing separation which seems to be taking place between manufacturers and operators of civil aircraft in this country. Handley Page Ltd became and still remain the largest shareholder in Imperial Airways, but unfortunately our promised representation on its Board has not been accorded, and we are considering what should be done in this matter. One hears of regrettable lack of interest in commercial aviation in this country, but this is not surprising when a concern which has shown both financial and technical interest in furthering the development of air transport is thus given the cold shoulder.' Nevertheless, Handley Page Ltd had made excellent progress during the year; all loans had been discharged; assets were totally unencumbered; trading profit maintained.

Supermarine also had consolidated as the result of profitable orders, and early in the year was registered as Supermarine Aviation Works Ltd, with capital of £300,000. Sqn Cdr James Bird remained the driving force, but now R. J. Mitchell was made a director and thus more authoritatively could influence policy. The original sheds were too small for Southampton flying-boat production, so the war-time flying-boat assembly shops of May, Harden & May at Hythe on Southampton Water were taken over.

From time to time seafarers might see the grey and silver prototype Supermarine S.5 contender for the Schneider contest being launched from the RAF hangars on the isolated Calshot Spit. A great roar told of considerable testing when the Napier engine was run. Rumour had it that on one timed run the S.5 had even beaten the world's speed record for land machines. Soon the two exquisitely finished Gloster IV biplanes and Colonel Bristow's Crusader from Felixstowe were assembled. At first the High Speed Flight was said to be a discontented group because of personal rivalries, and there were questions as to whether Service pilots, however skilled, had anything like the experience and aptitude of professional civilian pilots for racing – but when the Press were invited on Tuesday 9 August, all was peace. Said C. G. Grey: 'Whatever may have been the state in early days there is no doubt that the Flight now is a splendid example for any happy family. Everybody flies every machine quite cheerfully, including the C.O., whom one understands is not supposed to be a racing

'The primary object of lowering the wing of the Supermarine S.5 was to improve the pilot's view compared with the S.4, which had lower resistance due to fairing the cylinder blocks into the wing', explained R. J. Mitchell. By cranking the Gloster IV top wing down, Henry Folland saved some resistance, although the biplane arrangement intrinsically had greater drag, and view was bad. (*Flight*)

pilot but evidently believes he should know all about his command. The final selection will not be made known till six days before the race, when, according to the rules, the pilot for each machine must be publicly nominated. None are finding difficulty in cornering at highest speed. We have heard a good deal about the effects of centrifugal force on the human anatomy, and the need for special belts, but these pilots are an unusually tough lot and do not seem to find any trouble at all.'

The colours of the racers were beautiful. The Supermarine became bright royal blue with shining silver flying surfaces; the Gloster was a confection of darker blue with burnished copper wings and tail, the top wing roots faired like a dolphin's fin into the Lion's outer cylinder banks and the lower wing sweeping upward to the fuselage; the Crusader, un-helmeted because of cooling trouble but with fairings in front and behind each cylinder, was entirely white except for a blue fairing from central cylinder to cockpit and thence to the blue fin. All three had long white floats of Short's advanced design; the glittering propellers were Fairey-Reeds of twisted duralumin.

Two Supermarine S.5s, one Gloster, and the Crusader were scheduled to leave Southampton for Malta on 17 August aboard the Admiralty collier *Heworth*, and a third Supermarine and a reserve Gloster would go later. There they would be transferred to the aircraft carrier *Eagle* for transport to Venice where they were expected to arrive on 6 September.

Meanwhile the RAF was conducting the first of what were to be annual tactical exercises. With nothing much to report during the week preceding August Bank Holiday, the sensation-loving Press blazoned these as 'the air manoeuvres' and 'the air war' though the object was to simulate defence operations against an attack on Britain. Squadrons had to be ready 'to scramble'. Bombers were sent hunting their targets regardless of weather, and fighters searched for them while Territorials practised gunning with the help of searchlights.

After it was over Colonel Sempill wrote to *The Times* stressing that: 'An inferior attacking force was able to penetrate London's air defences in innumerable raids, and theoretically laid considerable portions of the city in ruins. If these results can be achieved, how much more effective would be a real enemy with forces such as any four European powers could send? We have not more than 500 machines wherewith to defend London, as about half our aircraft are on foreign stations. Yet we rest calmly in our beds without a gas mask among seven million of us.'

At a Press conference at Uxbridge, Sir John Salmond, as AOC Air Defences, stated: 'The exercises were only to find the weak spots of the defensive system, for it has only been going for two years and will take all of eight to complete. The organizers concentrated on essentials only, so some sides of defence had more attention than others. During the exercises only the telephone lines of the GPO could be used; in time of war we would have direct lines. There were also complications of co-operation between Air Force and sound indicators and searchlights. It takes a long time to train the men, who are Territorials and have much to do. The defence of London always has the shadow of the time factor: an invading aircraft would take only 25 minutes from the coast to London, so very high training is necessary to give defences warning of its approach.' What had become abundantly clear was the weakness of a warning system using sound as locator. There was need for a complex radio network instead of telephones, and 'some kind of wireless apparatus which will warn the pilot of one aeroplane when he is approaching another'. It was also found how widely differing were reports on any one action as individually given by bomber, fighter, or umpire.

Not unexpectedly the slow night bombers were easy prey to fighters in reasonably clear weather, but with clouds there was a fair chance of getting through. Fox bombers proved too fast to catch. A 200 mph fighter would be essential, and even more important was great rate of climb and high ceiling to enable diving at raiders if only they could be found. There seemed little chance of complying. Assessment of the latest fighters to specification F.9/26 had not been completed, but the Bulldog was favoured, closely followed by the Hawfinch, despite their inadequate speeds. As the Hawfinch had better spinning recovery, Bristols fitted the Bulldog with larger fin and rudder but this worsened weather-cocking when taxi-ing cross-wind, though aided spin recovery.

Because of these contradictory qualities it was decided to retain the small rudder and lengthen the fuselage instead. To that end a prototype was ordered by the Air Ministry in November for extended competition with the Hawfinch. In that form Uwins reported: 'It has no vices, view is excellent, and every moment with it is a pleasure. Controls are balanced and operative down to the stall of 55 mph, with no tendency to spin when ailerons and rudder are used. Even at highest speeds, the balanced surfaces, which are carefully sheltered to prevent "snatch", ensure lightness of operation, making the machine extremely untiring to fly. Stability in all senses is positive, so if the tail is adjusted to suit the cruising speed, the control column may be released and the machine will fly for considerable periods hands off. The "offset" fin ensures minimum foot load during cruising flight, and at certain speeds it is possible to remove feet from the rudder bar without swinging off course. The metal structure, with its high safety factor, ensures absolute rigidity under all manoeuvres, and even when diving at 270 mph terminal velocity the machine is as steady as cruising at 130 mph. There is no tendency to flutter, and the longest spin is immediately stopped by normal use of controls. Even inverted loops have been made from dives at over 200 mph – which illustrates the exceptional strength. In short, it is a machine with which a pilot is immediately at home and happy to fly anywhere under any conditions.'

By contrast the Martlesham report on the Boulton & Paul Partridge stated: 'Though lateral control is light and effective at all speeds, the elevators are very heavy unless trim is first obtained with the tail adjusting gear; directional control is effective but rather heavy, and there is a tendency to hunt directionally at top speeds. Aerobatic control is only fair generally and longitudinally is very bad; the machine is very unstable longitudinally, and force has to be used to pull out of a steep dive. If the tail be wound right back, engine off, it is almost impossible to control the aircraft longitudinally when the engine is opened out. Taxi-ing is poor; the undercarriage is too spongy and rather narrow. The aircraft is fairly easy to land, but difficult to get the tail down with the trimmer set for gliding. Take-off into wind is quite good, but even a slight cross-wind is difficult, full rudder being required.'

Why there should be such differences in two similarly powered

Boulton & Paul's Partridge was their first single-engined fighter since the Bobolink, and had a plump body like its namesake, but the bulge allowed the gun to be fitted low down to give breech access. (*Courtesy Air Marshal Sir Ralph Sorley*)

machines of identical size and weight may seem inexplicable, but despite careful use of wind-tunnels, interpretation was imprecise. Though difference between model and full-scale might only be second degree it resulted in test flying remaining that subtle game of trial and error. At Bristol, young Russell found that: 'On first flights, control balance was the dicey part. No one expected to make more than a rough shot at the right stick or pedal force the first time, much less harmonize them. To be on the safe side it was better to accept a strong arm rather than risk overbalance. Then the forward aerodynamic balance was altered until the test pilot was satisfied. All wind-tunnel testing was generally mistrusted, so efforts were made to standardize shapes and proportions of tailplanes and elevators, fins and rudders, wing-tips and ailerons. In this way accumulating flight experience was carried forward. Stability and control just happened – for the day of the aerodynamicist in design had not fully arrived. Perhaps the best identification of this era was that, throughout, it was possible to design, build and test a prototype for around 30s per lb all-up weight. Maybe the industry was unsophisticated, but variety was not lacking.'

Typically, a 46-ft span semi-cantilever monoplane, the Boulton & Paul P.31 Bittern, was proving a problem to its pilot, tall elderly Sqn Ldr C. A. Rea, AMI Mech E, an engineer of considerable experience. Designed expressly to break bomber formations, it was the forerunner of big twin-engined single-seaters for the RAF, and had a Lewis mounted each side of the nose in barbettes to enable the gun to be elevated through 45 degrees for attack from below, and the pilot was therefore located with unobstructed upward and forward view in an open cockpit just ahead of the centre-section. Here was an echo of Capt Albert Ball, VC, and the Austin

First twin-engined single-seat fighter was the Boulton & Paul Bittern – a very different but earlier conception than the Partridge, and relatively advanced aerodynamically for British practice.

fighter designed under North's jurisdiction to meet Ball's tactics of startling an enemy pilot with a burst from above, then diving beneath to use upward-firing guns. The result was a formidable looking machine, but as the operational requirement merely visualized slow-flying bombers like Virginias its speed of 145 mph was already outclassed by the Fox for the engines were only the 230 hp Lynx. Rea recorded: 'I found it to have practically no lateral control; yet the ailerons had adequate movement and there was no lost motion when tested on the ground, and various adjustments showed no improvement. I felt that if I could see the ailerons in flight matters might be solved, but the outer wings, which were cantilevers deep at the root tapering to a very shallow tip section, were obstructed by the engine nacelles, even with the seat in its highest position, so I added cushions which brought my head uncomfortably high above the wind-

Bittern II was a major modification of the prototype, with strutted wings. The gun mounting is at its elevated position enabling attack from beneath. (*Courtesy Air Marshal Sir Ralph Sorley*)

screen. Despite the battering slipstream I was then able to see that the wing was flexing so much that the control effect of the ailerons was almost completely neutralized by contrary wing twist. A ground test was therefore made with ailerons and wing loaded to produce the stresses of flight and it was at once evident that the flexing was of such an extent as to be dangerous. In the course of months a new set of wings with outboard strut-bracing was built and lateral control was then all that could be desired.'

Though the Bulldog was meeting the general fighter requirement, it was still considered that an interceptor was essential. Fedden had argued convincingly that a supercharged radial should be stipulated as it took less time to warm up than a liquid-cooled engine. An interceptor specification, F.20/27, followed which aimed at using the geared Mercury III radial currently being developed. Tenders were accepted from Bristol, Gloster, Hawker, and Vickers with biplanes, and Westland and Vickers for low-wing monoplanes.

7

August was not all holiday mood. There was increasing cohesion of Communism in worldwide class war in which far-fetched allegations of victimization were exploited to this end. Thus on the 22nd some 200,000 demonstrated in Hyde Park against a long delayed United States sentence of execution on two Italian Communists who had been in prison since 1921. It was the beginning of deliberate emotional exploitation by malcontents, and was accepted unquestioningly by a strange mixture of trade unionists, escapers from the hard world of facts, and those representing the Opposition whom H. G. Wells called 'leftists'.

Certainly aircraft factories had less unrest than in many other industries, for they attracted chiefly those who still saw romance in aeroplanes and found pleasure in craftsmanship, but undoubtedly the social texture was changing. Men were gravitating from work-short 'depressed' areas to London and surrounding counties whose population increased by a tenth. It was leading to problems of housing and ribbon development with low standard, unimaginative 'council' houses on the one hand, and on the other the rising domination of industrial consortiums among chain stores, motor companies, petroleum refiners, chemical and textile industries. When one looked at the aircraft industry there seemed little sense in having sixteen major airframe builders and four engine makers competing for orders of a score a time – yet the Air Staff considered it justified, for bitter war-time experience had shown production could not be stepped up quickly. That suited the single-minded pioneers who had managed to remain manufacturers; for at least, expansion of the RAF, small though it was, allowed them to continue in this fascinating game of designing yet another and another new type of aeroplane.

In France it was even more difficult, for there were thirty-six aircraft constructors and ten aero-engine manufacturers. As in Britain, to keep them going a small annual appropriation had to be spread over the whole industry – chiefly for experimental aircraft – but at least this led to many diverse techniques of metal construction. France, who had not attempted to repay her war debt of £623 million to Britain, was subsidizing aircraft constructed for overseas orders. She also had a stranglehold on Bristol engine sales, for the Gnome-Rhône concern held a constructional licence covering the whole of Europe except Scandinavia, and this prevented direct British sales to such countries as Poland where recently thirty-four Jupiter-powered twin-engined Goliaths had been sold; and Germany, who would have preferred buying direct from Filton, refused to purchase from their enemy France and were securing engines through the Swedish licensee. However, Bristol had been able to penetrate the Far East market, and appointed Nakajima Aircraft Co of Tokyo as licensed builders.

From England in August the four rival flying-boats built by the Short, Blackburn, Saunders, and Supermarine companies departed on a 3,000 mile cruise to the Scandinavian capitals. The all-metal Short Singapore, now with stiffened bottom and uprated Condor IIIA engines with four-bladed propellers, had been much improved by the outrigger Fletner on the main rudder, dispensing with the auxiliaries. The metal-hulled Blackburn Iris II still had wooden wings but was similarly re-engined, and, to give room for a tail gunner's station, the centre fin had gone and its rudder moved forward and converted into a servo-unit linked to the outboard rudders now devoid of balances and with bigger fins. Briefly test-flown at Brough on 3 August by Flt Lieut Sawyer of the MAEE, it immediately left to join the all-wood Saunders Valkyrie and Supermarine Southampton I at Felixstowe, where the Singapore arrived on the 6th with Oswald Short as passenger. On the 12th the squadron took off, Sqn Ldr Scott, the CO, piloting the Iris which was carrying Sir Samuel Hoare to the Copenhagen Air Traffic Exhibition. Almost immediately the Singapore returned with engine trouble but next day flew nonstop to Oslo, the others having stayed at Esbjerg overnight. On the 19th they made the 300 mile flight to Copenhagen despite a gale, disembarking Sir Samuel just in time for the opening next day, then continued their circuit of the Baltic – though the Iris had to dash back to Felixstowe with Sir Samuel on the 24th and back

Back in civil guise, with auxiliary rudders removed, the Short Singapore has a Fletner servo rudder just visible on arms behind the main rudder.

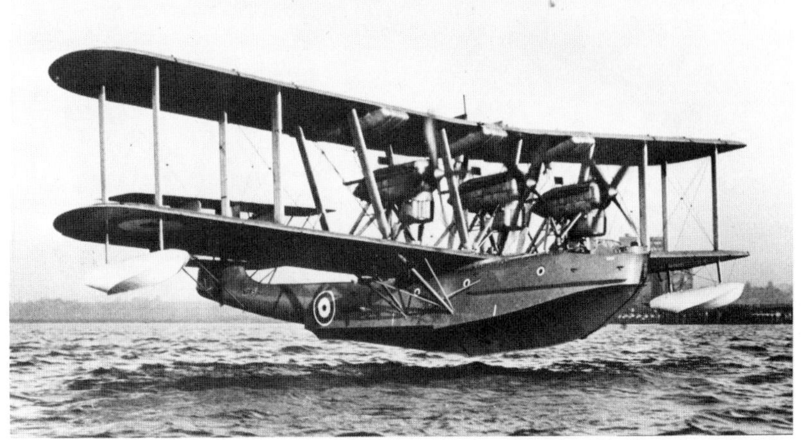

Now with a metal hull, the prototype Blackburn Iris becomes the Mk. II, and has had the centre rudder removed to give room for a tail gun cockpit. (*Hawker Siddeley*)

to Gdynia two days later. Engine repairs delayed the Southampton at Danzig on the 30th, and the Valkyrie alighted with engine trouble in the open sea and was towed to Königsberg, and at Copenhagen was damaged in attempting to take-off but eventually reached Felixstowe on 28 September. Splendid seaworthiness was shown by the Iris and Singapore, both completing the cruise without incident, their metal hulls fully vindicated. Of the two the Iris was preferred, for its deeper hull gave the lower wing and propellers much better water clearance, so it could cope with rougher seas, and the additional engine enabled quicker take-off and was an added safety factor on long patrols; nevertheless it was slower than the cleaner Singapore. Of the latter C. H. Barnes in *Shorts Aircraft since 1900* tells how 'Alan Cobham saw it and remarked "I could fly that round the world", and asked Oswald Short whether he thought the Air Ministry would lend it for that purpose, but they said it had cost taxpayers £20,000 and could not be lent to a civilian pilot for a stunt flight. Oswald referred this to Sir Geoffrey Salmond, who persuaded the Air Council to lend the Singapore to Short Brothers for one year, provided they insured it in favour of the Air Council for £12,000. Cobham then went into the details of his proposed round-the-world flight and found that some sectors would be beyond even the Singapore's range against the worst head winds, so he changed his plan to a 23,000-miles survey flight round Africa, the cost of which was £20,000 – made up by Sir Charles Wakefield with £12,000, Short Brothers who put up £3,000, Rolls-Royce with £1,000, and the remainder in small subscriptions from suppliers of materials and equipment.' What Cobham had in mind was the possibility of establishing an airline to carry gold and diamonds from the Rand to London by way of Cairo, and already an important link had been established from Khartoum to Kenya by Capt T. A. Gladstone who was associated with Robert

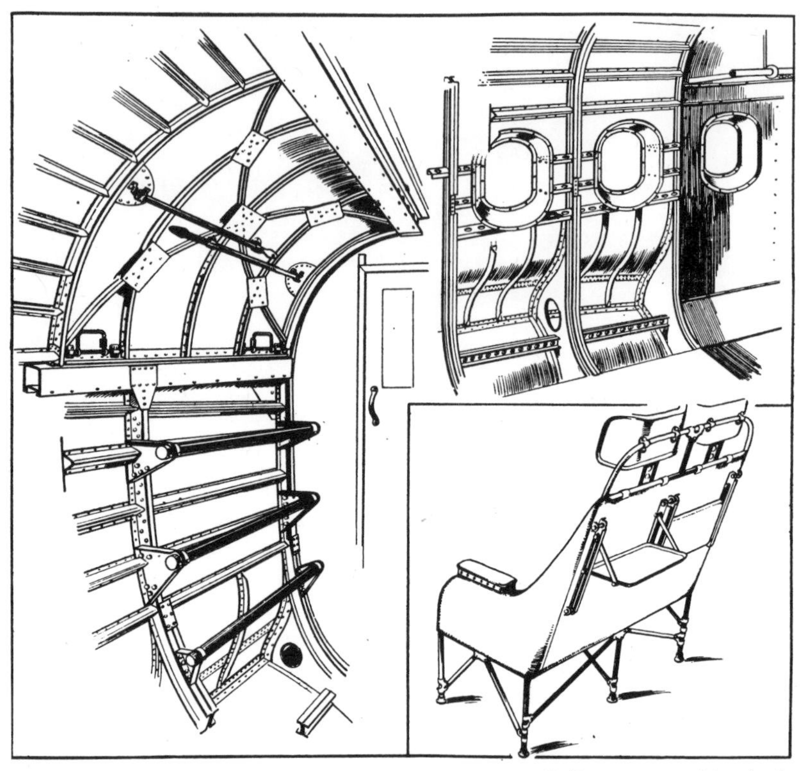

Construction of the Short Calcutta duralumin hull, showing (*left*) structure and hatch; the cabin (*right*) partly walled and floored, revealing integral chine; new form of seat (*below*) with headrests and tables. (*Flight*)

Blackburn. Meanwhile the two Short Calcutta flying-boat airliners were visibly taking shape.

The spate of other long flights continued with less justification than Cobham's proposal. The Americans, Brock and Schlee, undeterred by the perils of their Atlantic crossing, set out from Croydon on 29 August heading for Munich en route to Tokyo, and from there proposed returning to the USA by crossing the Pacific via Midway Island and Honolulu.

Two days later – after blessing by the Roman Catholic Bishop of Cardiff – a Jupiter-powered Fokker F.VII piloted by Lieut-Col F. F. Minchin and his friend Leslie Hamilton, with Princess Löwenstein-Wertheim, their backer, as passenger, laboured into the air from Upavon at 7.30 p.m. heading for Ottawa. They crossed the coast of Connemara at midday, and the steamship *Josiah Mason* saw the Fokker shortly before 10 p.m. that night – then silence. Months later a wheel was washed up on the shores of Iceland. Even before the war the Princess had sometimes chartered the Handley Page crescent-winged biplane owned by Roland Ding, and since the war often flew to public events in various aeroplanes piloted

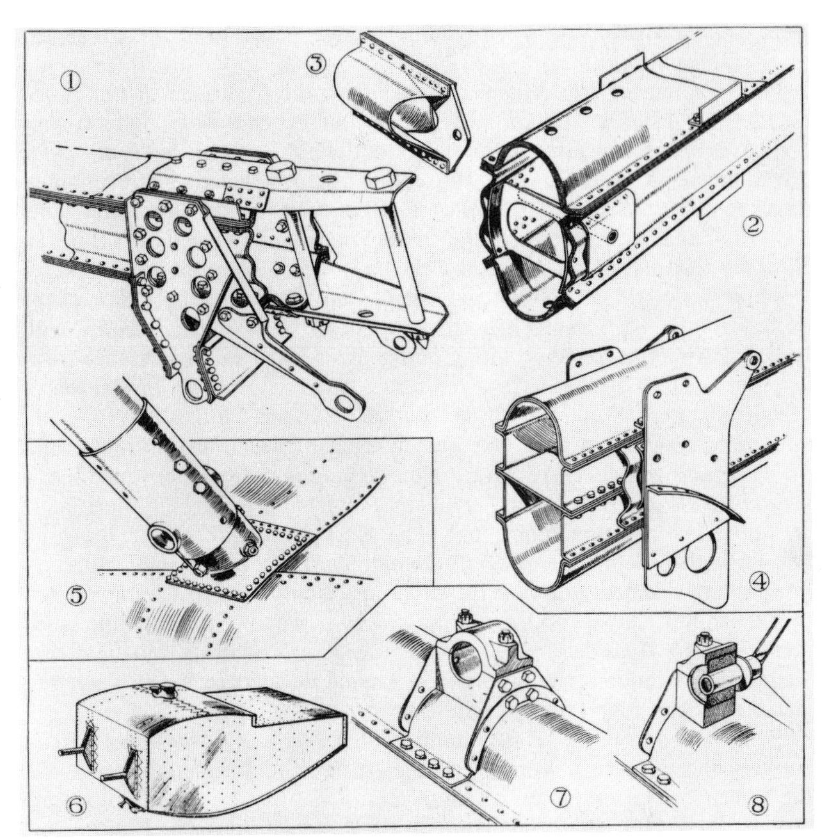

Calcutta wing structure was fabric surfaced. 1, strut and Rafwire attachment to spar. 2, simple dural top spar. 3, compression strut. 4, lower spar. 5, attachment of chine struts. 6, 7 and 8, tank with rubber-sprung mounting. (*Flight*)

by the handsome, skilful Leslie Hamilton who had been a Royal Air Force instructor and test pilot until he retired in December 1924 – yet neither he, nor Minchin, who resigned from the RAF in 1919, had found their niche in civil aviation. Everybody liked Minchin, that sad man of such charm and kindly nature who, after surveying the route to India for Imperial Airways in 1924, found the routine of airline flying too dull, and the endurance flying he conducted for Bristol when testing the Jupiter in the Bloodhound was too casual. Perhaps he missed the zest of war in which he had been so effective, and in the quest to find himself had made his disastrous Atlantic attempt.

Nor was he alone in desperate measures. Frank Courtney, backed by Charles Hosmer, a banker of Montreal, was attempting the east to west crossing from Plymouth in a Dornier Wal, with Flt Lieut Downer navigator and Bob Little as mechanic, accompanied by Hosmer's son. They took-off shortly after dawn on 3 September, making for the Azores, but battling against a head wind he found the fuel reserve too low so turned back for Corunna, and decided the attempt must be deferred as the machine was unsuitable.

On 7 September two Atlantic attacks from the American side ended in disaster, for a similar Fokker to Minchin's, named *Old Glory*, had left New York for Rome, but after fourteen hours the pilot sent an SOS and came down in rough Atlantic seas and could not be located, but eventually wreckage was picked up 100 miles north of the SOS position. That same day two Canadians attempted to emulate Brock and Schlee with a Stinson, *Sir John Carling*, but after leaving Harbour Grace no more was heard.

There were still others bent on similar 'do or die' attempts, for on the 16th Capt R. H. McIntosh, tired of endless cross-Channel flights with Imperial Airways, together with Commandant Fitzmaurice, a vividly restless personality of the Irish Free State Army, started from Baldonnel aerodrome near Dublin with yet another Jupiter Fokker F.VII in an attempt to reach New York, but after flying 300 miles in stormy weather, a 40 mph headwind forced them to return – but they were determined to try again. Meanwhile the erratic Mr Levine, returning to England, persuaded an unemployed airline pilot, one-eyed Capt Hinchliffe, to fly him and his embattled Bellanca, *Miss Columbia*, across the Atlantic, but as a preliminary decided on a nonstop flight to Calcutta. When they attempted to start from Cranwell on the morning of 17 September the machine failed to take-off. A week later they successfully got it airborne, but landed at Vienna in torrential rain, whereupon Levine decided to fly to Venice to watch the Schneider Trophy contest.

Venue for the Schneider was the Lido, the long island forming a barrier between the Lagoon of Venice and the Adriatic, and at its mid-point was the Excelsior Palace Hotel, headquarters of the British team. Competing aircraft were housed at San Andrea air station on an island within the north side of the Lagoon facing the main shipping channel for Venice. The contest would be along the 13½ nautical miles of the Lido bathing beaches. Unfortunately the US Navy Department had announced that Lieut Alford Williams, holder of the 1923 world record of 266.6 mph, could not participate as it was impracticable to complete adequate tests in time to qualify. The contest would therefore be Britain versus Italy.

The British racers arrived earlier than expected, and during the first week of September had been erected and were ready for test. Despite high hope, Calshot trials had shown the Crusader slower than the others, so it was designated as a practice machine. On 11 September, Flg Off H. M. Schofield, who was involved in several hair-raising experiences with it at Felixstowe due to engine cutting, took this racer out and headed for the Lido, where he opened up in a long take-off. As soon as the Crusader was clear of the water, onlookers were appalled to see it half roll to port and dive into the Lagoon upside-down. The impact tore off a float, and water rushed through a hole in the fuselage projecting Schofield out with such violence that his clothes were dragged off, and when the rescue boat reached him he was found vigorously swimming clad only in a shirt. On lifting the Crusader from the bottom after a week's immersion, the magnesium alloy crankcase had almost dissolved, and it was immediately evident that the aileron cables had been crossed to the wrong turnbuckles so that the movement was reversed. That neither pilot, firm's inspectors, nor AID had discovered this before flight seems inexcusable, yet it was no new thing; even eight years later I found the rudder of my experimental Lysander linked to the control column, and the elevator to the rudder bar, and it was longer still before there was mandatory instruction to make control fittings non-interchangeable.

On Friday 23 September, when the machines had to be presented to the *Commissaires Sportives*, for navigation and mooring tests, the skies were dull, the weather uncertain, but the sea held calm, and all completed the taxi-ing and take-offs without trouble except Flt Lieut S. N. Webster, who taxied too fast so that his S.5 lifted a short distance and touched down again. The six-hour mooring followed in a heavy thunder-storm, and not until it finished was Webster told he must repeat taxi-ing on the following day.

By Saturday night many visitors had arrived at the Excelsior Hotel. C. G. Grey was there with Lester D. Gardner and other aviation writers. Sir

Competing Italian (*top*) and British racing seaplanes moored in the lagoon at Venice, with Webster's S.5 in the foreground. (*Flight*)

Philip Sassoon arrived piloted by Flt Lieut A. G. Jones-Williams, universally known as 'John Willy'. Sqn Ldr the Hon R. A. Cochrane, with Flt Lieut H. V. Drew, landed with his Westland Widgeon. Alan Butler, Hubert Broad, and Miss Spooner came in Moths, and the irrepressible Mr Levine was there with Capt Hinchliffe, having asked nobody's permission to land at the Lido.

On the following day a tremendous crowd arrived to watch the race, but weather was hopeless. Nor was Monday morning encouraging, for, though the wind dropped, a swell rolled on the Adriatic. Slowly conditions improved, and at 12.30 the three British machines were towed down the canal on their own floats, followed by the three Italian racers on pontoons towed by steam tugs. The shore was black with spectators. By 2 p.m. excitement was intense. The Italian Crown Prince and high ranking officers arrived. There was a great hum of conversation, interspersed with loud-speaker announcements. Far at sea were HMS *Eagle*, two Italian battleships, a cruiser, minesweepers and coastal motor-boats, and overhead a Savoia S.59 flying-boat was circling to keep watch. Promptly at 2.30 loud-speakers announced that Flt Lieut Kinkead on the Gloster IVB had taken-off. Three minutes later he was seen approaching low down, and flashed past the starting line in a crescendo of sound. Five minutes later, to a great roar from the crowd, Major de Bernardi raced across the line with his red projectile-like Macchi M.52. Kinkead was seen returning, and made a beautifully judged semicircle with 50 degree bank around the acute-angled northern turning point. Five minutes later, flying higher, de Bernardi followed with dive and zooming turn, drawing further roars of applause. Immediately behind came Webster on the S.5 No. 4 with geared Napier, and two minutes behind him Kinkead had completed his second lap. But de Bernardi failed to appear, and was spotted through binoculars far down the seaward leg, his machine rolling quietly on the swell, his engine having failed through valve trouble. Then Lieut Guazzetti, the second Italian, crossed the line just as Webster completed his first lap at an average of 280 mph. Two minutes later Flt Lieut Worsley came into the race with the direct-drive second Supermarine, and behind was Kinkead on his third lap. A slight drizzle set in, making it difficult to see through the low windscreens of the racers. Just as Webster completed his next lap, Lieut Ferrarin, last of the Italians, started off, but before he had flown a quarter of a mile, clouds of black smoke and bursts of flame came from his machine, and he immediately pulled into a climbing turn, and rumbled back to the Lagoon, a piston having broken.

With Guazzetti the only Italian left, and all British machines lapping faster, it was certain the RAF would regain the Trophy providing the engines held out for the seven circuits – but on the sixth, Kinkead's Gloster had a spinner failure, a strip of metal wrapping round one of the propeller blades and setting up terrific vibration; but Kinkead made a successful landing. On his fifth lap Guazzetti ran into trouble, for he was nearly blinded with leaking petrol and suddenly disappeared behind the

Gone in a flash! Worsley's ungeared Supermarine S.5 streaks past. (*Flight*)

Lido to alight on the Lagoon. At this débacle, the crowd began to melt away, leaving the two Supermarines to battle it out. Webster was clearly some 10 mph faster than his colleague, and got round the seven circuits so much faster than expected that he continued for an eighth to make sure, then banked steeply over the hotel to the cheers of the dwindling crowd and headed back to the Lagoon. By averaging 281.65 mph he had beaten the world speed record by 3 mph, so at times he must have travelled at well over 300 mph. Boyish looking, sandy haired, young Webster from Walsall had more than vindicated the skill of Mitchell in devising the machine and Wilkinson in developing the original 450 hp Lion until it gave 900 hp for racing, nor could the contribution of Ralli be overlooked in designing the Fairey propellers. That evening the managing directors of Supermarine and Napier gave a party for both teams, but the British pilots had to leave early as Sir Philip Sassoon was taking them to Venice to meet Italian officials. Midnight had struck when two slight young men in dinner jackets came into the bar of the Excelsior. Someone recognized them as Webster and Kinkead, and asked what they were doing there instead of with Sassoon. Typically Webby replied: 'Just buying Kink a drink.' Next day Prince Scalea, President of the Italian Aero Club, formally handed over the Schneider Trophy, and the three British pilots received from the patriot poet Gabriele d'Annunzio handsome silver cigarette cases as mementoes, and for Webster there was also a gold ring in the form of an eagle set with a ruby.

A great British team: Flg Off H. M. Schofield, Flt Lieut O. E. Worsley, Flg Off S. N. Webster, Flt Lieut S. M. Kinkead, and the Commanding Officer, Sqn Ldr L. H. Slatter. (*Flight*)

Unfortunate controversy followed the Schneider contest arising from a comment in *Flight* that: 'The Macchi designer and Mitchell both took the Supermarine S.4 as a starting point, but whereas the Macchi was ready for, and won last year's race, the S.5 was but recently completed thus there can be no question of copying.' Immediately Italy's *Aeronautica* contentiously stated: 'As the M.39 appeared a year earlier it may be easily maintained that the S.5 is a copy. The Short Crusader also is wonderfully similar, for in July 1926 its builder, Mr Short, when on a visit to Macchi's workshops, inspected the M.39 in smallest detail.'

Oswald Short replied: 'When I read this I did not know whether to be annoyed at being labelled a spy, or flattered to be credited with so retentive a mind. The facts are that in 1926 I was invited to visit two Italian firms interested in my particular form of metal aircraft, and I was shown the M.39 undergoing its first engine test. I remarked I could have sworn the floats had been made in our own shops. One of Mr Macchi's staff replied, "Yes, when Mr Macchi Junior was in America for the 1925 Schneider no doubt he kept his eyes wide open." That year they had entered a flying-boat, so it was after its defeat that they turned to a twin-float seaplane similar to the British and Americans in that contest. Undoubtedly Mr Macchi observed the clean lines of the Supermarine and noted that the attempt to produce a comparatively thin wing in pure cantilever led to trouble; he also saw that the American float strut system was the cleanest. These facts no doubt led to the M.39 design, and to avoid trouble the wing was braced to the floats and top of the fuselage by wires, and this requirement alone, apart from good view, would suggest dropping the wing to obtain reasonable angles on the bracing wires. The M.39 of 1926 was therefore a combination of the good points of the 1925 British and American machines, and no one begrudges Messrs Macchi the credit of their achievement.'

8

Overshadowed by the Schneider was the successful conclusion of Richard Bentley's flight in a Moth to Cape Town, reached 27 days after leaving Stag Lane on 1 September, during which he covered 7,000 miles. 'His flight stamps Mr Bentley as being worthy to rank with the world's great pilots,' was the consensus of reported opinion. 'Once again the little de Havilland Moth with its simple A.D.C. Cirrus engine has proved worthy to rank with the world's great aeroplanes.'

Long-distance flights remained recurrent points of news. On 17 October four Supermarine flying-boats commanded by Grp Capt H. M. Cave-Browne-Cave, began a twelve-month cruise to the Far East with intention of visiting foreign ports in the manner of HM battleships as part of the Imperial Defence policy. Though a cruise of great importance, its routine nature had not the glamour of the widely publicized African flight on which Sir Alan Cobham was about to start, using the Singapore with which he and his second pilot, H. V. Worrall, had been practising from the Medway under the tutelage of Lankester Parker. C. R. Bonnett was again accompanying him as photographer, and as mechanics he had C. E. Conway and F. Green. On 17 November, after a farewell lunch in London at which Sir Charles Wakefield, the financial mainstay of the voyage, announced that Lady Cobham was going as well, 'the start was made from Rochester, despite the delays of the morning, promptly at the time fixed, and with punctuality characteristic of Sir Alan Cobham the machine passed over central London exactly at the appointed hour. He and his party flew up the Thames through London and cut across country to Southampton, where they alighted safely', though evidently the reporter

Oswald Short, that man of iron resolve and hidden kindliness, with Winston Churchill and John Lankester Parker, in the bows of a Calcutta.

had expected the splendid lunch to affect Cobham's judgment. Thereafter daily reports of progress were eagerly consumed by readers needing a figure of romance on whom to pin their sense of adventure.

Whatever the hopes of trail-making flyers, it was not machines or engines which would dictate success of eventual air routes but ability of pilots to fly through every kind of weather, trained in the assurance of radio bearings and gyroscopic indicators instead of dependence on natural horizons. To that end Sqn Ldr G. H. Reid, DFC, inventor both of a turn-indicator and a reaction testing machine for pilots, on retiring from the RAF formed the Reid Manufacturing & Construction Co Ltd to make and market such devices. Associated with him was Fred Sigrist of Hawkers. 'This is certainly a happy augury for the business,' reflected *The Aeroplane*'s editor, 'for in very nearly 20 years knowledge of Mr Sigrist one has never known him back the wrong horse.'

Far away at Long Island, New York, a man who helped pioneer the way nearly 30 years earlier died on 17 October; he was Charles M. Manly, who at 22 had been the brilliant engineer assistant of the late Professor Langley at the Smithsonian. By 1898 Langley had already spent a decade investigating the unknown subject of aerodynamics. Following success with his tandem-winged steam driven models, he was urged by the US Government to build a full-sized machine, and unable to find a suitable light-weight petrol engine among those with which early automobile builders were experimenting, he set Manly the task both of designing the 'aerodrome' as he called his machine, and producing a 52 hp five-cylinder water-cooled radial petrol engine weighing less than 3 lb/hp, which presently ran a type test of three 10-hour nonstop runs at full power. So convinced was Manly of the practicability of flying that he piloted the machine on both unsuccessful attempts at launching, but with cessation of Langley's work he turned to the American automobile industry, patenting many successful devices, and then in the early years of the Great War became consultant for the British Government in the purchase of USA aircraft and engines, and subsequently was assistant general manager of the Curtiss Aeroplane & Motor Corporation until he formed his own firm of consulting engineers in 1920.

British aviation suffered a more direct blow on 20 October with the death from pneumonia, following an operation, of Fairey's great friend and co-director Lieut-Col Vincent Nicholl, DSO, DSC. He and his fellow directors F. G. T. Dawson and Maurice Wright had been undergraduates at Cambridge in 1912, and as Maurice was a friend of Alec Ogilvie, the proud owner of a Wright biplane, they spent much time at Eastchurch and so met Dick Fairey who was then working on Dunne's inherently stable tailless biplane. All their lives they remembered those Eastchurch days as a joyous episode. Nicholl had a power of command that was his characteristic. When he joined Faireys, though he was not a trained engineer, what 'the Colonel' said went, before anything else. 'Nicholl was tall and very thin, a man of charm and irrepressible fun, much consideration for others, but with a body too frail to contain his spirit long,' wrote Norman Macmillan in valediction.

Fairey was involved in a most secret project. The two trail-blazing long-distance attempts by the Horsleys having proved unsuccessful, the monocled Air Member for Supply and Research, Air Marshal Sir John Higgins, decided a special aeroplane must be built to establish a prestige lead over the French, Italians and Americans. Fairey learned of this through his friend, Sqn Ldr Carr, one of the Horsley pilots, and offered to undertake the design. He received no specification or detailed requirement, only an official letter inviting him to build a research aircraft capable of flying long distance in expectation of an attempt in 1928.

Hollis Williams heard of it through the D.O. 'jungle telegraph'. 'With the Dove tucked in the experimental shop, I had spare time again, and wanted to make some return for C.R.F.'s generosity, so I started project work at home. When I saw a new requirement I tried to produce a more advanced design unofficially, sending a folder to Ralli, but they usually vanished without comment. When I realized work was being undertaken on a long range biplane I put in two designs – one a high-wing monoplane with fixed undercarriage and the other a low-wing with retractable undercarriage. A week or so later I was called into Barlow's office and told that C.R.F. had decided to build the high winger – designated the Fairey Postal, and for secrecy a separate design group would be set up under my charge. With F. H. Ordidge, we formed a team of about 30 draughtsmen and technicians housed in a half-section of the firm's cold and noisy transport garage.

'The idea was to provide the best possible L/D and fly at speeds slightly above it, using a special Napier Lion with cruising consumption promised at 0.52 pints/bhp/hour. As this speed was only 25 per cent above stall, we presumed high "g" would be avoided and we could use low strength factors and get a light structure. We were really providing a long endurance aircraft and hoped that "Met" would provide the right tail wind to waft it along.'

9

After three months' absence I had returned to Westland qualified as an RAF Reserve pilot on Jupiter-powered Bristol Fighter Trainers, and was given a miscellany of jobs such as liaison between D.O. and Works, mechanical testing, visits to Martlesham on Wapiti trials, and acting as observer to the recently appointed chief test pilot, Capt Louis Paget – an imperturbable, colourful 6½-footer who sported a monocle and elegantly cut clothes favouring blatant squares and checked patterns. Unlike Keep, Openshaw and Geoffrey Hill, he had no scientific training. His description of an unstable machine was 'it turns round and looks at you', and one with poor controls 'flew like a piece of blotting paper.' My task was therefore to record, for which my pay was increased to £5 a week and I was insured for £1,000. Paget had £650 and was insured for £2,500. Stuart Keep, who in effect was general manager, had a salary of £700, and Bruce, as managing director and chief engineer of Westland and director of Petters Ltd, received £1,000/annum compared with Oswald Short's £700 for the same task.

Westland had sold several Widgeons, and now had two demonstrators, G-EBRL and G-EBRO, which Keep allowed me to fly after a brief check from Paget. When their Lympne predecessor, the taper-winged Widgeon II, was sold to tall and handsome Dr Whitehead Reid, to my surprise I was told to fly it to his hangar near Canterbury. It was an October afternoon before the machine was ready, and by the time I had followed the railway to Farnborough, visibility had deteriorated, so I decided to stay there overnight. I had found the Airscrew 64 section wings of this Widgeon stalled very suddenly, unlike the R.A.F.34 of Widgeon III, so it was disconcerting that a north wind made it necessary to approach over the high ground of the adjoining golf course and land across the narrow width of this small aerodrome, heading at the tall buildings dominating the opposite side, but to my relief I discovered it could be done. Reassured at this success I went to the wooden hut labelled 'Pilots Report Here', and after booking in, told the sole occupant that it had been my ambition to land at Farnborough ever since war days, when as a schoolboy I cycled there from Reading to watch the flying. For a while we chatted about the early pilots.

'I've been here ever since the days of Cody,' said the man. 'Would you like to meet his son Vivian? He's foreman of the fabric shop here.'

I nodded. He lifted the phone. 'Viv, there's a young man interested in your father.'

Presently Vivian Cody, a stocky man of 40 or so, arrived with a box of photographs of his father's kites, gliders and power machines, including several of the first flight of Cody No. 1. 'That was in May 1907', said Vivian – and as this was only twenty years later, when his recollection was

The Westland taper-winged Widgeon II in final form with Genet engine, reveals neat wing folding, with permanent 'jury' strut which faired into the fuselage.

fresh and none had contested a date so memorable to him, it is likely to be true, nor did he differ on later occasions when he told me more.

Next day, above a countryside bronzed with autumn, I flew to the small, sloping, war-time field where Whitehead Reid housed his R.A.F.-engined S.E.5 alongside an oily Avro 504K. To my horror he climbed into the cramped little forward cockpit and asked me to fly him around. Except for my instructor I had never carried a passenger, and certainly not in this slightly dicey little high-winger. It took most of the field to take-off, but I was thankful that the uphill gradient braked the landing, for we ran much too close to the hangar door – though my surgeon passenger seemed unperturbed.

Still with ply-extended rudder, but now with horn-balanced skewed ailerons, the Westland Wizard rebuilt after its crash and fitted with a Rolls-Royce F.XI.

At Yeovil the private-venture Wizard had been rebuilt and one of the earliest pre-production Rolls-Royce F.XIs installed. In November it was flown by Paget. From the moment he took-off it was clear that climb was spectacular. Speed trials gave hope of a winner, for almost 190 mph was attained, but lateral control was still too heavy, so trial and error adjustment of the inset horn balance began. Despite a long sequence of wind-tunnel tests since February 1926, only lift, drag, and pitching moments for several wing-sections and body shapes had been measured. 'Ailerons and rudder always give trouble,' explained the aerodynamicist, 'so they might as well be done full scale.' Changes and performance testing would continue several months, during which Bruce endeavoured to interest the AMSR and the Director of Technical Development (DTD), and Paget talked about the machine enthusiastically to his friends in the RAF.

While the Wizard was proving the value of aerodynamic cleanness afforded by the narrow engine, Fedden was quick to seize the possibility of proving his supercharged Jupiter VII at extreme altitude by inducing his directors to have their private-venture Bulldog prototype fitted with 50-ft span wings. They were ready when it returned from Martlesham at the end of October, and on 7 November it was tested in this guise by Uwins, and then delivered to Farnborough for installation of special oxygen equipment so that Flt Lieut J. A. Gray could attempt a record climb – but before this could be made, Commendatore Donati raised the record to 38,800 ft for Italy flying a single-seat Dewoitine built under licence by Ansaldo and powered by a high-compression Jupiter IV. As in any case Bristols had achieved splendid advertisement, the Bulldog attempt was abandoned.

With long wings having ailerons only on the lower rather than reduce efficiency of the top wing, the supercharged high-altitude Bristol Bulldog awaits flight at Filton where the war-time hangars could still be seen.

As far as airframes were concerned the future of the Bristol Aeroplane Co was becoming critical, for they had been existing on overhauls of the obsolete Fighter. A production order either for the Bulldog or two-seat Type 101 was essential. Of other constructors, Beardmore, Parnall, Saunders, and Short Bros lacked production orders, though Beardmores were fully occupied in designing and building the 140-ft span all-metal Rohrbach Inflexible under Bill Shackleton's direction; Bolas was busy on naval requirements; Knowler at Saunders was designing a fighter and a simplified flat-sided metal hull for a standard Southampton superstructure; Shorts were considering a four-engined version of the Singapore. Other manufacturers were better placed. Armstrong Whitworth had both Siskin and the Atlas in squadron use, and production lines were full; Avro were busy with Lynx-powered 504Ns filling the Manchester factory; Blackburns were suffering a lull, but were finishing the last of the Darts and a small production batch of Bluebirds, and expected an I.T.P. for seven pre-production Ripon IIs and an experimental contract for three Iris boats to specification R.31/27. Boulton & Paul had six Sidestrand IIs; de Havillands their Moths. At Faireys were production IIIFs, differing from the prototype in having a rounded fin shielding a balanced rudder. Gloster had Gamecocks, but were becoming apprehensive of their rivals in the F.9/26 competition; Handley Page was slowly turning out Hyderabads; Hawkers were full with Horsleys; Supermarine making metal-hulled Southamptons; Vickers still turning out Virginias and Victorias; Westland had the Wapiti.

Like Westland, Blackburn had made a fighter meeting F.9/26 specification. Designed by B. A. Duncan, it was calculated capable of 200 mph at 15,000 ft powered with a supercharged Mercury – a figure beyond belief to the Air Ministry performance estimator – so the tender had not been accepted, but a private-venture all-metal version known as the Turcock

At first glance the metal-framed, fabric-covered Blackburn Turcock might have come from the Armstrong Whitworth stable, but its streamlining marked a departure from Bumpus's previous outlook.

was built with a supercharged Jaguar VI, and tested on 14 November, achieving 176 mph only to be destroyed in a crash two months later. Though no replacement was built it led Bumpus to consider a fast, small, light fighter which Petty began to scheme for him.

Speaking at a Conservative meeting at Chelsea on 14 November, Sir Samuel Hoare revealed that total recent production, together with a residue of obsolete types such as the Bristol Fighter and D.H.9A, amounted to 750 first-line aeroplanes – yet France, the greatest air power in Europe, had 1,350. Nevertheless our three Services totalled 284,000 officers and men compared with 320,000 before the war, but though £118 million was currently expended on Defence, it represented no more than the £80 million spent in 1914, as the £1 had so greatly depreciated.

Not French, but German aviation was the growing worry. When C. F. Snowden Gamble, author of *The Story of a North Sea Air Station*, extensively toured Germany that summer he was impressed by the State-operated landplane flying schools of Staaken, Stettin, and Schleissheim, with seaplane schools at Warnemünde on the Baltic and Sylt in the North Sea. Courses took four years, and pupils were housed in barracks and under strict discipline. In essence it was a military force in civil guise. Unlike my experience of the RAF Reserve, all pupils wore parachutes, and for initial training had crash helmets. After learning on 80 hp Udet biplanes, they passed to more powerful machines including the 240 hp B.M.W. Junkers-A 20 monoplane, and then bigger, quasi-military types such as the Junkers-F 13, the 600 hp Dornier Merkur and three-engined Junkers-G 23. Subsequently they received seaplane training on war-time Friedrichshafen two-seaters and the latest 500 hp Heinkel H.E.5s. All instructors were ex-war pilots, though care was taken to avoid addressing them by their late rank. Snowden Gamble also visited the Rohrbach works in Berlin, and at the Heinkel works alongside the Warnemünde school was shown round by Ernst Heinkel, but had insufficient time to visit the Dornier-Metallbauten part of the Zeppelin organization of thirteen associated firms, which was 'the largest aircraft manufacturing concern in the world, and undoubtedly one of the largest engineering units'. Dornier itself had seven factories – two in Germany, two in Italy, one in Holland, one in Spain, and one in Japan.

One of the smaller German companies, Focke-Wulf, who had produced several original designs including a tail-first monoplane, were interested in a licence to build the Autogiro. At Göttingen experiments were being made on models of it to determine disc coefficients for lift, drag and pitching moment. That the British Air Ministry still thought there might be other solutions was confirmed by their acquisition of design and construction rights of a rotary-wing machine known as the Helicogyre devised by an Italian engineer, Vittorio Isacco, as a variant of Brennan's tip-driven helicopter, with two wide-chord rotor-blades universally articulated to a common hub on a substantially vertical axle, each blade having an engine at its tip. Despite objections, it was to be built by Saunders Ltd under government contract – though the industry's technicians dismissed it as wasted money, for it seemed improbable that the engines would work under the centrifugal loads nor could the propellers be effective because of the varying translational velocity. 'Evolution, not revolution,' said Westland's designer to me.

Experiments with the Autogiros were continuing at Hamble and

Wimperis retained an open mind on rotary wing development despite failure of the Brennan helicopter, and recommended financial support for the somewhat similar Isacco Helicogyre.

Farnborough. The Viper-powered C.8V owned by A. V. Roe Ltd reverted – after trying it with different undercarriages, such as two wheels and skid, tricycle, and even four wheels – to its original two-wheeled undercarriage and was demonstrated at Croydon on 20 October by Bert Hinkler who was still organizing his proposed Australian flight. Recently he had tested the Cierva C.9 which Roy Chadwick had virtually devised by superimposing a pylon-mounted 30-ft four-bladed rotor above the second Avian prototype fuselage.

Chadwick had come to the fore as a virtuoso designer. His 29-ft span Avro Avocet, had a duralumin monocoque fuselage built by Short's method as a natural follow-on of the wood monocoque Avenger.

Ever active behind the scenes was John Lord, the force that drove A. V. Roe & Co Ltd. Stocky, short, outspoken yet kindly, unmistakably a Northerner, he had the business acumen which Alliott Roe lacked, and by his enterprise and foresight Chadwick had been encouraged to design many interesting aircraft. Latest was Type 584, the Avocet, a single-seat fleet fighter to specification 17/25 issued the previous June. Though it had a stressed-skin fuselage, its possibilities seemed limited, for it was only of 180 hp; but Vickers had gone even further by designing a similarly powered low-winger with Wibault-type all-metal construction.

Concurrently John Lord was elected Honorary Secretary of the SBAC, and Capt Peter Acland succeeded T.O.M. Sopwith as Chairman. 'Since the death of Major Wood,' wrote C. G. Grey, 'Acland has directed the fortunes of Vickers aircraft department with marked success, for it has steadily paid its way and is not responsible for the recent financial reconstruction of the firm. Capt Acland is one of the best liked personalities in the industry.' Tall, handsome, with wavy dark hair, his influence was through socialities and instinct for people and affairs rather than any pretension to technical knowledge of design and production. These he was content to leave to Rex Pierson and Percy Maxwell Muller assisted by Archie Knight. In the experimental shop was the Vireo low-wing cantilever monoplane competitor of the Avocet, with Wibault-type metal-skinned construction which made it seem a more advanced conception than any recent British aeroplane except the Supermarine S.5.

Model of the Blackburn Nile civil flying-boat. It would have had Jupiter engines and carried fourteen passengers. Only the hull was completed.

Certainly Grp Capt J. A. Chamier, the new DTD, was alive to the importance of monoplanes, for he authorized Blackburns to build a 100-ft span monoplane version of the Iris despite strong competition from Supermarine and Short Bros. At the latter factory day and night shifts were working on a replacement port lower wing for Cobham's Singapore. Overtaken by dusk on the evening of 24 November, he had anchored alongside HMS *Queen Elizabeth* in St Paul's Bay, Malta. Next morning he took-off for Kalafrana and alighted safely, but while being towed a heavy swell tore off the starboard wing float. His men climbed onto the port wing to lift the other clear, and he turned for the lee of cliffs and again anchored. On the 29th the Navy towed the Singapore to their base at Kalafrana, but it was illusory shelter, for heavy seas swept the sea wall and damaged the other float, so she was dragged onto the slipway – but even here the waves surged onto the port wing, which smashed, and only after partly hacking it away to clear a wall could the weighty flying-boat be dragged clear. Shorts were magnificent: the new lower port wing was delivered at Malta on 26 December, though slightly damaged in transit when a heavy weight was dropped on the packing case and several ribs smashed.

The Singapore's three-engined sistership, the Calcutta, specifically designed for Imperial Airways, was in the last stage of completion, but instead of the envisaged open central bay a vertical strut now stabilized the spars above the centre engine. In the 17-ft long cabin, with decor of blue leather wall-trim and upholstery, buff ceiling, and yacht-like white enamel, there were four rows of three seats, one of two, and a single seat at the back, together with galley and toilet rooms. That civil aviation seemed at the turn of the tide was indicated at the third annual general meeting of Imperial

Airways, when the first-ever profit was revealed, though only £11,000. Traffic and air freighting had increased steadily, achieving 92 per cent scheduled flights in Europe and 100 per cent in the Middle East where weather was more favourable, but unfortunately Persia had refused to ratify arrangements for extension to Karachi, so Basra was the present terminal.

Said Sir Eric Geddes: 'Imperial Airways have the knowledge and experience to run these Imperial routes, but a bolder policy on the part of the Government is essential or we shall be left behind by Continental competitors whose Governments give them much greater financial support.' This led to the resolution: 'That this meeting considers the present Government's subsidy inadequate for the services rendered and should be substantially increased.' Frederick Handley Page was prompt to second the motion. Already he had Volkert formulating still larger machines than the W.10s now operating on the cross-Channel route, and ideas were crystallizing as a four-engined biplane. 'If there is a limit in size,' said H.P.,

Near disaster for Cobham's Short Singapore at Malta, but saved by the determination of the RAF.

'it is not yet in sight, and I might say that it is more difficult to obtain orders for giant aeroplanes than to execute them when received.'

Undoubtedly H.P. would have whole-heartedly agreed with C. G. Grey that: 'Revolutionary ideas are not the result of direct inspiration. They are produced by a good mental process which absorbs all existing knowledge, rejects what is unsound, and then subconsciously arrives at something which looks new, but is probably only a modification in practical form of an idea which was absurd and unworkable years before. The world is waiting for air transport. And if we set to work the right way, we English will provide it, just as we provided the world with ships and railways.' But it was D.H., not H.P., who would have accepted C.G.G.'s further dictum: 'The present system by which the Trade exists solely on Government orders and regards manufacture of civil aircraft as a side-line is altogether wrong. If the Trade is wise it will lay its plans to get there without Government orders. The right way is for any firm which aspires to build commercial aircraft to set to in a little cheap workshop right away from the existing factory – at any rate as far away as the opposite side of the aerodrome.'

That was not quite what de Havilland had done, but his little factory had extended chiefly on civil orders, and with rejection of his fast and beautiful Hound because its slim fuselage could not contain the mass of gear, including a spare wheel, required for general-purpose aircraft, he had finally become disenchanted with military prospects, though did not altogether dismiss them. His interest essentially was in private flying and small commercial aircraft. He looked at design with a pilot's eyes, and still worked at his own drawing board on the preliminary design of every new type. As R. M. Clarkson, his young technician, observed: 'He believed a designer should have the closest possible contact and understanding with a prospective user, whether private owner, airline, or air force, and having established what the user wanted, he should be completely unfettered in interpreting those requirements and must thoroughly believe in what he was doing.' De Havilland himself wrote autobiographically: 'Although most designers would resent being called an artist and reply that they could not draw well nor paint at all, a designer must have much of the creative artist in him, backed by a lot of practical engineering experience.' Certainly in Arthur Hagg, his ambitious head designer, he had a man who was an artist by instinct and training coupled with more than a decade of aeronautical engineering experience under de Havilland's and Walker's guidance.

Their latest product, the D.H.61 Giant Moth, was typical of their combined outlook, though all details had been designed by Hagg and his team who migrated to a couple of rooms in his Bournemouth home for ten weeks' seclusion from the exigencies of Stag Lane. One of the draughtsmen was Harry Povey who, early in Moth production, privately designed at home the teak assembly jig for its fuselage, and it proved so effective that constructional time was cut from one week to one day. He was therefore favourably regarded by Hearle, who during the Bournemouth sojourn asked Hagg if a draughtsman could be spared to assist with production matters. Hagg told me: 'Of them all, Povey made the worst drawings, and the works had frequently complained they were undecipherable. So he was the one I could most easily spare – and thereafter he had the task of trying to discover what his own drawings meant! He never looked back, and next was made chief inspector with a long career of promotion stretching ahead.'

Flown in mid-September by Hubert Broad, the D.H.61 was typical of de Havilland's unaltering B.E./D.H.9A conception enlarged to 52-ft span and

Awaiting initial flight at Stag Lane, the prototype D.H.61 Giant Moth, although showing the typical hand of Geoffrey de Havilland, was designed in detail in ten weeks by Arthur Hagg and four draughtsmen at Bournemouth. (*Flight*)

6,200-lb all-up weight. A direct-drive Jupiter VI was fitted for the first flights, but the geared 500 hp Jupiter XI was its designated engine. Designed as a D.H.50J replacement to the order of the Commercial Aviation Co founded by H. C. Miller, and the Mac.Robertson Co of Adelaide, a big fruit-preserving firm, it had a generous cabin for six as a transport for those of the company who had to make long journeys. Landing speed was 46 mph, with top speed of 126 mph and range of 475 miles. Notably it was the first British aeroplane with crossed-axle undercarriage to enable landing in scrub and grass tall enough to overturn a machine using a conventional axle. Said *Flight*: 'It is one of the first of modern moderate-power machines suitable for general commercial service where heavy traffic could not be expected, and for this compares well with many much boosted American civil aircraft.' Even so, the *Canberra*, as the first was named, could not really show advance over the 68-ft span D.H.54 Highclere which had flown 2½ years earlier.

Design was proceeding somewhat slowly on two other projects, for these had to be all-metal structured – a new problem for die-hard specialists in wood. They were redesigns of the successful Hercules – made smaller as the twin-engined D.H.67 for aerial survey, and larger as the three-engined D.H.72 bomber to specification B.22/27 by interposing an additional bay outboard of the wing engines to give a span of 95 ft. To the same specification, John North at Boulton Paul was evolving the P.32, remarkably similar in shape, both it and the D.H.72 having the third engine on the centre-section above the fuselage as the RAF insisted on keeping the nose free for a gun. However, these machines were regarded only as supplementary to slightly lighter but more important bombers designed to specification B.19/27 as replacements for the Virginia, and design work was proceeding at Armstrong Whitworth, Avro, Fairey, Handley Page, and Vickers.

The year ended with great snow-storms, and the pilots of Air Taxis Ltd, W. L. Hope and G. Birkett, dropped hundreds of parcels of food by parachute to snow-bound villages and isolated houses.

CHAPTER IX

NEW TYPES FOR OLD 1928

'An important work monopolizes a man and, besides many other sacrifices, claims the whole personality. It fires the imagination, and gently approaches its elected disciple in an alluring way. Gradually it draws the soul more firmly into its golden net.'

Gustav Lilienthal (1891)

1

All firms had been instructed to develop automatic wing-tip slots for their Service aircraft. Concurrently with the trial installation by Handley Page Ltd on the Bristol Fighter, Westland, after many wind-tunnel experiments, had fitted a D.H.9A with slots sliding on cantilever supports, but the parallel motion gave an inefficient vent. Therefore, a swinging link motion was retrospectively fitted to the prototype Wapiti and satisfactorily tested in time to be incorporated in the pre-production batch initiated with J9078, to which had been fitted Frise ailerons, and a larger and rounded fin with rudder having a triangular horn balance. The slats had a torque tube within the wing to avoid twisting on opening. Decking was pale grey for all production models, and the internal dural tube structure was olive green.

The Witch high-wing bomber featured an enclosed bomb bay, and skewed ailerons modelled on the Wizard.

The Westland Witch was typical of its day, though regarded as advanced in being a monoplane with metal-framed fuselage and wood wings on steel-tube struts.

Early in January both the slotted Wapiti and rebuilt Wizard Mk I were being tested at Yeovil, and the big Witch high-wing bomber to specification 23/25 made its first flight at Andover on the 30th, piloted by Louis Paget, but proved unstable with C.G. aft. In the experimental shop the Pterodactyl Mk I's moulded nacelle of diagonal spruce strips and bird-like wings were being assembled, and the dural tube framework of the low-wing interceptor monoplane was taking shape. One corner of the erecting shop was the civil aviation department of which I was made manager at what seemed a

Unfortunately the Witch proved longitudinally unstable despite similar geometry to the Wizard. Because the fixed wing struts prevented an increase of sweepback, the engine was moved forward as indicated by this mock-up nose.

tremendous salary of £400/annum. A three-engined commercial monoplane, comparable with the Hamlet though greater in span, was being designed, and it had been agreed I could make a flyable mock-up cockpit enclosure for the Widgeon to see if cabin light-planes were feasible.

While the Witch was awaiting weather for climbs, the Hawker Harrier equivalent was already at Martlesham, but as a replacement for the Horsley it revealed minimal advance, though its single-bay wings of 10-ft less span looked neater, but even its all-metal structure was matched by the latest Horsley.

As first flown, the Blackburn Beagle's rudder had a triangular horn balance, but soon the point was removed to reduce its area, and the elevators gained an unshielded horn balance. (*Hawker Siddeley*)

At Blackburn, George Petty, recently appointed chief designer by Bumpus, had schemed the B.T.1 Beagle for the same purpose with swept-back, square-tipped wings, and a fuselage faired with stringers to rounded section conforming with its Jupiter VIII. When flown at Brough by George Bulman on 18 February, the rudder was overbalanced and elevators too heavy. The usual tests with varying balances began. At this juncture the Air Ministry accepted a tender from Blackburns for a somewhat similar, though smaller, biplane, the Nautilus, to specification O.22/26 for a carrier-borne Fleet spotter powered with a Rolls-Royce F.XI which it was hoped would give interceptor climb.

Like the Harrier, the competing Gloster Goring was at Martlesham, and at first might be mistaken for it, but instead of the elaborate split-axle undercarriage it had a simple cross-axle and the tail was angular like that of the Witch. Last of the four contenders was the distinctive Warren-strutted Handley Page Hare, but it was not until 28 February that Tom Harry England took it on its first flight. By the time Beagle and Hare were flying, experience with the Harrier and Goring indicated that the specification was outmoded. No machines were ordered. With only low priority, the Beagle remained on test for eighteen months before going to Martlesham, and the Hare for almost as long, and although the Witch was

The Gloster Goring, shown in original guise, was built of wood, had high-lift wings, and initially flew with a Jupiter VI engine. Ultimately it had a flat-sided fuselage, taller rudder and steeper fin.

satisfactory after moving the engine forward, it eventually went to the RAF parachute school at Henlow. Though this specification seemed money wasted on too conservative an outlook, it had been beneficial to each company by helping pay overheads of their design staff and bridging the frail financial path to the next State-supported design in hope of eventual production giving a few years security.

While manufacturers were busy with design and development to meet every military purpose, British civil aviation took a step forward with the preliminary availability of new buildings on the eastern boundary of the recently united Waddon and Wallington aerodromes re-named Croydon. There were four main blocks: to the north an hotel, then the terminal and main offices, and south of them two big sheds, each of which could house ten machines the size of an Argosy. Glass-domed booking and waiting hall were centrally placed but only 60 ft square, with similar areas for immigration and emigration and a still smaller Customs hall from which

Despite a cleaner appearance than the Horsley, the Hawker Harrier was disappointing and its lower aspect wings led to poor take-off at full load. (*Courtesy Air Marshal Sir Ralph Sorley*)

A single diagonal compression strut each side took the wing load of the 50-ft span, slotted Handley Page H.P.34 Hare. Like the Witch, it was slightly faster than the Beagle, though the Witch had better climb.

passengers emerged to the novelty of a concrete apron where previously embarkation had been from the muddy field. Dominating everything was the control tower, but the duty officer's room was only 25 ft square, though surrounded by windows giving unobstructed view of all approaching aircraft. A small compartment was used by a wireless operator 'for receiving and transmitting messages and for direction finding' using three Marconi units of 3 kilowatts on 900 metres for telephonic, the same for Morse, and the third on 1,400 metres for inter-aerodrome communications previously received by the Air Ministry and transmitted to Croydon by GPO telephone. Aerials were two and a half miles away on the edge of Mitcham Common.

Though all airline operators had transferred to the new buildings, they would not be officially opened until 2 May, but difficulty followed because the grass flying area became waterlogged and tailskids were broken on the edge of the concrete, necessitating a chalk entry platform. There were complaints that rents were too high, for the Treasury ruled they must give fair commercial return on capital value, though 'the premises are admittedly planned on a scale far larger than justified by the actual business at Croydon aerodrome.' Only visionary dreamers seemed to imagine that commercial aviation might become a very big business. Luckily three bays were available in the old Waddon aircraft factory, now mainly occupied by A.D.C. Aircraft Ltd, and there such firms as Surrey Flying Services established themselves. But the great boon was the hotel, for it quickly became the social focus not so much for travellers but for the pilots, particularly welcome after wind-battering flights in the open cockpits of those airliners.

Of other voyagers, Alan Cobham at last found calm enough sea to relaunch the Calcutta in safety, and on 21 January he took-off from Malta's Grand Harbour for Benghazi, thence to Tobruk and Aboukir. On the 27th he began flying along the Nile, completing the 390 miles to Luxor in six hours, and there explored the Valley of the Kings. Two days later he left in the relative cool of 7.30 a.m., reaching Wadi Halfa before midday, and on 30 January arrived at Khartoum. On the 4th he followed the rapids to Lake Albert, 2,000 ft above sea level, thence to Entebbe on Lake Victoria, 2,000 ft higher again, leaving on 6 February for Kisumu and Mwanza where he would make, at the Colonial Office's request, a survey diversion of 2,700 miles to Khartoum and back.

Next day it was Hinkler who took the news. He was off to Australia. He left Croydon at 6.48 a.m. on 7 February with his much modified Avian prototype, which had the wings from the high-altitude Avro 594C used by Mrs Elliott-Lynn for her record climb the previous October. Hinkler also fitted a special wide-track undercarriage he had devised which moved backwards on folding the wings to reduce weight on the tail and let him

Metamorphosis of Bert Hinkler's Avro Avian prototype with Cirrus II by fitting the special wings and modified undercarriage of the Avro Avian II with Alpha radial with which Mrs Elliott-Lynn attained a height of 19,200 ft in October 1927.

reach the engine – for he was a small man. Already an attempt to fly to Australia with a standard Avro Avian, the *Red Rose*, piloted by Capt W. N. Lancaster, with his sponsor Mrs Keith Miller as passenger, had come to grief. When they left Croydon on 14 October of the previous year their proposed journey was treated as a joke, but they pressed on and made good time until making a forced-landing at Ur, in Iraq, and then on 10 January the engine cut when taking-off at Muntok in the Netherlands East Indies, and the machine was smashed, necessitating shipment to Singapore for repairs. While they were waiting, the indomitable Hinkler passed on his swift passage to Australia, for he achieved Darwin in 15½ days, having covered 11,005 miles in 128 hours flying time with overnight stops each day. Not until 19 March did Lancaster and Mrs Miller arrive at Darwin after an anxious 550-mile crossing of the Timor Sea, but they were not expected, and nobody was on the landing ground to meet them. Meanwhile Hinkler was being greeted by crowds of thousands everywhere he landed during an extended tour of Australia with his wife as passenger.

Australia and Canada were already convinced of the value of light aircraft. The RAAF had decided to replace their Avro 504 trainers with D.H. Moths, and ordered twelve from England with the intention of building others in Australia. To further this, the energetic St Barbe formed de Havilland Aircraft Proprietary Ltd of Melbourne; and a visit to Canada by him led to an order from the RCAF for twelve and incorporation of the de Havilland Aircraft Company of Canada Ltd, though it was only a small factory at Leaside aerodrome, four miles from Toronto. R. A. Loader, who had been St Barbe's assistant for some years, was made Canadian manager. Elsewhere it was St Barbe's avowed intention to set up further servicing agencies wherever there was a reasonable market for Moths. A. V. Roe & Co Ltd followed suit, granting a manufacturing licence to the United States Aircraft Co of Spokane, Washington, and a battle of price cutting began. The Moth with 84 hp Cirrus, advertized as having accumulated the equivalent of five return trips to the moon in 2,250,000 miles of flying, was reduced to £650 painted to choice; promptly the Avian was reduced to £600; Blackburns responded with a cautious drop to £725 for the Bluebird. Nevertheless the Moth was sweeping the board. Every working day at least one emerged from the production line at Stag Lane.

A newcomer, the Parnall Imp, appeared in February, and was tested by its designer, Harold Bolas. Anything from his board was bound to break tradition, so the Imp was a Genet-powered cantilever biplane with sparless wings skinned with a tapering thickness of spruce, the top wing having acute sweepback and the lower straight, with an I-strut connecting them in the manner of the war-time Fokker Dr.I triplane. Soon Bolas found this unique little machine dropped a wing all too easily, so slots were fitted, but it still landed fast and all found it unpleasant to fly. Currently Bolas was

The angular Parnall Imp found no purchasers, though it was well demonstrated at flying meetings by a keen young Farnborough pilot, Flg Off D. W. E. Bonham-Carter. (*Flight*)

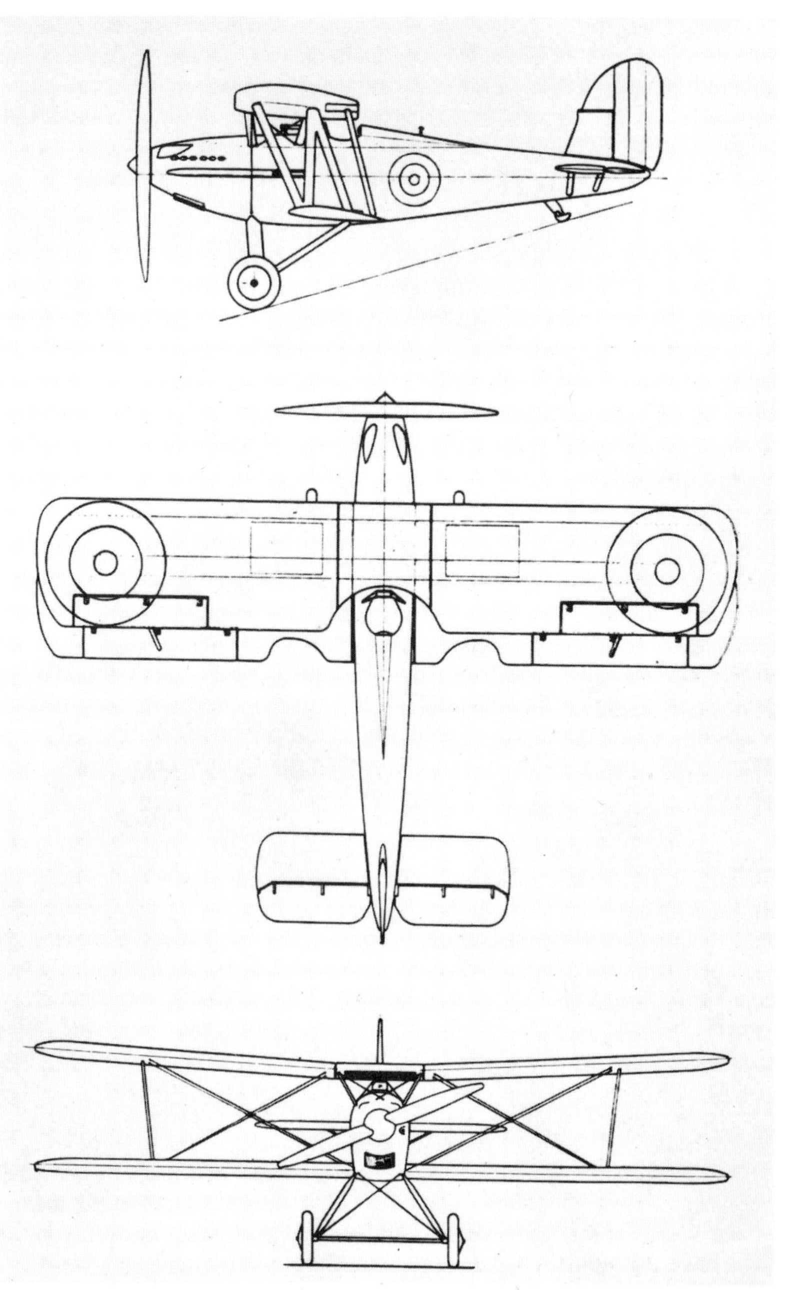

General arrangement of the Rolls-Royce XI engined Parnall Pipit designed as a Flycatcher replacement to specification N.21/26 in competition with the Hawker Hoopoe and Gloster Gnatsnapper.

designing the Pipit, a single-seat ship fighter structured entirely in metal, and powered with a Rolls-Royce F.XI. One of its minor features was inspired by the Wapiti 'Zipp' fasteners, for he used quickly detachable panels to cover the fuselage structure so that it could be completely revealed in a few minutes for inspection.

Meanwhile his essay with Autogiros had not proved the success he had hoped, the smaller C.10 having been written off at Farnborough in comparative trials with the Avro C.9, and the Airdisco-powered Parnall C.11 was crashed at Yate when Cierva attempted to take-off before the rotor attained sufficient speed, and the machine turned over and was wrecked. A few days later Cierva attempted to fly the Viper-powered Avro C.8V in a high wind to gain experience under bad conditions, but again took-off before the rotor was spinning fast enough and turned over. Subsequently Flt Lieut G. I. Thomson, chief instructor at the Hampshire Aeroplane Club at Hamble, was engaged as test pilot for the Autogiro company.

2

Press announcements that Orville Wright was sending his re-built, earliest biplane to the Science Museum at South Kensington, because the Smithsonian Institution had described the Langley tandem as the first capable of flight, revived old controversy and surprisingly little sympathy for Wright. C. G. Grey summed it up: 'The Wright machine was the first to fly, but was not a practicable flying machine. The Langley was the first practicable flying machine but did not fly because of errors in construction. At South Kensington the Wright will cause considerable interest, some amusement, and a good deal of admiration for those gallant fellows who survived taking the air in such a contraption. By 1910 the Wright aeroplane was a back number, so they abandoned the front elevators without tail, and took to the tail-behind type, and it became just like any other maker's in essential aerodynamic design, enabling anybody to fly them. But for many years the Wrights endeavoured to block progress by claiming that their patent for interconnected rudder and wing control covered the case of a man operating ailerons with his hands and a rudder with his feet, and consequent litigation did incalculable harm to the development of aircraft before the War 1914–18, both in Europe and America.' When Frederick Handley Page took issue with C. G. Grey he was sharply told: 'When one considers that the Wrights knew so little that they made no provision either for lateral stability or directional control it would seem almost ridiculous to hold them up as contributors to the science of aeronautics when Langley had already proved his superior science by providing precisely those aerodynamic qualities. Control by warping, in the manner of the Wrights, was soon abandoned as dangerous, so even in this matter they have not contributed anything to the modern aeroplane, for a well designed and inherently stable aeroplane to-day can be flown quite conveniently without using the ailerons, controlling by rudder and elevator alone.' In fact neither Langley nor the Wrights had the full picture, yet the honour and glory of making the first flight with an aeroplane which ultimately proved controllable, though unstable, rests imperishably with Orville Wright: but had the Wright brothers never flown, aviation would have developed in Europe precisely as it did.

It was a far cry from the Wrights to the multitude of RAF aircraft I found at Martlesham that February. I had flown there with Widgeon G-EBRL from Yeovil; there were no formalities of prior permission or radio contact – from Yeovil one headed eastward over quiet countryside; watched for RAF aircraft at Sarum and Andover; admired the soft smoke-clad vista of London; droned happily across East Anglia to Ipswich, and a few minutes later circled the pine-fringed heath of Martlesham. On the tarmac of the fighter Flight, I could see the Wizard, distinctive with silver parasol wing, shining pointed cowling and royal blue decking. Paget had flown it there on 29 January. Alongside were the Armstrong Whitworth Starling and Hawker Hawfinch. There were no Very's light signals of red

Compared with the Bristol and Hawker fighter biplanes, the Armstrong Whitworth Starling seemed clumsy and handling was heavier. (*Courtesy Air Marshal Sir Ralph Sorley*)

or green: if the approach was clear, we landed. Within the flight hangar I found the Gloster Goldfinch, Boulton & Paul Partridge, the Rolls-Royce F.XI Vickers 141 revamped from the Hispano-powered 123, and a dusty long-unused Gloster Guan with an unsuccessful supercharged Lion. While I was there the civil-registered Avro Avenger landed, piloted by helmetless F. L. Luxmore, who was assisting H. A. 'Sam' Brown, the amiable ex-instructor of the Lancashire Aero Club recently appointed as chief test pilot of A. V. Roe & Co at Woodford. Presently the new Bulldog II came circling down from an altitude test, piloted by ginger-haired Flg Off 'Pat' King, muffled to the ears but frozen after two hours in an open cockpit at well below zero. King and 'Poppy' Pope favoured the Hawfinch; Webster of Schneider fame and 'Mutt' Summers preferred the Bulldog; but as there was no consensus as to which was best, it was decided to send both aircraft

Entered as Type 141 for the fighter competition, the re-engined Vickers Type 123 had excellent controls which remained effective below the stall.

for Service trials to the five fighter squadrons of Kenley, Hornchurch, Northolt, Biggin Hill and Upavon.

The flat-gliding Wizard, though superior in performance, was criticized both for initial heaviness of the horn-balanced ailerons, and for obstruction to pilot's view by the top wing, though I could not see that this was worse than the Bulldog with its deep root ends of the top wing housing the petrol tanks. Certainly the Wizard was thought to have potential, for subsequently a contract was placed for redesign with metal-structured wings.

I still have the triplicate book in which I reported every day to Westland, and it has many sketches of fittings and details I thought of interest to the Yeovil-based designers. In a sense, all we manufacturers' representatives were not only salesmen but spies. Such people as burly Bennett-Baggs, bald-headed George Bulman, the elderly, quiet Rea of

Similar in appearance to the Lion-powered Gorcock, but without pointed nose, the Gloster Guan had greater span and an exhaust-driven supercharger.

Boulton & Paul, the rather withdrawn grey-haired Uwins, tall, twinkling but sallow Tom Harry England, and smooth young Oliver Vickers were frequent visitors to Martlesham, and from time to time came the chief designers.

Of absorbing interest at Martlesham, moored in the open, was the huge, triple Condor-engined, $157\frac{1}{2}$-ft span, khaki-painted Beardmore Inflexible monoplane which had been brought by sea to Felixstowe and transported with great difficulty by night to the A & AEE for assembly. Its construction of plain dural plates, channels and angles of substantial thickness was more akin to bridge and shipbuilding than aircraft, and not surprisingly was overweight, leaving negligible payload at the all-up weight of 37,000 lb. A hawser-like 6-in circumference funk-wire had been added each side to relieve the up load, though the cantilever strength of the big box spar was

With three Rolls-Royce Condors, the all-metal Beardmore Inflexible was probably the most impressive and useless aeroplane in the world in 1928, earning the nickname 'Impossible', yet it was a trail-maker.

adequate for down loads. Boyish-looking Sqn Ldr de Haga Haig had left the RAF to take charge on behalf of Beardmores, for Bill Shackleton was recovering from a serious operation; but with no prospect of immediate orders, the forceful Lord Invernairn had already decided to close down the aircraft department again, though would continue development of the R.101's engines under the management of Alan Chorlton. Disillusioned, Shackleton presently migrated to Australia and joined the Larkin Aircraft Co as chief designer. Yet when, after many taxi-ing and tail-up runs, the Inflexible was flown by Sqn Ldr Noakes, handling was found easy though somewhat stately.

The road bordering the aerodrome was closed for that flight, and Noakes taxied the big machine across it onto the unused heath the other side for a longer run. The huge wheels, specially built by Dunlops to Shackleton's design, were more than 6-ft diameter and took the rough ground easily. Invernairn was there to watch, surrounded by senior officers, and most of the station personnel, as well as many manufacturers' representatives who by remarkable coincidence seemed to have business in the vicinity which enabled them to watch. Take-off was almost anticlimax. There was a distant roar from the three engines as the machine trundled forward; the tail lifted and was held steady; speed seemed no more than that of a light aeroplane when the Inflexible skimmed across the road and within another 100 yards was airborne and climbing away at a shallow angle. Its giant silhouette made the half dozen biplane bombers parked around seem anachronisms – yet at least they could carry a big load of bombs to Düsseldorf and back if the wind was not too strong.

Most numerous of all versions of the lumbering Vickers Virginia was the metal-framed Mk. X, for its inertia and slow speed made it easy to fly and therefore highly regarded. (*Flight*)

Vickers Virginias were the mainstay. They had been modified again and again. The latest Mk X had twin 580 hp Lion VBs which doubled the ceiling and raised the speed from 97 mph to 108. At Martlesham, tests were in hand of a Mk VII with metal wings but there were problems with aileron overbalance and flying one wing low; its metal-structured tail eliminated the standard fins and had all-moving, balanced rudders. To give further improvement, trial was being made at Brooklands with Jupiter engines as the result of much eloquent salesmanship by Roy Fedden.

Even the day-bomber Boulton & Paul Sidestrand I had snags. I was given a flight by the RAF's oldest Flight Lieutenant, 'Bill' Markham, and allowed to try the controls. So great was the stability and so heavy the controls, it seemed like a train on rails, and to apply rudder without bank required full strength of the leg. With one engine out, trim was impossible unless flown with side-slip so that yaw from the fin amplified the rudder, but an outrigger servo Fletner, like that of the Singapore's, was about to be tried. The same type of rudder servo was also used for the prototype

'Light hands' was the old-time requirement for pilots, but the Boulton & Paul Sidestrand necessitated considerable muscle; yet its climb, ceiling, and speed of 130 mph, made it the best light bomber of the decade. (*Flight*)

Calcutta, G-EBVG, but the first test flight by Lankester Parker on 14 February, with Major Brackley as co-pilot and Eustace Short and the late Horace's son Francis as passengers, had to be abandoned after a two minute straight, because the elevators were overbalanced. After modification all went well. Air Vice-Marshal Sir John Higgins with Major Penny, the Air Ministry specialist on flying-boats, were prompt to take a ride, and on 10 March Sir Samuel Hoare was given a six-minute trip despite rough wind and whirling snow. Five days later Parker flew the Calcutta to Felixstowe for C of A trials.

Away for Felixstowe! The first Short Calcutta leaves the Medway for C of A trials after only twelve brief test and demonstration flights. (*Flight*)

While these were being conducted, Cobham, after overhauling the Singapore at Durban, reached Cape Town on 30 March. Compared with Dick Bentley's 28-day single-handed dash over the same route in September 1927 and Bert Hinkler's remarkable 15½ days to Australia, this slow journey of months had negligible public acclamation, but its practical experience was invaluable to Imperial Airways, the RAF, and the constructor. So also was the Far East cruise of RAF Southamptons which

One of the splendid metal-hulled Supermarine Southampton flying-boats hauled ashore at Karachi for hull-cleaning and general maintenance during the 1927–8 RAF cruise. (*Grp Capt G. Livock*)

arrived in fine fettle at Singapore at the beginning of March. The formation leader, Sqn Ldr G. E. Livock, DFC, in later years wrote: 'As our safe range was only 500 sea miles we had to alight 29 times for fuel between England and Singapore, partly because we had to fly round the coast of India instead of crossing direct. Refuelling was one of our main headaches. As we lay at moorings – often in swift-flowing rivers or rough water – the petrol had to be brought out in local boats handled often by native crews who knew nothing of flying-boats and their fragility. We never knew what to expect, and usually had to handle the drums or tins ourselves and hand pump the petrol to the top wing tanks – arduous work after several hours flying in the heat of the tropics.' They would remain until mid-May at Singapore, and then continue to Australia, visiting Hong Kong before returning.

The Press were more interested that a woman was attempting a lone flight to South Africa – for Lady Bailey, wife of millionaire Sir Abe Bailey, with characteristic lack of ostentation, had left Croydon on 9 March, heading for Johannesburg, sublimely unaware of the function of the verge ring on her compass. Inevitably she had difficulties, and on reaching Cairo on 19 March was informed that she would not be permitted to proceed unescorted across southern Sudan because of the danger of a forced landing, and her Moth was impounded.

Quiet Lady Bailey with extrovert Louis Paget, chief test pilot of Westland.

It so happened that Bentley was on his way northwards, en route for England in his Moth, formating with Lady Heath (formerly Mrs Elliott-Lynn) who had departed from Johannesburg for Cairo on 24 February in her Avian. Hearing of Lady Bailey's plight, he asked his sponsors, the *Johannesburg Star*, for permission to return southward from Khartoum to escort her. On that understanding she was permitted to follow the Nile

Lady Heath, an athletic tall Irishwoman, was never averse to publicity, but though elected Lady Champion Aviator of the World by the USA, British sporting aviation tended to belittle her achievements.

unaccompanied, flying through a severe sandstorm to Wadi Halfa, and on 29 April reached Khartoum where she met Bentley, who escorted her to Kisumu on Lake Victoria. She continued alone but unfortunately crashed at Tabora after losing her way because she had no local map. The aerodrome was at 4,000 ft, the atmosphere treacherous from heat, and she stalled at the last moment, hit the ground hard and turned over, breaking the fuselage and wings but not herself. At the behest of Sir Abe, a Moth owned by the Johannesburg Light Aeroplane Club was flown to her by Major Meintjes of the SAAF so that she could continue and finally attain Cape Town on 30 April.

Meanwhile Lady Heath had arrived at Cairo on 3 April, but here the RAF refused to let her undertake the Mediterranean section alone. On 15 April, as Bentley was still in Kenya, she decided to fly via Heliopolis and Benghazi to Tripoli where Mussolini promised that a seaplane would escort her to Italy. On 17 May the *Daily Express* Paris correspondent reported: 'Lady Heath stepped from her tiny airplane at Le Bourget this morning after her long flight from Cape Town as fresh as a daisy. "It is so safe," she said, "that a woman can fly across Africa wearing a Parisian frock and keeping her nose powdered all the way." ' That afternoon she arrived at Croydon, but Bentley had overtaken her and reached England five days earlier. Because of her forceful personality Lady Heath unfairly received little credit for her achievement compared with the diffident, dreamy Lady Bailey.

While these flights were in progress the Atlantic claimed more victims, for Capt W. G. R. Hinchliffe and the Hon Elsie MacKay – whose father, Lord Inchcape, was unaware of her participation – had not been heard of since they left Cranwell on 13 March, heading towards the USA in her imported Stinson monoplane. Hinchliffe had been unable to meet British medical requirements for a commercial licence because of his single eye and for months had been seeking some form of livelihood, such as a spectacular long-distance flight which might bring big financial returns to safeguard his wife and three children.

Others were more successful. On 12 April, flying a single-engined Junker-W 33 monoplane named *Bremen*, Hermann Köhl and the charmingly irrepressible Colonel Fitzmaurice, commander of the recently formed Irish Free State Air Force, accompanied by monocled Baron von Hünefeld, their sporting financier, took-off from Baldonnel to achieve the first eastward Atlantic crossing by aeroplane, their luck holding for a safe landing in a snow-storm on Greenly Island in the Strait of Belle Isle, their fuel almost exhausted. Said Fitzmaurice before he left: 'I am doing this because it will put our little Air Force on the map, and secondly I might lift myself out of a rut, for I want to get enough money to establish civil aviation in Ireland.'

A few days after this dangerous achievement, the greatest of world flights safely ended when the Frenchmen Capt Costes and Lieut le Brix landed back at Le Bourget having flown 35,000 miles by long stages in

330 flying hours with their veteran 500 hp Hispano-Suiza Breguet XIX which already had achieved over 40,000 miles in other long-distance flights. A tremendous crowd greeted them at Le Bourget and lined the long route to Paris where President Poincaré officially greeted them, and on 15 April they were accorded the insignia of Officers of the Legion of Honour. By contrast the Far East cruise of the RAF flying-boats and the Cairo to Cape formation flying of Fairey IIIFs seemed almost commonplace.

In unofficial Martlesham parlance, the advanced looking Vickers Vireo light fighter 'hurled itself at the ground' when landing. (*Courtesy Air Marshal Sir Ralph Sorley*)

Far from commonplace in Britain were the Lynx-powered 38-ft span Vickers Vireo cantilever low-winger and the 29-ft span Avro Avocet single-seat ship-planes to specification 17/25. Both were disappointing. Currently Scholefield was testing the Vireo, but it was not realized that its low top speed was largely due to the frictional drag of its Wibault-type corrugated skinning, for the main discouragement was an unexpected hard drop which often occurred when landing, due, it was imagined, to the wing section. The possibility of wing root interference seems to have been unsuspected by Rex Pierson.

Nor was the Avocet free from snags, for the rudder was badly deficient in power when initially flown at Hamble at the end of the previous year, tailplane/fuselage interference affecting the air flow over the lower third. By the time the Vireo was flying, the Avro's rudder had been made taller, and with horn balance added to lighten the load, retaining the original truncated fin as had been done with the earliest Wapiti, and in April it went to Martlesham for brief comparative handling trials with the Vireo. Subsequently both had power increased to 230 hp and were tested with floats. Although the Vireo was scrapped it held lessons for future designs,

Although the Avocet prototype rudder was theoretically large, it was unaccountably necessary to add $1\frac{1}{2}$ ft and a horn balance, as exemplified by N210, before it was sufficiently effective. (*Courtesy Air Marshal Sir Ralph Sorley*)

and the Avocet similarly was a useful experiment because it was Chadwick's first monocoque dural fuselage, based on Short patents. All designers have to be skilled adapters as well as cautious innovators, so the single lift strut in conjunction with 'N' interplane struts to transmit torsion was used by Barnwell for his next design, Type 118, for he was well aware of the advantage of a single braced panel as he had used a Raf-wire version in the 1927 Type 101; nor was strut bracing new, for Shackleton had introduced it in England with his novel W.B.XXVI built for the Latvians in 1925.

3

On 1 April the Royal Air Force was ten years old. Lord Weir, who was largely involved with its war-time conception, cabled to Trenchard: 'I regard it as incomparably the most efficient Air Force in existence, creditable alike to British organization and British character. It is a welcome portent to the Empire that this new fighting Service affords opportunity not only to British qualities of courage, determination, and enterprise, but also to the spirit of true scientific progress.' Yet when a war-time S.E.5a was compared with a Gloster Gamecock, or the Handley Page V/1500 with the latest Vickers Virginia, progress seemed mighty small. Clearly it was in engines that the great strides had been made, for Britain was well served by Rolls-Royce and Napier with powerful, reliable, water-cooled engines; and by Bristol and Armstrong Siddeley with relatively lighter but no less reliable radials costing almost half the price of their water-cooled equivalents, and easier to install and maintain.

The Mk. IX Vickers Virginia introduced 'tail-end Charlie' aft gunner. The nose was longer to get the C.G. forward, and the tail had increased span, less chord, and balanced elevators. Virginias were the mainstay of Bomber Command until 1937. (*Courtesy Air Marshal Sir Ralph Sorley*)

Of the status of the industry Fairey considered that: 'It has been nurtured by the Government with the specific objects of arming the Services and providing the necessary means for rapid expansion and high production in the event of national emergency. It has been encouraged to produce highly efficient machines of very robust construction, not only to give maximum performance but to supply the multitudinous demands of the various branches of the RAF, and I suggest it has been very successful. It does a larger export trade than America in military types, and if the necessary money was forthcoming could entirely re-arm the Air Force with aircraft of every type.

'I believe the extraordinary recovery of this country, which fought the Great War and paid for the larger part of its cost, which pays its international debts in full, which brings the pound sterling above par, and employs more people than have ever been employed before, is an effort not to be ashamed of. The standard of living is rising, and there is distinct hope of a general increase in the nation's prosperity. I suggest that this has not been the work of politicians or even the Press of the country, but of that much despised person, the British industrialist of which the British aircraft industry is a humble unit.'

Above all, that unit depended on the annual Air Estimates. Sir Samuel Hoare in his introductory statement on 12 March pointed out that the air vote now included the air expenditure hitherto carried by the Colonial Office, but in spite of this increase of about £2 million the net total of £16,250,000 was only £700,000 more than in 1927. Nevertheless the RAF would be stronger by the equivalent of four squadrons before the end of the year, and no less than 70 per cent of all squadrons would have new types of machines – meaning those of design not used in the war years. He stressed that since 1922 the number of squadrons had doubled; but none saw the weakness when he said personnel had increased by only a fifth – maybe he pinned hope on the fact that '75 per cent of RAF machines are now equipped with parachutes.'

As to airships, the R.100 and R.101 would be flying within the next twelve months, and he reaffirmed the Government's belief that though aeroplanes were invaluable for short distances, airships could travel several thousand miles without landing on foreign territory – though he cautiously added 'if proved safe and dependable'. The lesson had certainly been learned, he insisted, that policy should be mainly Imperial and not European; it had therefore been decided to institute a new subsidy agreement for airlines based on three principles: first, a weekly mail service to India; second, regular substitution of new types for old; third, the right of the State to share in any ultimate prosperity the company might achieve – for it was a hopeful sign that when Imperial Airways started, running costs were 4s 2d/ton-mile, whereas the new types operated at 1s 10d/ton-mile.

In concluding his speech Sir Samuel said: 'I may be pardoned for making an observation that I would not make upon any ordinary occasion, for I know how rightly sensitive the House is to the introduction in Debates of the names of any permanent official or serving officer, but there has never been a fighting Service so closely identified with its Chief as has the RAF during the ten years it had been identified with Sir Hugh Trenchard. I do not mention his name to bring it into the field of Parliamentary controversy, but solely that Members should give credit to whom credit is due, and remember his sound judgment and resolute purpose without which the RAF as it is would have been impossible within the first ten years of its life.'

The Debate proceeded in the usual garrulous manner of Parliament, with contention between Labour, Liberal and Conservative, yet with a notable warning from Sqn Ldr F. E. Guest, Liberal Member for Bristol, who quoted Marshal Foch: 'I must impress upon you, gentlemen, to forget the last war; not to imagine that the next will begin where that one left off. It will be as completely different from the last as the last one was from the one before.' This gave pretext for Lieut-Col J. T. C. Moore-Brabazon,

Conservative Member for Chatham, both to congratulate Sqn Ldr Guest on his work as commanding officer of one of the Reserve Squadrons, and to warn the Minister that technicians at the Air Ministry not only stifled development but were unfair in handing out contracts, for 'one firm in England developed the all-metal machine, but another has been given the orders along the lines of the firm with which the machine originated.' It was clear that Oswald Short had powerful friends. In support Mr L'Estrange Malone, now more than ever a Socialist doctrinaire, strongly advocated that the Air Ministry should spread orders evenly 'instead of filling up the staffs of the firms one part of the year and sacking them at the end of it', and accused Ministry technicians of being dominated by military minds, red tape, and red tabs. Inevitably it drew the riposte that this was better than domination by the Red Flag. As to that symbol, Sir Philip Sassoon, winding up the Debate, asked if it was surprising that more attention was paid to the long record of blood in Russia rather than that country's recent invitation to universal disarmament; it was unreal that the British Empire 'exposed in every part of the world to the envy and greed of the unscrupulous should be asked to accept this glorious and noble gesture to strip itself of its means of defence.' When the House went into Committee, Sir Samuel Hoare referred with special sympathy to the death of Flt Lieut S. M. Kinkead, DSO, DSC, DFC, a South African of the RAF, who on 12 March was killed when flying the Supermarine S.5 in an attempt on the speed record. 'Let us turn aside,' said Sir Samuel, 'before I deal with questions raised in this Debate, to offer sympathy to his family and pay tribute to one of the finest officers in the Force, a young man with unrivalled record who might in the ordinary course have reached the highest post in his great profession.'

Everyone loved 'Kink' – an inapposite nickname for a man who 'thought straight, talked straight and lived straight.' He had taken out the S.5 in an oily calm, and on his second attempt at getting off, flew to the Isle of Wight shore, turned towards Calshot for the measured three-kilometre course, slanting down from 250 ft, then at 50 ft suddenly disappeared in a shower of spray. Major J. P. C. Cooper, the Inspector of Accidents, came to the conclusion that Kinkead had decided to alight, but in the sunfilled haze of afternoon misjudged his height and flew into the water or stalled.

A fortnight later Major de Benardi, flying a Macchi float seaplane at Venice, achieved what Kinkead had attempted, beating the world speed record with an average of 512.776 km/h – but even racing cars were faster than the fastest fighter, for Malcolm Campbell with *Bluebird* on 19 February had set a record of 206.96 mph at Daytona Beach, Florida, beating that of Major Segrave's racer by 5 mph.

On the day following Kinkead's death, the ever present peril of stalling was the cause of a fatal accident to Flt Lieut I. E. McIntyre, CBE, AFC, at Melbourne. He had become prominent in 1924 as co-pilot with Wg Cdr S. J. Goble on the first seaplane flight round Australia, using a Fairey IIID in which they covered 8,568 miles in 90 flying hours; two years later he had made a remarkable 10,000 mile out-and-return flight from Sydney to British mandated territories in the Pacific using a D.H.50 seaplane.

Clearly even the most skilled were not immune from accidental stalls, and though by using Handley Page's automatic slots the danger could be ameliorated, it did not entirely remove the risk, as Geoffrey de Havilland had found. His company's demonstration Moth G-EBTD had been fitted with slots based on Handley Page wind-tunnel recommendations. D.H. had foreseen that though slots stabilized laterally, the rate of descent would be greater than a conventional straight-axle undercarriage could stand, so a split-axle type of wider tread and one-third greater shock-absorbing travel was substituted. Many journalists and club representatives were invited to

'Not wisely, but too well!' – Geoffrey de Havilland steeply stalls the prototype Moth while demonstrating its modified form with tip slats and balanced rudder. (*Flight*)

Stag Lane to see de Havilland himself demonstrate the Moth's control in stalled flight and its safety from spinning even when pulled violently up at 200 ft. On the next flight he showed a tyro's landing by flattening 10 ft too high, and the Moth pancaked safely down; then he let it dive into the ground without flattening so that it bounced 15 ft high, but again stalled down in safety. Finally he stalled down from 250 ft, but at the last moment the nose dropped slightly and the rate of descent increased. The Moth hit the ground tail first but laterally level, spread-eagled its undercarriage, and the fuselage snapped at the front cockpit, the nose sagging to the ground. Not surprisingly D.H. looked a little embarrassed, but St Barbe's publicity agility instantly made the best of it by saying how safe it showed the Moth to be, for no unslotted machine could have got away with it!

Nevertheless, the unslotted Widgeon was as good as the slotted Moth, largely because, like the Fokker, its elevator power was limited; so when Sqn Ldr England and his new assistant, the debonair Major J. L. B. H. Cordes, a few weeks later began to demonstrate stalled landings with Handley Page's own slotted Moth, Louis Paget and I sometimes took pleasure in countering it by flagrantly landing the Widgeon from a stalled glide, saving its frail little undercarriage at the last minute with subtle increase of power.

4

At the annual general meeting that spring, the chairman of Handley Page Ltd, Mr S. R. Worley, impassive as ever, announced that after allocating £15,000 to reserve and writing off £3,239 for depreciation, a credit balance of £31,638 remained, so a dividend of 10 per cent would be paid. He emphasized that the slotted wing patents were most valuable, but as H.P. explained: 'In this country the Government has the right to any patent without referring to the patentee, and the question of compensation has to be determined by the Royal Commission on Awards to Inventors, so it is impossible to forecast the extent of the award.' The United States Government was more acceptably commercial and had paid $125,000, and would continue until royalties amounted to $875,000 (£175,000 at that time) after which there was free use of the patent.

Handley Page's rivals in 'giant' aircraft, Vickers Ltd, at their annual meeting dealt collectively with the grouped business of ships, armament, and engineering, giving a gross profit of £1,275,995 – but the aircraft department must have been soundly financed and profitable for it was announced that it would be floated as a separate subsidiary to be known as Vickers Aviation Ltd, with Sir Robert McLean as chairman, and Capt Acland and Maxwell Muller as special directors. A few weeks after its formation Oliver Vickers, elder son of the chairman, and liaison director between the company and the RAF in general and Martlesham in particular, died from pneumonia.

Changes were also in the wind for A. V. Roe & Co Ltd, for their chief shareholders, Crossley Motors Ltd, revealed a loss of £65,518, attributed to the cost of developing a new 20.9 hp six-cylinder car, but there was comforting assurance that the accumulated reserve profits of the Avro aircraft section exceeded £37,000. On 25 May the *Financial News* stated: 'As recently foreshadowed, arrangements have been completed whereby Sir William M. Letts, Crossley's manager, has sold to the Armstrong Siddeley Development Co Ltd the whole of the share capital of A. V. Roe & Co Ltd. Sir William Letts has handed a sum of over £270,000 in cash to Crossley Motors Ltd for the acquisition of their shares, which shows them a profit of more than £200,000 from their original investment in A. V. Roe & Co Ltd.'

Alliott Verdon Roe, that early flyer, was 51, but remained optimistic as ever in his fussy, diffident, charming manner. Widely it was assumed that he would now sit back with his riches as the *doyen* of the aircraft industry. But A.V. had other ideas, and was considering how he could start again.

There was no question of dismissing the Avro staff. John Siddeley decided to run Avro and Armstrong Whitworth as complementary firms. Lloyd contined to design the A.W. machines and Hiscocks to produce them; at Avros, Chadwick remained chief designer with Parrott managing the Hamble factory and Roy Dobson running the Manchester factory with firm and practical hand. Lloyd was designing a new fighter and a much modified version of the Atlas. Chadwick was engrossed with a medium-powered five-seat high-wing monoplane similar to the US machines which had flown the Atlantic, but as he proposed a welded steel-tube fuselage structure, it was decided that Dobson and Broadsmith should visit the Fokker works at Amsterdam in order to learn the techniques, for British experience was confined to a few women welders making non-structural parts.

Coincidentally D.H. was toying with an almost identical machine for which Halford would redevelop the vee-eight Renault as an inverted engine to give 200 hp. Like Chadwick, de Havilland adopted welded tubular-steel for the fuselage and wooden wings, but with the insight of one who was both pilot and designer, so arranged the Triplex windows around the cockpit that they gave optimum view for an enclosed aeroplane. Provisionally the new D.H.75 was named Hawk Moth, with Ghost engine, though Lynx or Whirlwind were alternatives. What with this, studies for D.H.50 and D.H.66 replacements and a private-venture interceptor fighter to F.20/27 specification, design work was so overwhelming the D.O. that he passed completion of the D.H.72 three-engined bomber design to Glosters who already were tackling a metamorphosis of his projected D.H.67 twin-engined survey biplane. To ease the work load, W. G. Carter, freed of further work on the Crusader, was engaged as specialist designer for the D.H.77 interceptor – which was a low-wing monoplane, like the competing Westland and Vickers machines, but would have a novel supercharged air-cooled engine which Halford was designing in H arrangement to give low frontal area.

It was going to be an instructive match between monoplanes and biplanes for the F.20/27 competition. Bristol had their Bulldog-derived Bullpup, which made its first flight on 28 April with Jupiter VI, as the Mercury IIA was not yet ready. At Hawker, Camm was designing a somewhat similar but neater biplane based on the Hawfinch, but with single-bay wings, and this was leading him towards a slick Kestrel-powered version. There was also Fairey with a Mercury-powered Firefly, renamed Flycatcher II, but he too felt that a Kestrel installation was preferable, so Lobelle was making tentative designs, though major pressure was on the secret 'Postal' monoplane.

Hollis Williams recorded: 'This was the first time we had attempted a tapered and twisted cantilever wing. Although the NPL was doing the wind-tunnel tests we couldn't wait for them, but I found a paper by Glauert which offered a mathematical solution for such a wing. After burning much midnight oil, the main characteristics emerged and were used as the basis for strength and performance. By the time Relf of the NPL had his preliminary results for wing alone we were committed, and components were being built. When Ralli asked me to produce our calculations they agreed with the NPL figures except for half a degree in "no-lift" angle. Relf was surprised and Glauert delighted as this was the first time industry had

With no resemblance to the 1923 Flycatcher, Fairey's eventual Flycatcher II, intended for the same naval purpose, had a 540 hp Jaguar VIII in lieu of its delayed Mercury.

used his work. Glauert was killed by a silly accident shortly afterwards. While walking one Saturday afternoon on Farnborough Heath, tree stumps were being blown out of the ground, and Glauert got in the way of a lump of root sailing through the air, and it fatally injured him.'

The Air Ministry made a cautious announcement: 'Reports that the RAF has decided on an attempt this year to break the world's long-distance non-stop flight record are premature. An aeroplane is being built with the purpose of exhaustive tests at home to determine how long an engine will run in flight and will naturally be a machine which might beat the present duration record.' This only vaguely interested a public who demanded a hero to worship rather than speculate on hoped-for achievements. As C. G. Grey said: 'We have reached in aviation a period much like that of 1907 to 1909 in motoring. People no longer look at flying as a great adventure. A forced landing by an Imperial airliner is recorded in the papers because it is unusual and no longer described as a miraculous escape from death. Owner pilots are merely regarded as good sportsmen and not as would-be suicides.

'What with the new motorist and the old char-à-banc driver competing each week-end for every square inch of road, there is no pleasure for young sportsmen and sportswomen who hitherto have driven for driving's sake. More than ever this year there is reason to heed the de Havilland slogan "Get off those crowded roads." Pilots like Lady Bailey, Lord Ossulston, the Master of Sempill, and Sir John Rhodes, not to mention quite a number of other more or less well-to-do people, have already influenced in the right direction thousands of their own class by showing how an aeroplane can be used as an ordinary sporting vehicle. Such trips as those of Colonel Sempill from his home near Aberdeen to John o' Groat's and back in an afternoon, Lady Bailey's daily trips from London to Newmarket and back during the race week, and the remarkable tours by the Duchess of Bedford, with Charles Barnard as pilot, have shown what is possible for the private owner. These journeys have not been advertized as have minor flights by those who suffer a craving for publicity, but they have had far greater influence among people of the better class than any self-advertizing flights could have.'

But it was not merely the rich who engaged in the delights of flying, as the tremendous activity of the many clubs increasingly revealed. Their week-end meetings were delightful occasions, full of enthusiasm and the mutual understanding of people who found new freedom in the skies and delighted in the skill and self-reliance required for perfect turns and faultless landings or overcoming the minor hazards of navigation and the bigger ones of weather. Almost as carefree were those in the Auxiliary Air Force and Reserve, though they had at heart their more serious purpose as defenders of their country should there ever be a war.

In the background of the public's 'conditioning', publicity for airships steadily proceeded. Even that secretive, publicity shunning, brilliant engineer Barnes N. Wallis emerged from seclusion at Howden to lecture at the Royal Society of Arts in mid-May on *The Design and Construction of Rigid Airships*, and now that the structure of both R.100 and R.101 were assembled, disclosed some of the ingenious details of his airship, such as the spiral mesh net for the gas-bags and method of making large tubes for the framework from strips wound spirally and riveted. He said: 'There are only 42 dissimilar components in the entire 730 ft structure; apart from variations in the thickness of metal, every girder is made from only seven of them; thus we could order items in lots of half millions, and make even only one airship a mass-production job. Despite initial anxiety in attaining the necessary accuracy we got the main girders, each 43 ft in length, within limits of plus or minus .01 in.' His rival on the R.101, Lieut-Col Richmond, BSc, FRAeS, who in 1926 had lectured to the RAeS on *Airship Research and Experiment*, restrained himself to the comment that it was extraordinarily interesting to listen to one who was seeking ends similar to his own by somewhat different means. Nevil Shute Norway in *Slide Rule* throws light on how unco-operative was the Royal Airship Works in passing information to Howden: 'In the five years that elapsed before either airship flew, neither designer visited the other's Works, nor did they meet or correspond on common problems each had to solve. Each trod a parallel road alone, harassed and overworked. Towards the end I made contact with my opposite number, Dr Roxbee Cox, and visited Cardington to see their ship, but his chiefs prevented him from visiting Howden. If the Cabinet wanted competition they had it with a vengeance, but I would not say it was healthy.'

Many in Britain criticized airships, among them E. F. Spanner, an experienced naval architect, who wrote *This Airship Business* with a theme that R.100 and R.101 had been designed in defiance of sound engineering knowledge and experience, and construction undertaken in a spirit of ignorant and exaggerated optimism, so both were doomed to disaster.

The superintendent of production at the Royal Airship Works, A. H. Hall, a Clyde-trained engineer who had been war-time Director of Torpedo and Mine Production, had recently been appointed chief superintendent to the Royal Aircraft Establishment. Of his predecessor, W. Sydney Smith, one-time inspector of factories, C. G. Grey at least was prompted to say: 'As in so many Government establishments, the Augean stables of the RAE are in such a state as to defy any official Hercules, but Mr Sydney Smith certainly got on with his job in spite of difficulties, and promoted efficiency and economies. He has worked hard for good feeling between the aircraft industry and the RAE.'

The kind of thing which made the industry suspicious could be instanced by the Gloster variable-pitch propeller which at Burroughes' initiative had been developed by Dr Hele-Shaw and Mr Beacham and was tested by Saint on a Grebe. It was the first in England in which hydraulic reaction to the centrifugal governor could be adjusted by a lever in the cockpit to maintain any desired rpm. 'If the engine runs over the set speed, the governor will move the crank pin of the Hele-Shaw pump in such direction that oil is sucked from one side of the cylinder from the propeller and forced into the other side in the correct direction to increase the pitch of the blades and so pull down the engine rpm.' In dives from 19,000 ft to 6,000 ft, with governor set to maintain maximum engine speed, Saint found that the rpm had only momentarily risen by 20, whereas a normal fixed-pitch propeller would have raced and wrecked the engine unless the pilot closed the throttle.

Despite the money Glosters were spending on it, the Air Ministry doubted the value because of added weight, increased cost and possibility of mechanical failure, and Capt Lynham, the RAE's propeller expert, argued that if the increased weight of the variable-pitch propeller was used to provide an engine of increased power, a better performance would be obtained despite a fixed propeller. Thus were the purse strings tied against finance for development. 'Nevertheless,' said Hugh Burroughes, 'by dint of much effort, we secured some development contracts for propellers to suit Jaguar, Jupiter, Lion, and Kestrel engines.'

Of these engines it was the Jupiter which was becoming the world's chosen power unit. Fedden had delegated French-speaking Norman Rowbottom, a confident, energetic inspector, eyes gleaming through gold-rimmed spectacles, to aid Gnome et Rhône with manufacturing techniques, and now he was appointed general manager of the French company. Licences and sub-licences began to proliferate, with the result that the Jupiter was being built in France, Japan, Germany, Belgium, Italy, Czechoslovakia, Switzerland, Spain, Hungary, Jugoslavia, and Portugal.

5

France was mourning the passing of the industrialist Jacques Schneider, donor of that ardently sought international Trophy, whom Louis Blériot had taught to fly in 1909. 'By his death France loses one of her most honoured pioneers, and one who has contributed very largely to the progress of high speed aviation throughout the world,' recorded his obituary. The incentive remained, and was leading not only to big advances in aero-engine design by Pate and Wilkinson of Napiers as well as the shadowy, bespectacled, white-headed Rowledge of Rolls-Royce, but was resulting in a vital contribution by Rodwell Banks of the Ethyl Corporation, allied with much research at Farnborough, in avoiding fuel detonation to enable compression ratios twice as great as war days. Fortunately the FAI had agreed, at the insistence of the RAeC, that the Schneider should in future take place every two years, so the next would be in 1929; even so time was at a premium whether for engines, fuel, or the improved S.6 at Supermarines and the new and pretty monoplane at Glosters.

The antithesis of the Schneider racers was the pioneer new conception of an ultra-light, high performance single-seat fighter which Blackburns demonstrated at Brough on 15 May to a large gathering of Air Ministry officials and overseas and foreign representatives. A long line of current Blackburn machines was displayed on the tarmac – Bluebird, Dart, Ripon, Velos seaplane, Brough-built Siskin IIIA, but the star was the 22 ft 6 in span Lynx-powered Lincock biplane with monocoque wooden fuselage set midway between the wings. With a speed of 150 mph, tiny dimensions and low inertia giving fantastic manoeuvring power, the Lincock seemed to have considerable potential as an export fighter for the lesser air forces of the world, and it had immediate appeal to every pilot, as shown by Sqn Ldr Noakes's spectacular display of 'manoeuvres that one has never seen before, including an upward spin of about three turns, and a half loop with a sharp turn upside-down, and then a half roll, ending at least a hundred feet higher than it began.' By contrast, demonstrations of the Ripon and the Dart by the new Blackburn test pilot, Capt A. M. 'Dasher' Blake, formerly of the Air Survey Co, concentrated on the value of the slots with which

Pilots loved the tiny Blackburn Lincock, which Flt Lieut R. Sorley (later Air Marshal Sir Ralph) described as 'light as a Sopwith Pup, with controls beautifully harmonized and effective.' (*Courtesy Air Marshal Sir Ralph Sorley*)

Development of the Blackburn Beagle was so slow that not until the following summer was it delivered with altered rudder, elevators, and effective slots, and then further delayed for a supercharged Jupiter XF. (*Hawker Siddeley*)

each had been fitted, though the Ripon threw a sedate loop. Those were the saleable military products, for the Beagle was still being altered and evaluated, and the Nautilus 2F.I two-seat fighter was only in the mock-up stage.

Demonstration of prototypes had become a good excuse for any manufacturer to throw a party in hope of strengthening relationship with civil and military officials and ensuring that the latest aeroplane was known in reality rather than by report. There might also be gifts of trophies for squadron mess, or occasional personal bottle or box of cigars left as a friendly gesture by the firm's representative. That was the British way – but in some foreign countries the price was expected to be marked up and the additional percentage pocketed by the purchaser's negotiator; it was a method disliked by the industry's seniors, who preferred not to know how their aircraft were sold, and certainly no money was ever passed to an Air Ministry official, but occasionally a key RAF pilot might be offered a job.

Civil aviation interests in somewhat similar manner endeavoured to attract the public to increasingly elaborate flying meetings, often described as 'pageants', with races, aerobatic displays, and joy-riding. Most were supported by the industry's test pilots and demonstration aircraft, together with occasional donations and prizes. Though Moths preponderated, backed by a few Avians, Bluebirds and Widgeons, one might sometimes discover a fascinating old-time aeroplane, such as an archaic B.E.2e at Hadleigh, Sopwith Pup airframes in the hangar at Castle Bromwich, a Curtiss-powered D.H.6 at Leatherhead, Whitehead Reid's dusty R.A.F.-powered S.E.5 at Canterbury, a Le Rhône Grahame-White Bantam and an Avro Baby re-engined with Cirrus I flown at Shoreham by the pre-war aviator Cecil Pashley and the young Fred Miles, owners of the Gnat Aero Co and Southern Aero Club. Often there were foreign light-planes, of which I flew a number, including a fascinating 20 hp Klemm which was like a powered glider. Of RAF participation at meetings there was occasional criticism: 'One may argue that officers and gentlemen ought not to turn themselves into aerobats for amusement of the public. There is no doubt that by exhibition and competition flying, Service officers obtain newspaper notoriety which is not good for the tone of the Service and in some cases detrimental to the outlook of the individual officer. One might argue that government petrol ought not to be used, nor government machines risked to make a British holiday. On the other hand, the publicity given to the RAF at these local meetings certainly might be used for the good of recruiting.'

There was certainly big publicity for Cobham on 31 May when he left Bordeaux at 6.30 a.m., heading for Plymouth on the last stage of his 20,000 mile African flight; but he found the Channel blanketed by fog, and after flying back and forth for one and a half hours alighted on a small river to await clearer weather. At 3.30 p.m. he took-off again, and four hours later came into view over Drake's Island, circled the harbour three times and alighted off Cattewater Air Station – but clearing Customs and other formalities took so long that not until almost 10 p.m. could he and his party cross the harbour to the Mayflower Steps to be welcomed by the Mayor and Mayoress and Sir Sefton Brancker. Said a reporter: 'Despite the unfortunate mishap at Malta on the outward journey and long delay on the West Coast of Africa on the return owing to radiator trouble, Sir Alan has proved that airlines using flying-boats can reasonably be maintained through and round the African Continent. The whole performance emphasizes his value as our foremost commercial aviator. Whatever he does in the way of flying is with a definite commercial end in view, not for his personal gain but for the development of commercial aviation. One hopes that the enormous amount he has done in this cause will bring in the end the prosperity he so thoroughly deserves.' There were tributes to Lady Cobham, to Capt Worrall, who as an experienced air mariner had done so much of the flying, and to the diligent crew who had shared the perils of this notable journey.

Coincidentally, the Far East cruise of four Supermarine Southamptons arrived at the Netherlands East Indies on 30 May, and next day reached Broome, Western Australia, having flown from Kupang in just under seven hours, but they dragged moorings and had to taxi around until the wind lessened.

On that same day, Charles Kingsford Smith, with his co-pilot Charles Ulm and American crew of navigator and wireless operator, left Oakland, California, in his Fokker F.VII-3m, the *Southern Cross*, on the first flight across the Pacific. Next day at 12.15 p.m. they landed at Honolulu having flown 2,400 miles in 27½ hours. At dawn on 3 June they left for Suva in Fiji, where they landed at 2.20 p.m. after covering 3,200 miles through rain and tempest in 34½ hours, the longest over-seas flight ever made. In the final stage on 8–9 June they took-off in a strong cross-wind from a prepared strip of beach, and reached Brisbane, Kingsford Smith's birthplace, in 20 hours of battling through worsening weather. Throughout each stage their wireless messages were relayed to the public by the Sydney

The famous Fokker F.VIII-3m *Southern Cross* flown by the Australian ex-RFC pilot Charles Edward Kingsford Smith, widely known as 'Smithy'. (*Flight*)

Broadcasting Station with such distinctness that occasionally the roar of the three engines could be heard above the dots and dashes of miscellaneous Morse. As 'Smithy' made clear, his aeroplane was not named after the southern constellation, but for a Western Australian town which had raised sufficient money to finance this great flight. A few days after its completion the Attorney General of Australia announced in the House of Representatives, Canberra, that his Government had decided to grant £5,000 to Kingsford Smith and his associates in recognition of their achievement.

There was echo in the King's Birthday Honours of another pioneering effort, for a knighthood was awarded to Capt George Hubert Wilkins, navigator, naturalist and explorer, who recently had flown from Alaska over the North Pole to Spitzbergen investigating its possibility as the shortest airline route between America and Europe. He had made many other Arctic contributions. After learning to fly in 1910 he was with Vilhjalmar Stefansson's Canadian Arctic expedition of 1913, flew with the RFC and RAF, attempted the £10,000 flight to Australia, and since then had been involved with geographical discovery in the Arctic, where he operated Fokker tri-motors and Stinson biplanes. In a lecture to the RAeS

The clean Vega, built by the small new US company of Lockheed, owned by Victor Loughheed, makes striking contrast with the old-style Waco biplane.

on 14 June he described his recent Polar flight for which he used one of the sensational aerodynamically advanced Lockheed Vega monoplanes piloted by Lieut Eielson. Their special map, only 15-in square, was drawn by an Englishman employed by the American Geographical Society, and the navigational method was devised by A. R. Hinks of the Royal Geographical Society. A standard Pioneer compass backed by a star compass was employed, but they depended more on accurate piloting, keeping direction by observation of ice, snow drifts and sun, changing course every 100 miles.

Frederick Handley Page also had been lecturing to the RAeS. His was the sixteenth Wilbur Wright Memorial occasion. Inevitably the slotted wing was his lively theme. 'For a period of seven years Jacob laboured to gain the wife of his desire, only to attain Leah as consolation. However another seven years of toil won Rachel, so in another seven, though I hope not lean years, I might find the slot beginning to be used to full capacity for improving performance and control!' That stirred the DSR, Major Wimperis, to well directed witticisms in moving a vote of thanks, but on more serious note he warned: 'Britain used to be the industrial workshop of the world, carrying out all the mechanical work needed by other countries. To-day those countries can do their own, so if we are to hold our position we have to find out how to do the more difficult jobs for other nations – such as producing stainless steel, artificial silk, and automatic slots.'

Meanwhile General Nobile had been attempting a Polar expedition with the airship *Italia*, but had been missing since 25 May. On 10 June it was announced that wireless communication had been established, and that the airship had alighted near Foyn Island, knocking off the control car, only to rise and come down 20 miles further on. Relief parties of ships, seaplanes and flying-boats from Italy, Norway, Sweden, France, and Russia were proceeding to Spitzbergen.

Into the limelight next came a woman aviator, Amelia Earhart. On 17 June, with Wilmer Stultz as pilot and Lon Gordon as mechanic, she took off at 3.50 p.m. from Trepassey Bay, Newfoundland, in the Fokker trimotor seaplane *Friendship*, and landed at Burry Port, Carmarthenshire, next day at 12.40. It was a flight which particularly attracted attention because Amelia not only was the first woman to cross the Atlantic by air, but she had extraordinary resemblance to Charles Lindbergh and quickly was nicknamed 'Lady Lindy'. Few realized there was British participation in this flight, for it had been financed with £8,000 from the wife of Sqn Ldr the Hon Freddie Guest, MP.

A few days earlier the Cherub-powered Westland Pterodactyl IA had been completed after transporting the components to the RAF station at Weyhill, Andover, and there, on a calm afternoon, Louis Paget took it on its first flight. In this first form it had only a single central landing wheel and skid-like supports outboard. As the pilot's cockpit was above and forward of the wheel the ride felt as tricky as with the first Pterodactyl.

Birdlike Pterodactyl IA with horizontal 'electroscopic' rudder inboard of the 'controllers', as elevons were then called.

Louis Paget (*left*) and Geoffrey Hill aboard the Pterodactyl before a central tandem replaced the single wheel.

The Pterodactyl IC was the ultimate version with a tandem-wheel undercarriage torsion-bar sprung, with steerable frontwheel.

Some months later, after a flight with a tense moment when the differential gear ratio lever slipped from low to high, lurching the machine into an unexpected dive, Louis had made a perfect landing with me aboard, when suddenly the undercarriage collapsed, skidding us along the ground in showers of mud and turf. Eventually a stronger undercarriage was devised with wheels in tandem which gave a smoother ride, but for the time being it was rebuilt with mono-wheel.

6

'Taking the ninth RAF Display all round there was nothing very exciting about it compared with previous Displays,' wrote one critic, 'but it struck me that the whole standard of flying was much more finished and neater and cleaner than any previous occasion. That means we may pride ourselves on being streets in front of anybody, for we have undoubtedly got the finest Air Force and the best built aeroplanes and engines in the world.' As far as personnel were concerned, that was true, and in a measure it applied to the aircraft and even more to engines; but compared with other air forces, British machines obstinately remained brilliant compromises rather than aerodynamic advances. Thus the aeroplane in which the bearded, swashbuckling General Balbo flew nonstop from Rome to London, in order to attend the Display, was a fast, clean, high-wing two-seat reconnaissance monoplane far and away ahead in appearance compared with the Fairey IIIF or Westland Wapiti of unmistakable war-time ancestry. Yet when it came to flying the somewhat archaic-looking British biplanes, they handled with much more pleasant and reassuring response than most foreign aircraft.

At the Display on 30 June the King and Queen and Duke of York were the focal point in the Royal Box, attended by bowler-hatted officials and high-ranking officers. The sky was overcast although clouds were high, and the gusty wind made it overcoat weather for the watching 200,000. Such was the growing use of cars that for the first time over 1,000 were parked in the enclosure. In the great fly-past of experimental types the new Westland Pterodactyl IA led the way, its apparently rudderless configuration mystifying many, for it was steered in flight by split horizontal drag flaps inboard of the controllers. Then came the latest conventional aircraft: dumpy Boulton & Paul Partridge, slender Hawker Hawfinch, stocky and aggressive-looking Bristol Bulldog, and slim Westland Wizard which rocketed into a steeper and longer zoom than all. Nearly as spectacular was the prototype Fairey Fox, now three and a half years old, with which Norman Macmillan had been testing the prototype Rolls-Royce F.XI and its successor. 'We had a number of forced landings before the R-R specialists got them right,' he recorded, 'and I recall anxious looks when I glided down on one occasion with a connecting-rod sticking through the crankcase. The trouble was traced, cured, and these engines were fitted to the Fox Mk IA which replaced the Felix engine of the Foxes in No. 12 Squadron.' Yet what one missed at the Display was

Many a pilot preferred the appearance of the old-fashioned Hawker Hawfinch to the businesslike Bristol Bulldog.

the banshee scream of those Felix engines, for the Kestrel was remarkably quiet.

A strange pair followed: the Halton amateur-built light aeroplane and the hugely dominating Beardmore Inflexible which a reporter described as: 'Flying with a gravity and impressiveness almost ludicrous in its solemnity. With its enormously long fuselage and generally clean outline it somehow gives the impression of a witch's broomstick of enormous size which has been reformed and piloted by an archbishop.' By contrast the

At last with a British engine, the Fairey Fox with Rolls-Royce Kestrel IIA was supplied to only one squadron.

Extremely light wing loading of 6 lb/sq ft characterized the 76-ft span Vickers Vellore freighter, so it seemed to float from the ground at take-off but was risky to taxi in a strong wind. (*Courtesy Air Marshal Sir Ralph Sorley*)

Vickers Vellore biplane, with long fuselage midway between the wings and carrying a biplane tail with quadruple rudders, looked antique, though it lifted from the ground with the lightness of a balloon. The Blackburn Ripon, Boulton & Paul Sidestrand, and Handley Page Clive troop carrier derived from the Hinaidi, completed the circus with sound and stolid competence. Yet when four night-bombing squadrons of Virginias and Hyderabads took-off, the terrific noise and very massiveness of the machines was enormously impressive, for no other country had aircraft of this calibre. But it was the flying of Flg Off Dermot Boyle and Flg Off Richard Atcherley of CFS on two Genet-powered Moths which beat anything in the programme for virtuosity. Never before had outside loops, nicknamed 'bunts', been displayed in public. Using engines modified to run inverted by feeding from a small tank on the undercarriage, they flew in close formation upside-down for many minutes, and made banked turns around the aerodrome, or similarly formated with one the right way up and the other inverted. By contrast, the air battles, squadron manoeuvres, and various individual aerobatics had become so customary that they were regarded as normal for their respective squadrons. So to the finale portraying an attack on an oil refinery, with stutter of guns, bang of bombs, a balloon in flames, and ultimate grand conflagration of the set piece. Among participating pilots were several of increasing distinction – Flg Off C. S. Staniland, Flt Lieut A. H. Orlebar, Sqn Ldr K. R. Park, Sqn Ldr R. B. Mansell, Wg Cdr W. Sholto Douglas, Wg Cdr H. R. Busteed and Wg Cdr A. Portal.

The Paris Aero Show followed. Said C. G. Grey: 'One has seen worse Shows. The French aircraft industry seems to be doing fairly well from orders for assorted governments as well as its own. It may not be so rich and prosperous as the British industry – supported as it is by a fatherly Air

Ministry, let alone the business in military and civil machines obtained abroad by the more enterprising British manufacturers – but the chiefs of the French firms seem quite well fed, and most of them, like their British opposite numbers, drive large automobiles.' But except for the Bristol Aeroplane Company, the British airframe industry gave the Salon a miss. There was not even a representative light aeroplane to compete with the French, German, Italian and Czechoslovak – so Bristol scored with their 'L'Avion de Combat Monoplace shown entirely in structure on the port side but fabric covered on the other, and considered by visitors of all nations to be the finest example ever seen of all-metal girder construction.' In British eyes it was a Bulldog almost undisguised, so there was some surprise that the Air Ministry had permitted it to be seen in detail.

Engines worthily upheld British prestige. Centre of interest was the Jupiter already famous the world over, but the Bristol stand was hidden under the staircase leading to the gallery and only with difficulty were the Salon organizers induced to provide adequate lighting. One could imagine the shrug: was there not the stand most evident of the French licensee, Gnome et Rhône? They were rumoured to be producing 70 engines per week, including the five-cylinder 230 hp Bristol Titan, the first examples of which were revealed both on British and French stands. Its established competitor, the Lynx, was displayed with Genet, Mongoose, and Jaguar by Armstrong Siddeley on an equally inconspicuous small stand on the gallery. Not until the second week of the show was the 480 hp Rolls-Royce F.XI displayed on another minute stand in one of the sub-galleries. Among the overwhelming display of French, Italian, German, and American engines it was of significant interest that Junkers displayed a 600 hp vee engine with gear-driven supercharger resembling the established French-designed Rateau, but British designers were still cautiously awaiting results of supercharger experiments at the RAE. Less obvious, though it had been on display at every Paris Salon since the war, was a Panhard with Knight sleeve-valve on the Levasseur stand, but it certainly attracted Roy Fedden.

The anti-British attitude was even emphasized at the banquet arranged by the *Chambre Syndicale* at the Hotel Continental for exhibitors and important members of foreign governments. No British manufacturer was present; no Air Ministry official, except the British Air Attaché. Under the chairmanship of M Bokanowski, Minister of Commerce and Industries in the French Cabinet, the theme of aviation was voiced by France, Italy, Germany, and Czechoslovakia, each boosting their country's aircraft.

In international affairs the British public limited its attention only to spectacular tragedies, such as the disappearance of the millionaire Belgian financier Alfred Lowenstein, an ostentatious user of private aeroplanes for his business, who allegedly fell through an open door while flying across the Channel – or was he in hiding? Speculation on this affair matched continuing interest in the aftermath of the *Italia* disaster, for though the injured General Nobile had been rescued by the CFS-trained Swedish pilot Lieut Lundborg with a Moth and flown to Whale Island, his rescuer

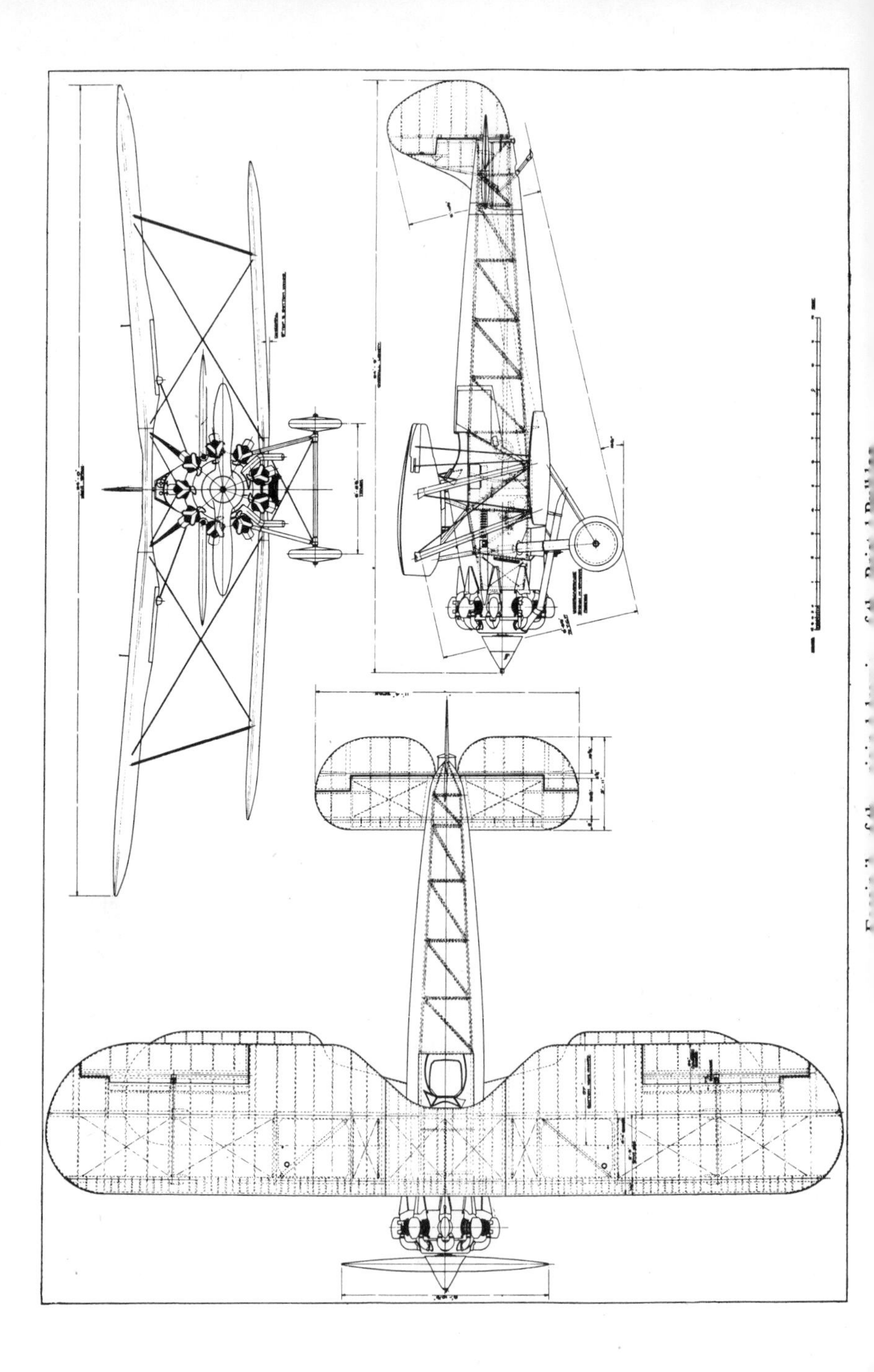

crashed on attempting to pick up five of the crew and was marooned until another Swedish Moth found them. Next came news that Capt Amundsen, Commandant Guilbaud and Lieut Dietrichson, had not been heard of since they set out in a Latham flying-boat on 18 June to search for General Nobile. Eventually a seaplane float was washed up near Tromsö, and it had to be concluded that Amundsen and his crew had all perished shortly after departing from that base. Meanwhile there was effective help from the Bolsheviks for the Fascists, for the Russian ice-breaker *Krassin* rescued two parties of *Italia* survivors while Swedish and Finnish seaplanes picked up others; but Dr Malmgren, with broken hand and both legs frozen, insisted on his Italian rescuers leaving him to die rather than hinder their search for others. It all provoked bitter controversy between the Italian newspaper *Impero* and the French and Swedish Press.

From tragedy to tragedy. The King's Cup race, starting from Hendon on 20 July and finishing at Brooklands next afternoon, not only had more than the usual crop of forced landings and accidents among the 30 starters of the 38 entries, but G. N. Warwick, flying the I-strutted A.N.E.C.IV *Missel-Thrush* fatally crashed into the Scottish hills at Broad Law near Tweedsmuir in Peebles. Of the mixed bag of aircraft that were entered for the stage-by-stage 1,000-mile circuit to the Solway Firth and back, it was believed that the Avro Avenger flown by Summers of the A & AEE would win, though it was expected to have doughty competitors in the Hawker Heron to be flown by Bulman, the Bristol 101 in the hands of Uwins, a Gloster Grebe flown by Atcherley, and the little Lincock raced by Noakes.

The Parliamentary Secretary, Ministry of Transport, Lieut-Col John Theodore Cuthbert Moore-Brabazon, pioneer aviator, is intrigued at Hendon by the Blackburn Lincock raced by Flt Lieut Noakes. (*Flight*)

On handicap some were prepared to back the re-powered Cirrus two-seat Avro Baby prototype, for it was capable of 98 mph. Another favourite was the Lynx-powered Autogiro C.8L Mk II flown by the Cierva Co's recently appointed test pilot A. H. C. Rawson, and boldly entered by Air Commodore J. G. Weir their chairman – but though credited with 120 mph, most people had yet to be converted to faith in its rotating mechanism, nor were they encouraged by seeing its rotor wound up with rope followed by a complete ground circuit of Hendon before Rawson was able to haul it off. Some thought that the legend on its rudder 'Beware of Rotors' was very sound advice.

Under a cloudy sky and into a chilly wind most competitors took-off without incident, but Bulman was eliminated when he taxied towards the starting line and blundered into Sydney St Barbe's Morris car. By evening it transpired that 24 had arrived at Glasgow. Of those out of the race, Yeatman damaged the undercarriage of his Moth at Newcastle; Stammers with a Moth also retired there having gone far off course; Trench with the Halton monoplane retired at Leeds with a broken magneto drive; Birt with a Bluebird forced-landed at Minworth; Whitehead with an Avro Baby turned over when landing in a field near Bury St Edmunds; Rawson with the Autogiro ran out of petrol near Nuneaton, windmilled into a small field and could not get out again; Smith with a Moth and Boyes with an Avian drove into each other at Nottingham; Soden with his Moth had the propeller detach itself and gently whirl into a field into which he came gliding, but could not re-attach it; Ragg with an Avian forced-landed near Atherstone; Jones with the speedy Nimbus-powered Martinsyde smashed his undercarriage taking-off from Castle Bromwich; and Warwick with the *Missel-Thrush* had not been heard of since leaving Newcastle, nor could he be found next day when Miss Leathart with the Grasshopper, which she now owned, searched the Cheviots for more than an hour, and the RAF later sent two more aircraft, though their quest was fruitless.

Next afternoon at Brooklands at 3.50 p.m. a Moth came into view flying low and fast, and was clearly W. L. Hope's. 'Thinking all was over he proceeded to loop and stunt before landing, and having landed switched on his well known winning smile. Suddenly there was terrific hooting, and Sir Francis McClean in his white Rolls-Royce came tearing across to tell Hope he had not crossed the finishing line. The winning smile switched off and the engine switched on in one movement, and within 30 seconds Hope was in the air again, discovered the finishing line, landed, and again switched on the winning smile *fortissimo*.' Within seconds Uwins appeared on the Bristol 101, crossing the line two minutes later, and then came the only lady entrant, the friendly and able Winifred Spooner. After which Moth after Moth seemed to appear, among which was an almost indistinguishably different machine – the home-built Simmonds Spartan prototype flown by Webster of Schneider fame.

Currently Oliver Simmonds announced his resignation from Supermarine in order to produce Spartans commercially. With a competitive

A Simmonds Spartan at Croydon – though forty-eight were built it was not popular. (*Flight*)

eye on the Moth's price of £650, it was advertized as 'the world's most economical aircraft' at £620. Though the idea of interchangeable components had first been formulated by Fritz Koolhoven for his war-time Baboon trainer, Simmonds had gone a step further by using a symmetrical-section wing so that it could be used on right or left, top or bottom, with main bracing wires all the same size and length and with identical struts and fittings; similarly an interchangeable component could be used for right or left elevator or for rudder, or for left or right aileron, and the fin and each half of the tailplane were identical. Said one of his admirers: 'Mr Simmonds is not well known outside the aircraft trade; he is a young man with very considerable experience. After the war he was at Cambridge, and then at the RAE where he was secretary of their flying club which built entrants for the earliest light aeroplane meetings. Then he went to Supermarine, where as a designer he has acquired valuable knowledge in such widely different types as the Southampton flying-boats and the Supermarine-Napier Schneider racers. It was his experience of club work, as chairman of the Hampshire Aeroplane Club, which first indicated to him the need to reduce the production cost of light aeroplanes.' Formation of Simmonds Aircraft Ltd followed, with capital of £20,000 in £1 shares, and with him as director was Lieut-Col L. A. Strange, well known before the war as pilot of the Grahame-White *Lizzie* at Hendon and subsequently for his war-time RFC adventures.

Regardless of competition the Moth was keeping well to the fore. More than 500 had been built, and production of one a day was being exceeded. On 26 July, Geoffrey de Havilland, with his wife as passenger, secured a light-plane world height record of 19,980 ft flying the latest Gipsy-powered Moth, beating the German record of a year earlier by only 1,280 ft, but making a telling advertisement feature which was supplemented three weeks later when Hubert Broad took-off at 5.30 p.m. on 16 August with the same Moth fitted with overload tanks giving 80 gallons total, and remained aloft 24 hours during which he read three novels as relaxation.

Geoffrey de Havilland, with Mrs de Havilland and their height-record Gipsy Moth.

On landing precisely at 5.30 the following day he still had 12 gallons reserve, so could have continued a further four and a half hours. Had the flight been in a straight line it could easily have beaten the world's distance record for light aeroplanes as he would have covered some 1,440 miles.

A long-distance attempt by Frank Courtney was more spectacular but less successful. Ever since his severance from Armstrong Whitworth he had been hoping to make a transatlantic flight, and though the Dornier Wal he bought with the backing of E. B. Hosmer, his Canadian millionaire friend, had made an unsuccessful attempt in 1927, he now made another assault from Lisbon, reached Horta in the Azores on 28 June, leaving for Newfoundland on 1 August – but that evening ships in mid-Atlantic picked up messages from his radio saying he had alighted on the open sea

There could be no greater contrast with British biplane flying-boats than German conceptions such as Frank Courtney's Dornier Wal monoplane with sponsons and central tandem engines.

due to fire in the aft engine housing. Six ships diverted to his rescue, and the liner *Minnewaska* picked him up with his crew of four, one of whom was Hosmer, and they were landed in New York – but their welcome lacked warmth, for ocean flyers had become too numerous for public interest in civic greetings. The abandoned flying-boat was found four days later by the Italian steamer *Valprato* and taken in tow as salvage, reaching Montreal badly damaged but repairable. Courtney was far from defeated, and declared: 'The flying-boat is one day going to be of the greatest use to shipping companies, though there must still be failures among those who try difficult experiments. We took every reasonable precaution, and I think we learnt more about ocean aviation than could have been learnt by any amount of success.' But had there been a gale when the Wal alighted the story would have been very different, for such a flying-boat would have been at the bottom long before any ship could have reached it. As things were, Courtney had achieved America and could see expanding prospects compared with Britain.

7

While the sea-sick crew of the Wal were anxiously awaiting rescue, the Committee of Supply in the Commons was debating civil aviation on the motion 'that a sum, not exceeding £415,000 be granted to His Majesty to defray the expense of civil aviation, which will come in course of payment during the year ending the 31st of March 1929'. A handful participated in the discussion, of whom Sqn Ldr the Hon F. E. Guest urged the Air Minister to press for increased expenditure on civil aviation to strengthen the potential reserve for war, pointing out that the Treasury were handing tens of millions to the unemployed, to railways, and for rating relief, but though it was vital to maintain an efficient aircraft industry and the highly skilled engineers on which its success was based, only paltry sums were given to aviation. Lieut-Col Moore-Brabazon took the contrary view that training for war should be divorced from development of civil aviation. Sir Robert Lynn took a domestic view, and appealed for subsidies for internal airlines from Southampton to Manchester and Liverpool, thence to Belfast and Glasgow, or between Liverpool and Belfast – an idea which Sir Harry Brittain advocated as even more applicable to flying-boats, and particularly for a service between the West Indies, Bahamas, and Florida. Rear-Admiral Murray Sueter, prodding at the Air Ministry as of old, and well informed that an indoor aircraft show had already been discussed by an SBAC committee chaired by Capt Peter Acland, stressed that 'a great air exhibition in this country should be held next year, financed with at least £20,000 from the Treasury, to attract leading people in aviation from all over the world.'

In closing the Debate, Sir Samuel Hoare, with customary lucidity, outlined the Government's view on the development of civil flying, particularly consolidation of Imperial Airways 'which was the envy of every country, with a remarkable record of reliability. We have now got to the stage when we can see clearly that with possibly two or three more changes from the present types to more up-to-date types, our aeroplanes will be covering expenses and our airline will be self-supporting.'

In the report for the year 1927–28 of the Aeronautical Research Committee there was indication of the envisaged technical trend, for it was stated that Sir Sefton Brancker as DCA had been consulted on his research requirements concerning civil aviation in the immediate future, and among these he stated: 'A high power heavy-oil engine to secure low fuel bills; thick winged monoplanes, which are believed to have low maintenance charges; use of geared engines and pusher propellers to reduce noise; development of all-metal structures.' As a further step, following a crucial report by Professor B. M. Jones on the importance of streamlining, a special sub-committee was formed to examine the problems of reducing

In characteristic fashion George Bulman displays the Hawker Hart prototype at Brooklands. (*Flight*)

This Fairey Fox IIM, J9834, proved to be 40 mph faster than the original Fox I. (*Flight*)

resistance in 'practical' aircraft – an aspect largely neglected during the past two years in favour of research on safety and stability.

All this was far from an immediate pointer to acceptance of monoplanes. True there was the Fairey long-range monoplane well on the way to completion; but the latest two-seat day bomber to specification 12/26, flown by George Bulman during June, was a metal-framed biplane, the Hart prototype J9052, with the Rolls-Royce F.XIB engine. Like earlier Sopwith creations, it had immediate pilot appeal, and Bulman was pleased with its smooth handling characteristics. When Dick Fairey discovered that Hawkers had this experimental contract, he angrily protested to the CAS that he had not been given a chance to tender, though the requirement must have originated from consideration of the geometrically similar Fox, and such was the strength of his position that he was invited to compete, though with little time in which to prepare a development of the Fox. Yet he achieved a much improved Mk IIM, with fuselage of steel-tubes and wings having spars of high-tensile strip drawn to double-lobe section riveted to webs stabilized by angle plates – superficially similar to Camm's 'dumb-bell' spars which had booms of steel strip rolled to polygonal section connected by a single plate web. Inevitably with two machines designed to the same specification they followed identical lines of thought and similar solution, for both had a top wing of greater span and chord than the lower, with outward raking N interplane struts and relative stagger, undercarriage disposition, crew location, and tail leverage – giving very similar profile.

A third contestant, the Avro Antelope, was almost the same dimensionally, and similarly with lower wing of smaller size, but it had a flat-sided fuselage of duralumin sheeting riveted to a frame of L-section drawn duralumin, and instead of a pointed nose there was a chinned radiator akin to the Horsley. Chadwick argued that such a fuselage was less liable to structural damage in fighting because it was virtually monocoque – while the other two designers championed the open structure of their machines as more accessible for maintenance. The Antelope also had duralumin-

Structurally the neat Avro Antelope was a step ahead of the Hart, but though it handled well RAF maintenance crews preferred space-frame construction.

structured, fabric-covered wings tapered in thickness at the central portion to give least obstruction to view, and supported by an inverted V. All three contenders had special balanced gun-rings to facilitate movement of the Lewis at acute angles, as greater speed had made Scarff rings unusable. The Hart had a wooden propeller, but the Fox IIM and Antelope featured the Fairey-Reed of twisted duralumin, yet Hart and Fox were equally matched at over 180 mph, but the Antelope was some 10 mph slower. Meanwhile they were almost the most secret machines in the country, though Fairey had corresponding single-seaters flying as a Firefly development for interceptor and ship-plane fighters as well as a two-seat fleet fighter reconnaissance ship-plane similar to the Fox IIM.

Top Secret in the Hawker D.O. was the rapidly advancing design of a single-seat fighter with Rolls-Royce XIS, based on the Hart arrangement, but, though 6 ft less span, was only 130 lb lighter. This was the Hornet as an answer to the slightly larger Firefly II. In the experimental shop was a Mongoose-powered two-seat trainer, the size of a Moth, which Camm had devised with standard Hawker metal construction; and at Brooklands, Bulman had just flown the F.20/27 Jupiter-powered biplane which looked

Rival of Bristol, Westland, and Vickers Interceptors was the 30-ft span Hawker F.20/27 which led the way for Camm's pointed nosed Rolls-Royce powered Hornet. (*Courtesy Air Marshal Sir Ralph Sorley*)

so like the Hawfinch, or its Naval variant the Hoopoe, with single-bay wings.

At Bristol, Uwins and his tough assistant Jock Campbell had been flying their interceptor, the Bullpup, with temporary Jupiter VI, since April. Just as Hawker or Fairey fighters bore unmistakable family characteristics, so did the Bullpup, and it was difficult to distinguish it from the Bulldog predecessor except that it had N interplane struts. Chances were small that another Bristol would be selected, because on 21 August it was finally confirmed that the Bulldog II had won the F.9/26 competition, and a contract for twenty-five production aircraft to specification F.17/28 followed. At last the company had a breakthrough. But it had been touch and go, for it was only the disadvantage of the additional bay of interplane rigging which ruled out the Hawfinch.

At Yeovil, Louis Paget had tested the Westland low-wing Interceptor, powered like the others with a Jupiter pending development of the Mercury. It was Rafwire braced, the landing wires going to the top longerons of the deep fuselage and the flying wires to the Vs of the Siskin-like chassis – a feature criticized by pilots, for if such an undercarriage was damaged, the wing structure would collapse, and even in normal flight it was at first disconcerting to see how slack the top wires became. Unfortunately there was a more serious snag. After a subsequent flight, Paget clapped his eye-glass firmly to his eye and said: 'The little beast turns round and looks at you if one attempts a loop – it's a flying corkscrew!' Further tests confirmed that odd things happened in a 'g' stall, though longitudinal control when landing was delightfully easy and precise. In the course of months a bigger tailplane was fitted but did not effect a cure. Hot wire explorations made in the wind-tunnel around model wings and tail revealed root interference at the fuselage. Inboard slats proved ineffective.

The 38-ft span Westland Interceptor low-winger was light on controls, delightfully easy to fly and gave perfect pilot view. It was the first of current designs to feature wheel brakes.

Presently it was found that the disturbed flow could be largely eliminated with a fillet, and a flight test at last showed the machine no longer made inadvertent rolls – but now that the next step could be taken there was doubt of recovery in spins and further long delay occurred in designing and trying a taller fin and rudder. By then the Hawker F.20/27 had long been assessed, and was known to be 10 mph faster than the Westland, but by that time both were obsolete, and so was the long delayed Vickers

In an attempt at countering the Westland Interceptor's wing/fuselage interference, automatic slots were tried at the roots, but the jerk on opening was disconcerting.

The Westland Interceptor was finally cleared for Martlesham with tall fin and rudder, and recently devised Townend ring which reduced drag by streamlining the flow around radial engines.

Jockey Type 151 draughted by John Bewsher, designer of the ill-fated *Missel-Thrush.* Like the Westland Interceptor, the Vickers was a low-winger with R.A.F.34 section wings, but though only moderately thick, the Wibault form of construction enabled it to be a cantilever.

Despite limited flying experience, I was impressed at the splendid all-round view of these low-wingers compared with contemporary biplanes, and certainly the Westland was easy to fly. Nevertheless RAF pilots preferred biplanes, nor did they believe these two interceptors were pointers to the future, though it seemed obvious when compared with a machine such as the Armstrong Whitworth Starling which was merely a clumsy single-seat derivation of Atlas geometry. Their conservatism was understandable. War-like purpose was hard to remember. All we really

Smaller than the Westland Interceptor, the 32 ft 6 in span Vickers Jockey Type 151 had its vicissitudes due to overbalanced rudder, torsional flexure of rear fuselage, root stalling, and dangerous spinning – apart from engine trouble.

cared about was the exhilaration of those open cockpits, with a heady wind rushing by, the mingled smell of oil and acetone, the sunny, peaceful world stretching far below, and the beauty of the towering clouds where we would swoop and zoom above their glittering chasms.

That flying had its war-like purpose became more obvious with such operations as the second annual Command exercise of the Air Defence of Great Britain which was held for three days beginning 13 August to determine the operational efficiency of individual units as much as actual defence of London. Eastland, commanded by Air Vice-Marshal Sir John Steel, represented a Continental Power with operational base in Northern Europe at war with Westland, commanded by Air Vice-Marshal Sir Robert Brooke-Popham, whose capital was London. It was bomber against interceptor.

Bombing forays from east and southern coasts were made by day mainly by those old warriors the D.H.9As, backed by Fairey Fawn, IIIF and Fox, together with Horsleys; at night they were chiefly Virginias and a squadron of Hyderabads. Opposing were nine squadrons of Siskin, two of Woodcock, and one of Gamecock. Such was the state of obsolete equipment. At noon each day an official communiqué described what had been going on in a great effort to condition the public to spending money on Defence – but interest was apathetic except for the excitement of a Siskin catching fire over London, the pilot parachuting and dropping on a roof, while the machine fell into Kew Gardens.

But what was the war-like result? The official summary stated: '(1) Approximately 250,000 miles were flown. The only mishap involving injury was to a fighter which turned over. Total weight of imaginary bombs carried was nearly 202 tons discounting aircraft judged shot down before reaching the objectives.

'(2) Day bombers made 57 raids and were attacked 39 times on the way in and 37 on the way out. Only nine raids succeeded in evading the defence both ways, and 151 day bombers were adjudged brought down by fighters and 20 by anti-aircraft, while 139 fighters were lost.

'(3) By day, clouds and strong winds high up favoured bombers on the whole. In operations after dark the weather favoured the defence on three of the four nights so that a large percentage of bombers was intercepted. This interception shows decided improvement since last year, as does the standard of air pilotage in use of clouds by bombers.

'(4) The Auxiliary and Cadre squadrons played a creditable part in these exercises as the first in which they have co-operated, and invaluable work was done by the Observer Corps and Searchlights. The AOC Air Defence of Great Britain is very satisfied with the efficiency shown by all units and thanks those who allowed searchlights within their grounds and the civilians who looked after the flares on emergency landing grounds.'

It seemed suspect that no mention was made of the number of raids carried out by night bombers, so the complacent assumed that only one or two got through and therefore the risk was not great if one could stand the

The three-engined Short Calcutta airliner moored on the Thames was visited by Colonel The Master of Sempill, who alighted alongside with his Blackburn Bluebird seaplane. There was no bar those days to waterway operations. (*Flight*)

loss indefinitely of thirty to fifty fighters a day. But could one? And if bombers flew higher and faster would they not get through?

Meanwhile, to more peaceful purpose, the Short Calcutta alighted on the Thames and was moored for several days off the Embankment for inspection by Members of the Lords and Commons.

8

Perhaps the undertones of war seemed subdued beneath the pursuit of peaceful purpose. In August, fifteen nations signed the Kellogg Pact in Paris outlawing war among civilized nations. In September, Professor Alexander Fleming made his discovery leading to production of penicillin and its remarkable curative properties. At Scapa Flow there began a great salvage operation to raise the sunken German warships so that their steel could be turned to productive use.

In his speech to the shareholders of Imperial Airways on 7 September, Sir Eric Geddes, massive of figure, was optimistic of the future. Not only did his balance sheet show £72,500 profit compared with £11,000 for 1927, but he was able to announce a new form of subsidy of £335,000 a year for the first two years; £310,000 a year for the next four; £220,000 for the seventh; £170,000 for the eighth; £120,000 for the ninth; and £70,000 for the tenth year – which, as he said, gave 'a total of almost £2½ million compared with under £600,000 remaining to be earned under existing agreements. So you will see that your company has a greatly extended security of tenure of Government support, representing an extension of five years for the existing European agreement and over seven years in the case of the existing Middle East agreement.'

It was felt very satisfactory that passengers had risen from 11,395 in the first year, to 11,703 in the second; 17,083 in the third year, and 26,479 in the fourth. The fleet comprised three Armstrong Whitworth Argosys, five de Havilland Hercules, two Handley Page W.8bs, one W.8f Hamilton, one W.9 Hampstead, three W.10s, two de Havilland 50s, two Short Calcuttas and one Supermarine Sea Eagle. Three further Argosys were on order.

'In the development of civil flying there appear to be two more or less inherent schools of thought,' said Sir Eric. 'One of them regards high figures of aeroplane miles flown, machines in existence, and pilots trained yearly, and sees a map with aeroplane routes spider-webbed across it, and says "That's fine"; but the other looks to the future commercial development of flying and its ultimate independence of subsidy. The British Government is undoubtedly adopting the second and wiser outlook; but no one is ever content with what a government does. I would like it to have said: "We give you this agreement now; we will shortly enter into a further agreement for you to go to the Cape or Singapore, and then a further agreement to go to Australia." They have not said so, and "not one penny more" has been their cry. Nevertheless I am sure that once this Australian route is organized, further assistance will be forthcoming. That is the future as I see it for Imperial Airways. There is another future for feeder lines, for special charters, for taxi work, for the use of aircraft by our large business firms, all of which will come in time; but the great mission of Imperial Airways is on Empire routes.'

To the bewilderment of those present, the meeting was then asked to accept his resignation, for he was anxious to devote his energies to further improvement of the Dunlop Rubber Co. Ltd. Said C. G. Grey: 'Like all strong men he has great enemies, but he has at any rate earned the admiration and esteem of people who have been forced, often against their will, to recognize that he has succeeded in every task he has undertaken. He has that talent, so necessary in commerce and war, of recognizing when a subordinate is a good man and giving him a free hand to do his job in his own way.'

That more than Government-sponsored heavier-than-air aviation was progressing was emphasized by an open day at the Royal Airship Works for the Press to see and publicize R.101. Within the towering shed the huge airship structure was nearly complete. Several sections were fabric covered, and one contained an inflated gas-bag to show the method of suspension, and it was said that a million bullocks provided the gold-beaters skin to make the sixteen gas-bags by the expensive and tedious method devised by James Templer for the Balloon School in the 1880s.

To those who had seen R.100, comparison with R.101 was of greatest interest, for there were striking divergences in solving similar problems – particularly the method of joining main rings and longitudinal members. Both Wallis on his own, and Richmond advised by North, swept away the mathematically unstressable tangle of girders, webs, and fish-plates used in early airships. The 'Capitalist' ship used a standard system of screwed collars to connect the helically seamed tubular framed girders at all junctions; the State ship, with Boulton & Paul patented girders of high-tensile steel strip rolled and drawn into thin tubes, used machine stampings bolted together at junction points and braced by diagonal wires in the bays. Both airships had more than 25 miles of steel or duralumin tube, but could be assembled like a Meccano toy. Each transferred the lift of the gas-bags by a different system identically intended to prevent abnormal loads falling on the main components when the ship was out of level or a gas-bag became deflated. They differed too in motive power, R.101 having, in order to reduce fire hazard, five Beardmore compression-ignition engines totalling 3,250 hp, using crude oil at £5 per ton, while the R.100 had six Rolls-Royce Condors totalling 4,200 hp and burning petrol at £25 per ton. The advantage seemed to be with R.100, for its power cars at 10 tons weighed only half those of R.101 whose less powerful engines were suffering snags. The long crankshaft of the eight-cylinder inline arrangement produced such torsional resonance that torque variations twenty times normal had been measured at 950 rpm. There was allied trouble with the reversible-pitch propellers which originally had hollow steel blades, but as these failed because of resonance it meant developing solid light alloy blades, though to guard against delay fixed-pitch wooden propellers were being built, one of which was solely for thrust astern when approaching the mooring mast.

The *Graf Zeppelin* made world news with its long-distance demonstration flights. (*Deutsche Lufthansa*)

Attempts at discovering how long it would take to complete either airship met with disarming reticence, but everyone knew that Germany had beaten both teams, for the Luftschiffbau Zeppelin GmbH had built an even longer airship, though of 1,300,000 cu ft less capacity. Named *Graf Zeppelin*, she was moved from her shed at Friedrichshafen in mid-September, and on the third flight was flown to Salzburg and back with 75 aboard, including Major G. H. Scott and Sqn Ldr R. S. Booth from the British airships. On 11 October, commanded by Dr Eckener, she set off for Lakehurst, New Jersey, with a crew of forty and twenty passengers comprising representatives of the German Reich and Prussian Governments, the late Count Zeppelin's son-in-law Count Von Brandenstein, several war-time Zeppelin captains and Lieut-Cdr

Rosendahl who had commanded the *Los Angeles* since her delivery to the USA, but the only British passenger was Lady Drummond Hay, the popular writer for popular papers. Over the Atlantic the weather was unfavourable and a southerly course was set, turning north-westerly after Madeira. At 11.25 a.m. on 13 October the US Navy Department received a wireless message from Rosendahl stating that the port horizontal fin had failed and asking ships to stand by. In fact a fifth of the fabric had stripped, resulting in such sudden violent pitching that the passengers were thrown from their seats. Three hours later a further message said assistance was not needed as temporary repairs enabled the airship to continue at reduced speed. At midnight she passed over Bermuda and, despite head winds, crossed the American coast next day at 3 p.m. and proceeded to Washington where she circled, then turned for New York before heading back to Lakehurst where she landed that evening and was 'walked' to her mooring mast for the night.

Coincidentally that early autumn, the man who in 1902 made the first British airship, Harry Spencer, contemporary of Willows, lost his life endeavouring to salve a balloon with which his son had descended among the chimney-pots of Rugby School. Overcome by escaping gas, Spencer slipped, rolled down the roof and fell to the ground with mortal injuries.

9

Long-distance aeroplane flights had become so commonplace that they received little attention – yet all were new adventure. Test pilots habitually made long European journeys to demonstrate company aircraft. Scholefield had flown the Vickers Vivid to Bucharest in 10 hours. Sam Brown had taken the Avro Avenger to Roumania to compete against French, Italian and German aircraft for military orders. Lieut P. Murdock, SAAF, piloted an Avian to the Cape and return. From India Charles Barnard and E. H. Alliott arrived back at Croydon on the evening of 5 September with the Duchess of Bedford's Jupiter-powered Fokker F.VII *Princess Xenia* in a record three days – which atoned for an outward journey, three months earlier, which had been dogged by misfortune. With the Duchess aboard they had originally left Croydon on 10 June, aiming at a record eight-day journey to India and back but at Bushire had to wait for a new engine. Karachi was reached only on 22 August, but as a new propeller was required, the Duchess returned by steamer.

Sir Philip Sassoon planned an equally ambitious tour to India and back, and at dawn on 29 September, piloted by Sqn Ldr C. L. Scott and Flt Lieut L. Martin with crew of four and Sir Philip's valet, left Plymouth Cattewater as passenger in the Blackburn Iris II, accompanied by Air Commodore A. M. Longmore. Officially they were testing the latest RAF flying-boat on a journey of 15,000 miles, but at Cairo, Sassoon would disembark and fly with Longmore in a Fairey IIIF along the Nile to Khartoum and back, then by Wapiti eastward to Baghdad and back, thence again by flying-boat to India, where Hinaidi, D.H.9As and Bristol Fighters

Major Sir Philip Sassoon, Bart, (*facing camera*) proved a brilliant Under-Secretary of State for Air under Sir Samuel Hoare on appointment in 1925, and qualified as an RAF pilot. (*Flight*)

would be used for local journeys. One outcome was a delightful book* describing this six-weeks adventure in fabulous lands. 'It is no small advantage to have personal experience of the mobility of air power, and to possess something more than theoretical knowledge of probable application and effect of decisions taken at home,' he wrote. 'Tours of this kind are not likely to become frequent – there is too much to do in Whitehall – yet they should at intervals be undertaken by one or other of those responsible for administration of the RAF. The whole Empire is the Air Service's field of operation. No doubt the first civilian passengers to use the Third Route will either be those pressed for time or frankly adventurous; but that stage will soon pass. Before many more years, long-distance air travel will stand on its own merits, as by far the most enjoyable method of seeing the world.'

As C. G. Grey sagely wrote: 'When Sir Philip first became Under-Secretary of State for Air some believed he was just a rising young politician who saw opportunity for political advancement. Those who have met him know his interest is much more personal than merely political, for he has an insatiable thirst for knowledge about all branches of aviation, and the information he demands shows he forgets nothing he has seen or heard since taking office. He has all the Sassoon family gift for assimilating information, and those who are well qualified to judge regard him as a most valuable asset to British aviation.'

Another great flight had also reached conclusion. The *Daily Mail* of 17 September reported: 'The 23,000 mile journey of the four RAF flying-

* *The Third Route* (Heinemann)

boats which left Plymouth last October under Wg Cdr Cave-Browne-Cave finished to-day when the aircraft alighted at Singapore. As a demonstration of reliability the flight will rank as one of the greatest feats in the history of aviation. Throughout the long journey each stage was accomplished to prearranged plan. The machines reached Karachi via the Persian Gulf, then made the first flight round the coast of India and reached Singapore by way of Rangoon. After flying through the East Indies they began their trip round Australia by Darwin, Perth, Adelaide, Melbourne, Sydney, Brisbane, and back to Darwin, and have just returned to Malaya by the same route as the outward journey. The flying-boats will now be put on permanent service at the Singapore Air Base, their duty being to assist the Navy to patrol our ocean trade routes.'

In England, what with private flyers setting out on long journeys, or flying to France for a brief week-end, there was increasing need of regulation, for they were becoming a nuisance at Croydon by neglecting Customs requirements and disappearing into the blue without notifying the Traffic Office of departure or giving details of itinerary. Many pilots ignored the instruction to make a warning circuit round Lympne and St Inglevert before and after crossing the Channel, with the result that many unnecessary wireless enquiries had to be made, for it was automatic that a Channel search would follow if machines known to be crossing did not report or signal at these two aerodromes.

Yet everywhere efforts were being made to encourage private flying. A string of little aerodromes was spreading all over Britain. Newest was by a consortium known as Airwork Ltd which H. N. St V. Norman had formed with F. A. I. Muntz, grandson of the original Muntz of the metal business, to acquire land near Heston in the angle between the Great West Road and the Bath Road. They intended making it a social focus for flying, with attractive club house and sheds for private and commercial owners; so the recently formed Household Brigade Flying Club, whose keen secretary was young Capt 'Mossy' Preston, expected to move there from Brooklands. Hoping to sell Widgeons for the Airwork Flying School, I landed with a demonstration machine on a 100-yard grass strip among the ploughed ground on which Norman and Muntz were working in Wellingtons to turn it into an aerodrome – but as almost always, the Moth was chosen.

The Widgeon was selling in smaller numbers than other light planes, partly because at £750 it was much more expensive, but also its narrow-track undercarriage had caused several nose-overs when taxi-ing across wind. One of the Westland pilots hated to fly Widgeons because of this, so I made a mock-up cross-axle undercarriage for Bruce to see, and he adopted it for future machines re-ciphered IIIA. Concurrently he and Davenport were designing their tri-motor, strut-braced, high-winger of similar carrying capacity to the Hamlet. The cabin was a ply box, with Warren girder spruce-structured rear and front fuselage, fabric covered throughout, and the R.A.F.34 Witch-type wing was similarly of wood with

The Westland C.O.W. gun fighter to specification F.29/27 was a scaled-up, similarly structured, Interceptor with 40-ft wings, and needed similar development treatment.

two deep box spars. In addition to the Wapiti, the military side of the D.O. was completing a larger version of the Interceptor to carry the formidable 8-ft long 37-mm diameter C.O.W. gun which I had mocked up on the starboard side of the cockpit, firing upward at 45 degrees. Vickers were also tendering for this same specification F.29/27 with a pusher I-strutted biplane and were advised to mount the gun identically in the cockpit, throwing away the advantage it could have had if mounted horizontally.

When the first post-war Internationale Luftfahrt Ausstellung, known as 'ILA', was opened in Berlin on 7 October it afforded a concentrated impression of the notable aeronautical progress Germany had made. One of the halls was exclusively for their products, the other for foreign exhibitors, among whom were France, Czechoslovakia, Italy, Belgium, Holland, and Britain who showed a slotted Gipsy Moth with split undercarriage, two Avians, a Bluebird, and a notable engine section with the

Last influence of war-time pusher design was the complex Vickers C.O.W. gun biplane with conical aft fairing astern the propeller. Nacelle slots show the gun position; the hemispherical nose is still to be fitted.

Lion Series XI shown for the first time, the Rolls-Royce F.XI, all four Armstrong Siddeley engines, the Bristol range, D.H. Gipsy, and the A.D.C. Nimbus and Cirrus. Unfortunately engine stands were too small and cheaply made to compare favourably with the splendid display of the German range of B.M.W., Junkers, Daimler-Benz, Argus, and Siemens Halske, ranging in size from 30 hp to 1,000. Even more extensive were the French range with the Renault, Lorraine-Dietrich, Salmson, Hispano-Suiza, Gnome-Rhône, S.F.F.A., Michel, and Farman; impressive also were the Italian Fiats and Isotta-Fraschinis; and Czechoslovakia completed the engine exhibitors with the Skoda and Walter. The aircraft shown by these countries were equally well represented.

Overwhelming everything in the German display were the enormous wings and towering bulks of the Rohrbach Romar, Junkers-G 31 and Dornier Super Wal, but new names were coming to the fore. Ernst Heinkel had become well established and was regarded as the leading builder of floatplanes on the Continent. Dr Focke and Herr Wulf had found a niche with smaller commercial machines, and Colonel Udet, a great fighter pilot and aerobatic master, had joined Messerschmitt in tackling a similar market with high-wing cantilever monoplane passenger-carriers.

Nothing could surpass the kindliness and hospitality of the German aeronautical people to their British visitors, despite uneasy rumours in the Press of all countries, and particularly Italy, of a secret agreement between British and French Governments for mutual co-operation in common measures of aerial defence; but the British Foreign Office issued prompt denial. In fact there was strong disagreement between M Laurent Eynac, the newly appointed French Minister for Air, and the Ministers of War and Marine, Messrs Painlevé and Leygues, concerning the extent of the Air Minister's control of Service aviation; but M Poincaré, President of the Republic, announced that all votes for aviation, whether military, naval or civil, would appear on the budget of the Ministry for Air, though personnel provided by the Air vote would be at the disposal of the Ministries of War and Marine, while the Minister for Air would direct what personnel were to be provided and how employed – a typical French compromise.

The best of both worlds had also been sought by Siddeley Proprietary Ltd, who, following the return of Chadwick and Dobson from Holland, announced that their subsidiary A. V. Roe & Co Ltd had reached agreement with Fokker to manufacture the Fokker F.VII-3m for sale in Great Britain, to the Irish Free State, and all Dominions and Colonies except Canada. Said C. G. Grey: 'Among our junior men there are plenty of brains far in advance of the foreigners to whom we go for our improvements. In due course these bright youngsters improve these imported designs and ultimately we arrive at the best products in the world. History will repeat itself, and thus through Anthony Fokker we shall arrive at better British aircraft. One can only hope that other firms will follow Mr Siddeley's example and import other promising designs which in turn they also will improve.'

British redesign of the Koolhoven Coupé was little more than stiffening, and then fitting a Cirrus-Hermes engine. (*Flight*)

Steps were taken at Westland towards a cabin Widgeon, such as this mock-up three-seat proposal, and they flight-tested a full-depth fuselage, leading to initiation of a new design with the pilot in front of two passengers.

That happened almost immediately. The famous 'Kully', now more generally known as 'Fritz' Koolhoven, had designed and built among his many post-war aeroplanes a cantilever high-wing Fokker-like monoplane with enclosed cabin for two – which interested Danish-born M. L. Bramson, an independent engineer in England who had been one of the Sky-writing pilots and was endeavouring to market an auditory stall warning operated by a free-flying small aerofoil. In October one of the FK.41 Koolhoven Coupés, powered with a 70 hp Siemens, was brought to Croydon for demonstration where I tried it, and reported to Westland that it could prove a competitor with the three-seat cabin monoplane which Toby Heal was designing after it had been proved in flight, to the surprise of Bruce, that the Widgeon with mock-up cabin was faster than the open machine.

Marcel Desoutter, the pre-war Hendon pilot whose crash with a Blériot had led him and his brother to develop light metal legs, decided to join Bramson in acquiring the British licence for the Koolhoven. Desoutter Aircraft Ltd was formed, with headquarters initially at Croydon, and presently George Handasyde, builder of beautiful early monoplanes and splendid fighters, joined as manager with the task of redesigning the FK.41 to British requirements. This led to a somewhat specious announcement of subsequent production headed 'Twenty years' experience built it'. As a three-seater with Hermes it was priced at £795.

Sensing the increasing interest in enclosed aeroplanes, Geoffrey de Havilland had designed a simple coupé enclosure for the Moth, foreshadowed by one which Sqn Ldr Probyn had fitted to the front cockpit of his Widgeon. The first of these enclosed Moths coincided with a new sequence of British aircraft registration letters, and it became well known as G-AAAA in the hands of D.H. himself. In effect it was a subtle method of introducing full enclosure, for unlike a full cabin, when sitting in either seat of the coupé Moth with everything closed, one felt as free as in an open cockpit and could see equally well. It merely required practice to wear a hat. Yet most pilots still preferred 'head-in-the-air'; but whether it was for that reason, or strong price competition of established fully-enclosed American monoplanes, there was little interest in their British replica, the D.H. Hawk Moth monoplane, and the similar Avro monoplane was abandoned. Meanwhile it seemed expedient to produce the Moth in an alternative version with fabric-covered metal-framed fuselage, of which the first, G-AAAR, was sent to Canada for trials on wheels, skis and floats by the RCAF. As production had attained forty Moths a month, it was urgent not only to have more floor area for Moth construction but a separate factory in which to manufacture Gipsy engines. Extensive building operations were therefore rapidly altering the appearance of the Stag Lane factory.

Almost finished in the old erecting shop were several production versions of the D.H.61 Giant Moth for Canada and Australia, and among them was scarlet-painted G-AAAN *Geraldine* with Jupiter XI, ordered by Associated Newspapers Ltd for the *Daily Mail* to help gather news all over

First of the new production Gipsy Moths bore the new national registration code of the 'A' series, and soon was fitted with a simple coupé top. (*Flight*)

the country, for which purpose it had radio and intercom, a motor-cycle, and folding desks for reporters. Capt Bernard Wilson, late of A & AEE Martlesham, was appointed pilot.

Significant changes were becoming the rule in the industry. While John Siddeley was setting his team of designers and builders to work on commercial aircraft with welded-tube fuselages, Alliott Roe had been negotiating with Sam Saunders with a view to purchasing the Cowes business, for he had long been interested in the possibility of flying-boats whether for commercial or military use. His close friend and business manager John Lord was partnering him in this project, and their Hamble factory manager, H. E. Broadsmith, was delighted to join them.

Pursuing the two-seat biplane theme, Vickers in effect scaled-up the Vendace from 44 ft 7 in span to 49 ft and doubled the weight to make the Vildebeest torpedo-bomber prototype. (*Courtesy Air Marshal Sir Ralph Sorley*)

To the surprise and interest of other manufacturers, Vickers (Aviation) Ltd broke precedent by obtaining the services of an Air Ministry official – the Director of Technical Development, Air Commodore J. A. Chamier, CB, CMG, OBE, DSO, who was appointed technical director although disclaiming any pretence of being scientist or engineer. As a commentator said: 'Many a time he has apparently ridden for a fall, as, for instance, when he came home from India in 1927 to take the job of DTD and forthwith proceeded to stake his reputation on having Handley Page slots fitted at once to all military aircraft except fighters.' Both within and without official circles most people thought the new marriage would soon fail, for Service outlook is not compatible with commercial necessities nor is the value of a temporary post-holder likely to be lasting – it would also take strong nerves to live with Chamier's tenseness and neighing laugh. Currently the Vickers Vildebeest Type 132, first flown in April by Scholefield, was despatched on 14 September to Martlesham for competition with the neater Blackburn Beagle which had suffered even longer delay before it was passable.

There followed an announcement that Vickers had acquired the entire share capital of Supermarine Aviation Works Ltd, and its managing director Sqn Cdr James Bird was given a seat on the parent Board. Pemberton Billing's little company had gone a long way, and so had his original partner and manager, Hubert Scott-Paine, who was building a great reputation for fast power boats at his marine company on the west bank of Southampton Water.

The name Supermarine was much in the news, for the RAF high-speed section at Calshot throughout early autumn made many practice runs with the Supermarine S.5 in expectation of breaking the world's speed record. It required ideal conditions of water surface and visibility. On 4 November, with little Mr Reynolds and Colonel F. Lindsay Lloyd of Brooklands track fame as official time-keepers and Harold Perrin observing for the Royal Aero Club, six runs were made by Flt Lieut D'Arcy Greig over the standard three-kilometre course, but though he attained 319.56 mph (514.285 km) it did not give the 5-mph margin required to break the existing record held by Major Mario de Bernardi with 318.464 mph. Still valuable for practice for next year's contest, the S.5 had reached its limit, for resistance could not be reduced nor more power obtained from the Lion or more thrust from the series of propellers which had been tried. Hope must be pinned on the Rolls-Royce powered Supermarine S.6 which Mitchell had almost finished designing.

10

While manufacturers were expanding, moves by Imperial Airways to extend activities with a combined air and rail transport system for goods failed to gain Ministry approval. Almost immediately authoritative statements appeared in several London papers that the Great Western, Southern, London and North Eastern, and London Midland and Scottish railway companies were applying to Parliament to operate their own air services. To many it seemed that an immense combine of railway interests was to be formed with intention of putting Imperial Airways out of business, nor was it lessened by Sir Felix Pole's comment as general manager of the GWR that: 'The railways will certainly not make the mistake made by many canal companies in adhering slavishly to one form of transport, but will adopt all forms according to circumstance or the demands of the future, whether by rail, road, sea, or air.'

While that seemed disconcerting, near panic struck the flying clubs and schools with the announcement of formation of National Flying Services Ltd, based on a scheme proposed by Sqn Ldr the Rt Hon Freddie Guest for a Government subsidized system of aerodromes throughout the country standardizing on only one type of instructional light aircraft so that it could buy more cheaply *en masse* than existing clubs could purchase machines singly. Though Guest specifically stated there was no intention of absorbing existing clubs or starting aerodromes where clubs exist, or of intriguing and taking away their subsidies, Alan Goodfellow of the Lancashire Club, a lawyer who was a key figure in North Country aviation, strongly opposed the idea, arguing that if the proposed company obtained a subsidy then all clubs should have the right to become associated and participate separately in the overall subsidy. Others were happy to continue as independents, such as Capt H. D. Davis, AFC, who had been chief instructor for Lieut-Col G. L. P. Henderson's activities, and recently took over his assets to form the Brooklands School of Flying. Enthralled by aeroplanes, he started his career by running from school and imposing himself on the late Samuel F. Cody at Laffan's Plain. When Cody was killed through structural failure of his last machine, young Davis joined the Sopwith Aviation Co as premium pupil, and on outbreak of war became an RFC mechanic, then as a flight sergeant early in 1916 he learnt to fly on Maurice Farman Longhorns, and less than two years later was posted flight commander at the Gosport Special School of Flying founded by the unconventional Colonel Smith-Barry who had raised the technique of instruction to such scientific art.

Sir Sefton Brancker liked Guest's scheme. The company was formed with Guest as chairman. Managing director was Lieut-Col Ivo Edwards who had been DD Air Transport under Brancker and later appointed Technical Adviser to him; his deputy was G. E. F. Boyes who had initiated the Seven Aero Club of the RAF. With them on the Board were Colonel The Master of Sempill, Sir Alan Cobham, their solicitor A. G. Hemsley, and John Graham Peel who put up much of the money. Hanworth Park was selected as main base, and the mansion in the centre of the grounds would be converted for club use. Currently Boyes was investigating the merits of every light aircraft, and it seemed the Spartan was attracting favourable attention because of the economy of holding interchangeable spares. The Desoutter monoplane also was making appeal as an economic taxi-plane.

A short distance north of Harmondsworth, the Fairey Aviation Co had bought some 150 acres of flat, well-drained land, free of obstruction on all sides between Harmondsworth and Heathrow, for it was becoming increasingly difficult to operate from the busy RAF aerodrome at Northolt, and this would be a good base as it was only three miles from their Hayes works and could be reached by good broad roads. But it was to Northolt that the fuselage and 82-ft span one-piece wing of the long-range monoplane, still known as the 'Postal', were taken by road at first light of dawn one September morning, with cars ahead and cars behind, for re-erection and first flight by Norman Macmillan.

There had been many problems, and Dick Fairey, anxious that the RAF should succeed, had remained somewhat concerned with the monoplane commitment. In earlier moments, when Hollis Williams and his men were bogged down by a design feature, he would anxiously say: 'Wouldn't we be more certain of success if we adopted a well tried biplane lay-out?' Particularly worrying was the Fairey-Reed fixed-pitch propeller which had to provide every ounce of thrust for take-off, but thereafter gave a cruising efficiency of only 57 per cent. Hollis Williams suggested that a French Ratier propeller should be ordered, for this had a one shot variable-pitch system in which it was pumped up for take-off to appropriate pitch and a controlled leak changed it to cruising conditions some five minutes later. There was a short and sharp reply from his big eagle-like chief that his aeroplane would fly with all-British equipment and particularly a Fairey propeller. Such things as a retractable undercarriage, flaps and slots had been discussed with Farnborough pundits and Air Ministry technicians but all sought to prove that neither these nor a variable-pitch propeller were worth the weight and complication. Outstandingly the torsional problems of that great wing had been solved by using an ingenious pyramidal system of four rods in each drag bay, devised and patented (No. 300,393) by Hollis Williams, so that the big wood spars took all the bending and the pyramid all the torsion apart from the lower flanges of the spars which formed the base of the pyramid. Quite independently, Danish born H. J. Stieger of the Beardmore D.O. had reached a similar solution applied to a single spar using pyramids of wire or tube on front and rear face, which he patented (No. 306,220) soon after the Fairey.

The 51-ft span Bristol long-range Type 109 had an enclosed cabin for two, with dual control, and range was estimated at 5,400 miles.

Competition was not lacking against the 'Postal', for though the Bristol Aeroplane Co had unsuccessfully tendered a long-range, 51-ft span, deep-gap biplane, the Type 109, with supercharged Jupiter XIF and pilot's enclosed cockpit, the directors decided in April to build it as a private venture to gain publicity for the engine – though not before the Duchess of Bedford had offered to back construction if she could be taken as a passenger, but as this could not be permitted they had loaned her a special engine for the Fokker F.VII *Spider* in which she and Barnard had attempted the record-breaking flight to India.

Because of its simplicity compared with the Fairey, the Bristol Type 109 was completed in July, temporarily powered with a Jupiter VIII; but there were problems with the petrol system. Only after limiting fuel to gravity tanks could Uwins make the first flight, on 7 September, but found deficiencies of longitudinal and directional stability. Modifications were put in hand, including moving the engine forward, but by now the Fairey was ready, so there was little point in pressing on with the 109 except for air endurance testing of prototype engines.

Explaining the emphasis on long range, Fairey said: 'Of contemporary designs, the maximum range of fighters does not exceed 400 miles; two-seater reconnaissance or light bomber classes will be limited to 600 miles; the longest range of troop carriers or bombers does not exceed 1,200 miles – so a 600-mile band of ocean is all that is at present necessary to protect one country against another, operating from its own territory. Increases in the range of any of these types must of necessity have a great effect on tactical considerations. However, if special need arose, much greater radius of action could be forthcoming, although it must for ever remain the case at any stage of development that as the radius of action becomes greater the effective load becomes smaller, but, incidentally, the greater is the chance of obtaining warning of an enemy's approach.

'Certainly in civil aviation an increase in radius of action is an outstanding necessity as airlines are extended throughout the world. In particular the problem of inter-communications throughout the British Empire appears to demand a machine with reasonable payload and nearly 2,000 miles range.

'Everything depends on engine development. If the future theoretically possible fuel consumption can be obtained in practice, the range of the Fairey "Postal" would be increased from the estimated 5,500 miles to 6,700 miles, and if its range was then reduced to 2,000 miles it could carry a useful load of 5,000 lb. If this same figure could be applied to an engine of 1,800 hp installed in an "ideal" aeroplane, then a total range of 10,000

Initial flight of the Fairey 'Postal' long-range monoplane which had a tapered, thick cantilever wing of 82 ft span employing clever torsional bracing.

miles is possible or alternatively it could carry a useful load of nearly 20,000 lb for a range of 2,000 miles.'

On 30 October the 'Postal' monoplane was cleared to fly. Macmillan – with O. E. Tips and another ground engineer aboard – made some fast taxiing runs and two short straight flights. This was the usual prelude to get the 'feel' before flight testing. No more was done that evening as it meant take-off and landing into the dazzling setting sun. A fortnight's delay followed, during which Macmillan tested a number of other aircraft. It had been intended that Sqn Ldr Noakes would be chief pilot on the record flight, but he was in hospital with a broken neck from attempting to land Bolas's latest fighter, the Pipit, after the leading edge of its tailplane collapsed at low altitude. In August Sqn Ldr A. G. Jones-Williams, MC, had therefore been substituted, with Flt Lieut Major as second pilot. In mid-November they made the initial flight. For the next three weeks or so Macmillan continued with preliminary test evaluation at Northolt. On one occasion

Marcel Lobelle, as chief designer, accompanied Macmillan and Tips. 'It was very nearly their last effort,' said Hollis Williams, 'because Lobelle felt off colour and opened a small hatch provided in the centre-section for astral navigation. This disturbed the airflow and the tail started to buffet wildly, but Macmillan managed to get the machine down safely.' Tips's diary merely comments: 'Fore and aft balance bad. Rudder wobble.'

On 7 December Macmillan and Tips took the machine to 10,700 ft. Next day it was flown to Cranwell, and there the programme of flight trials would continue for several months in the hands of Macmillan and Jones-Williams, with the assistance of Fairey technicians.

It is expressive of the industrial outlook of those days that Macmillan was never insured by his company. 'I only broached the question of my lack of insurance once with Fairey,' he wrote, 'and it was not at that time. He replied that he considered I could insure out of my salary if I wished to do so. I was never insured at any time during my career of civil flying. The insurance companies refused to accept test-flying risks when I first asked them about it.' Later they did so at what was regarded as a very high premium, so most manufacturers seemed to regard their early pilots as 'disposable', and unquestionably any insurances, and often pay, were remarkably mean for the risks undertaken in probing into the unknown of flight developments.

At the end of November death had struck two aviation pioneers: Frank Hedges Butler, the kindly hospitable chairman of Hedges & Butler who was 72 – and Samuel White, the 67-year-old chairman of the Bristol Aeroplane Co Ltd and Bristol Tramways Co Ltd. It was Butler, ballooning in 1901 with his daughter Vera, Charles Rolls, and Stanley Spencer, who suddenly thought of founding the Aero Club, which was registered on landing, and the first additional members were Stanley Spooner, editor of *Flight*, and H. Pollock. Many balloon voyages followed, some of which were of record distance, and in 1908, when Wilbur Wright demonstrated his aeroplane in France, Hedges Butler shared with Griffith Brewer the honour of being the first Englishmen to fly as passengers.

What Samuel White helped to found in partnership with his brother, the late Sir George White, was the British & Colonial Aeroplane Co Ltd in 1909 – precursor of the Bristol Aeroplane Co Ltd – an intensely family concern comprising Sir George's son Stanley White (Sir Stanley), and his nephews, Sydney White-Smith (Lieut-Col RAF), Henry White-Smith (Sir Henry), and Herbert Thomas. After the war Lieut-Col White-Smith turned solely to affairs of the Tramways Co, and a few years later Sir Henry White-Smith left the firm after disagreement; so control had reposed in a triumvirate – Samuel White, Sir Stanley White, and Herbert Thomas – who made the many bold decisions for private-venture aircraft and accepted the great gamble of embarking on engine construction. Spurred by achieving the Bulldog contract, they now authorized Barnwell to design for the next year's Olympia Aero show a four-passenger biplane to be powered with Fedden's projected 315 hp seven-cylinder radial.

The success of another great gamble was annually celebrated by de Havilland Aircraft with a dinner for employees. Five hundred were present on 7 December presided over by Alan Butler, the chairman. Recently the company had taken the unprecedented step of founding a superannuation scheme, which 40 per cent of its employees had joined. Butler referred to it in replying to Tom Clark's eloquent toast of 'The Company,' reminding the guests that Clark, as works superintendent, had seen every D.H. machine through the shops from D.H.1 to the D.H.75, and it was his long service which had prompted the company to safeguard the retirement of its employees. As to current progress, there would be 300 Moths made by the end of the year, against 185 in the previous year, and output was booked to the end of next March; engine output was well in its stride and 131 Gipsys had been produced – yet a year ago the company had been having a difficult time and had only 400 men, but now it gave full-time employment to 1,560, and as the new shops became established they would employ still more. The de Havilland School of Flying also was expanding, and 4,441 flying hours had been completed against 2,266 the previous year. Of world records held by British aircraft, all except one were with de Havilland machines, and recently the Hound gained three such records, yet despite superiority to competitors it unfortunately had not been selected for the RAF. However, the market for larger aircraft was expanding: five D.H.61s had been delivered, and four further Hercules were almost ready for West Australian Airways. What with branches overseas and the formation of the Moth Aircraft Corporation of the USA, the de Havilland Company was clearly playing a big part in overseas trade.

Of the problems of aircraft production few could speak with greater authority than Fred Sigrist. He told an RAeS audience that sixteen distinct classes of work had to be tackled at Hawkers, and about 45 per cent of the labour was skilled, 35 per cent semi-skilled, and 20 per cent unskilled – though special instruction was necessary for all categories before the individual could be usefully employed on aircraft. The eternal problem was that the year consisted of high peaks and depressions. For six or nine months a factory might be very busy, even working overtime, and then run into a slump during which there was every endeavour to retain a nucleus of skilled men even if it meant short time, though this led to the better men seeking work in industries offering more stable employment. Cracking at the Air Ministry's frequent change of mind in specifying requirements not only for prototypes but during production runs, Sigrist said: 'As modest laymen we cannot be expected to understand the considerations of high strategy and tactics which dictate these requirements and any criticism would therefore be ambiguous.' As to production in emergency, he said: 'The Staff view of aerial warfare is that in the first clash of hostilities casualties in machines and personnel will be heavy, and the side which can most rapidly equip will be the predominant party. In other words, the industry must be in a position to assume the rôle of a third line of defence. Frankly, we cannot accept this to-day, and until we are on a basis which ensures continuity of output the position will remain serious. We cannot expand immediately, neither can we find sufficient men to instruct allied trades which might be of assistance. Even if we did manage to produce the machines, the engine question would still remain, and in its way seems less capable of solution.'

December saw King George V so seriously ill after contracting a chill at the Cenotaph on Armistice Day that on the 4th a Regency Council was appointed comprising Queen Mary, the Prince of Wales, the Duke of York, the Archbishop of Canterbury, the Lord Chancellor and the Prime Minister. There was sense of grave anxiety affecting all activities, but to the Nation's relief there were signs of improvement in his general condition at Christmas.

The year ended with worrying news from Kabul, capital of Afghanistan. A force of 1,000 rebels led by an insurgent called Habibullah, captured the forts north-west of the capital, stole their stock of rifles and ammunition, and from a strategic position near the British Legation engaged King Amanullah's attacking troops. With the Legation cut off and in dire peril, the British Commissioner radioed for help. Air Vice-Marshal Sir Geoffrey Salmond, AOC India, decided on what seemed a fantastic course of evacuating British women and children by air, though it involved flying from Peshawar over 10,000-ft mountains often covered in cloud. First he sent unarmed D.H.9As to drop leaflets warning of the air lift, then on the 23rd a Wapiti with radio went as pathfinder through the Khyber Pass to check that machines could land at Cherpur aerodrome, and was followed by a Victoria of No. 70 Squadron which brought out 23 women and children, their baggage being collected by D.H.9As. On Christmas Day a Victoria, a Wapiti and eleven D.H.9As took out another 28, and by New Year's Day they were safely in India. An ominous lull followed.

CHAPTER X

ACHIEVEMENT AND DISASTER
1929

'There must be a beginning and end of any great matter; but the continuing until the end, until it be thoroughly finished, yields the true glory.'

Sir Francis Drake (1587)

1

On 1 January, George Holt Thomas, greatest of British pioneer aviation industrialists, died at Cimiez, in his 60th year, following a further operation for cancer. Distinctive with clipped beard, kindly, yet hard headed, he had been a practical visionary, understanding from the beginning the sky's potential, and with his wealth as owner and general manager of the *Graphic*, *Daily Graphic*, *Bystander* and *Empire Illustrated*, was able to give practical encouragement to early aviation. In the beginning a mutual motoring interest with the Farman brothers led him to engage Louis Paulhan, star of the great 1909 Rheims Meeting, to demonstrate a Farman box-kite at Brooklands and later to compete for the London to Manchester *Daily Mail* £10,000 prize, which Paulhan won after erecting his machine in a field at the end of Colindale Avenue which eventually became Hendon aerodrome – the greatest pre-war aviation sporting location in the world. Thereafter Holt Thomas formed the Aircraft Supply Co to build Farmans under licence at Hendon and, by vigorous policy, forced progress against official apathy and even hostility to British aviation. His business, renamed Aircraft Manufacturing Co Ltd, became the biggest of all aero companies during the Great War, building not only his famous line of D.H. aeroplanes but flying-boats, airships, propellers and aircraft engines in great number. However much he hid it, he felt that those in authority had not adequately recognized his work in developing British military aviation. Few even remembered that it was he who had established Geoffrey de Havilland. Though Holt Thomas had been the most influential aircraft constructor in Britain, he remained personally too remote to become well known, and with the advent of his illness, retired to the country where he bred cattle with record milk output much as he had produced aircraft of

George Holt Thomas, the one-time dictator of the greatest early aircraft empire in Great Britain.

record-breaking quality. In making farewell, C. G. Grey said: 'One of the greatest troubles of British aviation has been that men of wealth and position and influence have not taken an active part. Numbers of prominent men, some of them millionaires, have posed as patrons; but with the exception of the late Sir George White of Bristol, Holt Thomas was the only man of first-class mental calibre, and at the same time of good position, to take personal part in the development of aeroplanes from the beginning. Thomas was a man who in every way was worthy of a title, and he had done the service which deserved it. If he was a little disappointed, he was justified in being much more so.'

Throughout the hard winter, cross-Channel pilots were finding that triple engines, good instruments and adequate wireless, with excellent co-operation from the ground stations, enabled them to fly through weather conditions that in earlier days, with no such aids, were so fraught that successful accomplishment of a flight resulted in the pilot being 'treated rather like a survivor of some forlorn hope who had won through in the face of incredible danger. It was magnificent, but it was not commercial aviation.' To attract more passengers, the French Air Union was scheduling a new service named the *Golden Ray*, described as *de luxe*. Said one cynic: 'Presumably that means the possibility of obtaining a drink in mid-air to counteract feet wrapped in a rug, a cold nose, and a feeling of queeziness, and every prospect of the drink suddenly splashing one's face with violence. Ten years from now it may be permissible to employ the term *de luxe* in speaking of aircraft.'

The New Year had brought the inevitable honours, promotions, and changes of personnel. Air Chief Marshal Sir John Salmond became Principal Air Aide-de-Camp to the King; Air Vice-Marshal Sir Edward Ellington was appointed AOC Air Defence of Great Britain, with Air Vice-Marshal F. R. Scarlett commanding Fighting Area. There was promotion to Air Marshal for Sir John Higgins the AMSR, and to Air Vice-Marshal for Air Commodore H. T. C. Dowding as Director of Training. Sir Frederick Sykes, the great founder of the RFC, was appointed Governor of the Bombay Presidency. Bert Hinkler was elevated to Squadron Leader in the Reserve and gained the distinction of the Gold Medal of the Fédération Aéronautique Internationale for his flight from

Arthur Gouge was the rising star of Short Brothers, but ultimately experienced bitter hostility from Oswald Short.

England to Australia as the most meritorious in the world for 1928. There was also promotion within the aircraft industry, among whom Arthur Gouge, BSc, AFRAeS, was made general manager of Short Bros (Rochester & Bedford) Ltd. His was a brilliant instance of a manual worker, by sheer ability, hard work and tremendous energy, acquiring a degree through night-school study, and from tin-bashing and experimental metal work becoming chief draughtsman and then a director. Nor was his task easy, for grey-haired Oswald Short was so single purposed that he could be impossibly difficult, nor did he really like Gouge. In charge of structural engineering was C. P. T. Lipscomb; Nobby Clarke had become chief draughtsman, and D. H. Laver ran hydrodynamic research. Arthur Gouge neatly summarized the task of designing an aeroplane: 'Many think the designer sits in a sort of studio and sketches pictures of aircraft, then gives them to other people to make. The facts are very different. An aeroplane is the result of an immense amount of team work. The operational requirements are supplied by the Air Ministry from its various technical staffs, and are examined by the chief designer and passed to specialized assistants to study. Some points cannot be met, so suggested variations are put before the Air Staff for approval. Finally the general outline is agreed. Then the real work begins. A full-sized mock-up is made in wood and cardboard so that all may see what the aircraft will look like, and many discussions and conferences are held around it before every instrument and gadget is correctly placed and pilot's and bomb-aimer's view is approved. Once RAF and Ministry experts are satisfied that this mock-up represents the best that can be done, the various detail sections of the aircraft are handled by assistant designers who co-ordinate the efforts of the various specialists, section leaders, and draughtsmen.'

Short's D.O. was very busy. Success of the twin-engined Singapore and three-engined Calcutta had resulted in a contract for a strut-braced high-wing three-engined floatplane for flight comparison with the Calcutta. Nevertheless, production orders for a military version of the Calcutta were delayed pending completion and flight trials of alternative three-engined flying-boats tendered by Saunders with their A.7 sesquiplane having Jupiters on the top wing, Supermarine with the Southampton Mk.X of similar geometry but the Jupiters in mid-gap, and the Blackburn Sydney as a breakaway in the form of a Rolls-Royce powered 100-ft span monoplane with strut-braced R.A.F.30 wing, $2\frac{1}{2}$ ft deep, mounted on a turret cabane to give wave clearance for the propellers. Forcefully Oswald Short argued the merits of an alternative more powerful four-engined machine and successfully achieved a contract to build one prototype called Singapore II which was virtually a Calcutta of 3-ft less span, with improved hull having pronounced keel and deeper forebody, and the engines in tandem pairs daringly mounted with a single strut at front and rear spar thus offering lower drag than a conventional twin of half the power.

Not content with this, Oswald formulated a still bigger biplane of 120-ft span with three sets of tandem engines mounted similarly to those of the Singapore II but with power totalling 5,000 hp. He told me: 'Having prepared the outline drawings I took them to the DTD, who at that time was Air Commodore Chamier. I found him a man of vision. He was enthusiastic about the prospect. All looked well, but weeks passed (during which Chamier resigned) and I received no indication that I would get an order. Finally I went to Sir Hugh Trenchard. He spoke about the money wasted in attempts to build big aircraft in metal. I replied that Short Bros

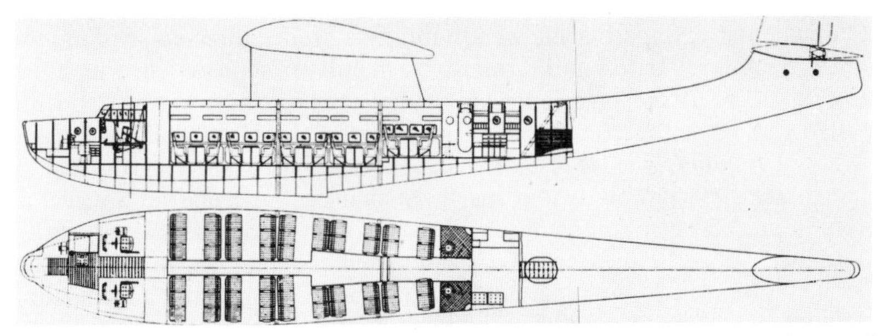

Every manufacturer developed many a design in detail only to be abandoned, as instanced by the six-engined Blackburn monoplane flying-boat proposed by Major Rennie, and which foreshadowed America's big Sikorsky boats.

had produced no failures. After half-an-hour's conversation I felt I was not making all the points for big metal flying-boats that I should, and asked if I could put my ideas in the form of a treatise, to which he agreed – but as I was about to leave he added: "Sit down, I'd like to talk to you." From so gruff a man as Sir Hugh was known to be, that was encouraging. I stayed an hour. In due course it was agreed to build the 33-ton Sarafand at a cost of £60,000 exclusive of engines and armament.' But it would be several years before completion, particularly as Oswald had argued the merits of big boats so conclusively that Supermarine and Blackburn were invited to tender for high-wing monoplane flying-boats of similar weight, equipped in civil form with 40 seats, though eventually their designs were not accepted.

Construction of the Blackburn Sydney monoplane with Iris-type hull had already begun, for it was ordered twelve months earlier, so the company decided to build alongside it a civil version, the Nile, to carry three crew and fourteen passengers, for the Alexandria to Cape route which Alan Cobham hoped to operate as a result of his survey with the Singapore, and for that purpose the Blackburn-owned North Sea Aerial & General Transport Co had combined with him as Cobham-Blackburn Airlines Ltd.

Supermarine's three-engined Air Yacht was an unexpected breakaway from their traditional beautifully curving hulls, but Mitchell was influenced by the Dornier Wal which the RAF at nearby Calshot had been trying. (*Vickers*)

Coincidentally, Supermarine were building a cruder three-engined monoplane flying-boat, based on Dornier-type cantilever sponsons to which the 92-ft untapered wing, on a cabane high above the fuselage, was braced with V struts. Built as an Air Yacht to the order of the Hon A. E. Guinness, owner of *Fantome*, the largest and most spectacular barque-rigged yacht in British registry, it was to join his three-engined Supermarine Solent for commuting between Southampton Water and the Mediterranean where he kept his yacht. However, the Supermarine rival of the Nile, the Southampton Mk X, was less bold in conception, despite the introduction of stainless steel as the hull skin, and had a long top overhang braced with struts from the stumpy lower wing at the outer engine mounting structures. Like the Air Yacht, its hull was derived from the all-metal Saunders A.14

Described as a sesquiplane flying-boat, the 79-ft span Supermarine Southampton X used the Saunders A-14 hull design rebuilt in stainless steel. (*Supermarine Aviation Works*)

flat-sided dural hull which had longitudinal corrugations instead of stringers, and was built as an experimental Southampton. The competing Saunders A.7 Severn not surprisingly had a somewhat similar hull, but scored with an outstandingly clean powerplant assembly. It was a design which had promising features, with propeller clearance higher than any other flying-boat, though it risked considerable change of trim between engines on and off. While construction of this 88-ft span flying-boat was being initiated, tests were in hand with the F.XI-powered Saunders A.10 single-seat fighter, based on specification F.20/27, which had something of

Henry Knowler's neat conception of the Saro A.10 fighter is little known but was distinctive with very small lower wing, biconvex aerofoil sections, and thin 'hollow ground' centre-section for upward view.

There was criticism that the moderately thick centre-section of the original Westland Wizard handicapped view. The all-metal version therefore had a thin centre section and conventional cabane struts – but this spoiled the performance.

the Bulldog's appearance but with pointed nose and chin-type radiator lacking the shapely grace of its rivals – the almost completed Hawker Hornet, the Fairey Firefly II, and the re-build of the Westland Wizard with dural-framed wings and thin centre-section to give better forward view.

2

Relative merits of monoplane and biplane were a continuing source of argument. Biplanes gave the ideal light structure; the strut-braced monoplane or sesquiplane was at best a compromise; the cantilever monoplane was clearly supreme aerodynamically but at cost of payload and increased landing speed as well as structural difficulty. A lecture to the RAeS by Professor Melville Jones, MA, FRS, on *The Streamlined Aeroplane* therefore attracted much interest, for his theme was to compare current resistances of an actual aeroplane with those theoretically possible, and through the ARC he had ensured experimental substance for his contentions. 'It is not suggested that it is easy to design a streamlined aeroplane which will be a practical machine,' said Melville Jones, 'but the immense saving in power, and therefore in fuel consumption, which would follow, causes me to believe that design will evolve steadily in this direction and that the ultimate aeroplane will be as well streamlined on the whole of its external surfaces as say the bottom of a racing yacht or the external shape of an albatross.' But he offered more than theory, for it could all be expressed in figures. As Major Green said: 'It meant more to a designer to be able to say that his aeroplane represented a definite advance towards a definite ideal, than to say merely that it was a little better than its predecessor because he did not know how much further he might have gone.' A valuable yardstick had been established, and its ready acceptance was expressed by many designers and scientists, such as Geoffrey Hill who forecast: 'When the present drag of our military and commercial aeroplanes has been halved, it will be time to set the standard more securely on its basis with the aid of aerodynamic knowledge which will then be available.'

A very different view to the Professor's was given a fortnight later in a lecture on *Monoplane or Biplane* by W. S. Farren, MA. This able, fast talking, former member of the RAE was now a lecturer at Cambridge and London Universities and argued from detailed calculations beyond comprehension of all except mathematical purists, that the essential difference between biplane and monoplane was the parasitic resistance of the wing structure – which meant that turning biplane into monoplane reduced payload by 20 per cent, and 'if anybody wants to bring the cruising speed up from 100 to 150 mph they would have to be very prosperous to pay for doing so.' Clearly he was in favour of biplanes, nor did he believe they would ever be more than 40,000 lb weight. Geoffrey Hill, who always smiled charmingly while disagreeing strongly, said Farren had merely instanced a bad monoplane and should blame himself for its inferior results and not monoplanes as a type. Major Buchanan of DTD's department tactfully suggested the choice was as much fancy as fact and had become almost a question of national characteristics, but he certainly considered British designers were clever, intelligent, very agreeable, and extraordinarily conservative. Tough-looking but kindly Charles Walker, with feet firmly on the ground, maintained that the question of monoplane or biplane depended purely on practical consideration of its intended job, but monoplanes scored in simplicity of design, though in practice had not the efficiency assumed for them; the same stalling speed as a biplane meant lower top speed; the de Havilland shoulder-wing monoplane with body of known characteristics was slower than the corresponding biplane, and the early Junkers monoplane of the same size as the D.H.50 had worse all-round performance. Roy Chadwick thought Farren's distortion of the aspect ratio of a monoplane was not fair, and it was working to British airworthiness requirements that made monoplanes at a disadvantage, for it demanded such things as too large a tail area. Handley Page roundly declared that monoplane or biplane was a matter of choice, but with his flair for instant publicity agreed that monoplanes implied heavy loading and high landing speed unless slots were used; nevertheless he was sure passengers felt safer with a biplane's rigid wings rather than with the more flexible wings of a monoplane such as the Junkers where the wing-tips flexed up and down several feet in bumps. Mr Trost of the Junkers company riposted that of 68 world's records recognized by the FAI, 38 were held by monoplanes, and that maintenance and other considerations might well outweigh the biplane. Duncanson of Gloster argued that if materials were put in the right place a monoplane could be stiffer than a biplane, particularly if the wing had a stationary centre of pressure; Major A. R. Low remarked that a biplane was really a cantilever about six feet deep and he did not see how designers were going to get over that; H. J. Stieger, late of Beardmores, and now endeavouring to exploit his monospar wing, shrewdly said that Farren had started with a conclusion and then written his paper to prove it, for he had taken an inefficient Argosy as basis and put a monoplane wing on it, resulting in wrong wing weights, for this could be a mere 8 per cent on big machines and 12 per cent on small ones. Finally the pilots spoke, Sqn Ldr England saying they favoured biplanes because of easier handling, and Bramson, with more enlightened outlook, emphasized that if flying from good surfaces there was no harm in high landing speeds.

Sir Sefton Brancker, who alone was in position to survey the whole field of British commercial aircraft development, said he certainly preferred monoplanes. 'We must come to higher speeds. America is travelling faster than we. My department is issuing a specification for a mail carrying machine cruising at 130 mph. One thing that is clear from Farren's paper is that in this country we do not know anything about monoplanes. Certainly I am going to have another dig at the limitations imposed by our airworthiness certificates on the terminal nose dive question, for the Fokker has an unblemished reputation in this respect. As an experiment the Air Ministry is going to have two machines built to identical specification, one a monoplane and the other a biplane.' That contract was secured by Blackburns who tentatively formulated a number of monoplanes, strut braced in similar manner to the Westland Witch and the numerous American high-wingers. The initial proposal, a 38-seater of 126-ft span, with four Bristol Jupiter IX geared radials, was too large and expensive for Brancker's project, so Bumpus designed three progressively smaller machines, and specification 6/29 was finally written around the C.A.15C with two Armstrong Siddeley Jaguar IVCs, interchangeably as an 86-ft monoplane or 64-ft span biplane with identical fuselage.

For those who flew commercial aircraft, 1 February saw the inauguration of the Company of Air Pilots and Navigators at the hostelry of Rule's of Maiden Lane, London, where some fifty met under the chairmanship of Sqn Ldr E. L. Johnston, the navigational expert of the Director of Airship Development. Stressing the need for a professional body of unquestionable integrity to safeguard the interests of professional aviators, he suggested that instead of a president, chairman, and committee, there should be a Court consisting of master, deputy master, and six wardens. Sir Sefton, considering this a far-seeing and ambitious scheme, advised the pilots to take as model the Company of Master Mariners. To this there was general agreement, except for cautious objection to a clause giving the Company power to co-operate with trade protection societies as this seemed undignified. A drafting committee was elected, with Sir Sefton representing civil aviation; Norman Macmillan for test pilots; A. S. Wilcockson for air transport pilots; Lieut-Col G. L. P. Henderson for instructors; V. H. Baker for flying club pilots; W. L. Hope for independent transport pilots; Major H. G. Brackley for the administrative section of air transport; Sqn Ldr E. L. Johnston for air navigators; and Lawrence Wingfield as legal adviser. C. G. Grey commented: 'When in due course the Company of Air Pilots and Navigators has qualified for the coveted prefix "Honourable" and has acquired status equal to other honourable companies of this nation, the name of Sqn Ldr Johnston must always stand as its founder.'

Concurrently a White Paper was published defining 'An agreement being made by the Air Ministry with National Flying Services Ltd, providing Government assistance during 10 years to that company in the form of a grant payable in respect of each club member who qualifies for the issue or renewal of a pilot's licence. The rate for the first three years will be £10 for each member so qualifying, subject to a maximum of £15,000 a year. The corresponding rate for the next seven years will be £5, the aggregate being limited to £7,500.' After all the fears, it seemed a damp squib.

There was news also that further evacuation of Kabul had begun on 29 January. The dangers were evident from a reference in the *Daily Telegraph* to the forced landing of a Victoria, and the aid given to its pilots, Flt Lieut Chapman and Flg Off Davis, by the Pretender to the throne, Ali Ahmad Jan: 'The landing was a feat which will be recorded in the annals of the RAF, for the engines failed over mountainous country, and there was only a small plateau in which to land. Had they overshot they would have plunged down an abyss of 2,000 ft to certain destruction.' Russian propaganda was quick to blame England for the Afghan revolt, and instigated worldwide rumour that Colonel Lawrence had been sent to Peshawar to stir up trouble in Afghanistan, and was responsible for the downfall of King Amanullah. Questioned in the House, the Secretary of State for Foreign Affairs merely said that A/C Shaw had been posted to No.20 Squadron in May 1928 and was carrying out the ordinary duties of his rank and had not been granted leave while with the Squadron. Meanwhile Habibullah had occupied Kabul and declared himself Ruler of Afghanistan. Complete evacuation of the Legation followed, and by 25 February, 586 people of thirteen nationalities had been flown from Kabul – the first great peace-time air lift in history.

3

The popular and technical Press, the RAeS, RAeC, Air League and SBAC, with the British public in general, united to acclaim Lady Bailey when she returned on 16 January from her excursion by Moth to Africa and back. Currently it was described as 'Not only by far the most remarkable feat ever achieved by a woman, but the greatest solo flight ever made by any pilot of either sex.' That was somewhat unfair to Lieut R. R. Bentley who had just completed his third flight between England and the Cape with his

Moth G-EBSO, and had escorted both Lady Bailey and Lady Heath over the difficult stages. But Lady Bailey, so modest, so vague, so charming, was surprised that anyone should make a fuss about her journey. Croydon was snow covered when she arrived. 'Those who welcomed her, belonging to a generation surfeited with miraculous happenings, were enthusiastic over her pluck, proud of her achievement and sincere in their admiration that a woman could accomplish the longest solo flight ever made – but they were not astonished,' said C. G. Grey. She felt it almost embarrassing when a lunch was given in her honour next day at the Savoy, with Brig-Gen Lord Thomson, chairman of the RAeC, in the chair. Proposing her health, he smilingly emphasized that there was no prejudice against the fair sex in aviation: women such as she, Winifred Spooner, Constance Leathart and Sicele O'Brien had entered aviation in a spirit and manner that was altogether admirable. He was unaware of a typist named Amy Johnson who was learning to fly at Stag Lane, nor did he mention Lady Heath who, these days, was somewhat cruelly referred to as Lady Hell-of-a-Din through her proclivities at self-advertisement, but whose courage and determination were equally indisputable. Of her own adventures, Lady Bailey had little to say, for she had no mission to convert the world to airmindedness or Imperialism or anything else, but merely remarked that she set forth to join her husband, and just sat and listened to the engine. Certainly she was sensitive to its change of tune, could feel the play of air in thermals, for I remember her landing at Yeovil one day and saying quietly: 'There's no lift in the air. It's rather dull. It seems a bit of a struggle to get to Cornwall.'

Despite the growing equality of women with men, and their attack on men's strongholds, it was only twenty-five years earlier that man had first managed to control a powered aeroplane and currently the Royal Aero Club had been investigating who was the first British subject to make the first powered flight in the British Isles. On the investigating committee were Lord Gorell as chairman, Capt de Havilland, Lieut-Col Lockwood Marsh, and Harold Perrin the secretary. They decided as criterion that the aeroplane must have been maintained in the air by its own power on a level or upward path 'for a distance beyond that over which gravity and air resistance would sustain it'. What they meant by that has never been understood, though presumably they knew. Samuel Cody and Hiram Maxim were eliminated as aliens at the time of their first flights, and in any case the latter's were contentiously dismissed as not true flight because a rail limited altitude to only a few inches. For Horatio Phillips and his Venetian-slat multi-plane his son could produce no corroborated evidence. Of A. V. Roe, despite his letter to *Flight* on 26 June, 1909, stating he had made dozens of short flights, and a supporting letter from Herbert Morris, that very upright director of the Palmer Tyre Ltd who saw the machine at Brooklands early in 1908 drop with enough force to break a number of straining wires, the claim was dismissed on the wild assumption that because it 'dropped' it was not landed and therefore had not actually flown! Yet in the summer of 1928 the RAeC had honoured Roe with a dinner to celebrate him as the first to fly in England. D.H. appears to have been in the minority in supporting Roe, and in later years wrote to the former's son, Geoffrey, confirming that he considered him to have the prior claim.

With Roe ruled out, the decision was that the first free flight was by Moore-Brabazon at Leysdown during the week-end 30 April/2 May, 1909. Discounting this judgment, C. G. Grey commented: 'Nevertheless in a general way the position seems to be that Mr S. F. Cody made the first flight in the British Isles.'

Alliott Roe could well afford to shelter behind his quiet smile, for a few weeks later he received the first knighthood given to a pre-war aeronautical pioneer. Those little 'hops' at Brooklands in 1908 and the straights in 1909 on Lea marshes and at Wembley had been the foundation of a great business, initially financed and owned by his brother Humphrey, but managed since mid-war days by John Lord. Yet it was Roe's initiative, his courage, his patient study and interpretation which had made it possible – and now he was looking to new horizons where flying-boats might play a great part. Yet it was strange there was no knighthood for that even earlier pioneer Oswald Short, who had maintained his lead by initiating the duralumin monocoque construction now beginning to be adopted by other aircraft builders at home and abroad.

Following the consolidation of Armstrong Whitworth by purchasing Avro, Vickers with acquisition of Supermarine, and Saunders reinforced with finance from Sir Alliott Roe, both de Havilland Aircraft Co and Fairey Aviation Ltd increased their capital. De Havillands privately limited their issue to shareholders for subscription in £1 shares in an authorized capital of £400,000 – still with Alan Butler as chairman, Geoffrey de Havilland as technical director, Charles Walker as chief engineer, Frank Hearle as general manager, and Thomas P. Mills as solicitor. Agreements were renewed for Francis St Barbe as business manager and Arthur Hagg as assistant designer. A financial commentator prophesied: 'No doubt the directors of the de Havilland Co, some years hence when there is a financial boom in aviation as now in the States, will float the firm on the public for a capital of some millions. When they do, the shares will probably be just as well worth buying as at present. That is, at any rate, so long as the present directors are still handling the firm.'

By contrast Richard Fairey decided that the Fairey Aviation Co Ltd would be floated with nominal capital of £500,000 for public subscription in 10s shares 'with the object of entering into seven agreements with Fairey Aircraft Holdings Co Ltd, C. R. Fairey, F. G. T. Dawson, M. E. A. Wright, A. G. Hazell, T. M. Barlow and W. Broadbent respectively to carry on the business of manufacturers of seaplanes, aeroplanes, aerial conveyances and aircraft of all kinds and components thereof including engines.' The money was fully subscribed before the formal prospectus appeared in the Press, where it was disclosed that £300,000 of the capital was held by the Law Debenture Corporation Ltd as a first mortgage debenture on the fixed assets. 'Charles Richard Fairey, the well known aircraft designer and founder of the business, has agreed to act as managing director for a period of not less than five years.' Profits for the year ending 31 March, 1928, were some £120,000 compared with a mere £19,701 from the equally busy but more modestly self-supporting de Havilland company. Fairey had orders in hand for £430,000 and contracts for a further £250,000 were being negotiated, some 40 per cent of the total being spares.

Engine companies were doing better than airframe manufacturers. Rolls-Royce, who had recently appointed Arthur F. Sidgreaves as managing director in place of the courteous and amiable but untechnical Basil Johnson, who resigned owing to ill health, had made £156,878 net profit in the past year on cars and aero-engines, and D. Napier & Son Ltd, depending solely on Lion engines, had a profit of £167,438. As Sir Harry Brittain emphasized at the Napier annual general meeting, it was H. T. Vane, the managing director, who had consolidated the business, for Montague Napier, chairman and joint managing director, was compelled by ill health to live in the South of France where he devoted much time and ability to the firm's problems 'and in large measure our success is owed to his foresight and genius.' Capt George Wilkinson was still chief designer, but George Pate as technical director had long been the key man, and was described as 'one of the leading authorities on the higher techniques of aero-engine design and construction. For he is a scientist in the true sense of the word and an extremely good production engineer.' But he was getting at loggerheads with the management because no new design of Napier inspiration was available for future manufacture, and instead the company had acquired rights of Halford's larger engines, though this did not affect de Havilland's ownership of the Gipsy.

Compared with the giants, a very small company, Comper Aircraft Co Ltd, was registered on 14 March, with capital of £15,800. Nick Comper had resigned from the RAF to devote himself to aircraft construction, and

Nick Comper (*second right*), his supporters and staff, and their first product, the Cherub-powered Swift single-seat racer.

Social occasion for Westland, with line-up of current types, including the latest geared-Jupiter Wapiti, to greet the Australian High Commissioner.

with him were associated Flt Lieut Bernard Allen the chief instructor at the Liverpool & District Aero Club, and Mr Dawson of Hooton who was financing the business.

Another newcomer was Cirrus Engines Ltd, formed as an offshoot of A.D.C. Aircraft Ltd, with A. H. Caple as technical director. He had designed a larger four-cylinder than the original Cirrus, known as the Hermes, rated at 105 bhp at its normal 1,900 rpm. One was being tested in a Widgeon that was demonstrated to guests at Yeovil on the occasion of acceptance by Lady Ryrie, wife of the High Commissioner for Australia, of the first of the Wapitis recently ordered by her Government. As Bill Gibson, now works manager, explained, Westland had converted to metal construction without dismissing a single woodworker. 'Our system of using square-section duralumin tubes held together by fish plates and

Westland in full production with fuselages of simple construction which were fitted with metal-framed wings sub-contracted to Gloster's subsidiary, The Steel Wing Co Ltd.

braced with wire is practically a metalization of wooden aircraft construction, so the carpenters and joiners have been able to turn to it without special training as metal workers. Few special tools have been used except simple forms of drilling and riveting machinery, and everything is built to jigs, the whole going together like a Meccano toy. We have also installed baths for full anodic treatment of dural parts and cadmium plating for steel, and are about to build large stove-enamelling baths so that a complete fuselage can be immersed for final anti-corrosion protection.' The presence on this occasion of Hugh Burroughes and David Longden of Glosters was explained by the all-steel wings they were producing for the Wapiti; Sir Stanley White also participated as supplier of the geared Bristol Jupiter which the Australians had selected.

'The baptismal ceremony being over, Flt Lieut Louis Paget proceeded to show what the Wapiti could do', reported *Flight*: 'Even allowing for flying light it certainly put up a fine performance. To anybody accustomed to ungeared engines the huge geared airscrew gave the impression that it was not doing its job, for it seemed to wander round so slowly. But the way the machine tore across the aerodrome and then stood on its tail in a vertical climb like a single-seat fighter was altogether uncanny. The way the pilot flew it about on its wing-tips quite explained the popularity of the Wapiti in the RAF as one of the nicest machines to fly. Evidently the controls are beautifully balanced.' So it seemed – but I thought them much too heavy, though that very feature made the machine safe.

The new Westland three-engined 'Limousine' monoplane with four-seat cabin was also shown for the first time, painted in the drab olive that Keep thought so workmanlike, though relieved by silver wings and deep blue top and nose. Sir Sefton Brancker immediately wanted to be flown in it to the Continent, but luckily it had only a temporary C of A. A modified Wapiti for Army co-operation with split-axle undercarriage particularly interested RAF, SAAF, and RAAF visitors. 'Of course we were not allowed to see the revised Wizard, in which the wings are being slightly raised and carried on splayed-out centre-section struts instead of on the stumpy pillars which it had at the RAF Display,' said C. G. Grey with tongue in cheek. 'It was

The all-wood Westland four-seat Limousine shoulder-wing monoplane, later known as the IV, had three Cirrus engines and similar wing structure to the Witch.

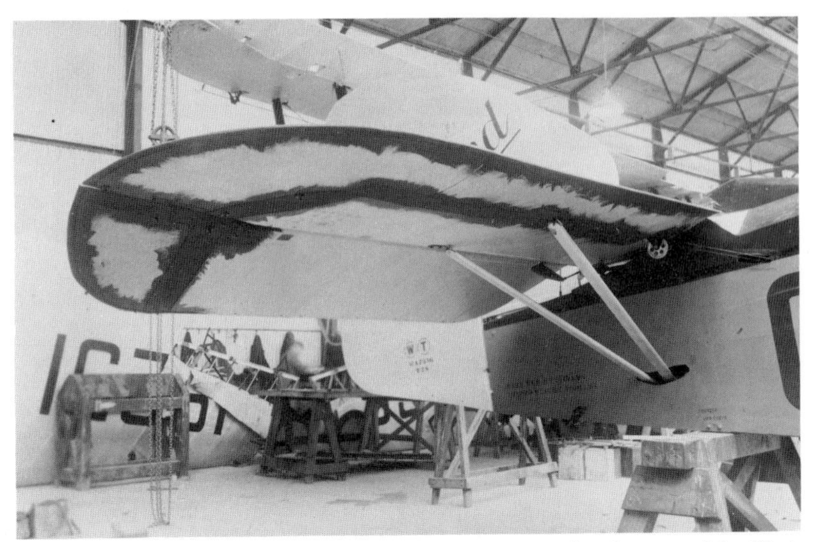

Despite wind-tunnel assurance of longitudinal stability, immediate increase of the Westland IV's tailplane area was necessary in the time-honoured way using ply.

in the particularly hush-hush experimental department along with the low-wing attack monoplane with the latest hotted up Rolls-Royce Falcon, about which everybody in the trade is talking so much though few have seen it. But the existence of these unseen machines shows how far ahead the chiefs of the firm are thinking.' Certainly there had been a scheme for the C.O.W.-gun fighter with an 'eversharp' nose containing a Rolls-Royce, but Air Ministry policy insisted on the even more secret, but greater drag, Bristol Mercury.

An example of how an aeroplane could be designed in a hurry was the metal-frame Blackburn Bluebird IV, for it was built in the experimental shop solely from sketches drawn by Norman Martin, and only after it flew on 23 February were they turned into production drawings. The 'new look' initiated by B. A. Duncan with the Turcock not only marked the Nautilus and Beagle of G. E. Petty as chief designer, but were apparent in the new Bluebird as the result of his overall responsibility. Tall and thin, with glittering glasses, small moustache and serious expression, Petty was regarded as unapproachable; but though the draughtsmen seemed nervous of him, he was one of the few designers who, like Geoffrey de Havilland, drew an entire project in substantial detail, locking himself away for a month at a time, and then handed it to his assistants to turn into working drawings. Major Bumpus, though chief engineer, was never seen in the D.O., for his primary concern was major policy decisions and technical contacts, leaving Robert Blackburn free to conduct the business in the ground floor offices below the D.O. The design executives were in offices forming a division separating the project department from the main D.O. which had a hull and float section controlled by the ever immaculate

The first fully metal-structured light aeroplane was the Blackburn Bluebird IV prototype which was a completely new design and slightly bigger than the original Bluebird. (*Flight*)

Rennie, his empty right sleeve tucked neatly in the pocket of his navy reefer jacket.

The new Bluebird had a Gipsy I instead of Genet, and construction was rushed largely because Blackburn had promised his friend Sqn Ldr L. H. Slatter, of 1927 Schneider fame, that he could fly the prototype to South Africa during his imminent leave. After only a week of testing by 'Dasher' Blake at Brough, Slatter flew it to Stag Lane for engine tuning on 3 March; five days later he took-off on the first of the stages to Durban – which he eventually reached on 15 April, but the Bluebird had to be freighted back because his leave expired.

Currently Bumpus and Petty were considering an enlarged version of the Ripon for Japan as the result of a visit by the war-time Sopwith designer Herbert Smith, who had retired to England and subsequent anonymity after completing his contract with Mitsubishi for whom he had designed a developing sequence of torpedo aircraft based on his original Sopwith Cuckoo. Because the Imperial Japanese Navy had decided to replace these machines with metal-structured aircraft of latest design, Mitsubishi, as the leading Japanese manufacturer, asked Smith to negotiate with Blackburn and Handley Page for suitable aircraft powered with a 600 hp Hispano-Suiza, and to negotiate full licence rights for eventual production. Through this, H.P. also re-established contact with Lachmann, who since 1926 had been technical design adviser to the Ishikawajima Aircraft Works in Tokyo. Though no order ensued for a Handley Page torpedo aircraft it at least resulted in Lachmann demonstrating a slotted Moth and Avro 504 to the Japanese Air Force, and soon after this, H.P. engaged him as technician in charge of aerodynamics and stressing at Cricklewood.

Meanwhile critical interest centred on the Air Estimates, and once again they showed a decrease of £50,000 compared with the previous year, for Parliament was asked to vote £16,200,000 net. MPs thought the method of presentation confusing except to an accountant, but at least it was clear that seven squadrons would be added to the seventy-five in existence.

However: 'Permanent officers are to be provided only in such numbers as will suffice to fill those posts which essentially need men who enter the RAF for a life career. This body of permanent officers will form the nucleus in peace and for its expansion in war, and will provide the specialists in engineering, wireless telegraphy, armament etc on whom the technical work of the Service depends.' The remainder would be Short Service entry as the only economic method of maintaining a supply of young pilots, but to meet criticism of a stranded career they would be assisted by the RAF educational organization to prepare for subsequent civil employment.

Despite the overall cut, the industry had an increased vote of £615,000 for technical and war-like equipment, and it was promised that nineteen squadrons, including two in India, and several training units would be re-equipped with aircraft of latest design – by which it was meant machines already obsolete in conception though adequate earlier in the decade. Research and experiment benefited with an increase of £102,500, of which £40,000 was for a new variable density wind-tunnel. Civil aviation gained an increase of £450,000, of which £349,000 would go to Imperial Airways and £16,000 to subsidize thirteen light aeroplane clubs, with £3,000 added for National Flying Services, who were expected to establish twenty aerodromes and eighty landing grounds within three years.

Opening the Debate on 7 March, Sir Samuel Hoare mentioned, without revealing the manufacturers' names, construction of the Blackburn monoplane/biplane, and the Short Sarafand as a 'larger flying-boat than any we have yet built.' He stressed that four years ago the Air Ministry were ordering only one metal machine to every nineteen of wood, but to-day it was seven to one of wood. Of his continuing concern over accidents, he said: 'Further effort is being made to apply the data accumulated on the risks of flying, whether due to structural or human element. The brilliant work of experimental pilots at Farnborough and Martlesham and the special efforts of the ARC are increasing our knowledge of wing flutter and the stresses to which high speed machines are subjected, and concentrated attention is being given to possible development of the slotted wing.' As to Air Defence, he said the most urgent problem was protection of British shores, but unfortunately the extension programme was not yet complete; we had only thirty-one squadrons out of the fifty-two proposed, but their standard was becoming more and more efficient compared with any other squadrons in the world, and the past few years had shown that the Auxiliary and Special Reserve more than justified the hope placed in them.

Caustically C. G. Grey described the ensuing discussion as 'hours of sheer nonsense punctuated by moments of common sense' – though Mr Wedgwood Benn struck a commendatory opening note, saying: 'I believe this is the last year in which the Air Force will have the services of Sir Hugh Trenchard. Remembering his work during the war, I would like to say that he is one of the greatest pioneers the RAF has ever known, and whenever people later speak of the origins of the Air Force they will say with very great truth of Sir Hugh that "There were giants in those days".' It was a sentiment echoed by other speakers. James Hudson made the telling point that Trenchard had clearly shown that aeroplanes were not weapons of defence but offence – and he added: 'Our discussion is on building up a weapon which can only be used to press the enemy civilian population as hard on their side as they will press our civilian population, making the next war infinitely more ghastly than the last.'

Though several members groped for methods of disarmament, there was little real criticism of the Estimates, and interest centred on concern with civil aviation, for Britain had only nineteen aircraft on regular services, while the USA had 250 and Germany 240. Although seventy-five British municipalities were considering the possibility of providing aerodromes for their towns and cities, all except Glasgow, Liverpool, and Birmingham were waiting to see what N.F.S. would do with their proposal to develop internal charter services and taxi flying. Lieut-Cdr Kenworthy wanted the Air Minister to look further ahead. 'The great route of the future,' he said, 'will be from New York to Galway, across Ireland to Wales, and by way of the Humber to Hamburg where it would join the great European and Asiatic systems onward to Peking.' He reminded members that the Air Ministry in 1926 said that regular air services from Egypt to Karachi with three-engined machines would start on 1 January, 1927 – it was now 1929 and we were no further forward. The Under-Secretary of State for Air quickly pointed out that the new service to India would start on 1 April, with stages by aeroplane, flying-boat, and train. On deeper note, Lieut-Cdr Burney stressed that the real security of the Empire rested upon the economic capacity of England, so the policy should be to create great airlines to all parts of the Empire as a foundation on which to build the Empire into an economic entity. 'A Europe divided is a condition which detracts from the development of our Empire, but the development of civil aviation would bring about a condition of affairs which would be a prelude to a United States of Europe.'

For readers of the popular Press, reportage of the Debate had little interest, for the masses wanted glamour, adventure, thrills, and football results, but all displayed enthusiasm at the successful attempt by Major Segrave in breaking the world land-speed record at Daytona Beach, Florida, with his beautiful *Golden Arrow* at an average of 231.36 mph. In many ways it was a by-product of aircraft development, for not only had this famous racing driver been a war-time RAF pilot, but the car was designed by Capt J. S. Irving, chief engineer of Humfreys-Sandberg of which The Master of Sempill was managing director, the engine was a Napier Lion developed for the 1927 Schneider, and the radiators, built by Glosters, were based on the wing radiators of their racing aircraft. 'Experts' had previously expressed belief that pneumatic tyres could not stand the strain, but tyres and wheels had been successfully developed by Dunlop, whose aircraft duralumin disc wheels everywhere superseded the old-fashioned wire-spoked wheels with which the Palmer Tyre Co had become paramount supplier in the Great War. For Segrave a knighthood followed.

On more intimate note, every man, woman and child was sympathetically aware of the illness of the King, and relieved that he was making progress in quiet convalescence at Bognor on the south coast.

4

Two tough officers of the RAAF, Flt Lieut James Moir and Flg Off Harold Owen, with Frank Hurley their cinematographer, had been attempting to fly from Australia to England, but crashed their overloaded Ryan at Athens when taking-off on the last stage, and Hurley was slightly injured. Capt Acland of Vickers heard they were seeking finance for a return flight with a British aeroplane, and found it suited his publicity plans to loan the Vellore I chosen by the Air Ministry as a possible standby if the long-range Fairey proved unsuitable. With its 76-ft span, relatively good aspect ratio, and wing loading of only 7 lb/sq ft, it had fantastically short take-off, and even with 513 gallons of fuel, spares and stores, giving 300 lb overload, it left the ground in a mere 100 yards when Moir and Owen on 18 March departed on the first stage of their return flight to Australia. Flying the well-tried route via Marseilles, Rome, Malta and Benghazi, the Jaguar engine began to misfire and in a forced landing at Mersa Matruh one wing brushed a house on the approach and they dropped hard, wrecking the undercarriage. New components were rushed out, and at last on 28 April they resumed flight across the Middle East, and over India and South-East Asia, meeting the worst of weather, but on crossing the Timor Sea the engine again began to fail 160 miles from Darwin. They staggered on, and managed to reach the lighthouse at Cape Don, crash landing among trees on a small patch of land with shark-filled seas on one side and crocodile-infested lagoons on the other, but eventually were rescued. Though such flights were uninsurable, and the aeroplane a loss, the cost was easily absorbed in overheads of a great firm like Vickers, as well as replacement

With a Jaguar VI replacing its original Jupiter IX, and using extra tanks in the cabin, the Vickers Vellore I attempted a record flight to Australia. (*Flight*)

To redesign the Vickers Vellore for twin Jupiters was no great task, but the mid-gap fuselage was made slab-sided, then led to a new version, the Vellox, with deep fuselage and cabin.

with a very similar Vellore having twin Jupiter XIF engines, which was put in hand at the Crayford Works founded by Maxim in the previous century, and was the last aeroplane built there.

A very different aeroplane made its first flight in April, for the single-seat, 30-ft span Hawker Hornet F.20/27 had been erected at Brooklands, where it was tested by George Bulman a few days later. He was a great little man, eager yet cool, speaking with conviction but never raising his light voice, eyes crinkled in a smile, bald head gleaming as he clambered helmetless into the cockpit. His new aeroplane seemed the quintessence of elegance, with polished pointed nose, and silver-doped wings and fuselage devoid of RAF tricolours. From the Hawker sheds, and the Vickers factory the other side of the track, a crowd gathered to watch. Powerfully the

Prettier than competitors, the Hawker Hornet fighter was regarded by the RAF as the quintessence of how an aeroplane should handle, though its twin-gun fire-power was considered too limited by the discerning. (*Hawker Siddeley*)

little machine tore across the grass and in six seconds was airborne, climbing swiftly, though Bulman was holding it down. It became a speck in the west and disappeared. There was the usual impeccable landing when he returned, and as he got out he said to Camm with quiet understatement: 'It's very good.'

All were soon aware of the beauty of the Hornet, for anyone could see it on test at Brooklands, so it was not surprising that in the course of weeks and months the Firefly IIM was gradually altered – slim streamlined steel struts replacing the wide balsa-faired tubes of the original, the straight-edged fin and rudder changing to curves like the Hawker, and when the coolant system proved unsatisfactory the wing radiators were discarded for a retractable radiator between the undercarriage legs like the Hornet.

Rival of the Hawker Hornet was the Fox-like, metal-framed Fairey Firefly IIM, as it initially appeared when tested by Norman Macmillan. Cooling problems caused substitution of an underslung radiator for the original retractable radiator and interlinked wing radiators.

In the end the two were remarkably similar, but by using a special Fairey-Reed propeller the Firefly was able to claim a speed of 217 mph at 13,000 ft compared with 205 mph attained by the Hawker, and alternatively it could climb slightly faster with a different propeller – though the speed then dropped to 200 mph!

While modification of the Firefly was proceeding, Fairey design was centring on a low-wing, twin-engined night bomber to specification B.19/27 which the AMSR, Sir John Higgins, was farsightedly supporting in view of Hollis Williams's technical achievement with the long-range monoplane. Rivalling it at Handley Page was an unconventional biplane being designed by Volkert and 'Star' Richards in which the fuselage was bolted to the underside of the top centre-section instead of on the lower wing, thereby giving better defensive view. Vickers also were competing.

The Fairey Long-range Monoplane at Cranwell throughout this time was being flown to check performance, stability, and adjust the engine.

The Air Ministry was restive at every hitch, and a panic tended to develop when things got out of hand. Fairey would be phoned at Hayes and two hours later he would arrive in his Rolls-Royce having averaged 60 mph to the detriment of all in his way, and would masterfully take command. Hollis Williams as design authority, Tips as experimental engineer, and Freddie Rowarth, the Air Ministry technical representative, were established there to help with the RAF's further tests, for though the Lion gave satisfactorily low consumption on the bench it would not do so in flight unless the air was calm. A special light-weight fuel, Swan spirit, had been provided, weighing only 6.8 lb per gallon, but evaporation losses forced change to a heavier mixture.

The Fairey Long-range Monoplane with Lion XI being refuelled at Northolt by the standard slow method of rolling up drums and hand pumping with a bowser.

'The last test before the record attempt was a 24-hour endurance flight at a starting weight well within the estimated maximum gross,' recorded Hollis Williams. 'As usual I flew with the crew. During the night the oil level in the service tank started to run low and we found it was not coming down from the reserve, so I had to open up the system and vent into the cabin with a sickly smell of hot oil; in addition we were getting carbon monoxide contamination. This was too much for Major navigating in the cabin, and he became very ill. On landing the doctors pounced on us, for such a long flight without sleep was unusual. John Willy came through with flying colours; I didn't do too badly; but Major was turned down, and a replacement had to be found at short notice. John Willy decided to substitute an instrument flying expert who could take over every few hours, abandoning the idea of astral navigation and relying upon occasional visual fixes. The choice fell on Flt Lieut David Betts, a magnificent young man, alert and intelligent; but during a medical the doctors found slight sinus trouble which they thought should be cleared before he set off. He went to the RAF Hospital at Henlow, but by the worst luck, picked up a germ of some tropical disease and within a few weeks died.

'Flt Lieut Jenkins, a Martlesham test pilot, was then selected – a slightly built man with tremendous powers of endurance. After a few hours 'Jenks' found snags which he wanted put right before the flight started. AVM Holt, the DTD, and Fairey rushed to Cranwell to find what caused the latest hold up. It was now spring. Weather had become unfavourable for the Cape route, so Rangoon was to be the target. John Willy was all for pressing on, and it looked as though Jenkins would be overruled, but Fairey took over: "The machine must only go when it is right and the weather is right. Jenkins, what are your points again – we will correct them." The main snag was carbon monoxide contamination. Long endurance checks of modifications were out of the question, for the special engine had been installed and we could not use up its time. A three hour test was agreed, checking contamination with a menagerie of canaries and white mice which remained so lively that the mice escaped and the wing and tail fabric had to be opened to ensure there wasn't a stowaway.

'Eventually all was ready. Take-off was to be down the slope from west to east during the dawn lull, using the hardest ground with smooth grass, for there were no runways anywhere in the UK. The crew turned out on the frosty morning of 31 March in tropical kit, clutching a bag of sovereigns for ransom money. But the wind was in the wrong direction and just too strong to risk a down-wind take-off. It was decided to taxi to the eastern end, but the starboard wheel sank into a rabbit hole and the machine had to be partially defuelled before it could be moved. Three hours elapsed, by which time the frost had thawed and the ground was soft. On attempting to take-off it was obvious they could not make it, and at the last moment the special ploughshare on the tail-skid was dropped, as they had no other brake, and the big monoplane came to rest with a propeller blade almost touching a low stone wall.

'Confronted with this awful fiasco, Mr Fairey was absolutely steady: no reproaches, no recriminations, but discussion with absolute composure on the basis that we must be prepared to take-off in either direction. In no time contractors were at work laying a runway at either end, though only 200 yards long, but enough to get rolling before hitting the grass. This made further delay, and not until three on the morning of 24 April were weather conditions satisfactory all along the route. In pitch darkness everything was prepared. This time there was no audience; the pilots got into the aircraft and took off without a hitch. I hopped into the back seat of a Bristol Fighter and stayed with them to the coast while they continued on and on, through day and night and the next day, until they ran into head winds which made Rangoon doubtful so they turned back and landed at Karachi after being airborne 50 hours 39 minutes, covering by Great Circle course 3,950 miles but failing to beat the record.'

Easter at least saw a record number of air meetings in England, for

For some years writing advertisements high in the sky with smoke proved profitable business using converted S.E.5as flown by St Barbe and Tait-Cox.

every club made it a special occasion. That week-end I visited several with a new demonstration Widgeon which had slots and a Gipsy engine. Lympne was the first port of call – a quiet grassy ridge-top expanse with some thirty visiting aircraft parked in front of the four war-time hangars. Among the light-planes were three Pander sesquiplanes from the Rotterdam Aero Club whose pilots had navigated with maps prepared by the Air Department of the Automobile Association which was run by the socially elegant Ivor McClure and his bespectacled cheery assistant Oliver Tapper. I flew one of these machines and found it very pleasant but more cumbersome than a Moth though with much gentler stall. From even further afield came Herr Frank Kirsch on a long-winged 45 hp Klemm monoplane from Stuttgart, 600 miles distant. In addition to the Moths, Avians, Bluebirds and the odd Widgeon or two, there was one of the earliest light aircraft, the Boulton & Paul P.9 flown by a charming couple wearing Bond Street cut flying suits, but their ancient R.A.F. engine eventually shed a cylinder, and in a forced landing the machine spread itself across the field in a complete write-off. From Martlesham came the keen Sgt Lowdell with the silver-painted Blackburn Lincock on which he put up a brilliant show featuring vertical upward rolls, matched only by Flg Off Atcherley on a Genet-engined Moth on which he made very long drawn out, smooth and beautiful rolls. By contrast Francis St Barbe's brother, Sydney, demonstrated skywriting with his S.E.5a, smokily pencilling *Shell* across the sky. Meanwhile I was trying the Ryan monoplane owned by parachute maker Leslie Irvin and thought it a solidly splendid machine.

In the evening I headed the Widgeon to Hadleigh in Suffolk, skirting London for the pleasure of looking at its softly smoke-veiled buildings and winding Thames – for no regulation denied such proximity provided we flew sufficiently high to make a forced landing on the outskirts – then turning at right angles, through the calm of evening came to Hadleigh 70 miles further on, a small field distinguishable only by a minute club house and shed just large enough for two folded Bluebirds. There we picketed the Widgeon over night. Next morning a full gale blew, but in company with the Linnell brothers in a Moth, and the club chairman and secretary flying Bluebirds, we headed towards Conington airfield midway between Huntingdon and Cambridge. At times we seemed merely hovering, and one of the Linnells afterwards remarked that his Moth was stationary so long over Cambridge that he and his brother had almost time to take their degrees. We landed like a helicopter at the Cambridge Aero Club. Some landings were even more spectacular, and it was a case of volunteers grabbing machines before they blew over. Though all aircraft had to be pegged down most of the afternoon, it did nothing to spoil the pleasure of meeting fellow enthusiasts, among whom was Lady Bailey. But as a fund-raising flying meeting it was a flop, for not until late could Lowdell and others make exhibition flights before the few hundred spectators who braved the gale. When the wind dropped, the Widgeon continued to Peterborough, landing in a park owned by friends, for another night-stop. Next day there was a brief visit to Norwich Aero Club, then a compass course to the London Aeroplane Club at Stag Lane, a passing call to the Phillips & Powis School of Flying at Reading to meet the CFI, Flg Off R. T. Shepherd, who would one day be first in the world to be lifted in a wingless machine sustained by vertical jet thrust. So to Hamble and the Hampshire Aeroplane Club, then home, having at each place demonstrated sky-clutching with slots, not so much as a labour of love but as excuse for five days of happy flying above the quiet beauty of a still unspoilt England.

Though slots were far from necessary for the Widgeon, inevitably they were a vital feature of Handley Page's entry for the Guggenheim Safe Aircraft competition and its attractive first prize of £20,000. So far there were twelve entrants, including five from Britain – de Havilland, Handley Page, Vickers, Gloster, and the Cierva Autogiro – yet only Handley Page had gone further than preliminary design, and was in the last stages of finishing his entry, which, to qualify, must be able to hold not more than 35 mph in level flight, glide at no faster than 38, and have adequate stability for five minutes without intervention by the pilot at all speeds from 45 to 100 mph. Should the engine fail when steeply climbing, the machine must glide down without pilot help; and with stick full back, nothing worse than a steep glide at 40 mph must ensue; landing must be possible from stalled flight, and take-off and minimum climb angle were strictly defined.

Volkert's solution was the two-seat H.P.39 Gugnunc, powered with an Armstrong Siddeley Mongoose radial, and characterized by a top wing of much greater span and chord than the lower, both having having full-length leading-edge slots and slotted flaps, with patented spoilers behind each outboard length of automatic slot, giving a powerful rolling moment as it angled up when an aileron was raised. A specially strong under-carriage was used to absorb the fast rate of descent when stalled, and wheel brakes were fitted to shorten the already short landing run. Tested at Cricklewood early in April, Jim Cordes found that with little practice he was able to hold the machine in such spectacularly steep attitude that, with the tail-skid dragging, the wheels were six or seven feet off the ground. Within a few weeks the machine went to Martlesham for C of A, and then was crated for despatch on the liner *Minnewaska* to New York, accompanied by Cordes, while Tom Harry England remained at Cricklewood to test the first Hinaidi having a metal-tube structured fuselage.

In the hands of monocled Major J. B. L. H. Cordes the Handley Page Gugnunc could put up a staggering display, but was far from ideal for private owners. (*Flight*)

Several private Moths and Avians were fitted with slots by Handley Page Ltd. On one of his journeys from Manchester with an Avian, Cordes earlier had run into thick weather, and followed the railway line, 'piloting by Bradshaw', but after an hour had to turn back a mile or two and landed in a scarcely discernible large field alongside the line. Once down he taxied cautiously through the fog until he saw a hedge, then turned along it until he found a gate where he switched off, got out, and found it led to a farm. In clearer weather next day he discovered that it was a very large triangular field between the LMS and Watling Street near Radlett, with other meadows beyond, which could be added to make an extensive level aerodrome. H.P. was interested. Cricklewood aerodrome was undulating, too small for fast landings, and fringed with new houses. He decided to purchase the farm, knowing he could sell Cricklewood at a profit of £100,000 to the Golders Green Development Corporation of which Barrett-Lennard, his former chairman, was a director; so he wasted no time on bargaining over the land value; within a few months it was his, and foundations were being laid for a large erecting shed at the apex of the triangle, though manufacture would continue at the old factory.

Of the art of test flying little had been written since its war-time analysis by Colonel Tizard, who currently was about to resign from the Department of Scientific and Industrial Research (DSIR), and accept Rectorship of Imperial College, London. To aspirant pilots, such as myself, a lecture by Cyril Uwins on *Experimental Test Flying* was particularly valuable because he dealt with prototype testing at the manufacturers rather than assessment in ultimate form by Martlesham where very few had either the necessity or technical background to investigate control response and stability scientifically, outstanding though these RAF pilots were on general handling and performance. Uwins indicated methods of investigating each control in turn, commencing with fin and rudder on the analogy of a weather vane. 'To arrive at fin area giving directional stability, you produce side thrust by side-slipping and use your rudder to keep the nose on the horizon: if you have bottom rudder the fin is insufficient; if top rudder at least there is enough; if no rudder at all then the fin must be slightly increased. The position now becomes more involved. Fin and dihedral are closely related, and the way to test is to trim, gliding hands off, then push on rudder and centralize. This causes the machine to bank violently to that side. If fin and dihedral are correct it will slowly right itself and glide in a normal path. If the fin is too big, bank is retained and a spiral glide develops which becomes steeper. If this happens, either reduce the fin or, if that is impossible, increase dihedral.'

After discussing rudder requirements he proceeded to longitudinal behaviour. 'To test pitching stability, adjust the tailplane to fly hands off, then disturb the flight path by pushing the stick forward until speed increases 10 mph, and release the stick; if stable, the nose will rise above level, stop, then fall below level but not so far as before, and so on in a series of decreasing oscillations. Should it be unstable the converse occurs.' Having adjusted tail areas, he described spinning and diving tests which must be explored step by cautious step. Referring to Bulldog dives, he commented: 'This test proved instructive, as when first carried out the fuselage fairings collapsed and needed considerable strengthening.'

Though Uwins gave the fundamental outline, the detail of prototype testing was involved and lengthy, but all manufacturers tried to cut down testing time because of the expense of alterations and delay which might give competitors a winning lead for contracts. All test pilots were made aware that because of this a compromise in aerodynamic characteristics was as far as one could go, provided the machine was made safe for others to fly.

As Professor Bairstow emphasized: 'The aeroplane is still far too complicated a structure to foresee from first principles all that might happen in the course of flying, particularly when diving at high speeds. Catastrophic accidents have then occurred, and it is not easy to be sure of the cause, for parts may be widely scattered and badly broken on impact. It requires much skill by the Inspector of Accidents to piece material together before making a guess at initial causes. If the conclusion is that none of the ordinary static failures could account for what has been observed, the matter is referred to the Accidents Investigation Committee. Thus in particular instances, the impression was reached that the wings were fluttering and it was partly up and down movement of the wings and partly angular movement of the ailerons. It was decided to carry out two independent lines of investigation – the first, to explain flutter in mathematical sense; the second, to reproduce on model scale, under controlled conditions, what happened full scale. This has led to simple precautions to avoid aileron flutter: they should be irreversible, with C.G. slightly ahead of the hinge, have small moment of inertia, and be underbalanced aerodynamically, with an appreciable part inboard of the outermost bracing.'

A Committee memorandum expressed a cautionary note that conscientious observance of every design recommendation would not provide absolute insurance against flutter, as Bolas had found, for the second Pipit crashed at Yate when flown by outstandingly tall Flt Lieut S. L. G. 'Poppy' Pope of the A & AEE on the grey Sunday morning of 14 February. Noakes, who at that time was still in hospital recovering from his injuries in the crash of the first Pipit, later gave me Poppy's account of the second accident: 'I received orders to proceed to Parnalls to test No. 2 Pipit.

The shipboard Parnall Pipit tested by Pope had a larger rudder than the one flown by Noakes, but confusingly carried the earlier number N232.

Taking my parachute I duly arrived and inspected the aircraft. On giving the fin and rudder post a shake I thought it felt a bit flimsy and told those at hand, including Bolas, but was assured all was in order, but wasn't too happy. So on taking-off I decided to fly on a straight climb to 5,000 ft, but hit cloud at 2,500 ft and put on rudder to steer away, and at that moment the rudder broke away and the aircraft turned on its back. I pulled the Sutton harness pin, fell from the cockpit, pulled the parachute release – and hit the ground in seconds after the canopy opened. I thought I had broken my back, then painfully found I could move first one leg then the other, released the harness and got up and ran around shouting "I'm alright", and went straight to the railway station and took the first train back.'

A pilot's report quoted in R & M 1247, which described flutter investigation of the Pipit by W. J. Duncan and A. R. Collar, seems to refer to an earlier occasion, possibly when flown by Saint of Glosters who, it is said, tested the prototype – for drawings of the wind-tunnel model show a small rounded rudder above the fuselage, typical of Bolas's outlook, yet Pope's machine certainly had a large rectangular rudder, braced tailplane instead of cantilever, and the ailerons were changed from four to a pair on lower wings only. The report ascribes the original violent flutter above 150 kt to rudder mass characteristics in conjunction with rolling of the tail due to insufficient torsional strength of the fuselage, but adds that flutter would not have occurred had not a tail-lamp been fitted to the rudder after the aircraft had been passed for flight – thus marginal is danger.

The Air Minister was always prepared to show by personal action his own confidence in flying and on 30 March, piloted by Capt A. S. Wilcockson, he left Croydon aboard the Argosy *City of Glasgow* on the inaugural scheduled air mail and passenger flight to India, accompanied by his Principal Private Secretary C. Lloyd Bullock, and Air Vice-Marshal Sir Vyell Vyvyan. In five hours they reached Basle, where they continued by train to Genoa and then embarked in a Short Calcutta, flying by way of Rome, Athens, and Tobruk to Alexandria which was reached on 3 April. From there Sir Samuel was making a 4,000 mile tour over portions of the Cairo—Cape air route, and Sir Vyell transferred with the mails to the D.H. Hercules *City of Jerusalem* which flew via Baghdad to Karachi, arriving on 6 April after delay due to a sand storm. Next morning the Hercules *City of Baghdad* left Karachi with the reciprocal air mail for England, carrying Lord Chetwynd, the vice-chairman of Imperial Airways, accompanied by his daughter, Sir Vyell, and Sir Geoffrey Salmond. At Alexandria, Sir Samuel Hoare rejoined for the next stage in a Calcutta, and after a tedious train ride from Genoa to Basle, completed the journey to Croydon in another Argosy piloted by Capt O. P. Jones, arriving despite thick mist, four minutes ahead of schedule, to be greeted by Sir Alan and Lady Cobham.

Meanwhile Kingsford Smith and Ulm, with two crew, had started from Sydney for England on 30 March with the *Southern Cross*, but failed to arrive at Wyndham, their first refuelling point on the north coast of

Refuelling an airliner abroad was usually by handing up 2-gal cans of Shell or BP. An Imperial Airways de Havilland Hercules is seen here.

Australia. Days went by. Apprehension mounted over the national hero. Despite an unhealed quarrel over financing 'Smithy's' enterprises, his former partner, Keith Anderson, accompanied by John Hitchcock, set out to look for him with a Widgeon, the *Kookaburra*, and soon they too were reported missing. On 12 April a D.H.61, fitted out by the Sydney Relief Committee, found the *Southern Cross* on a mud flat midway between Derby and Wyndham, and after dropping food, medical supplies, and mosquito nets, returned next day, accompanied by a machine owned by West Australian Airways, landed alongside, and took the crew off. On 21 April Lester Brain, flying a Q.A.N.T.A.S. aeroplane, found the Widgeon, but Anderson and Hitchcock had died from exposure. Harsh criticism followed, alleging that Kingsford Smith and Ulm had staged the affair for publicity. The Australian Government set up an inquiry headed by Brig-Gen L. V. Wilson who decided the landing was not deliberate, but was critical of Ulm and said his diary was 'obviously meant for publication'. Kingsford Smith was criticized for failure to carry emergency rations and having an inefficient wireless set and operator which prevented reception of adverse weather warning and contact after landing, nor had there been any attempt to burn the 18 gallons of engine oil to supplement fire signals; as to the unfortunate Widgeon, it was totally unfitted and unequipped to fly across desert. The Aviation Department was therefore recommended to prescribe regulations governing long-distance flights, with power to prevent them if necessary.

5

In April there was nation-wide relief that the King had sufficiently recovered to return to Windsor without further need of the Council of

Regency. It added to the buoyancy of the Stock Exchange, for aircraft exports were booming. The Greek Naval Air Service contracted with H. G. Hawker Engineering for six Horsley torpedo-bombers; a batch of Southampton flying-boats with 450 hp Lorraine-Dietrich engines was purchased by the Argentine Navy; the French Government ordered a Calcutta, and the Japanese were negotiating for a larger version with licence for construction by Kawanishi – of which Tony Fletcher was formerly chief designer but had been succeeded by Yoshio Hashiguchi who was currently visiting the Short factory. Blackburns were building the 3MR4 enlarged Ripon to the order of Mitsubishi for the Japanese Navy, and had interested the Royal Canadian Air Force in the Lincock. In that country Armstrong Whitworth had sold constructional rights of the Siskin and Avian to the Ottawa Car Manufacturing Co who were building both types for the RCAF. Westland also was trying to interest Canada in Wapitis, encouraged by the Australian purchase of them and the South African Government following suit, for it was profitable business, matched only by de Havilland sweeping the board with Moths for clubs and schools

The Blackburn designed and built, Japanese naval T.7B prototype powered with 600 hp Hispano-Suiza engine, managed to acquire a distinctly oriental appearance. (*Hawker Siddeley*)

and supplying hundreds for the RAF, RAAF, RCAF, SAAF, Irish Free State Air Corps, New Zealand Ministry of Defence, Chilean, Danish and Italian Air Forces, and the Governments of India, Sarawak, Sweden, Finland, Japan and Greece. Nevertheless the Avian was getting into the export market, and twenty had just been ordered by the South African Air Force and six by the Estonian Air Force. Robert Blackburn was hopeful of metal Bluebird IV sales in the USA where the Blackburn Corporation of America had just been formed, and two had been ordered from Brough by the Larkin Co of Australia – but with the shops busy with Ripon production, arrangements were being made for all Bluebird construction to be sub-contracted to Alliott Roe's new Saunders-Roe business at Cowes. Westland had considered doing the same with the Widgeon but prospects did not warrant this. More hopefully the new business of Simmonds

Aircraft Ltd leased part of the old Tube Mill by Southampton Water, and was selling the Spartan in America, Australia, Canada, South Africa and New Zealand, and in view of extending production Chris Staniland, whose Short Service commission had just expired, was appointed chief pilot – for he had proved a brilliant aerobatic pilot in the RAF, though even better known as a driver of racing cars and motor-cycles.

May proved a busy month of aircraft testing for every manufacturer. The first production Bulldog was being rigorously flown to check conformity with the prototype, and subsequent machines were receiving the mandatory 30 minutes acceptance test. At Lympne, the Rolls-Royce Kestrel powered Short Gurnard, N229, reassembled as a landplane with modified ailerons, after near disaster as a seaplane due to dangerous over-

Designed with interchangeable float or wheeled undercarriage, and Bristol Jupiter X or Rolls-Royce Kestrel, the Short Gurnard had a welded tube fuselage frame instead of the Sturgeon's monocoque, because of RAF insistence on easy access.

First flight of the Blackburn Nautilus prototype, competitor of the Gurnard, was made by Neville Stack in May 1929, but there were the invariably unexpected problems which took long to sort out.

balance when flown by Parker on 2 May, now showed excellent stability and manoeuvrability, endorsed a week later by its sister ship the Gurnard I, N228, with 525 hp supercharged Jupiter X. At Brough, the ever hatless 'Dasher' Blake, with Neville Stack as temporary assistant, was flying the Nautilus fleet spotter/interceptor, but found it had snags somewhat similar to those of the Westland Interceptor. A likely cause seemed to be the radiator conduit obstruction between fuselage and lower wing, so it was decided to change the rectangular section to an inverted triangular form and fit smaller-chord elevators – but these modifications would not be completed in time to exhibit the Nautilus at the Aero Show in July. Concurrently trials were still proceeding with the Jupiter XF Beagle, though it had first flown fifteen months earlier, but now that it had larger horn-balanced elevators and reduced horn balance on the rudder, it was almost ready for despatch to Martlesham for retrospective comparison with the Gloster, Handley Page, Westland and Hawker machines of specification 24/25, though by now theirs was an out-dated requirement.

There was similar concentration on civil aircraft. Campbell-Orde at Whitley was busy with the Mk II Argosys for Imperial Airways. They had Jaguars of greater power cowled with circumferential slot-like aerofoils, known as Townend rings, to reduce engine drag, and in addition to wing-tip slots had ailerons servo-balanced by small vertical surfaces at the trailing edge of the inboard wing bays, for which John Lloyd had obtained Patent 323,289.

At Vickers there was disaster. The Vanguard, G-EBCP, loaned by the Air Ministry to Imperial Airways, had been returned to Brooklands for experimental modifications, the latest of which was a Virginia X tail unit with finless twin rudders to enable easier trim with one engine out. On 16 May Tiny Scholefield, with Frank Sharrett as observer, took-off from Brooklands to test its directional characteristics, but when near Shepperton the Vanguard suddenly dived and crashed onto the Middlesex side of the Thames, caught fire, and both were killed. It seems probable

The civil Vickers Vanguard had been on route tests by Imperial Airways since May 1928, leading to the fatal modifications made by Vickers to improve control. (*Flight*)

that Tiny, with his immense strength, pushed the rudders to full lock and overstressed the tail so that it broke away. Significantly a compensating device was fitted to rudders of subsequent Virginias to prevent excessive application.

Edward Rodolph Clement Scholefield was 35, and had joined the RFC as a mechanic early in the war. After learning to fly in 1915 he was shot down eventually and imprisoned, but after the Armistice was posted to the RAE as a test pilot. I had first met him twelve months before at Lympne when I had just landed the Pander semi-cantilever biplane; as I came to a stop a resounding, bull-like roar demanded: 'Keep the engine running, young feller! I want to see if I can get into that little toy,' and turning my head I saw this great bear-like figure of a man lumbering towards me. 'His method of expressing himself was a standing joke among his friends,' wrote C. G. Grey. 'But what he said carried immense weight, for there are few, even among those exceptionally skilled test pilots we so fortunately possess, whose opinion on aeroplane or engine is as reliable or

Flt Lieut E. R. C. 'Tiny' Scholefield, like Harry Hawker, was an enthusiastic Brooklands racing driver at week-ends.

worth having as Tiny Scholefield's. In every way he was first class. He could handle a light aeroplane as delicately as needed, or wrestle with the heaviest machine with ease. He could go through a long, hard flight untired, or fly with the best possible judgment the fastest machine anybody could produce. His was a lovable disposition, and he could be at once amusing, illuminative, and instructive.'

Because there was surprisingly little information on rudder forces, particularly their intrinsic power in the form of lift coefficients, Miss F. B. 'Fanny' Bradfield, who had joined the RAE after her war-time mathematical work, was given the task of collating data, and found that in the case of machines with a monoplane tailplane, ranging from Bristol Fighter to Beardmore Inflexible, there was not much difference in the degree of rudder power – though it seemed anomalous that the Westland Dreadnought rudder which had looked so large was in fact the least effective, and the small circular rudder of the Avro 504K was among the best. However, if the F.E.4 could be accepted as criterion, then an all-moving rudder between biplane tailplanes was much more powerful. Ease of application was a different matter, governed only by the amount of balance, and the extent of yaw response depended on fin and fuselage areas, interference and tail lever-arm length.

Test flying could have its difficult moments. In this instance the RAE pilot crawled from the cockpit uninjured.

From Martlesham, on hearing of the Vanguard's crash, the ambitious, tough Flg Off Mutt Summers telephoned Pierson at Vickers expressing genuine dismay at the loss of Scholefield, whom he had always felt might take him as assistant pilot, and offered to carry on with test work in temporary or permanent capacity with permission of the Air Ministry. At this stage RAF Command was well aware of the industry's difficulty of time in training test pilots to standards such as those of Uwins and Parker in their long service with Bristol and Shorts – but George Bulman was an instance of how readily this could be done if an RAF pilot had been at Farnborough or Martlesham. The Air Ministry was currently disposed to release occasional test pilots provided they had no permanent commission. Vickers Ltd therefore made speedy moves to obtain Summers as chief test pilot, shortly followed at Hawker's behest by release of Flg Off P. E. G.

'Gerry' Sayer from the same squadron as Summers at the A & AEE in order to join Bulman as assistant. They were young men of good judgment and high skill, but of very contrasting temperament: Summers of average height, heavy in build, round of face, and friendly with the right people – Sayer more sparsely built though taller, with vivid personality and infectious grin. Mutt, I had known from earliest days of the Wizard, with which he was favourably impressed – and for Gerry, I had acted as observer when he was briefly attached to Westland to check the prototype installation of Frise ailerons on the first production semi-wooden Wapiti.

Of the arduous nature of flight research on spins and flutter, Flt Lieut Linton Ragg of the RAE said: 'In one test a modified tail caused the machine to spin at such high rate that it was impossible to take any observations after six or seven turns, the machine becoming completely out of control in a flat stable spin. After trying to recover from 18,000 ft to about 7,000 ft the pilot eventually succeeded by moving throttle and control column violently backwards and forwards together, thereby rocking the machine out of its stalled state. It was touch and go whether he remained conscious long enough to stop the spin, for he lost sight of everything, and it took 20 minutes gentle flying round the aerodrome before he recovered sufficient physical and mental condition to land. Similarly wing flutter has caused trying experiences, such as coming down with hands and knees badly bruised by the control column as it played hide-and-seek round the cockpit. Already large aeroplanes are becoming too much like hard work for one man to control. Coupled with servo controls, gyroscopes are being developed so that the crew of the aircraft of the future can let the machine automatically hold its course, right itself in bumps, and fetch up over its destination to scheduled time without the controls being so much as looked at from take-off.' But there was a long row to hoe.

6

Parliamentary session having ended on 10 May, Baldwin decided to go to the country without further delay on the major issues of world peace arising from the Briand-Kellogg Pact, and the steady increase in unemployment which now totalled some 1,200,000, counting only those within the insurance scheme. The Liberals pledged themselves to conquer unemployment; Labour claimed it could do it better. Baldwin said trade was recovering and the gravity of unemployment exaggerated, but promised more technical education, slum clearance, and better welfare schemes for mothers. 'Safety First' was his Conservative slogan.

A number of candidates used light aircraft in their campaign. Moths were in the ascendancy, for they were owned by Sqn Ldr the Hon F. E. Guest, the Member for Bristol North; Capt Harold Balfour for Thanet; the Marquess of Douglas and Clydesdale for the Govan District of Glasgow; and H. R. Murray-Philipson for Peebleshire, who arranged his meetings close to the few good landing grounds in Scotland. Sir Philip Richardson,

Conservative candidate for the Chertsey division of Surrey, used his own D.H.50, and W. E. Allen, Unionist candidate for West Belfast, commuted between Baldonnel and Aldergrove in a friend's Gipsy Moth.

Despite 29 million increased electorate due to the concession of votes to all women over 21, there was remarkable initial apathy, and an indecisive result from the poll on 30 May. The usual great crowd gathered in Trafalgar Square to watch electric lights flashing results. Excitement began to grow at the repeated 'Labour gain'. With 287 seats it became the largest party – but nobody had really won, for the 260 Conservatives and 50 Liberals could combine to defeat any Labour proposals. Inevitably Baldwin resigned, and on 4 June Ramsay MacDonald was invited to form a Labour Government. Snowden became Chancellor, J. R. Clynes was made Home Secretary, and Arthur Henderson took charge of Foreign Affairs, but the novelty was the first British woman Cabinet Minister when Margaret Bondfield became Minister of Labour. The tall and authoritative Lord Thomson was reappointed Air Minister. His Under-Secretary was Frederick Montague, 'self-educated newsboy, shop assistant, and subsequently parliamentary agent for the Labour Party.' Of Thomson, C. G. Grey said: 'In the first Labour Government he maintained the policy of expansion planned by his predecessor Sir Samual Hoare. Since then he has nursed British aviation as assiduously because of genuine interest in aviation. He has been an excellent chairman of the Royal Aero Club, and as an after dinner speaker at aviation banquets has been an outstanding success. As British representative at the International Conference at Washington he did valuable work and won personal popularity and respect to an unusual degree; so his reappointment will be a valuable international asset.' The burning question was whether the policy of expansion would be continued despite Labour's pacifist outlook. Said MacDonald: 'I wonder how far it is possible, without abandoning any of the party positions, to consider ourselves more as a Council of State and less as arrayed regiments facing each

Of the ten de Havilland Giant Moths manufactured, two were operated in Britain – *Geraldine* by the *Daily Mail*, and *Youth of Britain* owned by Alan Cobham Aviation Co, and later sold to Imperial Airways.

other in battle? So far as we are concerned, the co-operation of other parties will be welcome.'

Whitsun heralded the usual round of air meetings, and introduced Sir Alan Cobham on 19 May at Mousehold in new guise as the pilgrim of a 21-week aerial propaganda tour of towns and cities in an attempt to convince local authorities that municipal aerodromes were a necessity, but he also intended to give 10,000 school children free educational flights. Backed by Sir Charles Wakefield's finances, he had purchased for this purpose a D.H.61 Giant Moth, named *Youth of Britain*, but it attracted less attention at the Norwich Club than the giant Inflexible, piloted by Sqn Ldr E. S. Goodwin, OC of the bomber Flight at Martlesham, which came rumbling in beneath the low cloud ceiling, circled and landed, then parked among the light aircraft with an accuracy which all found amazing – for differential brakes were a novelty, the short landing run of contemporary military and civil aircraft rendering their weight and complication unnecessary despite difficulty of taxi-ing and turning in strong side winds.

Big aeroplane and little petrol cans! Topping up the Beardmore Inflexible at one of the flying meetings where it was sent to encourage airmindedness in the taxpayer.

Many crashes, some fatal, continued to beset the course of private flying and commercial aviation. A serious blow to the prestige of British air travel was the loss of the Imperial Airways Handley Page W.10 *City of Ottawa* on 17 June when it came down in the Channel, three miles from Dungeness, and seven passengers were drowned. It was yet another instance of a twin-engined machine unable to fly at full load with one engine out of action. In trying to touch down alongside a trawler at minimum speed, the pilot through ill luck hit the top of a swell, and the impact broke off the tail. As the machine nosed over, the passengers, who were not strapped in, fell heaped on top of each other in the front of the cabin, and only four could extricate themselves, though pilot and mechanic were swept from the open cockpit and saved by the trawler. 'Nose-dive to death' headlined the *Daily Telegraph* and other papers, causing sufficient outcry for the Secretary of State for Air to order a formal investigation which revealed that engine failure had been caused by fatigue fracture of bolts holding one of the connecting-rod bearings. For once the pilot was not blamed. Disregarding Imperial Airways assurances that 100,000 passengers had been safely carried since January 1925, the *Daily Express* vigorously opened a campaign for 'no more twin-engined aircraft to be used for Cross-Channel passenger traffic'. Nothing could be immediately done, but as the Olympia Aero show would shortly reveal, capacious four-engined aircraft were on the way.

H. A. Brown, ex-RNAS, joined Avros in 1919 and became chief test pilot after acting for them as chief instructor to the Spanish Naval Air Service from 1921–1926, followed by an interlude as instructor to the Lancashire Aero Club. (*Flight*)

Throughout June the light-hearted flying of light aircraft continued happily. Regular attenders, such as Neville Stack, George Lowdell, Hubert Broad, Sam Brown and others had become well known, but now Chris Staniland was noted flying the Spartan. Said a critic: 'One has heard people say that Mr Simmonds must have sacrificed desirable aerodynamic properties for the sake of his interchangeable surfaces, but Mr Staniland certainly proved the Spartan is well up to standard when it comes to stunting in such expert hands as his, but one would suggest that now he has become commercial he must consider his demonstrations from the view-point of spectators as well as that of pure flying. The other day he made all his evolutions directly in the line of the sun, and we certainly lost some of the enjoyment of his show on that account.' It was a point Chris never forgot.

Neville Stack, universally known as 'Stacko', was appointed chief instructor to National Flying Services. (*Flight*)

Private owners remained devoted to the conventional British biplane. Westland continued to find the Widgeon monoplane so slow a seller that the benignly autocratic Bruce decided to withdraw its production and offer manufacturing rights to any takers at home or abroad. All design work ceased on the cabin three-seat version, unaware that de Havillands already were constructing a very similar machine, the D.H.80, which they referred to as the Moth Three. However, Bruce and Keep were hopeful that the Westland Four would find favour as a small feeder airliner, for numerous enquiries seemed a good pointer.

Of passing historical interest was news on 29 June that the Spanish conqueror of the South Atlantic, Major Ramon Franco, with three companions, had been found by HMS *Eagle* after they had been missing for a week since attempting to fly the Atlantic with their Dornier Wal, for they overshot the Azores in the dark, ran out of fuel, and alighted in the open sea. There were public demonstrations of gratitude in front of the British Embassy in Madrid, and the Cross of Naval Merit was awarded to the captain of the *Eagle*; but Ramon the hero was court-martialed and found guilty of negligence, and this turned him to revolutionary activities – an attitude undreamed of by his Royalist brother Francisco despite his destiny as eventual Dictator.

On the same day that Franco was rescued, some twenty British light aeroplane pilots flew to Rotterdam for the Concours Voor Sportvliegtuigen. The ever energetic Sir Sefton Brancker went there too, for here was great opportunity to inspect Dutch, German, and French light aircraft, and towering among them was Fritz Koolhoven explaining in guttural English the merits of his new open two-seater derived from his FK.41. Hospitality was lavish, and even extended to payment of a fine for one inebriated British pilot who went to sleep on a doorstep. At the airport there was all the fun of the fair, with aerobatics, races, and joy rides, but a unique event was the first towed launch anyone had seen – for a two-cylinder, ultra-light monoplane built by a rough and enthusiastic German carpenter, Herr Espenlaube, towed a similar glider laboriously to 1,000 ft, where its pilot released and came circling down. In Britain there were no gliding clubs despite the 1922 competition, but here was a method which might well be used for practice instead of launching from a hill where the wind happened to be blowing up it.

7

Not only were the next few months aeronautically the peak of the year but of the entire decade. Events opened with the King's Cup race on 5 July, starting from Heston Air Park which, breaking established aeronautical tradition, was ready on time. There was no doubt that the partners Norman and Muntz were going to make it a social centre with its comfortable club house, pleasant lounge, bright bar, and overnight accommodation. Splendid workshops had been built and lock-up sheds for light aircraft. Landing approaches were easy, and the grassy aerodrome seemed large enough to operate any airliners of the near future. Efficiently managing the day-to-day business was dynamic little Susan Slade; and Capt V. H. 'Laddie' Baker was chief pilot, with John Parkes, recently an instructor at de Havillands, as his assistant.

Heston aerodrome and its club became a social focus. (*Flight*)

The Minus was a conversion to a parasol monoplane of the biplane designed by C. H. Latimer-Needham for the Halton RAF apprentices to build, and it gained 10 mph. (*Flight*)

There were sixty entrants for the race, though many withdrew, and on this cold, windy, and rain-threatened morning there were more to see the start than ever before, and still more came to the finish next day. Aircraft ranged from the ultra-light Halton Minus parasol monoplane and original two-seat Avro Baby, G-EAUM, which young Miles at Shoreham had resuscitated and fitted with a Cirrus I, through an array of Moths, Avians, Widgeons and Spartans, a Bluebird, two S.E.5as, two Grebes, and the Vickers 141 on loan agreement from the Air Ministry flown as scratch machine by their new pilot Mutt Summers. A familiar figure was missing: R. J. Goodman Crouch, who had handicapped all the races since the war, feeling frustrated in his Air Ministry job had resigned to become director of the American company manufacturing the Avian. His plump and good-natured assistant Wilfred Dancy, assisted by the very able and friendly Freddie Rowarth, were now official handicappers. The course was longer than previous circuits of Britain, and was flown clockwise instead of anti-clockwise, competitors heading from Heston to Norwich, then south to Lympne, west to Filton, and north to the old racecourse at Blackpool, staying at the *Queen's Hydro* before continuing next day to Moor Park at Glasgow and then back to Heston. There was every degree of skill among the pilots, many of whom were professionals, and included Service officers – for Flg Off Atcherley masquerading as R. Llewellyn, with Flt Lieut Stainforth as navigator, piloted a Grebe entered by Sir Walter Preston, MP for Cheltenham, and Mr E. H. 'Mouse' Fielden, late RAF, flew a Grebe entered by the elderly, craggy faced, Freddie Guest who flew in the rear cockpit as his passenger.

I was particularly interested in the prospect of swarthy teenager Carol Napier, a Westland apprentice who was flying his Gipsy-powered Widgeon. A few months earlier his tall, black-bearded, silent father, Montague Napier – that great but rarely seen invalid head of the engine company – had visited Westland.

'Are you sure this machine is safe?' he asked. 'I don't want my son to

The Gipsy-powered Westland Widgeon III purchased for Carol Napier had automatic tip-slats, and was the last to have the original type of narrow undercarriage.

break his neck.' I explained that the Widgeon was safer than most, and exceptionally well behaved at the stall, though all aeroplanes could be crashed by the foolhardy.

'What is a stall?' asked Montague Napier. I explained, and compared the Widgeon with a slotted aeroplane.

'Let's be doubly safe. I'd like it to have slots as well. Now take me up and show me,' said this man who never flew.

I gave him a cautious circuit and a semi-stalled descent using a little power for gentle landing and shortest run, then flew solo, with some crazy-flying to show what excesses one could go to with the Widgeon. Afterwards Napier said: 'Very well, Carol may have one if you will undertake to see that he flies it properly.' So at this King's Cup race I was a little anxious for the inexperienced youngster, who was somewhat clumsy – but to my relief he abandoned the race at Hamble.

The start of that race had its emotional moment when Flg Off H. H. Leech with the Avro Baby failed to take-off, bouncing and trickling the length of the aerodrome, but after taxi-ing back and working on the engine, he got away. Most departures were sedate, but Flt Lieut Armour, with B. E. Lewis' Gipsy Moth, was sensational, for he held it down to gain speed, then pulled up, banking steeply round to turn down-wind for Norwich. A model of good judgment was Winnie Spooner who made a smoothly beautiful turn at exactly the right moment with her specially streamlined Gipsy Moth. Most accomplished and spectacular of all was Wally Hope, winner of the previous two races, who was flying an even more streamlined scarlet-painted Gipsy Moth which had minute wheels, for he immediately put his right wing-tip nearly to the ground, and after travelling 20 yards spun round, latched on to the air, and went straight off on a steady climb dead on course at 1,000 ft to gain benefit from the following wind. By contrast the take-off of a commuting Handley Page W.8b, chartered by millionaire Van Lear Black, was like a lumbering elephant, its tailskid ploughing a great furrow in the recently sown grass.

There were the usual dismays and alarms and forced landings, commencing with Armour when his engine failed between Henlow and Norwich, and R. W. Jackson with a Spartan got lost. At Mousehold, which had a great dip in the centre, there were extraordinary landings, for the air was very bumpy, and Bailey's Gipsy Moth stalled down so heavily, with slots wide open, that a longeron broke. At Lympne H. V. Ashworth's Avian blew over on a bouncy landing and broke the propeller, and Leech retired because his Avro Baby was too underpowered to cope with the gale that was blowing. Meanwhile Harold Balfour, MP, retired at Hornchurch, and Maurice Brunton had been forced down at Harrow. Between Sittingbourne and Maidstone the ever genial Tommy Rose had engine trouble with his yellow S.E.5a and landed on the Downs, parking it behind a haystack while he made his happy way to the local and his usual sustenance. R. G. Cazalet had engine trouble with his Widgeon and landed near Hamble. On the northward run Allen Wheeler with an S.E.5a retired at Blackpool, and so did H. R. Law, the lisping but able son of Bonar Law, with his D.H.60X Moth. Then it was found that Sempill in his Bluebird IV had landed with engine trouble between Leeds and Birmingham, and Boyes in similar trouble with a Spartan landed near Lydney in the Forest of Dean. Flt Lieut Bruce with another Spartan retired at Blackpool as he was eighteenth and obviously out of the race. On the return from Glasgow, Hubert Broad ran into a heavy storm between Wooler and Alnwick and, unable to see the high ground ahead, landed near Chillingham Castle, the home of Lord Ossulston, another enthusiastic Moth owner. Summers with the Vickers 141 retired with engine trouble at Sherburn.

At Heston for the finish, C. G. Grey delightedly reported: 'One learned that the Air Park secretariat, which consists chiefly of Miss Slade, herself a pilot of considerable ability who has her own Moth, had invited the entire House of Commons as guests, to teach them something about aviation. Whether many were there seemed doubtful, but the spectators were very much like an Ascot crowd in sports clothes rather than an ordinary flying or aerodrome collection. What had helped a good deal is that the Household Brigade Flying Club has adopted Heston as its home aerodrome and installed itself in the best room of the social focus as its own club premises. Aviation is obviously going to be the done thing within the next 12 months.'

Diligent supporter of many flying meetings was the tireless Director of Civil Aviation, Sir Sefton Brancker, here piloting his official Moth appropriately registered. (*Flight*)

When Sir Sefton Brancker arrived in a coupé Moth, his official Moth G-EDCA being under repair, he joyfully declared himself converted to travelling in an enclosed machine. He was followed by Lord Thomson in another Moth, piloted by Major Travers of the London Aeroplane Club, and in a few well chosen words through microphone and loud speaker the Minister declared Heston Air Park officially opened and wished it long life and prosperity. By that time the loud speakers were announcing that the leading competitors had left Birmingham, with Neville Stack well ahead with his specially streamlined Hermes Avian, pursued by Richardson,

Winners of the 1929 King's Cup – Flg Off R. L. 'Batchy' Atcherley and his navigator Flt Lieut George Stainforth.

Atcherley, and Fielden. Promptly a great rain cloud blew across the northern horizon, but suddenly the first machine appeared, yet instead of the Avian it was Atcherley's Grebe, and behind was Richardson's Moth. Hardly had they landed than two light aircraft showed close together, the bright scarlet of the leader revealing it was Hope's, and the beige with red stripes that of Alan Butler. Soon Winnie Spooner's Moth appeared, having averaged at 105 mph the fastest speed of any light aircraft except those deliberately converted for racing. Sixth was a disappointed Stack with an engine obviously firing on only three cylinders – a literal knock for Cirrus Engines Ltd. Then came the other Grebe piloted by Fielden, who explained he had flown slowly because a flying wire had broken. Eighth was Le Poer Trench with an Avian; ninth was Geoffrey de Havilland in his beautifully cleaned-up Gipsy Moth. Tenth was Mr 'Parachute' Irvin with his Gipsy Moth; eleventh, Jackaman with another; twelfth was Bill Thorne, Avro's assistant pilot, flying an Avian; thirteenth came the doughty Sqn Ldr Probyn with his Genet Widgeon, which he claimed with pride had maintained 90 mph on four out of five cylinders, a rocker-arm having broken at Bristol on the first day. Fourteenth was the beautiful Mrs Alan Butler with her enclosed Moth; fifteenth was Wg Cdr Manning and his Widgeon; sixteenth Flg Off Bonham-Carter, the newest member of the Caterpillar Club, who was flying the Gipsy Moth of D. S. Schreiber, an inexperienced owner who could not get insurance cover; seventeenth came Mr and Mrs J. W. P. Chalmers, keen amateurs who now had a Gipsy Moth; eighteenth was A. F. Wallace in a similar machine; nineteenth came P. P. Grey in another; and last was Hubert Broad.

'The King's Cup is open for anybody to win with a British machine and British engine. To see people entering ancient things like Grebes and S.E.5as would be funny if it were not so tragic,' said C. G. Grey dispeptically. 'If the wealth of the British aircraft industry, which is spending its tens of thousands on showing its machines at Olympia, cannot put up a decent speed show in a race for the King's Cup then it is quite time that somebody in all due humility suggested to His Majesty that in future the race should be made international.' But the British spirit was not so serious as that. Here was a typical occasion where amateurs could compete with professionals, all enjoying the challenge of navigating without radio aids, and ever hopeful that the pretty little fields of England would at least be big enough in which to land should their engine fail. In any case we had the forthcoming Schneider Trophy contest as the great international event, and competitive entries from the USA, France, and Italy.

The many failures of engines, flogged to their limit in the King's Cup and other races, had concerned Geoffrey de Havilland as being detrimental to even more extensive sales of the Moth as a reliable vehicle. He therefore instituted a reliability test with a sealed Gipsy in one of his Moths, ostensibly to get data on wear and tear, but actually as evidence for setting its major overhaul at 600 hours, which would approximate to a year's hard flying. By the time of the King's Cup race it had already achieved 485 hours at 88 mph, and during the next few months the hours steadily increased until the target was attained, and double-page advertisement spreads emphasized that the Gipsy was 'the most reliable light aero engine in the world'.

A week after the King's Cup race came the Royal Air Force Display – the tenth of its kind. There was one immediate visible social change, perhaps emphasizing that we had a Labour Government, for instead of the regimented bowler hats in the enclosures, almost every man wore a grey trilby and informal clothes – and this was significantly underlined when their Royal Highnesses Grp Capt the Prince of Wales and the Duke of York entered the Royal Box as the King's representatives clad in similar clothes, the Prince having flown to Hendon in his special double-cockpit Westland Wapiti piloted by Sqn Ldr Don. Among those in the Royal

HRH The Prince of Wales had long used aircraft for transport. The Royal Bristol Fighter was now superannuated and replaced by a special Westland Wapiti built under specially rigorous inspection.

Box were King Alphonso and his son; the Prime Minister and the Labour Cabinet comprising Lord Parmoor, father of Stafford Crips and Lord President of the Council, Philip Snowden, Arthur Henderson, J. H. Thomas, Emmanuel Shinwell. There were the French, German, and American Ambassadors, High Commissioners of the Dominions and Commonwealth, Maharajahs and Sultans, diplomats, leading members of the Lords, Admirals, Generals and Air Marshals.

Felicitously the editor of *The Observer* penned that day: 'We write with noise of planes in our ears. To-day is Air Force day, and never have conditions been more favourable to spectacular effect. The very condition of our Empire, its long and disturbed Asiatic frontiers, its vast half-pacified African possessions, insist upon development of this Service as a fighting arm. Nothing short of air equality will serve our national and Imperial needs, and to attain it we are ready to face some abasement of Naval estimates. Flight has become an integral part of our civilization. We begin to realize the vast cultural importance of the air bringing continents into touch. As an instrument of economic progress air survey has proved its value in Canada and tropical Africa; by its aid preliminary work of years is reduced to weeks, and accomplished with greater certainty and fractional cost. Flying is widening the range of man's knowledge, and it is significant that a new achievement which looks so confidently to the future should potently contribute to rediscovery of the historic past, for air archaeologists have taught us that where man's hand has once been busy its touch is never wholly obliterated. Let us not shut our eyes to the moral of these facts. They proclaim that if we are to retain our high place in the world we must be pioneers in the use of air machines. The Prime Minister has set a notable example. The attacks upon him for his use of Service machines have happily miscarried, and have indeed brought us nearer the day when a plane will be permanently at a prime minister's disposal.'

With that reference to Ramsay MacDonald all cordially agreed, for though he was thought rather elderly to have taken to flying, people admired him for trusting his personal safety to the RAF. Certainly at the RAF Display he was much at ease and enjoying the occasion as he watched from his seat in the front row behind the familiar white-painted railings of Hendon. Said a spectator: 'At one time a curious ripple, like a wave slowly moving the heads of the crowd, turned out to be the raising of hats extending from one side of the promenade to the other, and then we recognized the lonely form of Sir Hugh Trenchard strolling down the hill and being greeted by all past and present Air Force people. Somehow it reminded one of a battleship and its bow wave.'

Because of the imminent Olympia Aero Show there was no parade of prototypes, but in any case the participating machines revealed the new look of the RAF. The morning passed pleasantly with races and lesser exhibitions as a warm-up for the main programme which started at 3 p.m. with three squadrons comprising twenty-seven Siskins taking-off in formation and playing follow-my-leader around the sky. Two Gamecocks flown by the winning pilots of the Area aerobatic competition displayed loops and rolls. Three ancient Vimys rumbled off with a parachutist on each lower wing, clinging to the mid-interplane strut, and after sedately touring the local countryside all six men simultaneously dropped, followed by a gasp from the crowd as one landed on the canopy of another, though both came down safely. Then it was the turn of five Supermarine flying-boats from Calshot to drone into view, one leaving the formation to skim ponderously across the grass and zoom up in front of the Royal Enclosure in salutation to the Prince of Wales. There came a thrilling bomb attack by three Flights of Siskins diving from three directions, so that it seemed they must collide, though it was the standard exercise of each Flight covering the attack of the next. By contrast, a Boulton & Paul Sidestrand and two Siskins performed a somewhat tame fight which reinforced the impression that the Sidestrand had heavy controls and was not so manoeuvrable as the Bugle of old. Army Co-operation next had its share with three Atlas at a time using their undercarriage fishing-rod hooks to pick off streamered messages suspended between paired poles – then, wheeling round, took them to a gun base.

With long hook dangling, the slotted Armstrong Whitworth Atlas dashes in to snatch up a message tied to a line stretched between two posts.

This year three Genet Moths instead of two did the inverted flying, piloted by Flg Off Dermot Boyle, Flt Lieut J. S. Chick and Flg Off W. E. P. Johnson, who curved around in formation upside-down, or with one upside-down and two right way up, and then two upside-down and the other right way up. Next came low level attacks by screaming Felix Foxes of No. 12 Squadron and roaring Bulldogs of No. 3 Squadron as the latest example of fighters. As *The Aeroplane* recorded: 'Though hardened by over 20 years of aviation one confesses to being thrilled by the howl of those diving Foxes. One is convinced that the sound alone would be

Nose gunner looks at tail gunner as the heavy Vickers Virginia Xs slowly taxi out in formation at the 1929 RAF Display. (*Flight*)

enough to quell any tribal war in five minutes. And no matter what alteration in the fashion of aircraft happens in the future, one holds that we shall have to go a very long way before producing an aeroplane so classically beautiful as the Fox.'

With a solemnity of their own, an armada of the green-painted night bombers thundered off, comprising No. 7 Squadron of Virginias and Nos. 10 and 99 with Hyderabads, and vanished into the distance in stately fashion to await the set-piece. Meanwhile the universally popular crazy-flying of two Avro 504Ns thrilled the crowd for, time and again, it seemed they must crash head on, yet skidded past at the last moment with wing-tip clearance. More air combats followed, again with a Sidestrand in leading rôle, but escorted by three Siskins which, to the mystification of many, were attacked by another Siskin of the same squadron, resulting in a

As one of the major time-honoured items at the RAF Display there was the usual breath-taking crazy-flying by two Avro 504Ns. (*Flight*)

fight with one of the escort which was quite exciting. Last came the great aerial attack on a foreign port, the picturesque set-piece waiting on the aerodrome with quay and Customs house at one end and fort at the other. On a sea made from blue parachutes billowing in the wind, a transport steamer was embarking troops and stores from lorries when a photographic reconnaissance Atlas came over and was attacked by 'Archie'. Up went a kite balloon to scan the sea for an aircraft carrier which must have been hull down, for presently a formation of antique Fairey Flycatchers shot down the balloon and machine-gunned quay and transport, and as they climbed away a squadron of Siskins dived to defence of the fort and attacked a formation of Fairey IIIF bombers and the squadrons of night bombers which had at last reappeared. Guns rattled, bombs thundered, targets flashed with explosions, and even when bombs were supposed to have fallen in the sea the RAF had cunningly arranged for a splash and fountain of mud. Finally the Customs house went up in flames to the joy of every schoolboy. And so home, the long queues good humoured, and the RAF delighted that everything had gone so smoothly. And as the visitors departed I saw the tall stern figure of Air Marshal Sir Hugh Trenchard, hat in hand, head inclined towards the slight, youthful figure of the fair-haired Prince of Wales. What was that great soldier thinking, I wondered: these men, those aeroplanes, that flying, were the children of his mind.

8

In the following week the arrival of Kingsford Smith and his crew at Croydon with the weather-worn *Southern Cross*, having flown from Australia in a mere thirteen days, attracted little attention, though Sir Sefton Brancker was there to welcome him. 'Smithy' merely stated he was 'here to purchase four Avro-built Fokkers of the large type which would be on display at Olympia'.

When on 16 July this seventh International Aero Exhibition was opened by HRH The Prince of Wales it appeared that the British aeronautical industry must have reformed, for instead of feverish work on uncompleted machines, every stand was ready, somewhat incongruous though aero-

One of the Lynx-powered British Fokker Avro Tens of 71-ft span ordered by Sir Charles Kingsford Smith for Australian National Airways.

planes seemed in such Victorian surroundings of cast-iron balustrades and bunting-decorated domed glass roof. It was a formal occasion of morning dress for senior members of the industry, followed by lunch somewhat better than tea-shop standard, at which HRH proposed a Toast of success to the exhibition, pointing out that this same month would see the 20th anniversary of the Channel crossing by Blériot.

Judging by attendance in the next few days, interest of the British public at first seemed negligible, though to all engaged in aeronautics the opportunity of closely studying so embracing a collection of aircraft made it a fascinating affair. C. G. Grey's assessment was, as always, shrewd: 'The outstanding feature of the Show is the competition between steel and duralumin and between air-cooled engines and water-cooled. You can see what can be done with steel by such firms as Bristol, Armstrong Whitworth, Gloster and Boulton & Paul; and you can see what can be done with duralumin by Vickers Ltd and Short Brothers. The Fairey and Hawker firms seem strictly neutral, and timber is almost dead as a material.

'Without mentioning names, there are two makes of aircraft in the Show which probably have higher performance with identical engines than anything else. One is designed by an engineering team of which each member is a brilliant specialist and the lot are overlooked by one man whose only form of argument is to say "That will do" or "Take it away".' That in fact was Fairey. 'The other is designed chiefly by a man who started life as a working carpenter and educated himself at night schools to become a scientist with real practical knowledge. His immediate boss, when one first

Olympian pre-opening unreadiness viewed from the Fairey stand, with the Long-range Monoplane on high (*right*), and Vickers and Shorts confronting each other in the distance. (*Flight*)

The far end of the Main Hall at Olympia with the steel-framed Gloster A.S.31 and the Gnatsnapper fighter in the foreground. (*Flight*)

met him, was one of the finest engine mechanics who ever lived, and to-day is a man of wealth, vast business sense, a thorough knowledge of practical mechanical engineering, and a natural eye for an aeroplane which prompts him to go round and knock off the corners.' Those two were Camm and Sigrist. 'In each of those firms the scientist is in his proper place. He is used in order to tell the practical engineer why his practical ideas work or do not work. And because the practical man is at the top, he has in each case managed to collect scientists whose calculations come out with amazing accuracy. But when once the scientist is at the top, things go to pieces because he is always ready to prove by mathematics that low performance is due to mistakes on the part of the practical engineer.' Grey proceeded to comment that: 'There seems to be a tendency these days to make a low-wing single-seat tractor fighter. The fighter on the Vickers stand is one example. And machines of that type have been built, though they are not being shown, by Boulton & Paul, the Westland people and de Havilland. If your machine is fast enough, there is no need to worry what is underneath your wing. All that concerns you is what is in front of your gun sights and whether anything is likely to pounce from above.'

Not fighters, but what seemed a scaled-down single-seat edition of the Westland cabin Widgeon displayed on the A.B.C. Motors stand as 'the Robin, price £375 with 40 hp Scorpion' attracted my initial attention, for its Martinsyde characteristic of raked wing-tips and tail led to discovery that Tony Fletcher had designed it – so there was joyful reunion before inspecting other aircraft.

The 25-ft span, folding shoulder-wing single-seat A.B.C. Robin clocked over 100 mph, but found no purchasers. (*Flight*)

Sir W. G. Armstrong Whitworth Aircraft Ltd had a combined stand with their Avro division, and exhibited the well-tried Atlas and Siskin IIIB; an A.W.XIV single-seat fighter with a Townend ring around its Jaguar; a slick Avro Antelope; the impressively large three-engined Fokker-type Avro Ten for eight passengers, and the Avro Five miniature version for four; and a Hermes and a Genet Avian having welded steel-tube fuselage frames. Said a critic: 'The two Fokker type machines, strike one as among the prettiest and most efficient looking in the Show. One gathers the price is a trifle high, but the superlative workmanship and finish will certainly bring orders.'

By scaling down the Fokker F.VIIb-3m, Chadwick devised the 47-ft span Genet-powered four-passenger Avro Five, of which two were purchased by Mrs Wilson of Wilson Airways in Kenya.

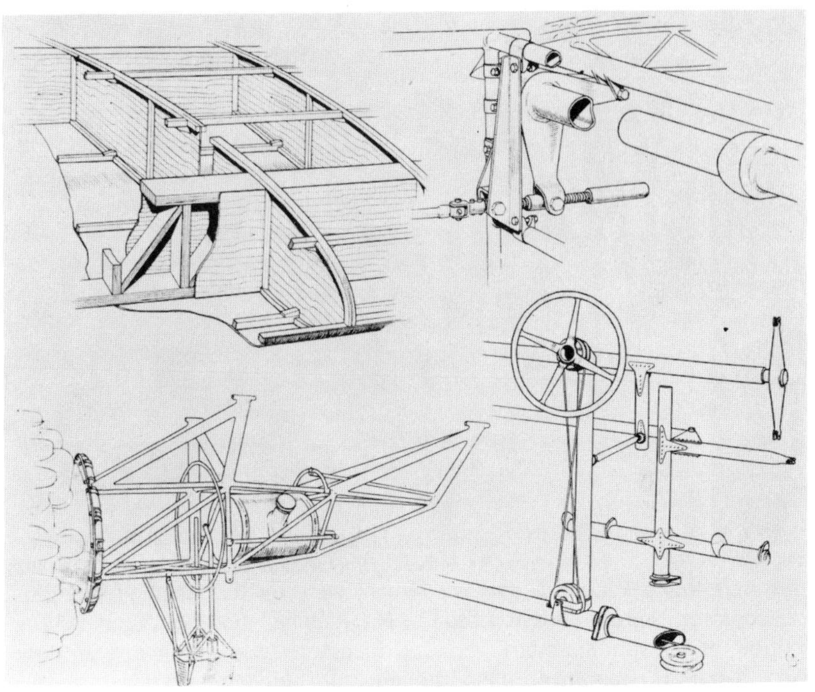

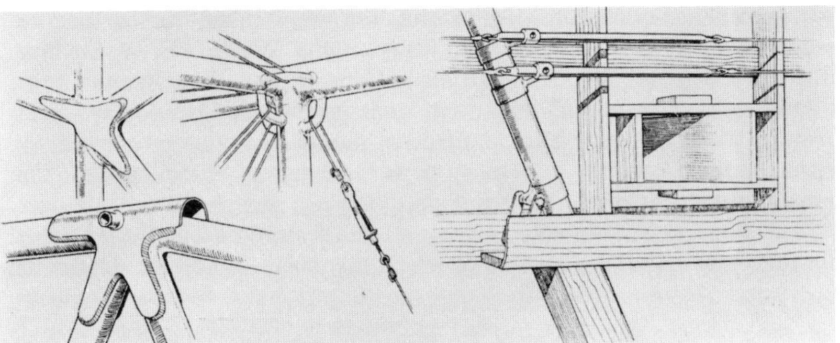

Details of the Avro Ten showing wooden wing, engine mount, control details and typical fuselage joints. (*Flight*)

The Blackburn Aeroplane & Motor Co Ltd overwhelmed not merely their stand but the entire Show with the huge hull of the Nile civil transport flying-boat, though the streamlined central cabane, which was such a feature of the design, was omitted, nor was the hull painted but shown with anodic finish. Those allowed to view the interior found an air of spaciousness in strong contrast to the average airliner currently in use. The fourteen pairs of facing seats each side of the central gangway were deep and had comfortable pneumatic upholstery, with a table between each pair and hand luggage space behind. Ventilation and heating were individually controlled, and bell-pushes communicated with the pantry. Almost hidden by

The 65 ft 7 in hull of the uncompleted triple Jupiter-powered Blackburn Nile high-wing commercial flying-boat. (*Hawker Siddeley*)

this hull was a Bluebird, and two others were on the stand of the concessionaires, Auto Auctions Ltd. The Nautilus had also been intended for exhibition but modifications were still incomplete, so a standard production Ripon II was substituted, though outdated compared with recent machines. Attracting considerable attention was a metal-framed Lincock, with geared Lynx, designed at the instigation of Wg Cdr E. W. Stedman of the RCAF, and intended for trials in Canada, though at Olympia it was unfinished and the special long-travel undercarriage had temporary wheels from a Ripon. Much play was made of its possibility as an inexpensive light fighter of half the customary power, but it was becoming clear that most governments preferred a full-sized first-line aeroplane.

The Boulton & Paul stand was small, but featured a Sidestrand poised on high, shown half uncovered to reveal the metal framework. Under the

Built for full-scale research, the Salmson-powered Boulton & Paul P.41 parasol Phoenix was used as a useful commuter by Sqn Ldr Rae their test pilot. (*George Swain*)

bottom longerons of the rear gunner's cockpit a special Boulton & Paul mounting had been fitted to enable a prone position with extensive field of fire. Within the spar bay of the fuselage, clear of slipstream but not enclosed, were racks for four 112-lb bombs, and under both lower wing roots were two external 230-lb bombs. Beneath this wide-winged biplane was a little open two-seat high-wing monoplane, the Phoenix, powered like the Robin with a 40 hp A.B.C. Scorpion and identically priced. Designed by Capt W. H. Sayers, formerly technical editor of *The Aeroplane*, it had a fuselage and tail somewhat similar to the Eastchurch Kitten which competed against his Grain Kitten in the war days.

Using similar strip steel construction to the sturdy Bulldog, the Bristol 110A, powered with a 220 hp Bristol Titan or 315 hp Neptune, carried four passengers but found no purchasers.

The Bristol Aeroplane Co Ltd displayed a Bulldog, half covered like the Sidestrand, and the new four-seat biplane, the 110A, painted palest blue but as yet untested. At first sight the 110A appeared retrograde compared with American high-wingers, yet its single-bay bracing was cleaner than their multi-struts though wing drag was greater. Construction was similar to the Bulldog, but the fuselage was Warren braced, and a special feature was the recently devised NACA cowling which completely enclosed a mock-up Neptune engine and would give lower drag than a Townend ring. The pilot had a roomy enclosure beneath the top wing leading edge, where view was good except to the rear, but in deference to open-cockpit pilots the topsides were permanently open and somewhat draughty. The main cabin was larger than in most American equivalents, and seated two passengers at the back and two forward on separate seats which were movable sideways and had folding backs to aid access to the cockpit. To compete, it was essential to have reasonably low price, yet estimated production cost was already £3,500, so prospects did not appear even as good as the D.H. Hawk Moth's which at £2,000 had already proved difficult to sell.

De Havilland Aircraft Co Ltd showed a cream and black Hawk Moth with geared Lynx and one wing folded to reveal the automobile-like saloon which was externally covered with leather fabric, though the rest was ordinary fabric. However, the centre of their display was an all-white enamelled Gipsy Moth coupé mounted on a pedestal show-case containing over seventy cups and trophies, including five of the eight King's Cups.

Though the de Havilland Hawk Moth with Halford's 200 hp V8 Ghost engine had no purchasers, eight were powered with the 250 hp Lynx and one with a 300 hp Whirlwind and sold to Canada and Australia.

The name *Moth* was strikingly embossed in gold letters each side of the fuselage. Two other Gipsy Moths, one in yellow and white, the other a blue and silver seaplane with white floats, had metal-framed welded fuselages made with square-tube longerons and circular spacers similarly to the Hawk Moth. All had lavishly cadmium-plated exhaust pipes and Raf-wires, giving an opulent impression.

That superfine quality was lacking in the blue and silver three-seat cabin monoplane shown by Desoutter Aircraft Co Ltd, though it unashamedly represented production standard. The Handasyde touch was apparent, for the windscreen had been improved, the tailplane lowered, a different fin fitted, and the undercarriage strengthened – but it remained substantially a stiffened version of Koolhoven's all-wood machine with Fokker-like wing.

From anywhere in the Main Hall one's gaze was drawn to the Fairey Long-range Monoplane mounted high above the white office on the stand of Fairey Aviation Co Ltd, for the great spread of the cantilever wing was dominant. Here at last was aerodynamic elegance that beat any other large machine – and yet, one reflected, this was because its special purpose enabled so slender a fuselage to be practicable; but surely that great undercarriage had more drag than all the rest of the machine? Why could it not be retractable without undue complication and weight? Why? Because except for oleo legs, development of hydraulics had been shunned by every designer.

Like Fairey's outlook, his stand was greater than all the others. Each corner was emphasized by a Fairey-Reed propeller. Facing the Addison Road entrance was the Firefly II single-seat interceptor, and along the centre aisle a Lion-powered IIIF in skeleton revealed that the fuselage was framed in welded steel-tube, with duralumin-structured wings; matching it was a three-seat seaplane counterpart with folding wings. Round the corner a single-seat ship-plane Firefly Mk II displayed little difference from Mk I

It was Fairey's extensive and profitable production of the IIIF which justified the size of his stand at Olympia.

except for a longer-stroke undercarriage and structure stiffened for catapulting. On the far side was the Fox, now standardized with Rolls-Royce F, but despite spearheading the new look for the RAF it still had no further Service orders and Fairey was still building the two-seat replacement rivalling the Hart. Next to it a IIIF with Armstrong Siddeley Jaguar was shown as a general-purpose two-seater, originally intended to compete with the Wapiti – but whatever the guises of the IIIF it remained basically a polished and streamlined development of the original war-time, slab-sided Fairey seaplanes, yet as one visitor put it: 'The engine cowlings of all

The Fairey Ferret II remained yet another variation of the well-proved Type III series airframe, but with a radial engine.

Fairey machines form a striking feature. They appear to have the engine poured in. Only close examination shows the way the panels are secured with patented Fairey clips to the main structure.' They had been adopted by every manufacturer, superseding traditional skewered hinges because of low drag, lightness and simplicity, for they consisted of a small circular stamping with spring-loaded plunger locked by a half turn with a screw-driver after pressing into a dished stamping on the cowling support frame. At 2s a time, stamped out by the hundred thousand, they were proving most profitable.

Whereas the prototype Gloster Gnatsnapper had flaps, rounded fuselage, and Mercury IIA, the second had a Jaguar VIII, and was flapless, had flat sides, and, instead of the rudder balance exposed above a low fin, it was shielded by a tall, curved fin. (*Courtesy Air Marshal Sir Ralph Sorley*)

There was a tendency to dismiss the Gloster S.S.18A as 'just another dated fighter' because of the two-bay wings, but none could guess it would prove precursor of a famous fighter in the next decade. (*Flight*)

Effective because designed for a single purpose, the prototype Gloster survey biplane was ordered by de Havilland's chairman, Alan Butler, for another of his interests, the Aircraft Operating Co Ltd. (*Flight*)

A very different school of design was represented by the five machines on the stand of Gloster Aircraft Co Ltd – the Gnatsnapper single-seat ship-plane fighter; an unnamed single-seat interceptor fighter biplane which was the two-gun S.S.18A; the pretty little Gloster IV Schneider racer in attractive blue and bronze; and the pale blue, tiny Gannet single-seater of 1923. Dwarfing them all was the black-enamelled structure of a twin-engined 61-ft span, metal-framed survey biplane. Though known as the A.S.31, this big machine was little Joey Preston's redesign of the D.H.67, and its prototype sister ship had been built for the Aircraft Operating Co and first flown in June at Bockworth by Saint who found that it had a remarkably short take-off run and that it could maintain height at 10,000 ft with full load on one engine so was safe to fly over dense jungles. The stand also featured the Hele-Shaw/Beacham variable-pitch propeller, and displayed a progressive series of high-tensile steel wings ranging from those built at the end of the war for the B.E.2d and Avro 504K to the Avro Aldershot, Gloster Gamecock, Westland Wapiti, and de Havilland Moth.

A corner stand between the two Halls featured the toy-like 25-ft span 40 hp Glenny and Henderson Gadfly, a low-winger of such simple design that it would suit the home-builder – but no demand followed, nor for a five-seat pusher tail-boom monoplane of which there were large photographs.

Handley Page Ltd had drawn an unlucky position, half under the gallery of the new Hall, for in the typical frugal manner of their founder they had requested the smallest area on which their large machines could be exhibited. An olive-painted Hinaidi fuselage rigged with the central portion of the wings tended to be passed unnoticed in the gloom, but attention was certainly riveted to the long, white-painted, elegantly furnished mock-up cabin portion of the new H.P.42 four-engined airliner ordered by Imperial Airways. It was like a Pullman of considerable height and width, with accommodation in two sections, each seating 20 passengers in pairs with central gangway, the remarkable chintz-covered seats rivalling the Nile's

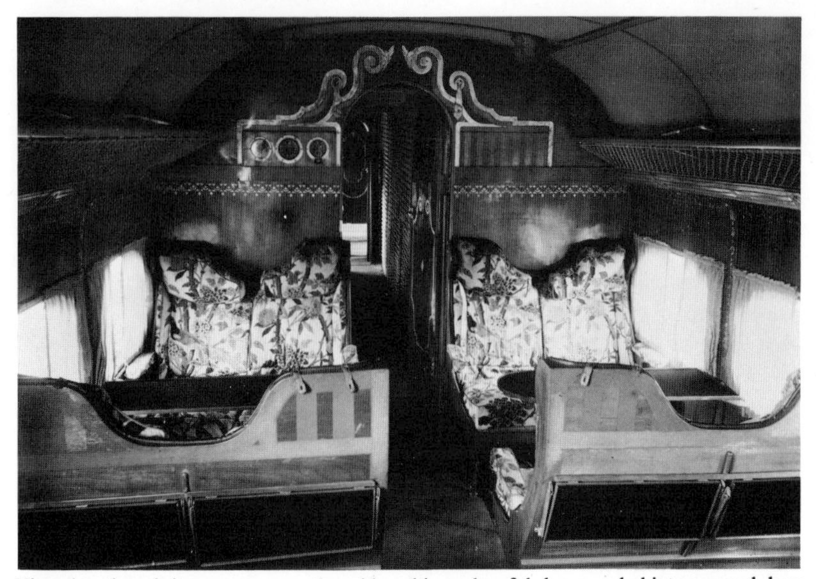

Victorian though it now appears, the wide cabin, colourful decor and chintz-covered deep double seats of the projected Handley Page H.P.42 were acclaimed as setting new standards of airliner comfort. (*Flight*)

for depth and leg room, and similarly had individual warm and cold air control and electric bell-push. In the ceiling above each cabin was an emergency rip panel disguised in the *décor*. Between the cabins was a cocktail bar to port, and two toilet rooms and a baggage compartment opposite. So that visitors could peer into this splendid airliner, an elevated gangway had been rigged outside.

The famous Hawker Hart prototype, J9052, outpaced established fighters so easily that it led to RAF production orders for some 1,000, plus 1,600 fighter, army co-operation, and bomber versions. (*Courtesy Air Marshal Sir Ralph Sorley*)

With its establishment of clear cockpit access the Hawker Tomtit marked a new outlook for specialist design of trainers, and its controls seemed ideal for this purpose.

Much better situated was the neatly efficient stand of H. G. Hawker Engineering Co Ltd in the same Hall, displaying three machines – the Hart two-seat day bomber, of which fifteen had initially been ordered; the new and beautiful Hornet single-seat interceptor, now with a 480 hp Rolls-Royce F.XIS replacing the XIA with which it made its first flight in the spring; third was a Moth-size but more rugged two-seat trainer with wings slightly swept back so that the two cockpits could be placed abaft the centre-section for unobstructed parachute escape. Named Tomtit, and powered with a 150 hp Mongoose, it had been flying since the beginning of the year, and so immediately satisfactory had the RAF found it and the incorporated Reid and Sigrist blind-flying panel, that an order for ten had been placed in March with the intention of adopting it as basic elementary trainer in place of the outmoded Avro 504N. Like wasps to a honey-pot came every visiting pilot and technician to see the Hawkers.

Down the aisle towards the annexe was the stand of George Parnall &

However experienced the designer there was no guarantee of success. Of the Parnall Elf a Martlesham pilot said: 'Hadn't a hope; directionally unstable, bad take-off, poor elevator and aileron.' (*Flight*)

Co showing their two-seat, Warren-braced, folding wing Peto, designed to fit the minute sealed hangar on a submarine's deck; and alongside it was Bolas's latest two-seat light aeroplane, the Elf, intended as a competitor to the Moth, but of similar pattern to the Peto, and flown only a fortnight before the Aero Show.

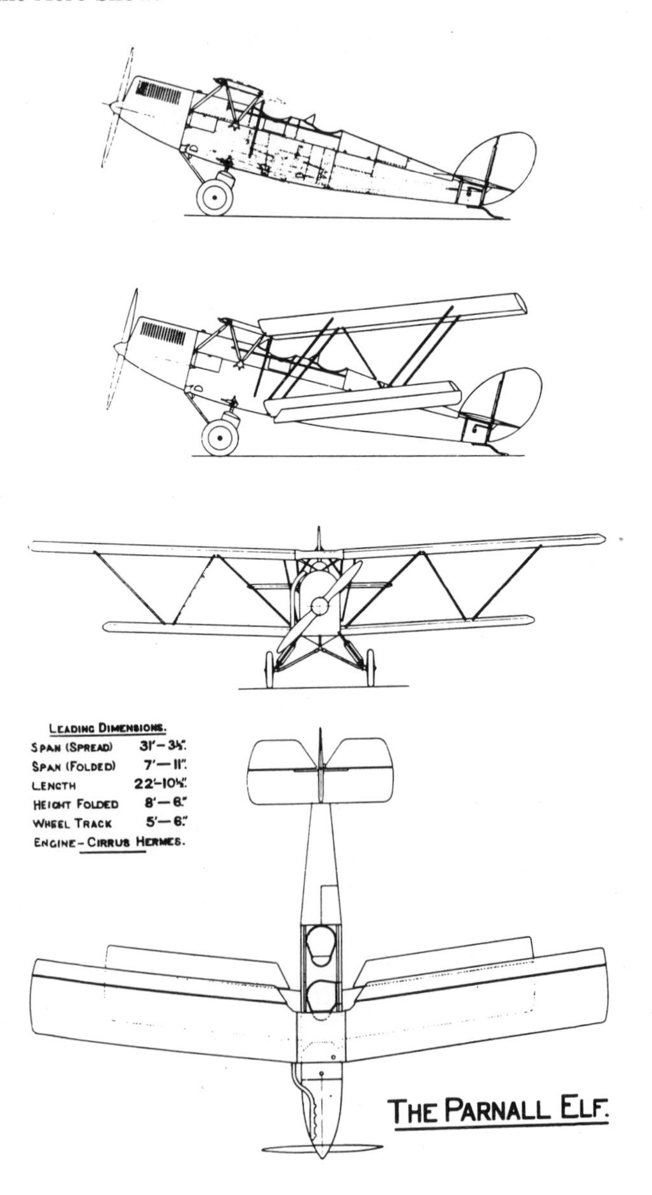

The Parnall Peto's deck hangar on the submarine M.2 seemed impossibly small, and ultimately led to catastrophe.

Hard by the Addison Road entrance the stand of S. E. Saunders Ltd featured Knowler's fascinating little twin-engined cantilever monoplane flying-boat, named Cutty Sark and painted blue, with metal hull based on the Saunders A.14, using flat sheets of duralumin to simplify manufacture, and a Fokker-type wing, containing the fuel tanks, was covered with three-ply on which were trestle-mounted twin Hermes. A Heywood compressed-air starter was fitted, as the propellers were out of reach. The cabin, located ahead of the wing and fully enclosed, had four seats arranged in two pairs, with low passage way between to give sufficient head room. The price was £3,500, or in amphibian form an extra £250, and it attracted considerable attention as a possible flying-boat trainer.

Nearby was the stand of Short Bros (Rochester & Bedford) Ltd, dominated by the Singapore, which, despite RAF tricolours on hull, wings and rudder, was the rebuilt machine used by Sir Alan Cobham on his African flight – the planing bottom having been modified to the lines of the Calcutta, and Rolls-Royce H.10 engines (later known as the Buzzard) substituted for the Condors, and Lamblin radiators under the top centre-section instead of nose radiators, resulting in a much neater nacelle. The

The prototype Saunders-Roe Cutty Sark, at Cowes, being prepared for Olympia. The high engine position was a gamble, but did not create undue trim problems.

public, with their hero Cobham in mind, thronged to gaze at his machine in admiring wonder that man dared to fly so huge and wonderful a creation; but of greater interest to overseas clubs and operators was a Moth with single central float carrying fork-mounted wheels on a rotatable axle linked to a screw mechanism to tilt them clear of the water or lower them for land use. As interesting comparison the Short Mussel Mk II was shown alongside with twin floats, but could take the same single float as the Moth. The original Mussel had been crashed the previous year by that keen novice pilot, Eustace Short, for he hit the mast of a barge with his wing-tip when gliding in, and cart-wheeled into the water. The new Mussel not only had a metal-skinned fuselage but metal wing spars and ribs of M.12 section, and had just returned from airworthiness trials at Felixstowe. Shorts also displayed a set of floats for a Wapiti recently tested as a seaplane at Felixstowe, and indicative of the shape of things to come was a model of the three-engined Valetta seaplane whose construction was sufficiently advanced for one of its duralumin box wing spars to be used as a balustrade of the stand.

Although displayed on a Moth at the Short stand at Olympia in 1929, the Short-built single-float gear was also fitted to the Mussel and carried a rotatable axle so that the pilot could swing the wheels up for water operation – but it was an untidy adaptation.

Like Desoutter, Simmonds Aircraft Ltd was unfortunate in having to be in the annexe – which otherwise contained only foreign exhibits. Two Spartans were shown, one a silver and blue three-seater with Hermes; the other, in the orange and black of National Flying Services Ltd, was a two-seater with Cirrus III. To emphasize interchangeability there was an ingenious model of a Spartan fuselage with a single lower wing which could be swung on rails across the fuselage until inverted the other side, where it could be employed because of the symmetrical section.

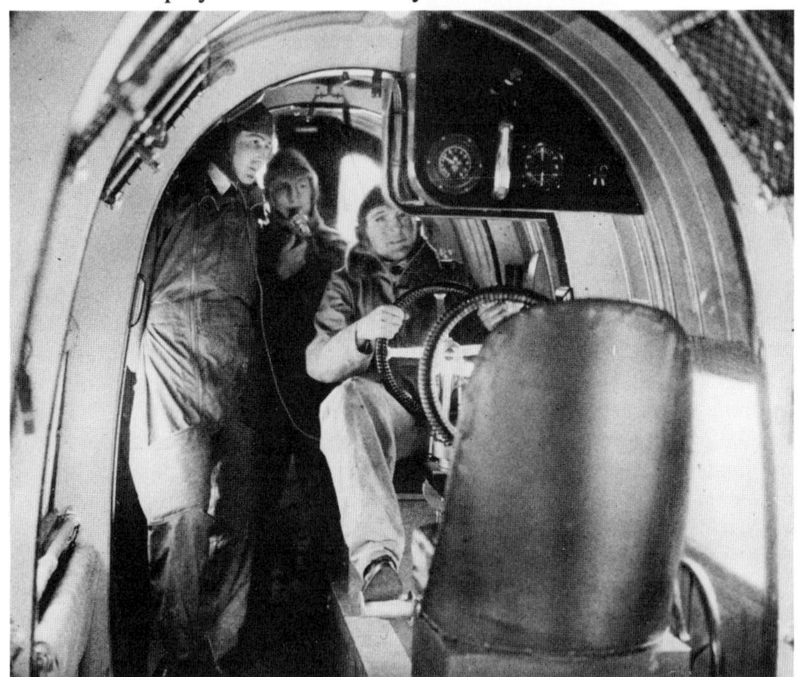

Although unobstructed, the cabin of the Vickers Victoria was relatively narrow though tall, as can be gauged from this interior of a blind-flying trainer version. (*Charles E. Brown*)

Now that Supermarine Aviation Works Ltd and Vickers (Aviation) Ltd were united they had a common stand, and the S.5 winner of the 1927 Schneider was mounted above as though diving at great speed. Almost filling the area was one of the metal-hulled Southamptons of the Far East cruise alongside a Vickers Victoria shown with lower wings uncovered to emphasize it was now metal-structured, with dural spars having channel-section flanges and zig-zag riveted webs. In the corner in front, the Vickers F.20/27 Jockey low-wing interceptor fighter was also mounted in a banking airborne position which revealed that the duralumin wing panels were attached Wibault fashion by upturned edges sandwiching the rib and held by U-section capping strips, and though the forward fuselage was smooth metal monocoque it was fabric covered over stringers mounted on the usual metal-framed structure abaft the pilot's cockpit.

Though hardly a clean design, the Hermes-powered Westland IV for Imperial Airways performed more effectively than its neat rival, the Avro Five.

Finally came the Westland Aircraft Works with a stand at the end of the Main Hall showing the latest Westland Four with Hermes engines, now known as the Wessex and built for Imperial Airways. It made interesting comparison with the much cleaner Avro Five, but Imperial had found the handling characteristics of the Wessex better, particularly at the stall because of the R.A.F.34 wings with negligible c.p. travel, and it took-off and landed in smaller space. Tucked partly under the balcony was a Jupiter VIII Wapiti and stripped Wapiti and Widgeon metal fuselages, but the Air Ministry would not give permission to show later aircraft.

Though fixed-wing machines completely dominated the exhibition, the Cierva Autogiro Co Ltd exhibited their latest C.19 Mk I Genet powered two-seater, which had a remarkable box tail which could be turned upward

Development of the Cierva Autogiro made a bound forward with the self rotor starting C.19 which was widely demonstrated by Flt Lieut A. H. C. Rawson. (*Flight*)

to deflect the slipstream onto the rotor blades, spinning them to 130 rpm in 45 seconds, at which the tail was returned to neutral, wheel brakes released, and the final vital 20 rpm attained during the take-off run of 30 yards.

While export prospects for aeroplanes were limited, those for British engines opened far and wide, and every manufacturer exhibited his full range, but there were also new engines: A.B.C. Motors Ltd featured their opposed flat four-cylinder air-cooled 80 hp Hornet which was about to be air tested in the Widgeon; A.D.C. Aircraft Ltd had a 300 hp six-cylinder inline; Armstrong Siddeley revealed their 700/805 hp fourteen-cylinder Leopard; Bristols displayed the Mercury though still on the Air Ministry Secret List; de Havillands featured the sealed Gipsy engine which had been duration tested in the Moth; Napiers displayed the 875 hp Lion which won the Schneider; Rolls-Royce had their 830 hp Type H; and, to the surprise of many, the Sunbeam Motor Car Co Ltd endeavoured to stage a come-back with a 1,000 hp twelve-cylinder vee water-cooled engine and a six-cylinder 104/112 hp compression-ignition engine. Parnalls also showed the geared Pobjoy. A novelty known as a Lever engine was staged by Redrup, and had seven cylinders disposed horizontally around the crankshaft giving minute diameter – but it still had to pass airworthiness tests.

Additionally there was a wide range of foreign engines and aeroplanes, including an all-metal Ford Tri-motor airliner which successfully amalgamated Junkers corrugated skinning with Fokker geometry.

The Junkers K-47 gave an impressive display at Heston, revealing Germany's potential ability if it came to another war. (*Flight*)

Throughout the week demonstration aircraft were available at Heston, and on 20 July Norman and Muntz gave an aerial garden party attended by 67 aeroplanes. As people in thousands began to arrive, a Junkers K-47 single-seat strut-braced low-wing fighter with supercharged Jupiter VII, together with a Junkers K-37 twin Jupiter-powered low-wing multi-seat fighter began a demonstration. Though nominally of Swedish design the Junkers fighters represented the veiled threat of Germany, and as *The Aeroplane* put it: 'The whole effect was grim and beautiful as they flew round and round in front of the enclosure a few hundred feet high, with the little machine sitting on the big machine's tail, and every now and then some curious opening through the gun tunnel of the big machine caused a sudden glare of light that one would have sworn was the flash of a machine-gun.' Both gave an astonishing display of aerobatics, looping, rolling, and spinning, but in landing floated as they had no air brakes and tended to swing after 100 yards because the rudders were no longer effective and the wheels had no brakes.

The duet was followed by the Hawker Hart flown by George Bulman, described as 'quite the most able demonstrator in this country. One has not met any other pilot who can show off an aeroplane at every conceivable angle at a certain place on the ground in the same way as he can. And he certainly showed off the Hart.' But the star turn was Cyril Uwins with the supercharged Bulldog, who 'even when he does an almost vertical dive and then goes up almost vertically shows no jerk at the bottom of the dive, and when he does an upward spin on his climb the machine seems to be doing it quite easily and naturally without any of the flick common with this trick.' Yet in an aerobatic competition it was Hubert Broad with a Moth who won, for the judges not unjustly considered that it was more difficult to present an effective performance with a lower powered machine than the spectacular effects available to those with greater power.

By the end of the week, warmed by the ever-flowing hospitality of his friends in the industry, *The Aeroplane*'s editor declared: 'Taking it all round the Aero Show is a really fine effort by the aircraft industry. If there is nothing startling or revolutionary in the design of any of the aircraft it at any rate shows we can make aeroplanes against anybody in the world. After all, if you want to get the best things in the world, whether a motor-car or a pair of boots or a suit of clothes, you have to go to England for it. Despite our bleats about bad trade we are the best fed and best housed and best clothed and best amused people in the world. We may have more unemployed than in other countries, but there is far less poverty, and altogether, jolly old England is very much to-day the Merrie England of the spacious days of Good Queen Bess. And regardless of all the wailing of the Trade about lack of support, Olympia was simply packed with people throughout Saturday, right until closing time, in spite of the blazing heat and all the other attractions of a London holiday afternoon.'

But was Britain really in the van of progress? True a good machine such as the Hart handled with perfection, and certainly our metal hulls were seaworthy, well proved, and elegant – but as engineering feats could they compare with Germany's giant creation, the 157-ft span Dornier Do X monoplane with its ship-like hull and six rows of tandem Siemens Jupiters mounted on an auxiliary wing above the gigantic rectangular 2½-ft thick wing which was strut braced to the sponson? The hull had three decks, with the topmost for captain, second officer, two pilots, engineer officer, four mechanics, telegraphist, cook and steward. Stairs led down to the main deck which had four- and six-seat cabins. Beneath were fuel and oil tanks. The nominal disposable load was 15 tons at 36½ tons all-up, but on

Many thought that Dr Dornier had overstepped the mark with his twelve-engined Do X, but it was a splendid experiment.

21 October she took off in 50 seconds at the tremendous over-load weight of 52 tons, flying for an hour with 10 crew, 150 passengers, and nine stowaways. Here was the greatest indication of all time of the future.

Nevertheless at Blackburns, Major Rennie had already investigated a high-wing flying-boat nearly as big, with all-up weight of $33\frac{1}{2}$ tons and six engines totalling 4,800 hp, but though the payload was 12 tons at a range of 250 sea miles, only 2.5 tons could be carried for 1,000 miles, equivalent to 10 hours cruising flight which Rennie said was the practical limit, for 'as a commercial proposition, a non-stop transatlantic flying-boat service is still beyond the science and technique of flight.' Though the design was established in considerable detail, the Air Ministry would not support so ambitious a project for what they considered was merely a prestige race.

Germany had not confined herself to maritime experiments. There was also the remarkable Junkers four-engined landplane, the G 38, which had a tapered cantilever wing 8-ft deep at the root, and, like the Westland Dreadnought of five years earlier, had part of the passenger accommodation within the wing. Range was estimated at 2,000 miles with 3 tons of useful load at an overload all-up weight of 24 tons, but it was indicative that take-off then required what seemed the fantastic distance of 1,500 metres – for there were no aerodromes in England even approaching that size. We were still wedded to the war-time outlook of little English fields. Thus the Westland Wizard took-off in a mere 300 ft, the Wapiti in less, the Virginia under 600 ft, as did the contemporary Vireo monoplane. No wonder the development of flaps and variable-pitch propellers was not being pursued with vigour.

On the last day of the Show, Louis Blériot, now 57, flew to England as assistant pilot of a twin-engined Blériot 127 in celebration of his flight across the Channel twenty years earlier. A great crowd watched at Croydon as his machine approached escorted by other aircraft, and Lord Thomson, Frederick Montague, Sir Sefton Brancker, and George Woods Humphery were there to welcome him. That evening the SBAC held a banquet in his honour, chaired by the Minister, and Brancker in his clipped, characteristic manner, toasted 'The Sport of Flying', saying that he found governments not so stupid as they look; consequently we had light aeroplane competitions officially promoted and even mediocre support for flying clubs as a sporting movement run by young men and women in their spare time – with the result that we had more private owners and more sporting flying than the rest of the world put together. Said Sir Sefton: 'The great thing to promote progress is to get away from rules and regulations. Personally I am all in favour of letting people break their necks if they wish. Sportsmen are usually poor, consequently we owe much gratitude to supporters of sporting flying such as Jacques Schneider, Gordon Bennett, Pulitzer, and Sir Charles Wakefield whom I always call the Patron Saint of Aviation.' Capt Peter Acland, toasting 'Lord Thomson', said the occasion of Blériot's visit was a happy culmination of one of the happiest periods in the history of British aviation.

Certainly it had engendered a renewal of enthusiasm, for the following morning a party of SBAC guests embarked in two *Golden Ray* aeroplanes of Air Union, and accompanied by Blériot in his big monoplane bomber, left Croydon for St Inglevert, and over the Channel were met by a replica of the original epic monoplane. At St Inglevert there was tremendous welcome for Blériot, and then the party drove to the monument at Les Baraques marking the spot where his flight had started in 1909. A civic banquet followed, attended by a thousand notables, and a great display of flying was held before the party left with M Blériot for England to circle the slope where he had landed at Dover twenty years earlier. Next day he gave a short lecture to the RAeS on his early experiences and showed a unique film of his flight attempts from the time of collaboration with Voisin onwards – for Blériot had always been a keen amateur cinematographer, and every machine he built was recorded on film that still exists for the benefit of posterity.

9

Airships increasingly became the background. On 29 July a week-long inflation of R.100 began at Howden, supervised by Lieut-Cdr Burney. It was hoped that within two months the airship would take the air. Vickers Ltd had long been anxious about the delay, and at the annual general meeting the chairman, General the Hon Sir Herbert A. Lawrence, GCB, said: 'It may not be inopportune to mention that the Vickers group accept contracts for work of national importance and of an experimental nature on which losses are incurred, as for example the giant airship R.100 now being constructed by the Airship Guarantee Co – one of our subsidiaries. This contract was accepted in 1924, and it has been necessary to provide a considerable sum to cover the estimated loss on completion because the airship has taken much longer to erect than originally estimated. A similar airship is being constructed by the Air Ministry at Cardington, the whole cost of which falls on the taxpayer, but Vickers, as private contractors, have to shoulder the loss themselves.'

Part of the passenger accommodation in the *Graf Zeppelin.*

Meanwhile the *Graf Zeppelin* was in the news. An attempt in May to fly to New York had been abandoned over Spain when three of the four engines failed due to synchronous vibration which broke the crankshafts, but on 1 August, commanded by Dr Eckener, another attempt was made carrying a crew of 42 and 19 passengers, among whom were Sir Hubert Wilkins, the Chinese film star Anna May Wong, and Susu a gorilla. In the general cargo of some 200 tons were a valuable Rubens picture, a Bechstein grand piano, 600 canaries, and a bust of the late Baron von Hünefeld who had died from cancer a year after his Atlantic flight with Fitzmaurice. Proceeding by way of Gibraltar and the Azores, after which radio contact could be maintained with the USA, the *Graf Zeppelin* was next sighted over Cape May, New Jersey, at 9 p.m. BST on 4 August and landed at Lakehurst an hour later. That the safety of such journeys was regarded with doubt was apparent from the relief with which President von Hindenburg and Lord Thomson immediately cabled congratulations. On 10 August the *Graf Zeppelin* landed back at Friedrichshafen on the 61st birthday of Dr Eckener. Five days later he began a world flight, heading across Russia and Siberia to Tokyo which he reached in three days, departing for Los Angeles on 23 August and arriving after a flight of 5,800 miles in 68 hours, but awaited dawn to land his 19 passengers before a throng of 10,000 spectators. The airship left on 27th, and two days later, after a stormy passage, reached Lakehurst where 16 passengers and 60,000 letters were landed. After the time-honoured ticker-tape greetings as crew and passengers were driven up Broadway to New York City Hall, *Graf Zeppelin* left on 1 September for Friedrichshafen commanded by Capt Lehmann, one of the war-time raiders on London, while Eckener remained in the USA to inspect two rigids being built by the Goodyear-Zeppelin Corporation. On 4 September the airship arrived safely in Germany having encircled the world in 20 days 4 hours, thus putting Jules Verne's dream of *Round the World in Eighty Days* forever in the shade.

Less spectacular was an aeroplane which was ancient history, for on 2 August the Duchess of Bedford, piloted by Charles Barnard, with R. F. Little as engineer, left Lympne in an attempt to fly to India and back within a week. The aeroplane was the Fokker *Spider*, built in 1926, with which Barnard and Alliott attempted the same flight in 1928, and had also been used by McIntosh and Bert Hinkler earlier that year with the same intention, but they had become lost and landed in Poland. Earlier still, named *Princess Xenia*, it had been lent by KLM to McIntosh when he tried to fly the Atlantic from Dublin in 1927. In the latest record attempt, re-engined with a geared Jupiter VIII loaned and fitted by Bristol for publicity, this Fokker covered 1,350 miles on the first day, arriving at Sofia in the evening. On the following day it reached Aleppo in Syria after a further 850 miles, and next day achieved 1,090 miles arriving at Bushire in Persia, thence on Monday, 5 August, 1,060 miles to Karachi. On Tuesday morning it began the return, and reached Croydon on Friday 9 August at 5.30 in the afternoon amid tremendous acclamation because a woman had shown what could be done in speeding Imperial communication. Bristol celebrated that the engine came from that city with a civic luncheon in honour of the Duchess and her crew. 'The task had called for physical endurance which any young man might have shirked,' said the Lord Mayor. 'Throughout the flight the Duchess had been an active, useful, and efficient member of the crew.' In reply she said that for her two companions it had been strenuous, but for herself it was a pleasure cruise. She had had the best of pilots, the best of mechanics, and the best of engines. Like Robert Bruce they had taken their lesson from the *Spider* and though at first they had not succeeded, they had tried again and spun their web across the Continent.

But now came the excitement of the Schneider preliminaries. On 12 August the Press visited Calshot Air Station to view the British competitors – and they were very impressed. The Rolls-Royce powered Supermarine S.6 seemed the last word in stark efficiency, and the Lion-powered Gloster VI was the most beautiful monoplane yet built. A glance at the previous year's S.5 monoplane and Gloster IV biplane, parked on the slipway, showed a great step forward by Mitchell and Folland. Though it was clear to technicians that the silver and blue S.6, which had 700 hp more than the fully exploited Lion's capacity, offered the best chance, reporters were more excited by the Gloster which was variously described as 'a winged meteor of gleaming gold', 'the Golden Arrow of the air', 'Venus of the air'. Apart from engines there was a special difference in the two contenders, for the S.6 had an all-metal wing in which even the flush radiators, consisting of two thicknesses of duralumin with water space between, helped take the torsional loads – whereas the Gloster had a

Reginald Joseph Mitchell (*left*), Supermarine's chief engineer and designer, who virtually designed the S.6 fuselage as the smallest envelope to take the 1,900 hp Rolls-Royce 'R'. Earlier Schneider aircraft had used the sequence of Lions originated by Montague Stanley Napier (*right*). (*Royal Aeronautical Society*)

number of spruce spars and stringers skinned with ply to form a torsion-resisting cellular assembly on which was superimposed a radiator of thin flat brass tubes sweated together and of greater resistance.

Of the British team, 35-year-old Sqn Ldr A. H. Orlebar, AFC – tall, thin-faced with firm jaw and sharp nose – had commanded the High Speed Flight since January. Next in service seniority was 29-year-old Flt Lieut D. D'Arcy Greig, DFC, AFC, of somewhat Mephistophelean countenance with narrow moustache and frown of concentration, who not only had been the first to make a full terminal velocity dive, but had become the first Caterpillar member of the RAF by escaping by parachute from a Grebe that was uncontrollably spinning, and subsequently became known to the public at the 1927 RAF Display by introducing inverted flying with Genet Moths in formation. Among his star pilots at Wittering were Flg Off Atcherley and Flg Off Waghorn; it was therefore not surprising that those

The 1927 Gloster Schneider biplanes were used as trainers in 1929, with top wing raised for better view and the cruciform tails changed back to the original Gloster design. (*Flight*)

two also were included in the Schneider team. Both were 25 years old, but tall Dick Atcherley, long known as 'Batchy', looked boyish in comparison with the heavily moustached Waghorn, a burly Rugby footballer who looked so fierce on the field. Completing the team was Flt Lieut George H. Stainforth, contemporary with them at Wittering, but 30 years old, tall, smooth faced and with a clipped triangular moustache, a man so quiet and withdrawn that some said he was dim, though in fact he was master of his art, carefully thinking every move, and he had been attached to the High Speed Flight longer than any other except Flg Off T. H. Moon who had joined the MAEE in June 1927 and was in charge of the technical organization of the successful British team that year, and now engaged on the same work.

Italy suffered a sad setback on 22 August when Lieut Motta fatally crashed at his training base with the tandem-engined Savoia Marchetti low-wing tail-boom seaplane, and the Fiat C.29 had been similarly lost a few days earlier though without killing the pilot. Consequently Italy asked for postponement, but the Royal Aero Club took shelter behind the FAI rules which precluded delay except due to bad weather; nevertheless there was precedent to agree since in 1926 the Americans had postponed the race from 23 October until 11 November because the Italian aircraft had not arrived, and that gesture enabled Major de Bernardi to win. On this 1929 occasion the USA was unlikely to be represented, for Lieut 'Al' Williams was having trouble with his Mercury mid-wing, wire-braced monoplane. France had already withdrawn. Nor was Britain without current trouble, for one of the S.6s was almost wrecked by swell from the *Mauretania* steaming up Southampton Water, and then a sequence of strong winds prevented flights even with the Avro Avocet and Fairey Flycatcher practice machines. On 25 August Sqn Ldr Orlebar, after getting accustomed to the Gloster VI by taxi-ing, was airborne a few seconds, but petrol feed was faulty and he was towed in again but later took the second S.6 for four circuits of the speed course.

Great hope was pinned by Folland on his attractive Gloster VI Napier-powered Schneider monoplanes, each of which cost £25,000 compared with £3,000 five years earlier for the Gloster II. (*Flight*)

The magnificent blue and white Supermarine S.6s had the appearance of utmost efficiency, but design of their fixed-pitch propellers had been a headache for Ralli of Faireys. (*Flight*)

On 28 August General Balbo informed the world's newspapers: 'The Italian team is going to England merely as a gesture of chivalrous sportsmanship to avoid the appearance that withdrawal would have seemed a counter move to the refusal to postpone the race. The probability of victory by our team, already hard pressed by the very great delay of constructors in delivering the machines and engines, is almost annulled by the death of Capt de Motta which not only deprived us of our best pilot but has cost us the engine and machine perfected for the best speed. We thus present ourselves in London with two seaplanes, one of them so new that it has never touched water. We have comfort only in the fact that we leave at Desenzano two new machines conceived on the boldest plans on which we place full hope for the record we intend to keep in the hands of our country.' Almost immediately the first batch of Italians arrived. 'The smartness of their uniforms and their excellent cut making our RAF men look quite shabby,' commented a journalist. 'Another interesting thing is the thoroughly practical type of water-proof suits supplied to those handling the machines, whereas our men merely wear waders, and if they are a little over-energetic, the water gets over the top and makes them even wetter than if they used bathing suits.'

Next day Sir Philip Sassoon, as current chairman of the RAeC, visited Calshot with Sir Sefton Brancker and Air Commodore F. W. Bowhill, the Director of Organization and Staff Duties. The S.6. was having cooling trouble, and additional radiators were fixed on the floats for Atcherley to test. Three little ducts had been added beneath each wing-tip: when Mitchell was asked what they were, he explained poker-faced that a pilot got his money back there if he made a good landing. That evening General Balbo arrived in London as authoritative spokesman of the Italian team, who were somewhat worried by the amount of casual shipping traversing the race course at Spithead.

On 31 August one of the Gloster VIs flew for the first time, tested as usual by Orlebar before letting his team try it; but clearly there was some trouble, for he only flew eight minutes. Waghorn then took the S.6, but failed to take-off, and the machine listed heavily as though sinking, but when retrieved it was found that 30 gallons of water had leaked into one of the floats through a defective seal around a strut. By now five Italian machines had arrived, and the 1927 M.52 was having its engine tested on the slipway; two new Macchi M.67s were locked in secrecy in their shed, and in the next was a little Fiat in disfavour and a tandem-engined Savoia Marchetti regarded with doubt. Late on Sunday, Waghorn again took the S.6 out but floated patiently so long for the swell to flatten that it was nearing dusk when a patch of white in the distance showed the Supermarine trying to get off in its typical cloud of spray, and it required four or five attempts before he was airborne. Quickly he was joined by Orlebar in the Gloster, and both streaked around until heavy clouds loomed in the west, and they landed in the glow of the setting sun.

On Monday Atcherley flew one of the S.6s for 10 minutes, but felt something was wrong, and on alighting found a strut fairing badly torn, and its fierce vibration in the slipstream had bent the petrol pipe between engine and float tank. It was assumed he hit a seagull. Then D'Arcy Greig took the Gloster VI for another 10 minute flight and there still seemed trouble. An Italian next took up the Macchi practice machine, but there was alarm when onlookers saw water pouring from a float and assumed he must have hit some obstacle. Rescue boats rushed out, but he alighted safely, though with a few bounces, and remained floating placidly. Later the little Fiat seaplane was brought out, and the engine run in the dark to judge carburation by colour of the exhaust flame. The next few days were crammed with practice flying, and it became clear that the Italians were depending solely on the two scarlet Macchis. Already the Gloster team was doomed to disappointment, for there was continued trouble with the Napier engines; three times in three days the entire petrol system was changed because of fuel starvation, yet when the engine was run on the slipway it seemed perfect, but as soon as it was airborne the popping began again. Both Greig and Stainforth said the two Gloster monoplanes were as nice to handle as any aeroplane they had ever flown. On Thursday night a last test was made by Stainforth. The Gloster climbed fast and looked fast, but as it came over Calshot all could hear the misfiring. Definitely it was out of the competition.

Friday, 6 September, was the great day of qualifying trials. Each competitor had to take-off, fly round a short course, alight, taxi between two buoys, take-off and fly round, and again alight; thereafter the machine had to be moored for six hours, and if it had not sunk by then, it was passed fit for next day's race. In the misty light of that early morning the three Italian machines and the three British were towed on pontoons to moorings in the middle of the Solent north of Cowes. All passed the tests. The British pilots had a last run round the course with the Avocet seaplane – Greig ending his flight with masterly loops, rolls and spins.

Forty-two-year-old Charles Richard Fairey on the Royal Yacht Squadron terrace at Cowes, with his French friend, the pioneer Louis Breguet, await the afternoon's Schneider contest in 1929.

Next day came the contest. Most of the morning was spent in getting the machines to the starting base. The sea was calm. The pilots sat around on shore, waiting. Stainforth had lost the draw and was reserve with the S.5. Groups of privileged guests walked the lawns of the Royal Yacht Squadron, Cowes waterfront was thronged. Tugs, fishing boats, and private craft kept getting into the contest area, and RAF patrol boats raced up and down chasing them off. The Prince of Wales arrived in the Saunders-Roe Cutty Sark. In Cowes Roads lay great steam yachts and large sailing vessels; beyond were four liners, the aircraft carrier HMS *Furious*, a light cruiser, a destroyer, and two Italian cruisers with their square-rigged training ship *Cristoforo Colombo*. Sopwith and Fairey had their own yachts – no new thing for Sopwith, but Fairey had bought the 12-metre yacht *Modesty* as relaxation on medical advice, and when he found it slow he built a small wind-tunnel to test the aerodynamics of sails, and had just commissioned Charles Nicholson to build to his requirements a new 12-metre named *Flica.*

Exactly at 2 p.m. the first starting gun boomed from Calshot. From no single vantage point, nor even from the specially chartered steamer *Orford*, could the race be fully witnessed, but from mid-Solent a wisp of smoke, far away off Cowes, headed towards Ryde pier and gradually lengthened until a tiny dot showed in front which grew with amazing rapidity into the Supermarine S.6 flown by Waghorn. It vanished into the distance, then was heard going along the Hampshire coast, having made the eastern turn, and presently could be seen speeding low down to Gosport, past Lee-on-Solent, then bank acutely round towards Cowes. Excitement was intense. What would be the speed of the S.6 on its first lap? Soon it was announced as 324 mph – 43 mph faster than the 1927 speed and 6 mph faster than the world speed record!

By that time W/O Dal Molin in the Macchi M.52bis was away, and there was a thrilling moment when Waghorn overtook him on the Cowes-Ryde leg. D'Arcy Greig on the outdated S.5 followed, and went round the course like clockwork, his engine full out. Next came Lieut Cadringher flying the first of the new Macchi M.67s, emitting an ear-splitting roaring whine, but he turned wide at Cowes, blinded by fumes which fogged the left side of his minute windscreen. Presently he was observed flying low and fast past Southsea streaming exhaust and evidently in trouble, for he skidded round the mark boat barely banking, tore across the mouth of Southampton Water, lifting just in time to clear the hill above Castle Point, then dived down out of view, but was found to have alighted safely. Meanwhile Waghorn was flashing past on lap after lap with wonderful regularity, followed by the Macchi M.52 and the S.5.

By the time the gun went for Atcherley to take-off there was considerable swell, and his S.6 porpoised badly in cascades of spray at the beginning of the run, but steadied and took-off smoothly. As Waghorn had been instructed to nurse his engine to ensure completing the course, everybody expected Atcherley to beat his time by going all out, but when the speed of his first lap was announced at 302 mph it seemed certain something was wrong. Typically, he had lost his goggles and mistaken one of the turning points, and was still tearing round the course when news came through that he had been disqualified. Last away was Tenente Monte in the second Macchi M.67, but after a single lap at 301 mph he alighted off Hayling Island with a broken oil pipe, and badly scalded on arms and legs. The three British continued steadily lapping. Waghorn flashed past on his last lap, but continued round only to turn suddenly with engine throttled, gliding steeply down. He had run out of fuel, and when picked up by his

The Rolls-Royce power pack that led the way to Schneider Contest victory and the speed record in 1929, and ultimately to saving Britain in the next World War. (*Flight*)

RAF tender was tremendously relieved to discover he had been flying an additional lap by mistake. D'Arcy Greig's S.5 completed her sixth and then seventh lap, and disappeared into history, and it became clear from the broadcast times that Waghorn had easily won. At first sight it seemed hollow victory against the sole Italian pilot flying the slowest of his country's racing seaplanes – yet surely that starkly beautiful creation of Mitchell's must point the way to future fighters from his board? And what triumph for the Rolls-Royce engine and its diffident, bespectacled designer, Rowledge, who had extracted more power per litre than ever known before – but even he could not have produced 1,900 hp from his 900 hp engine had not 30-year-old Rodwell Banks been called in to advise on fuel. 'With the engine operating on 100 per cent benzol,' he wrote, 'troubles began with exhaust valve distortion and burning, and plug sooting on idling, sometimes preventing the engine being opened to full throttle. Vic Haliwell was in charge at Derby of engine testing, assisted by the energetic Ray Dorey. With only a month to go before the Contest I had no time to do very much, but by the simple expedient of diluting the benzol with "light cut" gasoline it was possible to get the engine through its test and race-ready.' Blissfully unaware of technical difficulties, the great British public were content that it had been a beautiful day in the sun, and that the Italians had failed to make the grade, sporting though their effort was.

The general feeling of achievement was endorsed a few days later when the RAF beat the world speed record by a handsome margin over a straight three kilometres at Calshot, timed by an automatic camera recorder devised by the RAE. Incongruously there was considerable international latitude in the method of entering such runs, for it was only stipulated that the aeroplane must not exceed 400 metres high at 500 metres from the start, so it was permissible to dive to that level from a reasonable altitude and thus start with considerable kinetic energy in terms of excess speed. On this occasion the attempt was without diving as visibility was poor. The Gloster VI was given first chance at the record piloted by Flt Lieut Stainforth, who took off in a run of only 12½ seconds aided by a 12 mph breeze, and flew the course five times, averaging 336.3 mph, thus exceeding Bernardi's record by 18 mph. He was followed by Sqn Ldr Orlebar in the same S.6 which won the Schneider, though with a new engine as the 'life' was reckoned to be only 1½ hours. After two attempts to unstick he made four runs, but had difficulty in getting his bearings although Atcherley in a Flycatcher marked the approved height. Triumphantly the S.6 achieved an average of 355.8 mph, but Mitchell considered it could go faster, and decided to fit a different propeller. Two days later, after waiting throughout the afternoon for haze to lift, Orlebar headed straight from the slipway, unsticking in 40 seconds in dead calm, and made six runs over the course, but could only add 2 mph to his previous speed. After he had alighted, Stainforth took the Gloster, but after a short trial landed with engine trouble. But England at last held the record as well as the Trophy.

What almost escaped notice was the newspaper report that a man in Germany, Fritz von Opel, had managed to fly a rocket propelled aeroplane for $1\frac{1}{2}$ miles, attaining 85 mph before he crashed – but clearly he must be a lunatic to believe in so crazy a method of propulsion.

10

Throughout the autumn the many flying clubs continued to cash in on the good weather not only with increasing instruction but every week-end two or three would hold a pageant or a display at their home aerodromes. Lancashire, Yorkshire, Northamptonshire, Leicestershire, Cinque Ports, Clacton, Filton, Haldon, the Household Brigade Flying Club at Heston all had their day, and Hanworth Air Park was opened on 3 August by Her Grace the Duchess of Bedford impressively attended by Air Marshal Sir Edward Ellington, Sir Sefton Brancker, The Master of Sempill, Sqn Ldr Guest, a thousand onlookers and fifty visiting aircraft. It seemed an even better centre than Heston, yet could not attain the intimate and happy social atmosphere which Norman and Muntz had woven into their venture, though Hanworth had its splendid and dignified manor as clubhouse.

This air view shows the pleasant location of the National Flying Services' Manor clubhouse at Hanworth Air Park. (*Flight*)

Behind its large garden, screened by the trees, were two hangars for commercial use, and NFS had one by the south boundary. On the west of the aerodrome were several acres of factory buildings which had been the war-time home of the Whitehead Aviation Co, and Marcel Desoutter had his eye on their adjacent flight sheds for his new company. Neville Stack had been appointed chief pilot of NFS, with H. M. Schofield of Crusader fame as chief assistant, and they had a staff of eleven pilots who had made their mark in Service and private flying. To add emphasis they were attired in dark blue uniform of a pattern reminiscent of the RFC.

By contrast the Rally held that same day by the Suffolk Aeroplane Club at Hadleigh was poorly attended because of the counter attraction at Hanworth, but one Belgian visitor, M Morse, came from Brussels in his Hubert monoplane. Nevertheless it was an intimate and delightful meeting. Friendly George Lowdell, erstwhile Sergeant-Pilot at Martlesham where he was jealously restricted almost solely to a 504 even when revealed as a brilliant aerobatic exponent of the Lincock, had been appointed instructor for the club Bluebirds. It was here that I met Robert Blackburn, to whom I was introduced by Dr Sleigh, chairman of the club. Despite the matured

Robert Blackburn, aviation pioneer (*second from left*), at the Suffolk Aeroplane Club's small aerodrome at Hadleigh. George Lowdell, the club's chief instructor is on the extreme left.

figure and scanty light hair, this tall man in light grey double-breasted suit was still recognizable as the slim youngster who had pioneered monoplanes twenty years ago. He seemed amused in a kindly way that I knew the story of his early efforts at Marske Sands in Yorkshire. We talked of Benny Hucks who had flown his aeroplanes, and Harold Blackburn, currently commanding officer at Martlesham, who was unrelated but had been in charge of the pre-war Blackburn Flying School. As a Yorkshireman, Blackburn was pleased to insist that it was another Yorkshireman, Sir George Cayley, who led the way to world conquest of the air both by formulating the principles and making practical experiment. As Laurence Pritchard said of Blackburn: 'Behind his easy friendliness he had the Yorkshire obstinacy which did not let him be easily swayed, yet he inspired the loyalty and affection and team spirit of a cricket captain.'

Yorkshire also owned the distinction of operating the first municipal aerodrome in the British Isles, when Hull opened Hedon, six miles from the city centre, having beaten Manchester which had boggy ground that would take a long time to clear. HRH Prince George arrived at Hedon in the Royal Wapiti piloted by Sqn Ldr Don, commanding officer of RAF

Communications at Northolt. A welcoming crowd of 35,000 pressed so hard on the entrance barriers that the gates were opened to let them in free lest they were crushed from the pressure of those behind. Both the Prince of Wales and Prince George by now had flown quite considerably, and recently had been receiving instruction on D.H. Moths at Northolt. The Prince of Wales had purchased a Gipsy Moth, G-AALG, upholstered in scarlet leather and painted in the red and blue of the Household Brigade Flying Club, which he would operate from his new residence at Fort Belvedere in Windsor Great Park where a landing ground was being prepared at Smith's Lawn close to Virginia Water. He was being instructed by Flt Lieut 'Mouse' Fielden; Prince George was using a Moth loaned by de Havilland, and Don was instructing him.

The last competitions of the year, marking the effective close of the flying season except for the hardy, were held in rain and blustering wind at Cramlington, near Newcastle, on 5 October for Challenge cups respectively presented by the SBAC and that big, amiable pre-war pilot, the late Edward Grosvenor, who after a long illness had died in August. The same Cirrus II Moth, G-EBPT, won both events, each flown by different pilots. This near-antique had been tuned so well by the ground engineer of the enterprising venture Cramlington Aircraft Ltd, recently formed by Constance Leathart and W. L. Runciman, that it was faster than the current Gipsy Moth. Such was the popularity of this meeting that some 20,000 paid for admission, and another 30,000 'hedge-guests' lined adjacent roads as they 'hadn't a bob to pay for entrance, but had tramped miles to see the flying' – and a splendid show it was, for the RAF participated with Siskins in formation and there were the usual friendly displays of aerobatics, air fighting and comic turns. Unusual were the Hornet-powered Martlet with Avro Baby wings and ply fuselage built by big and confident Fred Miles, and a 'Blimp' non-rigid constructed by the Airship Development Co Ltd in the old RAF airship shed there, and it circled the aerodrome bearing a vast label on its side: 'This space to let'.

It was a reminder that soon the skies would see the R.101, which on 2 October the Press were invited to inspect in its shed prior to the programmed launch on the 7th, but on the latter day a 10 mph wind was too great to risk bringing it out, for it was calculated that this zephyr could crack the airship against the side of the shed with a force of four tons. Five days later, in the first faint glow of dawn, it took only six minutes to walk her clear and across the aerodrome, crabbing markedly in the 4 mph breeze until she reached the mooring tower. Aboard were Major R. B. B. Colmore the Director of Airship Development, and Lieut-Col V. C. Richmond her chief designer, Major Scott her commander, and a picked crew. Ballast was dropped, and with three cables at her nose, she rose to the securing cone. At 10.30 a.m. on 14 October the engines were started, and 40 minutes later the mooring cable was slipped, ballast dropped, and she lifted, drifting back, and then under power sailed away against the wind to 1,500 ft, and after a turn or two headed towards London, several years late.

Britain's officially favoured airship gamble, R.101, with midship engines running, slips her cables for the maiden flight. (*Flight*)

The 'capitalist' airship, R.100, safely moored at St Hubert, Montreal, after its transatlantic flight in 1930. (*Douglas Robinson*)

On the 18th, Lord Thomson was carried as passenger, but three days later R.101 was cautiously moved back to her shed on the excuse of predicted gusty winds, but in fact for adjustments. On 1 November she flew with Sir Samuel Hoare, passing over Sandringham for the King to see, and on the 2nd carried out speed trials over the Channel; then on the 9th she flew with 82 on board, but passed her stiffest test on the afternoon of 11 November when she rode at her mast in winds averaging 70 mph with gusts at 83 mph, the maximum load on the coupling rising to somewhat over a third of its breaking load of 30 tons. On the 17th she left Cardington for the longest flight so far, going to Edinburgh and Belfast, returning after 30 hours 35 minutes. Such success was somewhat galling to the Vickers team working long hours on R.100, but on 28 November they were able to emulate their rival with the Press visiting Howden. To enable her to be moored after flight testing, the R.101 was taken from the masthead at Cardington and housed in her shed, giving opportunity to correct control servo-motors and slack gas-bag netting. In the early morning of 16 December R.100 was taken out. Major Scott was in overall charge, and Sqn Ldr Booth, her commander, at the controls. After handling trials she proceeded to Cardington, but as she passed over it was noticeable that the fabric of the aft part of the hull and the bottom fin pulsated in the slipstream, and the low winter sun, casting shadows along the hull, showed all the fabric pressing inwards against the frames, unlike R.101 where it was kept flat by air pressure. On arriving at 12.15 the mooring coupling was dropped and linked to the tower, and 50 minutes later she was safely tethered. Everyone seemed pleased. On the 17th her second flight was made, largely to investigate the tautness of the fabric, and on the 18th she was taken from the mooring mast to the second shed for adjustments. Meanwhile in Parliament there were the inevitable questions. Sir John Porter asked the cost and was told that expenditure on R.101 had been estimated at £527,000, but would slightly exceed this amount, and the contract price of R.100 was £350,000, though in fact it cost much more. There were even pertinent questions whether royalties were paid for patents held by Major Scott and Lieut-Col Richmond, but the House was assured nothing had been paid. Meanwhile the Government offered a ballot to select eighty MPs for the thrill of a flight.

That autumn C. G. Grey rekindled interest in equally silent flight with a splendidly illustrated article on German sailplanes at the Wasserkuppe where soaring of considerable duration had become the rule, and sailplanes were beautiful long-winged creations of polished ply and transparent fabric. 'Therefore one suggests the time has come when something ought to be done about it in this country,' he said. 'If readers want to be put in touch with others with ideas on the subject they are invited to write to *The Aeroplane*. One feels this is a movement likely to grow very considerably, but one does not undertake to do anything more than put enthusiasts in touch.' This prompted Douglas Culver to organize lunch for sixty of them on 4 December at the Comedy restaurant in London with The Master of

Amid the beautiful Rhön hills, slope soaring and distance flights had become well established with perfected, long-winged sailplanes. (*Flight*)

Sempill in the chair. Mr R. F. Dagnall, owner of the R.F.D. company at Guildford, though primarily a fabricator of balloons and flotation gear, had aeroplane experience, and offered to build one or two gliders free provided members would give the designs, and if nobody felt like designing one he would even buy a German glider. Adroitly the RAeS arranged for Dr Georgii, head of the engineering department of Darmstadt University, to lecture on gliding early in the New Year.

11

The fifth ordinary general meeting of Imperial Airways revealed steady progress. Profits had increased from £132,000 to £140,000. There had been 75 per cent utilization of capacity in Europe and 61 per cent in the Near East; 35,522 paying passengers and 873 tons of mail and freight had been carried. Almost 94 per cent of scheduled flights had been completed. The Board was negotiating with the Government of India to extend an Imperial service to Calcutta, and were awaiting ratification for a service between Egypt and South Africa, for which Sir Alan Cobham and his colleagues would join the Board of a subsidiary company to operate this route.

Nevertheless there were still grievous setbacks. Although the recent Channel accident had resulted in the decision to use only three-engined aircraft on such services, these machines were not immune from disaster. On 6 September the *City of Jerusalem*, a D.H. Hercules, crashed on a night landing at Jask, the wing-tip flares setting fire to the petrol vapour, and the pilot, Capt A. E. Woodbridge, and his two passengers were killed, though the wireless operator and engineer escaped with burns. 'Woodie' was a pilot who in his quiet way gave complete confidence, so it was accepted that this crash was not fallibility of judgment but the inevitable risk of flying. This was re-emphasized on 26 October when the Short Calcutta *City of Rome*, on the London—India air mail route, ran into a gale exceeding 70 mph after leaving Naples, and the captain, L. S. Birt, was forced to alight at sea 10 miles south of Spezia. An SOS was picked up by the Italian steamer *Famiglia*, whose captain soon found the Calcutta riding the seas quite well. 'A cable 250 yards long was thrown, and the pilot fastened it to the bows. The tug made towards land at slow speed, the crew of the flying-boat helping by keeping their engines going. After a quarter of an hour the cable snapped. The tug crew were able to see the *City of Rome* for a few minutes more and heard the noise of her engines, but then the navigation lights disappeared and it was no longer possible to fix her position. Three times the tug circled the spot where the flying-boat was last seen, but nothing was found or heard.' Search continued the following day with aircraft and Italian naval vessels, and in the afternoon the floating bodies of pilot and wireless operator were discovered. Soon it became certain that the flying-boat had sunk and her crew and four passengers drowned. It dinned home that although hulls might be perfection, wings were relatively fragile and once a lower wing was buckled by heavy seas the entire biplane structure would collapse, and, though a cantilever monoplane might be safer, its wing-tip floats were equally vulnerable. Yet Air Ministry technicians had turned down earlier proposals by Oswald Short, and independently by Sam Saunders, to build twin-hulled flying-boats which would be stable with damaged wings, as recently proved by the Italian flying-boat used by the Marchese de Pinedo in his flights around the world.

There had also been disaster and narrow escapes among the industry's pilots. De Havilland's popular instructor Flt Lieut A. S. 'Jimmy' White, who had been so outstanding in instilling confidence even to the most nervous pupils, was gliding in to land with a Gipsy Moth carrying a potential customer, and, as he turned, a single-seat Moth, flown by Capt G. F. Boyle, approached slightly lower in the opposite direction and their left wings interlocked, both aircraft falling into a field and bursting into flames. Succeeding Jimmy was A. P. Eden who had piloted the sealed Gipsy Moth G-EBTD for much of its 600 hours, though this was shared by several of the keen young men Francis St Barbe had acquired to promote his sales campaign, and included Geoffrey, the elder son of D.H., fat and genial Ronnie Malcolm, and the elegantly handsome W. T. W. Ballantyne and Hugh Buckingham, sometimes known as 'the beautiful twins'.

At Avros, their chief pilot, H. A. Brown, had been badly injured when the first Avocet crashed at Woodford aerodrome through engine failure. Then Louis Paget at Westland had a near call when diving the Interceptor with its new tall fin and rudder, for on landing it was found that the fuselage fittings of the tailplane front struts had failed, so if he had dived again the structure would have collapsed. This was followed by trouble at Bristol when C. R. L. Shaw, an outstanding instructor at the Bristol Flying School, was diving the wooden Bristol Type 101 and suddenly the centre-section swung back over his head, a steel fitting in the cabane having failed; but such is strength in emergency that he pushed the structure forward to scramble out, and parachuted from a mere 1,000 feet. Even the two most brilliant recent exponents of aerobatics in the RAF were killed in separate accidents – Flg Off C. H. Jones who stalled through pulling round too steeply after taking-off, and Flg Off J. Clarke, when he flew into the downdraught above the Blackburn factory and crashed onto the concrete apron where the machine caught fire.

Certainly Harry Guggenheim knew the dangers of flying for he had flown as an American Naval pilot during the Great War, and consequently had launched aeronautical schools and flight research – for like his father, he declared that the inherited wealth of his family's mining fortune should be used for the progress of man and not for mere self-gratification. But at Mitchell Field, New York, his 'safety aircraft' competition was leading to bitter contention. The passing months had seen entries rise to twenty-five, but nine were withdrawn without submitting a machine, four without completing eliminating trials, and one was eliminated in these tests. Of the rest it was not known whether the Autogiro entered by Cierva, or another from his American licensee, the Pitcairn Autogiro Co, would be present for they had a habit of turning over in a side wind because the rotor could not be quickly stopped. However, nine aeroplanes were ready, and three had completed tests – the 170 hp Curtiss Tanager enclosed cabin biplane, which had leading-edge automatic slots on each wing and manually-operated flaps, and was fitted with floating ailerons similar to the controllers of the Pterodactyl; the 150 hp Handley Page Gugnunc in which all slots and flaps were automatic but with normal ailerons; and the standard 100 hp Fleet two-seat open biplane, somewhat similar to a Moth, but with manually operated trailing-edge flap.

At the conclusion of the Guggenheim competition the Handley Page Gugnunc was purchased by the Air Ministry and extensively evaluated at Farnborough.

From the USA came a private report: 'The Handley Page ship is, to my mind, the most promising, because everything is purely automatic, but the flap of the Tanager is manually operated. I am pretty sure there is quite an argument between Handley Page and the Guggenheim Fund as to the minimum speed of his machine. They have been trying to measure it by towing a pitot below it; but I have found by experience that this is not accurate as one gets varying readings. Mr Handley Page is backed by Martlesham Heath tests, so should be fairly able to prove his case.' The difficulty was that Cordes could not get the re-rigged machine to glide at the 33½ mph which he and Martlesham had attained, whereas the Tanager was indicating a lower speed. The witness continued, 'I only hope they do not break either machine during testing because when landing over the barrier they come down with an awful bang and all have broken wheels and undercarriage struts so far. I do not believe in this slow landing and gliding because it necessitates both Tanager and Handley Page entries being led onto the field with a man at each wing if the wind is over 20 mph. This is much the same trouble we have with Avians and Moths, and makes an airplane decidedly impracticable because sooner or later one will land under conditions where men are not available and the machine will turn over. Practically everybody who has flown a light plane to any extent in America has had this happen.'

The real row was over slots. The US Navy had purchased the right to use Handley Page slot patents early in their development, and to get experience of slotted flight had given Curtiss an order for an experimental machine with automatic slots, though the company itself was not licensed. When Cordes cabled news of the Tanager's slots, H.P. immediately took ship to New York, full of righteous anger to get the Curtiss declared ineligible under the rule stating: 'The employment of design features, which, in the opinion of the Fund, are copied from the design of another competitor may render the aircraft ineligible for entry,' and he instituted a law suit against Curtiss for infringement. The American firm claimed that, as a supplier to the US Navy, it was entitled to use the slot, and invoked various American patents they said the Gugnunc had infringed. So when H.P. further claimed that the tandem lower wing of another American competitor, the Cunningham-Hall with interceptor and flap, contravened his patents, its designer, Major Schroeder, also retaliated that the Gugnunc used 153 American patents and one further British infringement did not worry him. Unfortunately for Handley Page Ltd, although the writ might have succeeded for machines made commercially for profit, there was not likely to be legal refusal for sporting trials such as the Guggenheim Competition.

More serious for Britain was repercussion of Wall Street panic selling which began at the end of October. A month previously there had been financial disaster to many London investors through the crash of the Hatry investment 'empire'. When general investment confidence in the USA followed suit and heavy selling began, rapidly becoming a blizzard, British exports started to fall. Largely as an economy, British troops were withdrawn from the Rhine. Unemployment was steadily increasing. An Unemployment Insurance Bill was passed in November extending transitional benefits, and providing training for unemployed young people. But the aircraft firms seemed on the road to prosperity, for all had lucrative contracts except Parnall who were too under-capitalized to risk putting the Elf into serious production, and Beardmore had closed because Lord Invernairn had again decided the game was not worth playing.

12

In October Uwins tested the Bristol 110A cabin biplane, provisionally powered with a 220 hp Titan, and reported favourably, but advised that it would be much improved with greater power, so Barnwell substituted the originally proposed 315 hp Neptune. For the first time since the war the Bristol aircraft department was really busy. Delivery of the first production batch of Bulldogs had been completed, and a contract followed for a further forty. Meanwhile Barnwell and his team were busy with yet another private-venture two-seat fighter-bomber biplane, Type 118, of much the same size as the 110A but powered with a geared and supercharged Jupiter XFA.

At Brough on 21 November the impressive Iris III was launched. It differed from the prototype in having duralumin-structured wings and somewhat better accommodation. Dasher Blake had not flown flying-boats, so for the acceptance flight Boulton & Paul loaned Sqn Ldr Rea as an experienced marine pilot, and he was accompanied by N. H. Woodhead as second pilot, with Major Rennie as observer. The contract required only a 30-minute test, and then the machine was handed to RAF pilots who flew it to Felixstowe for performance trials.

Next with a new machine was Vickers with their Type 177 ship-plane, flown by Mutt Summers at Brooklands on 26 November. Basically it was a Jupiter XF version of the Type 143 ordered by the Bolivian Government, which in turn was derived from the Type 141 of the single-seat fighter competition won by the Bulldog. It was the last of Vickers fighter biplanes, and the last distinctly showing the hand of Arthur Camden Pratt, who before the war had been the armament specialist of Short Bros.

Four days after the first flight of the Vickers Type 177, Mutt Summers flew his biggest new prototype to date – the 76 ft 6 in span twin Rolls-Royce F.XIV bomber built to specification B.19/27 as a replacement for the Virginia. With a laden weight of 13,820 lb, carrying 1,546 lb of bombs, it had a range of 920 miles at an estimated cruise of 115 mph. The 12-degree swept-back R.A.F.31 single-bay outer wings had automatic slots, and like its predecessors it had a boxed biplane tail, but this time fully floating. However Pierson was not really convinced that a biplane met the case, and had also submitted a 100-ft span monoplane; but the Air Ministry insisted, as they did with Handley Page, that a biplane was required – for their technicians had already agreed that Fairey might try

In first form the Vickers B.19/27 had Rolls-Royce F.XIVs but their steam cooling gave prolonged trouble, and the aeroplane had a baffling rolling wallow eventually attributed to excessive sweepback.

the long-prejudiced monoplane geometry in view of success with the long-range machine. As yet neither rival was ready.

That month Avros sprang a surprise with a Mongoose-powered replacement for the 504N. The new 621 Tutor was a boldly staggered, N-strutted, conventional biplane with fabric-covered welded-tube fuselage, and steel-structured rectangular wings having automatic slots on top wing and Frise ailerons on the lower, though top ailerons were later added. It had one eye-catching difference from previous Chadwick designs, for the rudder had distinctive de Havilland shape, though he said it had merely the curve used on the Fokker-like Avro Ten. The Tutor's handling earned immediate approbation, and early in December the prototype, G-AAKT, was submitted to Martlesham. In *Pilot's Summer* F. D. Tredrey, an RAF pilot, described it as: 'A gentlemanly little thing from the flying point of view. Nicely balanced controls, light and smooth to handle, beautiful for all

The Lynx-powered Avro 621 Tutor was identical with the Mongoose prototype which went to Martlesham in December 1929.

manoeuvres, no tricks or vices, sweet glide at 70, and as easy to land as buttoning up your coat.'

That month a V-strutted monoplane, which at first sight seemed a Widgeon, circled and landed at Yeovil, and was found to be the new and almost secret D.H.80 three-seat cabin monoplane, piloted by Hubert Broad. It was more like the rejected cabin Widgeon than any yet built, employing the same method of folding, but the design was neater, with typical de Havilland ply-covered fuselage and characteristic tail. Because of its aerodynamic efficiency the glide was very flat, so the long undercarriage legs could be turned broadside as air brakes. Louis Paget took it up, and I went as passenger, finding the view far better than expected because of the boldly inverted Gipsy. As Alan Butler said at the de Havilland party on 14 December: 'The Puss Moth, as we shall call it, gives me an entirely different outlook on flying.' Broad had first tested it on 9 September, registered as E-1 under the new approval system by which manufacturers

To the surprise of its designers the neat D.H.80 Puss Moth was 7 mph above estimate, clocking 128 mph – which was 30 mph faster than the latest Moth and equalled the Siskin. A production aircraft is illustrated.

used an individual maker's letter until completion of airworthiness trials. Since then everyone with a pilot's licence at de Havilland had been flying it with such unanimous approval that it was clear it should go into production, and to that end Hagg was redesigning it with welded steel-tube framed fuselage accommodating two passengers in staggered tandem seats instead of a bench across the back of the cabin.

In the week following the D.H. Christmas party, Hubert Broad made the initial flight of the dainty and very functional D.H.77 semi-cantilever low-wing fighter, powered with the new Rapier H engine which Napiers had been developing. On this engine the whole future of Napiers seemed to depend, for the high-powered Rolls-Royce engines had made the Lion obsolete. Despondently Montague Napier wrote to the managing director, H. T. Vane: 'We have nothing in progress but Halford's H engine,' for his proposed eight- and twelve-cylinder vee engines had been abandoned, the chief engineer, George Pate had gloomily resigned, and the newly engaged works manager was sacked. Only Montague Napier's farsighted policy of investment was likely to carry the company through until new develop-

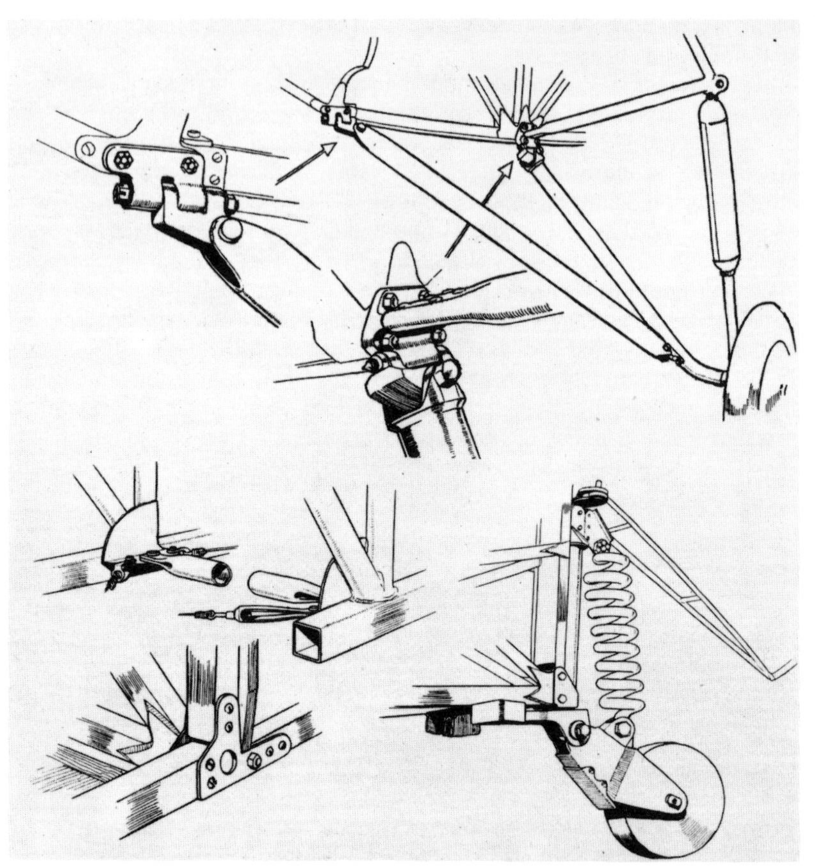

For the Hawk Moth, the de Havilland team came to terms with metal construction, using a welded-up variant of the Hawker and Westland square tube system. (*Flight*)

ments, such as a compression-ignition engine, or an intended six-cylinder for light aircraft, became profitable. 'I think,' wrote Henry Cooke, the co-director and close friend of Montague Napier, 'we're in for a difficult time.'

Last of the aircraft to fly in this decade was the Hispano-powered Blackburn T.7B, flown by Blake in the presence of Mitsubishi representatives at Brough on 28 December. Handling was sufficiently like its Ripon predecessor for tests to be cut to a minimum, though it was flown with and without torpedo, and thereafter hurriedly packed for shipment to Japan where George Petty would accompany it.

In the interim there was disaster. Ever since the Fairey Long-range Monoplane had returned, Norman Macmillan had been conducting numerous tests to improve longitudinal and directional stability. It now had a massive fin and taller rudder with relatively smaller horn balance. At 8 a.m. on 17 December the Fairey monoplane took-off from Cranwell again in the hands of Sqn Ldr A. G. Jones-Williams and Flt Lieut N. H.

Jenkins in another attempt to break the world's long distance record of 4,912 miles made by Costes and Bellonte in their flight to Manchuria. To beat the record they must fly at least 100 kilometres further, measured on a Great Circle course. That meant achieving a point in south-west Africa halfway between Benguela and Walvis Bay. Four hours later the machine passed Nevers in France, but after contacting Malta by radio, giving a position just north of Sardinia, no more was heard, and there was worry when it became overdue. Then came a report from the Consul-General in Tunis briefly stating it had crashed in the hills some 30 miles south, and that both pilots were killed. In the House on 19 December, Mr Montague

The improved Fairey Long-range Monoplane sets off again in an attempt to break the record.

said he was not yet in a position to add materially to the information, but a guard had been placed over the machine and the Air Ministry had despatched a technical officer to Tunis to make the fullest expert enquiry on the spot. Hollis Williams told me: 'Freddie Rowarth had the job of going out, hiring a team of Arab guides and mules, and they eventually found the wreck at a moderate altitude of 4,000 ft. It had flown straight into rising ground and was on track. What went wrong was never finally determined, although the cabin altimeter was the most likely cause. The flight plan was to reach 5,000 ft four hours after the start in order to cross the Rhône watershed, and then continue climbing slowly to 8,000 ft. The flight log was recovered, also two sealed barographs. The last entry read "Sighted Tunis 5,000 ft", but the sealed barographs showed no greater than 3,000 ft after crossing the French coast. All that could be done was to recover the remains of the crew and burn the wreckage. Jones-Williams was buried at Newtimber, Sussex, and Jenkins at Martlesham.'

On ceremonial inspection, 'Boom', Marshal of the Royal Air Force, Hugh Trenchard, GCB, DSO, DCL, LL.D. (*Flight*)

The year ended with the departure of Marshal of the Royal Air Force, Sir Hugh Trenchard, Bart, GCB, DSO, from his post as Chief of Air Staff which he had held since 1919. In the New Year's Honours his promotion to the peerage would be announced as Baron Trenchard of Wolfeton in the County of Dorset, and he would be succeeded by 49-year-old Air Chief Marshal Sir John M. Salmond, KCB, CMG, DSO, who in the war days of early 1918 had similarly taken over from Sir Hugh Trenchard the command of the RFC in the Field. But Hugh Montague Trenchard, now 56, would forever be remembered as Father of the Royal Air Force. He had led the way from chaos, and developed a Service capable of protecting not only the British Isles but the entire Empire against the next biggest Air Force in the world, and this he had done on the most limited money. Time after time the Sea Lords, backed by the whole weight of the Admiralty, had tried to dismantle his Air Force in order to maintain full command of Naval flying, but Trenchard saved it as the Third Service, one and indivisible, yet paying scrupulous regard in equipment and methods for the specific needs of Navy and Army. Moreover he ensured the future technical skill of the RAF by establishing first-class training at Halton for aircraft apprentices, building splendidly equipped barracks for the boys, and quarters for their instructors, together with excellent workshops and lecture rooms to replace the leaky war-time huts hitherto used. Having established training for the mechanics, he organized great Flying Schools, and a Cadet College so that young men could enter directly from school to become permanent officers, for that purpose using the former RNAS establishment at Cranwell in Lincolnshire, while for ultimate training in the higher arts of this great aeronautical military Service he established the Staff College at Andover. None could have done better. He was the man for the time. For his successor there was the task of leading this specialized Service into the future, confident that its equipment had all the latest technological advances, so that his men could beat the world maybe single-handed. In the ability of the British aircraft industry to fulfil these needs the entire Royal Air Force had every confidence.

INDEX

CONSTRUCTORS

Page numbers in italics refer to illustrations.

CONSTRUCTOR CLUBS

CONTINENTAL AND USA CONSTRUCTORS

ENGINE MANUFACTURERS

PEOPLE

GENERAL